A Tribute to Paul Erdős

A Tribute to Paul Erdős

Edited by

A. Baker
Trinity College, Cambridge

B. Bollobás
Trinity College, Cambridge

A. Hajnal
Hungarian Academy of Sciences

CAMBRIDGE UNIVERSITY PRESS
Cambridge
New York   Port Chester
Melbourne   Sydney

CAMBRIDGE UNIVERSITY PRESS
Cambridge, New York, Melbourne, Madrid, Cape Town, Singapore, São Paulo

Cambridge University Press
The Edinburgh Building, Cambridge CB2 8RU, UK

Published in the United States of America by Cambridge University Press, New York

www.cambridge.org
Information on this title: www.cambridge.org/9780521381017

First published 1990
This digitally printed version 2008

A catalogue record for this publication is available from the British Library

ISBN 978-0-521-38101-7 hardback
ISBN 978-0-521-06733-1 paperback

Contents

Preface

This volume is dedicated to Paul Erdős, who has profoundly influenced mathematics this century. He has worked in number theory, complex analysis, probability theory, geometry, interpolation theory, algebra, set theory and, perhaps above all, in combinatorics. His theorems and conjectures have had a decisive impact. In particular, he, more than anybody else, is the founder of modern combinatorics, he pioneered probabilistic number theory, he is the master of random methods in analysis and combinatorics, and he has created the fields of Ramsey theory and the partition calculus of set theory.

Paul Erdős is the consummate problem solver: his hallmark is the succinct and clever argument, often leading to a solution from 'the book'. He loves areas of mathematics which do not require an excessive amount of technical knowledge but give scope for ingenuity and surprise. The mathematics of Paul Erdős is the mathematics of beauty and insight.

One of the most attractive ways in which Paul Erdős has influenced mathematics is through a host of stimulating problems and conjectures, to many of which he has attached money prizes, in accordance with their notoriety. He often says that he could not pay up if all his problems were solved at once, but neither could the strongest bank if all its customers withdrew their money at the same time. And the latter is far more likely.

Ever since he was greeted at the station in Cambridge by Davenport and Rado in October 1934, on his way from Hungary to Manchester to work with Mordell, Paul Erdős has travelled the world constantly: he is the archetypal peripatetic mathematician. As the 'Professor of the Universe', he travels light, saying that 'private property is a nuisance'; wherever he arrives his 'brain is open' for new problems and ideas: 'another roof, another proof'. He is fond of idiosyncratic expressions (bosses, slaves and

Jacques Dixmier	11bis rue du Val de Grâce, F-75005 Paris, France
Alan Dow	Department of Mathematics, York University, North York, Ontario, Canada M3J 1P3
R. Durrett	Department of Mathematics, White Hall, Cornell University, Ithaca, New York 14853, USA
P. D. T. A. Elliott	Department of Mathematics, University of Colorado at Boulder, Boulder, Colorado 80309-0426, USA
Paul Erdős	Mathematical Institute of the Hungarian Academy of Sciences, Reáltanoda utca 13–15, H-1053 Budapest, Hungary
T. I. Fenner	Department of Computer Science, Birkbeck College, Malet Street, London WC1E 7HX, UK
Matthew Foreman	Department of Mathematics, Ohio State University, 231 West 18th Avenue, Columbus, Ohio 43210, USA
A. M. Frieze	Department of Mathematics, Carnegie Mellon University, Pittsburg, Pennsylvania 15213-3890, USA
Fred Galvin	Department of Mathematics, University of Kansas, Lawrence, Kansas 66045-2142, USA
R. L. Graham	AT&T Bell Laboratories, Murray Hill, New Jersey 07974, USA
K. Győry	Mathematical Institute, University of Debrecen, PO Box 12, H-4010 Debrecen, Hungary
Roland Häggkvist	Mathematiska Institutionen, Stockholms Universitet, Box 6701, 113 85 Stockholm, Sweden
A. Hajnal	Mathematical Institute of the Hungarian Academy of Sciences, Reáltanoda utca 13–15, H-1053 Budapest, Hungary

R. R. Hall	Department of Mathematics, University of York, Heslington, York YO1 5DD, UK
W. K. Hayman	Department of Mathematics, University of York, Heslington, York YO1 5DD, UK
D. R. Heath-Brown	Magdalen College, Oxford OX1 4AU, UK
Thomas Jech	Department of Mathematics, Pennsylvania State University, 215 McAllister Building, University Park, Pennsylvania 16802, USA
I. Juhász	Mathematical Institute of the Hungarian Academy of Sciences, Reáltanoda utca 13–15, H-1053 Budapest, Hungary
H. Kesten	Department of Mathematics, White Hall, Cornell University, Ithaca, New York 14853, USA
D. J. Kleitman	Department of Mathematics, M.I.T., Cambridge, MA 02139, USA
P. Komjáth	Department of Computer Science, Eőtvős Loránd University, Múzeum hőrút 6–8, H-1088 Budapest, Hungary
János Komlós	Department of Mathematics, Hill Center, Rutgers University, New Brunswick, New Jersey 08903, USA
K. Kunen	Department of Mathematics, University of Wisconsin, 480 Lincoln Drive, Madison, Wisconsin 53706, USA
Attila Máté	Department of Mathematics, Brooklyn College of CUNY, Bedford Ave. and Ave. H, Brooklyn, New York 11210, USA
E. C. Milner	Department of Mathematics and Statistics, University of Calgary, 2500 University Drive NW, Calgary, Alberta, Canada T2N 1N4
H. L. Montgomery	Department of Mathematics, University of Michigan, Ann Arbor, Michigan, USA

Z. Nagy — Mathematical Institute of the Hungarian Academy of Sciences, Reáltanoda utca 13–15, H-1053 Budapest, Hungary

Jean-Louis Nicolas — Département de Mathématiques, Université Claude Bernard (Lyon I), F-69622 Villeurbanne Cedex, France

Paul Nevai — Department of Mathematics, Ohio State University, 231 West 18th Avenue, Columbus, Ohio 43210 USA

P. Révész — Institut für Statistik und Warscheinlichkeits-theorie, Technische Universität Wien, Wiedner Hauptstrasse 8–10/107, A-1040 Wien, Austria

A. Schinzel — Instytut Matematyczny PAN, Skr. pocztowa 137, 00-950 Warszawa, Poland

Wolfgang M. Schmidt — Department of Mathematics, University of Colorado at Boulder, Boulder, Colorado 80309-0426, USA

Saharon Shelah — Department of Mathematics, The Hebrew University, Jerusalem, Israel, and Department of Mathematics, Hill Center, Rutgers University, New Brunswick, New Jersey 08903, USA

T. N. Shorey — School of Mathematics, Tata institute for Fundamental Research, Homi Bhabha Road, Bombay 400005, India

L. Soukup — Mathematical Institute of the Hungarian Academy of Sciences, Reáltanoda utca 13–15, H-1053 Budapest, Hungary

Joel Spencer — Courant Institute of Mathematical Sciences, 251 mercer Street, New York, New York 10012, USA

J. Szabados — Mathematical Institute of the Hungarian Academy of Sciences, Reáltanoda utca 13–15, H-1053 Budapest, Hungary

Endre Szemerédi	Department of Mathematics, Hill Center, Rutgers University, New Brunswick, New Jersey 08903, USA
Z. Szentmiklóssy	Mathematical Institute of the Hungarian Academy of Sciences, Reáltanoda utca 13–15, H-1053 Budapest, Hungary
Gérald Tenenbaum	Département de Mathématiques, Université de Nancy I, BP 239-54506 Vandœvre les Nancy Cedex, France
A. R. Thatcher	129 Thetford Road, New Malden, Surrey KT3 5DS, UK
Carsten Thomassen	Mathematical Institute, The Technical University of Denmark, Building 303, DK-2800, Lyngby, Denmark
R. Tijdeman	Mathematical Institute, RU Leiden, Postbus 9512, 2300 RA Leiden, The Netherlands
R. C. Vaughan	Department of Mathematics, Imperial College of Science and Technology, Queen's Gate, London SW7 2BZ, UK
A. K. Varma	Department of Mathematics, University of Florida, Gainsville, Florida 32611, USA
Péter Vértesi	Mathematical Institute of the Hungarian Academy of Sciences, Reáltanoda utca 13–15, H-1053 Budapest, Hungary
W. Weiss	Mathematics Department, University of Toronto, Toronto, Canada

Acknowlegements

The Editors are grateful to Miss Ellyn Dalitz, Dr I. Leader and Dr A. Thomason for their assistance in the preparation of this volume.

Generating expanders from two permutations

Miklós Ajtai, János Komlós* and Endre Szemerédi

Abstract

Given $\alpha > 0$, we say that the sequence $\pi_1, \pi_2, \ldots, \pi_r$ of permutations of $\{1, 2, \ldots, n\}$ has the expanding property if, for all $X \subset \{1, 2, \ldots, n\}$ of size at least αn,

$$\left| \bigcup_{i=1}^{r} \pi_i X \right| > (1-\alpha)n.$$

We show that there exist pairs (σ, τ) of permutations such that the generated permutations $\pi_i = \sigma^i \tau^i$, $1 \leqslant i \leqslant r$, are expanding, where $r = O(1/\alpha^3)$. In fact, most pairs of permutations have this property.

No construction of such pairs of permutations is known.

0 Introduction

Expanders are of ever-increasing use in computer science. The random generation of expanders (Pinsker [8], Pippinger [9] and others) has been replaced by explicit constructions (Margulis [6], Gabber–Galil [4] and, recently, Lubotzky–Phillips–Sarnak [5]).

On the other hand, the recent characterization of expanders through eigenvalues by Alon [1] provides efficient testing for the expanding property, thus reviving interest in the random generation of expanders.

Here we offer a compromise approach: generating only a fraction of the bits at random and then extending them deterministically.

* Work supported by the NSF grant NSF-CCR 8505053.

Definitions

In the following, we will always assume that $0 < \alpha < 1$, $A > 1$ and $A\alpha < 1$ and, for simplicity of notation, we will also assume that αn is integral.

A *weak α-expander* on $2n$ vertices is a bipartite graph $G = (U, V, E)$ such that $|U| = |V| = n$ and, for any pair of subsets $X \subset U$ and $Y \subset V$, $|X| = |Y| = \alpha n$, there is at least one edge between X and Y.

A strong *(A, α)-expander* is a bipartite graph $G = (U, V, E)$ such that $|U| = |V| = n$ and, for any $X \subset U$, $|X| \leqslant \alpha n$, we have that $|N(X)| \geqslant A|X|$, where $N(X)$ is the neighbourhood-set of X:

$$N(X) = \{v \in V;\, (x, v) \in E \text{ for some } x \in X\}.$$

We will use the parameter value $A = (1-\alpha)/\alpha$, so that $\alpha = 1/(A+1)$, and write *strong α-expander* for strong (A, α)-expander.

Connections

We will only deal with regular bipartite graphs $G = (U, V, E)$, where we identify both vertex sets U and V by $[n] = \{1, 2, \ldots, n\}$. By König's theorem, an r-regular bipartite graph $G = (U, V, E)$ consists of r one-factors between U and V, i.e. permutations π_i, $1 \leqslant i \leqslant r$, of $\{1, 2, \ldots, n\}$, where, within the ith one-factor, vertex $u \in U$ is connected with vertex $v \in V$ such that $v = \pi_i u$.

Since very often the connections represent expensive modules (e.g. comparator switches), an alternative way to represent one-factors between U and V is to use two extra sets U' and V' of in-between vertices with fixed connections (the jth element of U' is connected to the jth element of V'). The set U is then connected to the set U' using a permutation connection σ, and similarly V is connected to V' by using some permutation τ.

This is, of course, equivalent to using the permutation $\pi = \tau^{-1}\sigma$ directly between U and V, but, while σ and τ may vary, the identity connection between U' and V' remains the same. The connections σ and τ are only used to communicate data from U to U' and from V to V', and the expensive part of the work is done along the fixed connection between U and V.

Thus, given permutations σ_i and τ_i, $1 \leqslant i \leqslant r$, we define the corresponding bipartite graph as

$$G = (U, V, E); \qquad |U| = |V| = n,$$

$$E = \{(\sigma_i^{-1} k, \tau_i^{-1} k);\, k \in [n],\, 1 \leqslant i \leqslant r\},$$

which is the same as the graph

$$G = (U, V, E); \quad |U| = |V| = n,$$
$$E = \{(k, \pi_i k);\ k \in [n],\ 1 \leq i \leq r\},$$

where $\pi_i = \tau_i^{-1}\sigma_i$.

Notation

$c_0, c_1, c_2, \ldots$ are *absolute* constants, while c is a 'generic constant' with possibly different values at each occurrence.

log stands for the natural logarithm. $\log n$ and $\log\log n$ are truncated from below, so that their value is at least 1. $+$ will always mean addition modulo n. For simplicity of notation, we will always assume that αn is integral.

$[n] = \{1, 2, \ldots, n\}$. $[a, b]$ denotes the set of integers in the interval (a, b).

$\bar{A}$ denotes the complementary of the set A.

We will use the entropy function $h(\alpha) = -\alpha \log \alpha - (1-\alpha)\log(1-\alpha)$, together with the inequality $\binom{n}{\alpha n} \leq \exp h(\alpha) n$.

Random generation of expanders

It is known that, for most r-tuples of permutations $(\pi_1, \ldots, \pi_r)$, the corresponding bipartite graphs are strong α-expanders, where

$$r = r(\alpha) = c_1 \frac{1}{\alpha} \log \frac{1}{\alpha}.$$

(Note that Zarankiewicz's theorem [12] implies that *any* α-expander (even in the weak sense) has a degree of at least $c(1/\alpha)\log(1/\alpha)$.)

On the other hand, it is easy to see that, if we select only one permutation and generate the others from it as $\pi_i = \sigma^i$, $i = 1, 2, \ldots, r$, then, for large n, the corresponding graph is *never* an expander. (Here 'large' means large in terms of r, which, in turn, is determined by α.) The reason is that these permutations have the same cycle structures.

We are going to show, however, that **two permutations can already generate expanders.**

In what follows, $r = r(\alpha)$ will always be a fixed function, upper-bounded by a polynomial of $1/\alpha$. The proofs will show that

$$r(\alpha) = c_2 \left(\frac{1}{\alpha} \log \frac{1}{\alpha} \right)^2$$

suffices.

1 Formulation of results

We now define the crucial notion of *mixing* permutations. For the sake of better understanding, we define it in two equivalent ways.

Definition We say that the pair (σ, τ) of permutations of $[n]$ is *mixing* if

(first form): for all $X \subset [n]$, $|X| \geqslant n^{1-c_3}$, we have

$$\left| \bigcup_{1 \leqslant i \leqslant r(\alpha)} \pi_i X \right| > n - |X|, \tag{1}$$

where $\pi_i = \tau^{-i}\sigma^i$ and $\alpha = |X|/n$: in other words, for all $\alpha > n^{-c_3}$, the permutations $\pi_i = \tau^{-i}\sigma^i$, $1 \leqslant i \leqslant r(\alpha)$, form a weak α-expander;

(second form): for any two sets $X, Y \subset [n]$, $|X|, |Y| \geqslant n^{1-c_3}$, there exists an i, $1 \leqslant i \leqslant r(\alpha)$, such that

$$\sigma^i X \cap \tau^i Y \neq \varnothing, \tag{2}$$

where

$$\alpha = \frac{1}{n} \min\{|X|, |Y|\}:$$

in other words, for any $\alpha > n^{-c_3}$, the following graph G_α is a weak α-expander:

$$G_\alpha = (U, V, E); \qquad |U| = |V| = n,$$
$$E = \{(\sigma^{-i}k, \tau^{-i}k);\ k \in [n],\ 1 \leqslant i \leqslant r(\alpha)\}.$$

Probably the second form is the most illuminating.

Theorem 1 (a) *For every n, there exist mixing pairs (σ, τ) of permutations of $\{1, 2, \ldots, n\}$. In fact, most pairs (σ, τ) are mixing.*

Here 'most' means that the proportion of pairs that are not mixing approaches to zero as n tends to infinity. (It does so at an exponential rate.)

(b) *The same remains true if we restrict σ and τ to cyclic permutations (in which case, of course, proportions are calculated within the class of cyclic permutations of size n).*

(c) *The same remains true for most cyclic τ if we fix σ to be the right shift ($i \to i+1$, $n \to 1$).*

It is clear that (c) implies (b) and it is not hard to see that (b) implies (a). We will only prove (c).

Remark It is easy to see (by substituting $\frac{1}{2}\alpha$ for α) that (2) can be replaced by

$$|\sigma^i X \cap \tau^i Y| > \delta(\alpha) n,$$

where $\delta(\alpha) = \frac{1}{2}\alpha / r(\frac{1}{2}\alpha)$.

It is very likely that (2) can actually be replaced by

$$|\sigma^i X \cap \tau^i Y| > \tfrac{1}{2}\alpha^2 n.$$

Theorem 2 *Theorem 1 remains true if, in the definition of mixing, we replace the words 'weak α-expanders' by 'strong α-expanders'.*

In other words, most pairs (σ, τ) have the following property: for all $X \subset [n]$ and α, $\alpha \geqslant \max\{n^{-c_4}, |X|/n\}$, we have that

$$\left| \bigcup_{1 \leqslant i \leqslant r(\alpha)} \pi_i X \right| \geqslant \frac{1-\alpha}{\alpha} |X|, \tag{3}$$

where $\pi_i = \tau^{-i}\sigma^i$.

Remark Sarnak [10] has a simple construction (with a hard proof) for two permutations which generate expanders. Namely, if n is prime, then the graph corresponding to the permutations

$$\sigma k = k+1 \pmod{n}, \qquad \tau k = k^{-1} \pmod{n}$$

is expanding with a small factor. However, repeated application of the same permutations does not improve expansion, so the large amount of expansion required by our definitions cannot be achieved.

Using the above-described way of realizing connections through two extra layers, the theorems translate to generating expanding connections networks by using only five (fixed-wired) permutations, which give *3-regular fixed connection networks*: σ between U and U', τ between V and V', and the identity permutations between U and U', between U' and V' and between V and V'. U sends information to U' using σ and similarly V to V' using τ. Next, U' and V' communicate using the identity, then U' recycles information to U using the identity; same with V.

The process repeats a number of times, and we use the same network to get a better expansion by simply running it a few more times (increase r).

2 Proof of the theorems

For the proof of Theorem 1 we will use four lemmas, whose proofs we present in a separate section.

The following lemma has been conjectured by Minc [7] and proved by Brègman [3]. (For a simple proof, see Schrijver [11].)

Lemma 1 *The permanent of an $n \times n$ 0-1 matrix does not exceed*

$$\prod_{i=1}^{n} (r_i!)^{1/r_i}, \tag{4}$$

where r_i are the row-sums of the matrix.

Corollary 1 *The permanent of an $n \times n$ 0-1 matrix is at most $n!p^{c_5 n}$, where p is the proportion of ones in the whole matrix.*

We also use two lemmas that say roughly the following: There are two small families F_1 and F_2 of random-looking sets such that every set contains – as a subset – a member of F_1 and is contained in a member of F_2.

Definition A set R is called d-random if, for any set B of size at most d,

$$|R+B| \geqslant \tfrac{1}{2}|R||B|. \tag{5}$$

($R+B$ is defined as the set $\{r+b;\, r \in R,\, b \in B\}$; in particular, it is empty if either R or B is empty.)

Lemma 2 (Random-looking sets) *If $N > c_6 d \log d$, then any set A of size $c_7 dN$ contains a d-random subset R of size N. In fact, most subsets of A of size N are d-random.*

The lemma is true in any additive group; in particular, it is also valid modulo n.

Remark The lemma offers only a probabilistic construction. One may be tempted to exhibit such subsets R by constructing sets with small discrete Fourier transforms. This, however, presents unsurmountable technical difficulties.

For proving Theorem 2, the following counterpart of the previous lemma will be used.

Lemma 3 (Covering k-sets with random-looking sets) *Given n, k and $d > c_8$, let $l > c_9 k$ be such that*

$$\frac{l}{n} < \left(\frac{k}{n}\right)^{c_{10}/d}.$$

Then, there exists a family $\{S_1, S_2, \ldots, S_N\}$ of l-sets such that

(a) the sets S_i, $1 \leqslant i \leqslant N$, cover all k-subsets of $[1, n]$;

(b) for any S_i and any d-set B,

$$|\overline{S_i + B}| < c_{11} k.$$

Furthermore,

$$N = 2\frac{\binom{n}{k}}{\binom{l}{k}}\log\binom{n}{k}$$

and most families of N l-sets satisfy the above two conditions.

Proof of (c) of Theorem 1 Fix a set X, $|X| = \alpha n$, and the permutation σ, and let Y, $|Y| = \alpha n$, vary together with the random permutation τ.

Define the sets $I(k) = \{i \leqslant r; \sigma^{-i}k \in X\} = (k-X)\cap[1,r]$, $1 \leqslant k \leqslant n$. If $\sigma^i \cap \tau^i = \emptyset$ for all $i \leqslant r$, then, for all k and $i \in I(k)$, we have $\tau^{-i} \notin Y$.

Let us represent Y by its location on τ, that is, define $\tilde{Y} = \{j : \tau^j 1 \in Y\}$. We now fix $\tilde{Y}$ (rather than Y) and compute the probability of the event

$$C = \{\tau : \text{for all } k \text{ and } i \in I(k),\ k \notin \tau^i Y\}.$$

(The probability of C is defined as $|C|/(n-1)!$.)

Now,

$$\begin{aligned} C &= \{\tau : \text{for all } k \text{ and } i \in I(k) \text{ and } j \in \tilde{Y},\ k \notin \tau^{i+j}1\} \\ &= \{\sigma : \text{for all } k,\ k \notin \tau^{\tilde{Y}+I(k)}1\}, \end{aligned}$$

where $+$ is meant modulo n.

Let us represent τ by the $(n-1)\times(n-1)$ permutation matrix

$$b_{ij} = \begin{cases} 1 & \text{if } i+1 = \tau^j 1, \\ 0 & \text{otherwise,} \end{cases}$$

where $1 \leqslant i, j \leqslant n-1$. Then any event C means that τ is chosen as a permutation from the 0-1 matrix A, whose $(k-1)$-th row, $2 \leqslant k \leqslant n$, has zero in the locations $\tilde{Y}+I(k) \pmod{n}$.

This, by Lemma 1, has a probability less than

$$\left(\frac{1}{n^2}\sum_{k\in[2,n]} |\overline{\tilde{Y}+I(k)}|\right)^{cn} < \exp\left\{-c\sum_{k\in[2,n]}\frac{|\tilde{Y}+I(k)|}{n}\right\} < \mathrm{e}^{-c\alpha^2 n}, \quad (6)$$

since at least αn of the sets $I(k)$ are non-empty.

Unfortunately, the number of possible sets Y or, which is the same, the number of $\tilde{Y}$, is too large (about $\exp h(\alpha)n$). To cope with this, we select a small, random-looking subset Y' inside Y. This will drastically decrease the number of possible choices for Y', but, if r was chosen large enough, most of the sets $\tilde{Y}' + I(k)$ will still be as large as before.

More precisely, let us select a d-random subset $\tilde{Y}'$ of $\tilde{Y}$ of size

$$\frac{1}{d}|\tilde{Y}|, \qquad \text{where } d = c_{12}\frac{1}{\alpha}\log\frac{1}{\alpha}.$$

By Lemma 2, such a $\tilde{Y}'$ exists if $|\tilde{Y}| = |Y| = \alpha n \geqslant c_{13}d^2 \log d$. Now, since $\tilde{Y}'$ is d-random, we have

$$|\tilde{Y}' + I(k)| \geqslant \tfrac{1}{2}|\tilde{Y}| \min\{d, |I(k)|\}. \tag{7}$$

We have to estimate from below the sum $S = \sum_k \min\{d, |I(k)|\}$, since equation (6) estimates the probability in question by $\exp\{-c|\tilde{Y}|S/n\} \leqslant \exp\{-c\alpha S\}$.

Another problem we have to deal with is the large number of choices for X. Just as for the set Y, we can represent X by its location on σ, that is, define $\tilde{X} = \{i : \sigma^i 1 \in X\}$ and, similarly, let $\tilde{k}$ be such that $k = \sigma^{\tilde{k}}1$. Thus, $I(k) = [1, r] \cap (\tilde{k} - \tilde{X})$.

To combine the two remaining tasks (shrinking X and estimating S from below), we use the following simple result.

Lemma 4 *Let $r = d^2$. For any Z, $|Z| = \alpha n$, there is a subset $Z' \subset Z$, $|Z'| = |Z|/d$, such that*

$$\frac{1}{n}\sum_{k=1}^{n} \min\{d, |Z' \cap [k+1, k+r]|\} \geqslant c_{14}\alpha d.$$

Using Lemma 4 and inequalities (6) and (7), we get the bound

$$\text{Prob}(C) \leqslant \exp\{-c_{15}\alpha'\alpha d n\} = \exp\{-c_{16}\alpha^2 n\}.$$

The number of choices for X' and Y' is at most $\exp 2h(\alpha')n$, where $\alpha' = c_{17}\alpha/d$. Hence, the probability that C occurs for some X and Y is less than $\exp\{-c_{18}\alpha^2 n\}$ as long as $h(\alpha') < c_{19}\alpha^2$.

This is satisfied by the choice

$$d = c\frac{1}{\alpha}\log\frac{1}{\alpha}$$

(with large c), which leads to

$$r = c_{20}\left(\frac{1}{\alpha}\log\frac{1}{\alpha}\right)^2. \qquad \square$$

Proof of Theorem 2 Since the proof is very similar to that of Theorem 1, we will only sketch it.

By changing α to $\frac{1}{2}\alpha$, we can reduce the problem to the following. Show that almost all pairs (σ, τ) of cyclic permutations satisfy the following condition.

For all $X \subset [n]$, $|X| \leqslant \alpha n$, we have

$$\left| \bigcup_{1 \leqslant i \leqslant r} \pi_i X \right| > A|X|, \tag{8}$$

where $A = 1/2\alpha$ and $\pi_i = \tau^{-i}\sigma^i$.

Or, equivalently, for any two sets $X, Y \subset [n]$, $|Y| = n - A|X| \geqslant \frac{1}{2}n$, there exists an i, $1 \leqslant i \leqslant r$, such that

$$\sigma^i X \cap \tau^i Y \neq \emptyset. \tag{9}$$

The proof is similar to that of Theorem 1. We select a small, random-looking set inside X using Lemma 2, and a random-looking set inside Y from the small family of such sets guaranteed by Lemma 3. For this latter selection, we choose an l such that

$$\left(\frac{k}{n}\right)^{c_{21}/A} < \frac{l}{n} < \left(\frac{k}{n}\right)^{c_{22}/d}.$$

The rest of the proof is standard calculus. □

Remark The proofs would be greatly simplified if we only set out to prove that the larger family of permutations $\tau^{-i}\sigma^j$, $1 \leqslant i, j \leqslant r$, is expanding.

3 Proof of the lemmas

Proof of Lemma 2 Write $n = c_7 dN$ and $p = c_{23}/d$, where $c_7 c_{23} > 1$. We choose a random subset of A by making independent randomizations for each point in A whether to include it in R. If each point has a probability p to succeed, then, with a very large probability, the obtained subset will be of size greater than N. Delete arbitrary points to make the size equal to N.

Given an arbitrary set $B = \{b_1, b_2, \ldots\}$ of size d, we write $R_i = R + b_i$. We have

$$|R+B| = \left| \bigcup_{i=1}^{d} R_i \right| \geqslant dN - \sum_{1 \leqslant i < j \leqslant d} |R_i R_j|. \tag{10}$$

Let us write $\|R\|$ for the maximum number of times a non-zero integer can appear as the difference of two elements in R:

$$\|R\| = \max_{\Delta \neq 0} |\{(r, r'); r, r' \in R, r - r' = \Delta\}|. \tag{11}$$

Clearly, $|R_i R_j| \leqslant \|R\|$, and thus

$$|R+B| \geqslant dN - \binom{d}{2}\|R\|. \tag{12}$$

We show now that, with a large probability, $\|R\| \leqslant N/d$, which implies that $|R+B| \geqslant \frac{1}{2}dN = \frac{1}{2}|R||B|$. (Since $\|R\| \leqslant N/d$ is the only property we use, it is clear that the proof also applies in the case $|B| < d$.)

Let us fix a non-zero integer Δ, and define a graph G on the points of A by connecting two elements if their difference is equal to Δ. It is clear that G is a union of (vertex-) disjoint paths and cycles (additive group!) and isolated vertices.

The subset R defines a (spanned) subgraph H of G, which is also a union of disjoint paths and cycles. We have to estimate the probability that H has more than $x = N/d$ edges.

For a specific choice of x edges from G, the probability that they all get into H is p^m, where m is the number of vertices these edges cover. (Actually, it is even less than that, since some of the vertices have been thrown away.)

If $N(m,x)$ is the number of ways to choose x edges from G which cover m vertices altogether, then we want to estimate the quantity

$$Q = \sum_m N(m,x)p^m.$$

Now, one can specify the obtained paths (and cycles) in H by specifying their 'left' endpoints and their lengths. If the x edges form k such intervals then $m = x+k$. Since the sum of the lengths is x, we get

$$N(m,x) \leqslant \sum_{k=1}^{x} \binom{n}{k}\binom{x-1}{k-1}$$

and obtain the estimate

$$Q \leqslant \sum_{k=1}^{x} \binom{n}{k}\binom{x-1}{k-1} p^{x+k} < \binom{n}{x} 2^x p^{2x} < \left(\frac{2enp^2}{x}\right)^x$$

$$= (2ec_7c_{23}^2)^x < \frac{1}{n^3}$$

if $2ec_7c_{23}^2 < 1$ and c_6 was chosen large enough.

Since there are only $O(n^2)$ potential values for Δ, the lemma is proved. □

Proof of Lemma 3 Let us choose N sets of size l at random. For a fixed k-set, the probability that it is not contained in any of the l-sets is

$$p = \left[1 - \binom{l}{k} \Big/ \binom{n}{k}\right]^N.$$

Our choice of N guaranties that $\binom{n}{k}p$ is small, thus proving (a).

Now, let us assume, contrary to (b), that for one of the selected l-sets S_i there is a set A of size $c_{11}k$ and another set B of size d such that A is disjoint from $\overline{S_i} + B$, in other words, that $\overline{S_i}$ is disjoint from $A - B$. (Of course, $A - B = \{a - b : a \in A, b \in B\}$.)

For a fixed choice of i, A and B, this has a probability

$$\binom{l}{|A-B|} \Big/ \binom{n}{|A-B|} \leqslant \binom{l}{|A|} \Big/ \binom{n}{|A|}$$

and the number of choices for i and B is $N\binom{n}{d}$.

Unfortunately, the number of choices for A is too large. As before, we can get round this problem by selecting inside A a d-random subset A' of size $c|A|/d$. This cuts back the number of choices from $\binom{n}{|A|}$ to $\binom{n}{|A'|}$, but we still have $|A' - B| \geqslant c|A|$. The rest of the proof is standard calculation. □

Proof of Lemma 4 *Let* $m_k = |Z \cap [k+1, k+r]|$. *Then,* $0 \leqslant m_k \leqslant r = d^2$ *and* $\sum_k m_k = r|Z| = r\alpha n$.

Let us choose the subset $Z' \subset Z$ at random. Then $|Z' \cap [k+1, k+r]|$ is a binomial random variable with parameters m_k and $1/d$, and with mean $m_k/d \leqslant d$. It is easy to see then that (writing $\boldsymbol{E}$ for expected value)

$$\boldsymbol{E}(\min\{d, |Z' \cap [k+1, k+r]|\}) \geqslant c_{24} \min\{d, m_k/d\} = c_{25} m_k/d.$$

Hence the lemma follows. □

References

[1] N. Alon, Eigenvalues and expanders, *Combinatorics*, **6** (1986), 83–96

[2] N. Alon & V. D. Milman, Eigenvalues, expanders and superconcentrators, *25th IEEE Symposium on Foundations of Computer Science*, 1984, 320–2

[3] L. M. Brègman, Certain properties of non-negative matrices and their permanents, *Soviet Math.Dokl.*, **14** (1973), 945–9 (translated from the Russian)

[4] O. Gabber, & Z. Galil, Explicit construction of linear superconcentrators, *J. Comp. and Sys. Sci.*, **22** (1981), 407–20

[5] A. Lubotzky, R. Phillips, & P. Sarnak, Explicit expanders and the Ramanujan conjectures, *Proc. FOCS* (*Berkeley*), 1986, 240–6

[6] G. Margulis, Explicit constructions of concentrators, *Problems of Information Transmission*, 1975, 325–32

[7] H. Minc, *Permanents*, Addison–Wesley, Reading MA, 1978. Reissued by Cambridge University Press, Cambridge, 1984

[8] M. Pinsker, On the complexity of a concentrator. *7th International Teletraffic Conference, Stockholm*, 1973, 316/1–318/4
[9] N. Pippenger, Superconcentrators, *SIAM Journal of Computing*, **6** (1977), 298–304
[10] P. Sarnak, Personal communication, 1988
[11] A. Schrijver, A short proof of Minc's conjecture, *J. Combin. Theory*, **A25** (1978), 80–1
[12] K. Zarankiewicz, *Colloq. Maths.*, **2** (1951), Problem 101 on page 301

Sum-free subsets

N. Alon* and D. J. Kleitman*

Abstract

A subset A in an Abelian group is called *sum-free* if $(A+A) \cap A = \emptyset$. We prove that for every finite Abelian group G, every set B of n non-zero elements of G contains a sum-free subset A of cardinality $|A| > \frac{2}{7}n$. The constant $\frac{2}{7}$ is best possible.

1 Introduction

A subset A of an Abelian group is called *sum-free* if $(A+A) \cap A = \emptyset$, i.e., if there are no (not necessarily distinct) $a, b, c \in A$ such that $a+b = c$. There is a considerable amount of results concerning sum-free subsets of Abelian groups. Many of these appear in the survey article [14] and some of its references. Our research here was motivated by a question we heard from Y. Caro, who asked if there is a positive constant c, such that any set B of n positive integers contains a sum-free subset A of cardinality

$$A > cn.$$

We have found a very simple proof of the following statement, which answers this question. Not surprisingly we learned later that almost the same result, without the strict inequality and with a rather similar proof, had been proved by Erdős more than twenty years ago (see [7]).

Proposition 1.1 *Any set B of n non-zero integers contains a sum-free subset A of cardinality*

* Work supported in part by a grant from the United States–Israel Binational Science Foundation and by a Bergman Memorial grant.

$$|A| > \tfrac{1}{3}n.$$

We can show that the constant $\frac{1}{3}$ cannot be replaced by $\frac{12}{29}$ (or any bigger constant), improving the result in [7], which asserts that the constant $\frac{1}{3}$ cannot be replaced by $\frac{3}{7}$. Although this is a very modest improvement, we believe it is worth mentioning, as it suggests that $\frac{1}{3}$ may actually be the best possible constant.

For a subset B of an Abelian group, let $s(B)$ denote the maximum cardinality of a sum-free subset of B. Similarly, for a sequence $A = (a_1, a_2, \ldots, a_n)$ of (not necessarily distinct) elements of an Abelian group, let $s(A)$ denote the maximum number of elements in a sum-free subsequence $(a_{i_1}, a_{i_2}, \ldots, a_{i_k})$ of A, i.e., the maximum k such that there are $1 \leqslant i_1 < i_2 < \ldots < i_k \leqslant n$, where the set $\{a_{i_1}, a_{i_2}, \ldots, a_{i_k}\}$ is a sum-free set. In these notations, Proposition 1.1 is simply the statement that for every set B of non-zero integers, $s(B) > \frac{1}{3}|B|$. Its proof applies to sequences as well, and establishes the following result (which is clearly stronger than Proposition 1.1).

Proposition 1.2 *For any sequence B of non-zero integers, $s(B) > \frac{1}{3}|B|$.*

On he other hand, we construct sequences B with $s(B) < \frac{11}{28}|B|$. Moreover, we show that, for every sequence A, there is a sequence B such that

$$\frac{s(B)}{|B|} \leqslant \frac{s(A)}{|A|} - \frac{1}{(|A| - s(A) + 1)!\,\mathrm{e}|A|}.$$

Therefore, the infimum of the ratio $s(B)/|B|$, as B ranges over all sequences of integers, is not attained.

Babai and Sós [4] raised the problem of estimating the maximum size of the sum-free subsets of n elements of general groups. Our main result in this paper is the following theorem, which settles this problem for finite Abelian groups.

Theorem 1.3 *For any finite Abelian group G, every set B of non-zero elements of G satisfies $s(B) > \frac{2}{7}|B|$. The constant $\frac{2}{7}$ is best possible. Similarly, every sequence A of non-zero elements of G satisfies $s(A) > \frac{2}{7}|A|$, and the constant $\frac{2}{7}$ is optimal.*

Our paper is organized as follows. In Section 2 we present simple proofs of Propositions 1.1 and 1.2 which slightly improve Erdős' result. We construct sets B and sequences A of non-zero integers with relatively small values of $s(B)/|B|$ and $s(A)/|A|$.

In Section 3 we consider general finite Abelian groups and prove Theorem 1.3. Finally, Section 4 includes several extensions and consequences of the above results and a few open problems.

2 Sum-free subsets of integers

We first prove Proposition 1.2 (which implies Proposition 1.1). Let $B = (b_1, b_2, \ldots, b_n)$ be a sequence of n non-zero integers. Let $p = 3k+2$ be a prime, which satisfies

$$p > 2 \max_{1 \leqslant i \leqslant n} |b_i|,$$

and put $C = \{k+1, k+2, \ldots, 2k+1\}$. Observe that C is a sum-free subset of the cyclic group Z_p and that

$$\frac{|C|}{p-1} = \frac{k+1}{3k+1} > \tfrac{1}{3}.$$

Let us choose at random an integer x $(1 \leqslant x < p)$ according to a uniform distribution on $\{1, 2, \ldots, p-1\}$, and define $d_1, \ldots, d_n$ by $d_i \equiv xb_i \pmod{p}$ $(0 \leqslant d_i < p)$. Trivially, for every fixed i $(1 \leqslant i \leqslant n)$ as x ranges over all the numbers $1, 2, \ldots, p-1$, d_i ranges over all non-zero elements of Z_p and hence

$$\Pr(d_i \in C) = \frac{|C|}{p-1} > \tfrac{1}{3}.$$

Therefore the expected number of elements b_i such that $d_i \in C$ is more than $\frac{1}{3}n$. Consequently, there is an x $(1 \leqslant x < p)$ and a subsequence A of B of cardinality $|A| > \frac{1}{3}|B|$, such that $xa \pmod{p} \in C$ for all $a \in A$. This A is clearly sum-free, since if $a_1 + a_2 = a_3$ for some $a_1, a_2, a_3 \in A$ then $xa_1 + xa_2 \equiv xa_3 \pmod{p}$, contradicting the fact that C is a sum-free subset of Z_p. This completes the proof. □

Next we show that the constant $\frac{1}{3}$ in Proposition 1.1 cannot be replaced by $\frac{12}{29}$. Put $B = \{1, 2, 3, 4, 5, 6, 10\}$. If $A \subseteq B$ is sum-free, then $|A \cap \{1, 2\}| \leqslant 1$, $|A \cap \{3, 6\}| \leqslant 1$ and $|A \cap \{5, 10\}| \leqslant 1$, Consequently, if A has more than 3 elements, then A has precisely one element from each of the 3 pairs $\{1, 2\}$, $\{3, 6\}$ and $\{5, 10\}$. However, in this case, since $4 \in A$, $2 \notin A$ and hence $1 \in A$. As A is sum-free, $3 \notin A$ and $5 \notin A$ and hence $6 \in A$ and $10 \in A$. This is a contradiction, since $4+6 = 10$. Therefore, for the above set B, $s(B) \leqslant 3 = \frac{3}{7}|B|$. In fact, $s(B) = 3$ since, e.g., $\{1, 3, 10\}$ is sum-free. A similar simple case analysis shows that, if $A \subseteq B$, $|A| = 3$ and $A \cup \{8\}$ is sum-free, then $\{1, 10\} \subseteq A$.

Indeed, $4 \notin A$ and hence A contains precisely one element from each of the pairs $\{1,2\}$, $\{3,6\}$ and $\{5,10\}$. If $2 \in A$ then $6 \notin A$ and $10 \notin A$ and hence $\{2,3,5\} = A$, contradicting the fact that it is sum-free. Thus $2 \notin A$ and $1 \in A$. If $5 \in A$ then $3 \notin A$ (as $5+3=8$) and $6 \notin A$, which is impossible. Hence $10 \in A$ and $\{1,10\} \subseteq A$, as claimed. We next apply these properties of the set B to construct a set C with $s(C) \leqslant \frac{12}{29}|C|$ $(< \frac{3}{7}|C|)$. Put $C = B \cup 7B \cup 8B \cup 9B \cup \{64\}$. Clearly $|C| = 29$ and $s(C) \leqslant 4s(B)+1 = 13$. We claim that in fact $s(C) \leqslant 12$. Indeed, suppose this is false, and let $A \subseteq C$ be a sum-free subset of C of cardinality 13. Then clearly

$$|A \cap B| = |A \cap 7B| = |A \cap 8B| = |A \cap 9B| = 3$$

and $64 \in A$. For each $i \in \{1,7,8,9\}$ define $A_i = A \cap iB$ and $A_i' = \{a/i : a \in A_i\}$. Clearly each A_i' is a sum-free subset of B of cardinality 3. Since $64 = 8 \cdot 8$, $A_8' \cup \{8\}$ is sum-free and hence $\{1,10\} \in A_8'$. Therefore $8, 80 \in A$. Hence $A_1 \cup \{8\}$ is sum-free and thus $1, 10 \in A$. It follows that $6 \cdot 9 = 64 - 10$, $1 \cdot 9 = 8+1$, $2 \cdot 9 = 10+8$ and $10 \cdot 9 = 80+10$ are not in A. Thus $A_9' = \{3,4,5\}$ and hence $27, 36, 45 \in A$. Consequently $7 \cdot 1 = 8-1 \notin A$, $7 \cdot 5 = 45-10 \notin A$, $7 \cdot 4 = 27+1 \notin A$ and $7 \cdot 10 = 80-10 \notin A$. Thus $A_7' = \{2,3,6\}$ and $21, 42 \in A$, contradicting the fact that A is sum-free. Therefore $s(C) \leqslant \frac{12}{29}|C|$ as claimed and the constant $\frac{1}{3}$ cannot be replaced by $\frac{12}{29}$. Notice that the same estimate holds for each of the sets

$$C_m = C \cup 1000C \cup 1000^2 C \cup \ldots \cup 1000^{m-1} C.$$

Hence, for every positive integer m there is a set of $n = 29m$ positive integers such that $s(C_m) \leqslant \frac{12}{29}|C_m|$.

In the case of sequences, we can obtain better upper bounds than the above one. Here we need the following well-known theorem of Schur [13].

Theorem 2.1 (Schur [13], see also, e.g., [14] or [10]) *For every $k \geqslant 2$ there exists a finite smallest possible integer $f(k) \leqslant k!\,e$, such that there is no partition of the integers $\{1,2,\ldots,f(k)\}$ into k sum-free sets. In particular, $f(2) = 5$ and $f(3) = 14$.*

The next lemma provides a way of constructing sequences B with a relatively small value of $s(B)/|B|$.

Lemma 2.2 *Let $A = (a_1, a_2, \ldots, a_n)$ be a sequence of n non-zero integers and put $s = s(A)$. Suppose there exists a set $C = \{c_1, \ldots, c_k\}$ of k integers such that every sum-free sequence of A of cardinality s contains* *at least one term that is equal to a member of C. Let $f = f(k)$ be the number given in Theorem 2.1, and let B be the sequence $(b_{ij} = ja_i : 1 \leqslant i \leqslant n, 1 \leqslant j \leqslant f)$. Then $s(B) \leqslant fs-1$. Hence*

$$\frac{s(B)}{|B|} \leqslant \frac{s(A)}{|A|} - \frac{1}{nf}. \tag{2.1}$$

Proof The sequence B is a union of the f sequences $B_j = (ja_1, ja_2, \dots, ja_n)$ for $1 \leqslant j \leqslant f$. Since each such sequence is simply a product of the members of A by the constant j, we have that $s(B_j) = s(A) = s$. Consequently,

$$s(B) \leqslant \sum_{j=1}^{f} s(B_j) = sf.$$

To complete the proof we must show that the last inequality is strict. Assume it is not and let D be a sum-free subsequence of B of cardinality sf. Clearly D must contain precisely s elements from each B_j. We now define a partition of $\{1,2,\dots,f\}$ into $|C| = k$ subsets as follows. For each j $(1 \leqslant j \leqslant f)$ D contains a sum-free sequence of s elements of B_j. Let $(d_1,\dots,d_s)$ be this subsequence. Clearly $(d_1/j, d_2/j, \dots, d_s/j)$ is a sum-free subsequence of A of cardinality s. By the definition of $C = \{c_1,\dots,c_k\}$, there are i and l $(1 \leqslant i \leqslant s,\ 1 \leqslant l \leqslant k)$ such that $d_i/j = c_l$. Choose, arbitrarily, such i and l and assign j to the lth class of the partition. Since $f = f(k)$ was chosen according to Theorem 2.1, there is an l $(1 \leqslant l \leqslant k)$ and there are (not necessarily distinct) j_1, j_2 and j_3 such that $j_1 + j_2 = j_3$ and j_1, j_2 and j_3 all belong to the lth class in the partition defined above. Consequently, there are $d_1, d_2, d_3 \in D$ such that $d_i = c_l j_i$ for $1 \leqslant i \leqslant 3$. However, in this case,

$$d_1 + d_2 = c_l(j_1 + j_2) = c_l j_3 = d_3,$$

contradicting the fact that D is sum-free. Hence, our assumption that $s(B) = sf$ is false and $s(B) \leqslant sf - 1$. This completes the proof. □

Corollary 2.3 *For every sequence A of non-zero integers there is a sequence B such that*

$$\frac{s(B)}{|B|} \leqslant \frac{s(A)}{|A|} - \frac{1}{(|A| - s(A) + 1)!\,\mathrm{e}|A|}.$$

Proof Suppose $A = (a_1, a_2, \dots, a_n)$ and $s = s(A)$. Let C be the set of all values of the first $n-s+1$ members of A. Clearly the set C has $k \leqslant n-s+1$ members. By Theorem 2.1, $f(k) \leqslant k!\,\mathrm{e}$. Thus, the assertion of the corollary follows from Lemma 2.2. □

Corollary 2.4 *Let S be the following sequence of 140 elements: $S = (ijl : 1 \leqslant i \leqslant 2, 1 \leqslant j \leqslant 5, 1 \leqslant l \leqslant 14)$. Then $s(S) \leqslant 55$. Hence $s(S)/|S| \leqslant \frac{11}{28}$.*

Proof Let A be the sequence $A = (1,2)$. Clearly $s = s(A) = 1$ and s, A, $k = 2$ and $C = \{1,2\}$ satisfy the hypotheses in Lemma 2.2. Since, by Theorem 2.1, $f(2) = 5$, Lemma 2.2 implies that, for the sequence $B = (ij : 1 \leqslant i \leqslant 2, 1 \leqslant j \leqslant 5)$, $s(B) \leqslant 5-1 = 4$. In fact, since, e.g., $(1,3,8,10)$ is sum-free, $s(B) = 4$. Put $k = 3$ and $C = \{1,2,4\}$. One can easily check that C, k, $s = 4$ and $A = (ij : 1 \leqslant i \leqslant 2, 1 \leqslant j \leqslant 5)$ satisfy the hypotheses of Lemma 2.2. Since, by Theorem 2.1, $f(3) = 14$, the lemma implies that, for the sequence S defined by the corollary, $s(S) \leqslant 4 \cdot 14 - 1 = 55$, as needed. □

Remark 2.5 Lemma 2.2 (or Corollary 2.3) clearly enables us to construct from S sequences B with $s(B)/|B| < \frac{11}{28}$. Since it does not seem that this method suffices to close the gap between the lower bound in Proposition 1.2 and the upper bound in the last corollary, we omit the detailed computation of $s(B)$ for the resulting sequences B,

3 Sum-free subsets in finite Abelian groups

In this section we prove Theorem 1.3. We first show that, for any finite Abelian group G and every sequence A of non-zero elements of G, $s(A) > \frac{2}{7}|A|$. This clearly implies a similar inequality for subsets of G. The basic method in the proof is similar to that used in the proof of Proposition 1.2, but requires several additional ideas. We start with the following simple observations concerning the cyclic group Z_n. Define

$$I_1 = \{x \in Z_n : \tfrac{1}{3}n < x \leqslant \tfrac{2}{3}n\},$$

$$I_2 = \{x \in Z_n : \tfrac{1}{6}n < x \leqslant \tfrac{1}{3}n \text{ or } \tfrac{2}{3}n < x \leqslant \tfrac{5}{6}n\}.$$

One can easily check that both I_1 and I_2 are sum-free subsets of Z_n.

For any divisor d of n, let dZ_n denote the subgroup of all multiples of d in Z_n, i.e., $dZ_n = \{0, d, 2d, \ldots, n-d\}$. Clearly dZ_n has n/d elements. In our proof we need the values of the fractions $|dZ_n \cap I_j|/|dZ_n|$ for all divisors d of n and $j = 1,2$. Clearly these can be computed by a straightforward case analysis, which is summarized in the following statement, the easy detailed proof of which is omitted.

Lemma 3.1 *The table contains the quantities $|dZ_n \cap I_j|/|dZ_n|$ for all possible $n \geqslant 2$ and $d|n$ $(1 \leqslant d < n)$ depending on the value of $|dZ_n| = n/d$ modulo 6.*

$\frac{n}{d}$	$\frac{\lvert dZ_n \cap I_1\rvert}{\lvert dZ_n\rvert}$	$\frac{\lvert dZ_n \cap I_2\rvert}{\lvert dZ_n\rvert}$
$6k$	$\frac{2k}{6k} = \frac{1}{3}$	$\frac{2k}{6k} = \frac{1}{3}$
$6k+1$	$\frac{2k}{6k+1} \geqslant \frac{2}{7}$	$\frac{2k}{6k+1} \geqslant \frac{2}{7}$
$6k+2$	$\frac{2k+1}{6k+2}$	$\frac{2k}{6k+2}$
$6k+3$	$\frac{2k+1}{6k+3} = \frac{1}{3}$	$\frac{2k+1}{6k+3} = \frac{1}{3}$
$6k+4$	$\frac{2k+1}{6k+4} \geqslant \frac{1}{4}$	$\frac{2k+2}{6k+4} > \frac{1}{3}$
$6k+5$	$\frac{2k+2}{6k+5} > \frac{1}{3}$	$\frac{2k+2}{6k+5} > \frac{1}{3}$

In particular, for all admissible values of n and d,

$$\frac{4}{7}\,\frac{|dZ_n \cap I_1|}{|dZ_n|} + \frac{3}{7}\,\frac{|dZ_n \cap I_2|}{|dZ_n|} \geqslant \frac{2}{7}. \tag{3.1}$$

We can now prove the lower bounds in Theorem 1.3. Let G be an arbitrary finite Abelian group and let $B = (b_1, b_2, \ldots, b_m)$ be a sequence of m non-zero elements of G. As is well known, G is a direct sum of cyclic groups and therefore there are n and s such that G is a subgroup of the direct sum H of s copies of Z_n. Thus we can think of the elements of B as members of H. Each such element b_i is, in fact, a vector $b_i = (b_{i1}, b_{i2}, \ldots, b_{is})$, where, for each i, $0 \leqslant b_{i1}, \ldots, b_{is} < n$ and not all the b_{ij} are zero. Let us choose a random element $(x_1, x_2, \ldots, x_s)$ of $H = Z_n^s$ according to a uniform distribution and define m elements $f_1, f_2, \ldots, f_m$ of the cyclic group Z_n by

$$f_i = \sum_{j=1}^{s} x_j b_{ij} \pmod{n}.$$

Notice that, for every fixed i $(1 \leqslant i \leqslant m)$, the mapping

$$(x_1, x_2, \ldots, x_s) \mapsto \sum_{j=1}^{s} x_j b_{ij} \pmod{n}$$

is a homomorphism from H to Z_n. Moreover, if d_i is the greatest common divisor of b_{i1}, b_{i1}, $\ldots$, b_{is} and n then $d_i < n$, $d_i | n$ and the image of this homomorphism is just d_iZ_n. Consequently, as $\boldsymbol{x} = (x_1,\ldots,x_s)$ ranges over all elements of H, f_i ranges over all elements of d_iZ_n and attains each value of d_iZ_n the same number of times. It follows that, for each $j = 1,2$,

$$\Pr(f_i \in I_j) = \frac{|d_iZ_n \cap I_j|}{|d_iZ_n|}.$$

For each divisor d of n $(1 \leqslant d < n)$, let m_d denote the number of elements b_i in B such that $\gcd(b_{i1}, b_{i2}, \ldots, b_{is}, n) = d$. Clearly $\sum_{d|n, d<n} m_d = m$ and, for $j = 1,2$, the expected number of elements b_i such that $f_i \in I_j$ is

$$M_j = \sum_{\substack{d|n\\ d<n}} m_d \frac{|dZ_n \cap I_j|}{|dZ_n|}.$$

Moreover, since $\boldsymbol{x} = (0,0,\ldots,0) \in H$ maps every b_i into $f_i = 0 \notin I_1$, it follows that there is an $\boldsymbol{x} = (x_1,\ldots,x_s) \in H$ and a subsequence A of strictly more than M_1 elements of B such that each $\boldsymbol{a} = (a_1,\ldots,a_s) \in A$ is mapped by $\boldsymbol{x}$ into $\sum_{i=1}^{s} x_i a_i \pmod{n} \in I_1$. Clearly this A is a sum-free subsequence of B (since I_1 is sum-free) and thus

$$s(B) > M_1 = \sum_{\substack{d|n\\ d<n}} m_d \frac{|dZ_n \cap I_1|}{|dZ_n|}. \tag{3.2}$$

Similarly, since I_2 is sum-free,

$$s(B) > M_2 = \sum_{\substack{d|n\\ d<n}} m_d \frac{|dZ_n \cap I_2|}{|dZ_n|}. \tag{3.3}$$

Combining (3.1), (3.2) and (3.3), we obtain

$$\begin{aligned} s(B) = \tfrac{4}{7}s(B) + \tfrac{3}{7}s(B) &> \tfrac{4}{7}M_1 + \tfrac{3}{7}M_2 \\ &\geqslant \sum_{\substack{d|n\\ d<n}} m_d \left(\frac{4}{7}\frac{|dZ_n \cap I_1|}{|dZ_n|} + \frac{3}{7}\frac{|dZ_n \cap I_2|}{|dZ_n|} \right) \\ &\geqslant \sum_{\substack{d|n\\ d<n}} m_d \times \tfrac{2}{7} = \tfrac{2}{7}m. \end{aligned}$$

Therefore B contains a sum-free subsequence of more than $\frac{2}{7}|B|$ members, completing the proof of the lower bounds (for sequences and sets) in Theorem 1.3.

The fact that $\frac{2}{7}$ is optimal (for both sets and sequences) follows from the following result of Rhemtulla and Street [12].

Theorem 3.2 (Rhemtulla and Street [12]) *Let $p = 3k+1$ be a prime and let G be the elementary Abelian group Z_p^s. Then $|G| = p^s$ and the maximum cardinality of a sum-free subset of G is kp^{s-1}.*

In view of this theorem, if we choose $G = Z_7^s$ and $B = G\backslash\{0\}$, then $|B| = 7^s - 1$ and $s(B) = 2\cdot 7^{s-1}$. Since s can be arbitrarily large, this shows that the constant $\frac{2}{7}$ in Theorem 1.3 is optimal (for sets, and hence for sequences too). □

4 Extensions, concluding remarks and open problems

1. One can easily generalize Proposition 1.2 (and 1.1) to the case of real numbers. This was done by Erdős in [7], where he proves the next statement for sets B.

   Proposition 4.1 *For any sequence B of non-zero reals, $s(B) \geqslant \frac{1}{3}|B|$.*

   The proof is similar to that of Proposition 1.2. If $B = (b_1, b_2, \ldots, b_n)$ is a sequence of non-zero reals and $\epsilon = \min_{1 \leqslant i \leqslant n} |b_i|$, we choose, randomly, a real number x according to a uniform distribution on $[1/\epsilon, 10n/\epsilon]$ and compute the numbers $d_i = b_i x \pmod 1$. One can easily show that the expected number of $d_i - s$ that belong to $[\frac{1}{3}, \frac{2}{3})$ is more than $\frac{1}{3}(n-1)$ and hence there is an x and a subsequence A of at least $\frac{1}{3}n$ members of B such that $xa \pmod 1 \in [\frac{1}{3}, \frac{2}{3})$ for each $a \in A$. This subsequence is sum-free, since $[\frac{1}{3}, \frac{2}{3})$ is sum-free with respect to addition modulo 1.

   In fact here also we can improve on Erdős's result and prove that strict inequality holds.

   Proposition 4.1′ *For any sequence B of non-zero reals, $s(B) > \frac{1}{3}|B|$.*

   To prove this fact, we apply Proposition 1.2. Given any arbitrary sequence $B = (b_1, \ldots, b_n)$ of reals, we claim that there is a sequence $C = (c_1, \ldots, c_n)$ of integers such that, for any $\epsilon_1, \ldots, \epsilon_n \in \{\pm 1, 0\}$, the sign of $\sum_{i=1}^n \epsilon_i b_i$ (which is 0, +1 or −1) is equal to that of $\sum_{i=1}^n \epsilon_i c_i$. To prove this we argue as follows. For each of the 3^n possible vectors $\boldsymbol{\epsilon} = (\epsilon_1, \ldots, \epsilon_n)$, let $E(\boldsymbol{\epsilon})$ be an equation or an inequality with the n variables $x_1, \ldots, x_n$ defined as follows: if $\sum_{i=1}^n \epsilon_i b_i = 0$ then $E(\boldsymbol{\epsilon})$ is the equation $\sum_{i=1}^n \epsilon_i x_i = 0$. If $\sum_{i=1}^n \epsilon_i b_i > 0$, let q be a positive rational so that $\sum_{i=1}^n \epsilon_i b_i \geqslant q$

and let $E(\epsilon)$ be the inequality $\sum_{i=1}^{n} \epsilon_i x_i \geqslant q$. Similarly, if $\sum_{i=1}^{n} \epsilon_i b_i < 0$, then $E(\epsilon)$ is the inequality $\sum_{i=1}^{n} \epsilon_i x_i \leqslant q$, where q is an arbitrary negative rational satisfying $\sum_{i=1}^{n} \epsilon_i b_i \leqslant q$. Consider the linear program in the n variables $x_1,\ldots,x_n$ consisting of the 3^n constraints $E(\epsilon)$. This program has a feasible real solution $(b_1,\ldots,b_n)$. Since all the constraints have rational coefficients it also has a rational solution $(d_1,\ldots,d_n)$. By multiplying all these numbers d_i by a suitable integer we obtain a sequence of integers $(c_1,\ldots,c_n)$ such that, for any $\epsilon_1,\ldots,\epsilon_n \in \{\pm 1, 0\}$,

$$\operatorname{sign}\left(\sum_{i=1}^{n} \epsilon_i b_i\right) = \operatorname{sign}\left(\sum_{i=1}^{n} \epsilon_i c_i\right).$$

Returning to Proposition 4.1′, let $B = (b_1,\ldots,b_n)$ be a sequence of non-zero reals. By the above discussion there is a sequence $C = (c_1,\ldots,c_n)$ of integers satisfying

$$\operatorname{sign}\left(\sum_{i=1}^{n} \epsilon_i b_i\right) = \operatorname{sign}\left(\sum_{i=1}^{n} \epsilon_i c_i\right)$$

for all $\epsilon_i \in \{\pm 1, 0\}$. In particular, no c_i is zero and $s(B) = s(C)$. By Proposition 1.2 $s(C) > \frac{1}{3}n$ and hence $s(B) > \frac{1}{3}n$, completing the proof of Proposition 4.1′. □

2. Motivated by Schur's theorem (stated in Section 2) and by the problem of estimating Ramsey numbers (see, e.g. [14] or [10]), various authors considered the problem of partitioning all the non-zero elements of a group into the minimum possible number of sum-free subsets. As noted by Abbott and Hanson [1], the original argument of Schur easily implies that the non-zero elements of no finite Abelian group of order n can be partitioned into less than $c_1 \log n/\log\log n$ sum-free subsets, where c_1 is an absolute constant. On the other hand, as is also observed in [1], the non-zero elements of any finite Abelian group of order n can be partitioned into $O(\log n)$ sum-free subsets. By repeatedly applying Theorem 1.3 (and Proposition 4.1), we clearly obtain the following more general result.

Proposition 4.2 *Any set of n non-zero elements in an arbitrary finite Abelian group can be partitioned into* $O(\log n)$ *sum-free subsets. Similar statements hold for any set of non-zero reals.*

3. A close inspection of the proof of Theorem 1.3 shows that the constant $\frac{2}{7}$, although the best possible for the general case, can be improved for many groups G. Thus for example the proof of Proposition 1.2 shows that, for any sequence B of non-zero elements of a cyclic group of prime order $p = 3k+2$,

$$s(B) \geqslant \frac{k+1}{3k+1}|B|$$

(and this is best possible for each such p, as can be easily shown using the well-known Cauchy–Davenport theorem [5], [6]). Similarly, for any sequence B of non-zero elements of Z_p, where $p \equiv 1 \pmod 3$ is a prime, $s(B) \geqslant \frac{1}{3}|B|$ and this is best possible for each such p. The proof of Theorem 1.3 easily gives that, for any sequence B of non-zero elements of Z_2^s

$$s(B) \geqslant \frac{2^{s-1}}{2^s-1}|B|.$$

This again is best possible for each s. By modifying the proof of Theorem 1.3, we can also improve the constants for various cyclic groups. in particular, we can show that if n is not divisible by any prime congruent to 2 modulo 3 then, for any sequence B of non-zero elements in Z_n, $s(B) \geqslant \frac{1}{3}|B|$. The proof is similar to that of Proposition 1.2; we multiply all the elements of B by a random member of Z_n^* (i.e., by a random number which is relatively prime to n) and compute the expected value of the numbers that are mapped to a certain sum-free subset of Z_n. The situation is more complicated when n is divisible by primes congruent to 2 modulo 3. A fruitful approach here is to multiply, for each divisor d of n, the elements of B by a random element of dZ_n^* and obtain a lower bound for $s(B)$ by computing the expected number of elements of B mapped to a certain sum-free subset of Z_n, (e.g., the subset $\{x \in Z_n : \frac{1}{3}n \leqslant x < \frac{2}{3}n\}$. This lower bound can be expressed as a linear combination of the quantities $m_d = \{b_i \in B : \gcd(b_i, n) = d\}$. All these lower bounds and the constraints $m_d \geqslant 0$ and $\sum m_d = |B|$ define a linear program from which a lower bound to $s(B)$ can be extracted. Using this method we can prove, for example, the following.

Proposition 4.3 *For any prime $p \equiv 2 \pmod 3$ and any $s \geqslant 1$, every sequence B of non-zero elements of the cyclic group Z_{p^s} satisfies $s(B) > \frac{1}{3}|B|$.*

We omit the somewhat tedious (though not too complicated) details of the proof. □

4. The proof of Proposition 1.2 (or 4.1) can be easily modified to show that, for any $r \geqslant 2$, any sequence B of n non-zero reals contains a subsequence A of size $\Omega(n/r)$ such that there are no $a_1, a_2, \ldots, a_r, a_{r+1} \in A$ such that $\sum_{i=1}^{r} a_i = a_{r+1}$. A similar statement for general Abelian groups is false. In fact, in Z_2^n the set of all non-zero elements B has cardinality $N = 2^n - 1$. Trivially, any subset A of B of cardinality $|A| = x$ with $\binom{x}{2} > 2^n - 1$ has $a_1, a_2, a_3, a_4 \in A$ such that, in Z_2^n, $a_1 + a_2 = a_3 + a_4$ and hence $a_1 + a_2 + a_3 = a_4$. Some related results have recently been obtained by Zs. Tuza.

5. Let us call a subset A of an Abelian group G weakly sum-free if there are no three *distinct* elements $a_1, a_2, a_3 \in A$ such that $a_1 + a_2 = a_3$. For a set $B \subseteq G$, let ws(B) denote the maximum cardinality of a weakly sum-free subset of B. Since each sum-free set is weakly sum-free, Theorem 1.3 implies that, for every Abelian group G and for every set B of non-zero elements of G, ws$(B) > \frac{2}{7}|B|$. Using some of the methods of [2] we can show that the constant $\frac{2}{7}$ is optimal here too. Indeed, take $G = Z_7^s$ and $B = G \backslash (\mathbf{0})$. Then $|B| = 7^s - 1$. Let $A \subseteq B$ be a weakly sum-free subset of B. To prove the optimality of the constant $\frac{2}{7}$ we show that, for every fixed $\epsilon > 0$, $|A| < 2 \cdot 7^{s-1} + \epsilon \cdot 7^s$, provided s is sufficiently large. Call an element a of A *good* if there are two distinct elements $b, c \in A$ such that $2a = b + c$. Otherwise it is *bad*. We claim that the number of bad elements is smaller than $\epsilon \cdot 7^s$, for $s > s_0(\epsilon)$. This is because otherwise, by the main result of [3] (see also [8] for a short proof and [9] for a much stronger result), there are three distinct bad elements a, b and c such that $2a = b + c$, contradicting the fact that a is bad. Therefore, if $|A| \geqslant 2 \cdot 7^{s-1} + \epsilon \cdot 7^s$ and $s > s_0(\epsilon)$, then A contains more than $2 \cdot 7^{s-1}$ good elements. It follows from the result of [12] (stated in Theorem 3.2 in Section 3) that there are three not necessarily distinct good elements d, e and f of A such that $d + e = f$. Clearly $d \neq f$ and $e \neq f$ (since $0 \notin A$). If $d \neq e$ then A is not weakly sum-free, contradicting its definition and completing the proof. Otherwise $2d = f$ and, since d is good, there are two distinct elements a and b of A such that $a + b = 2d = f$. Hence, in this case too, A is not weakly sum-free, completing the proof. □

6. An analogue of Proposition 1.2 can be established for measurable sets in the torus. Recall that the (one-dimensional) torus T is the group of real numbers x $(0 \leqslant x < 1)$ with addition modulo 1. Let μ be the usual Lebesque measure on T with $\mu(T) = 1$. We can prove the following.

Proposition 4.4 *For any measurable $B \subseteq T$ and for any $\epsilon > 0$, there is a (measurable) sum-free set $A \subseteq B$ such that $\mu(A) > (\frac{1}{3} - \epsilon)\mu(B)$. The constant $\frac{1}{3}$ is best possible.*

To prove this proposition, we use the fact that $f\colon T \to T$ defined by $f(x) = 2x \pmod 1$ is ergodic. For $i \geqslant 1$, put

$$B_i = \{x \in B : \tfrac{1}{3} \leqslant f^{(i)}(x) = 2^i x \pmod 1 < \tfrac{2}{3}\}.$$

Clearly B_i is sum-free. Since f is ergodic,

$$\lim_{n\to\infty} \frac{1}{n} \sum_{i=1}^{n} \mu(B_i) = \tfrac{1}{3}\mu(B),$$

implying that $\mu(B) > (\frac{1}{3} - \epsilon)\mu(B)$ for some i. The fact that the constant $\frac{1}{3}$ is best possible is proved by showing that, for each sum-free $A \subseteq T$, $\mu(A) \leqslant \frac{1}{3}$. Indeed, suppose $A \subseteq T$ is measurable with $\mu(A) \geqslant \frac{1}{3} + \epsilon$, where $\epsilon > 0$. Let p be a large prime, and call an element $i \in Z_p$ *A-full* if

$$\mu\left(A \cap \left[\frac{i}{p}, \frac{i+1}{p}\right]\right) > 0.9\frac{1}{p}.$$

For sufficiently large p, the cardinality of the set $B \subseteq Z_p$ of all A-full elements of Z_p is clearly bigger than $\frac{1}{3}p+1$. Consequently, by the Cauchy–Davenport theorem ([5], [6]), $|B+B| \geqslant 2|B|-1 > \frac{2}{3}p$. Hence $(B+B) \cap B \neq \emptyset$ and there are $b_1, b_2, b_3 \in B$ such that $b_1 + b_2 \equiv b_3 \pmod p$. One can easily check that this implies that A is not sum-free, as needed. □

By replacing the Cauchy–Devonport theorem by Kneser's theorem [11], we can show that the maximum-possible measure of a sum-free measurable subset of the n-dimensional torus is also $\frac{1}{3}$ for all $n \geqslant 1$.

7. Our proofs of Proposition 1.2 and 1.3 are probabilistic. In particular, they clearly supply an efficient randomized algorithm which, given a set of n non-zero integers B, finds, in expected polynomial time (in the length of the input), a sum-free subset of it of cardinality $\Omega(n)$. It would be interesting to find an efficient

deterministic algorithm for this problem. It would also be interesting to determine the best-possible constants in Propositions 1.1 and 1.2.

Acknowledgement

We would like to thank Y. Caro and I. Krasikov for helpful comments.

References

[1] H. L. Abbott & D. Hanson, A problem of Schur and its generalizations, *Acta Arithmetica*, **20** (1972), 175–87.

[2] N. Alon, Subset sums, *J. Number Theory*, **27** (1987), 196–205

[3] T. C. Brown & J. P. Buhler, A density version of a geometric Ramsey theorem, *J. Combinatorial Theory, Ser. A*, **32** (1982), 20–34

[4] L. Babai & V. T. Sós, Sidon sets in groups and induced subgraphs of Cayley graphs, *European J. Combin.*, **6** (1985), 101–14

[5] A. L. Cauchy, Recherches sur les nombres, *J. Ecole polytechn.*, **9** (1813), 99–116

[6] H. Davenport, On addition of residue classes, *J. London Math. Soc.*, **10** (1935), 30–2

[7] P. Erdős, External problems in number theory, *Proc. Symp. Pure Maths.*, Vol. VIII, AMS (1965), 181–9

[8] P. Frankl, R. L. Graham & V. Rödl, On subsets of Abelian groups with no 3-term arithmetic progressions, *J. Combinatorial Theory, Ser. A*, **45** (1987), 157–61

[9] H. Furstenberg & Y. Katznelson, An ergodic Szemerédi theorem for IP-systems and combinatorial theory, *J. d'Analyse Math.*, **45** (1985), 117–68

[10] R. L. Graham, B. L. Rothschild & J. H. Spencer, *Ramsey Theory*, Wiley, New York, 1980

[11] M. Kneser, Abschätzung der asymptotischen Dichte von Summenmengen, *Math. Zeit.*, **58** (1953), 459–84

[12] A. H. Rhemtulla & Ann Penfold Street, Maximum sum-free sets in elementary Abelian p-groups, *Canad. Math. Bull.*, **14** (1971), 73–80.

[13] I. Schur, Uber die Kongruenz $x^m + y^m = z^m \pmod{p}$, *Jahresbericht der Deutschen Mathematiker Vereinigung*, **25** (1916), 114–17

[14] W. D. Wallis, Ann Penfold Street & Jennifer Seberry Wallis, *Combinatorics: Room Squares, Sum-free Sets, Hadamard Matrices*. Lecture Notes in Mathematics, **292**, Springer, Berlin–Heidelberg–New York, 1972, Part 3, 123–277

Is there a different proof of the Erdős–Rado theorem?

James E. Baumgartner*

Abstract

We define the notion of an Erdős–Rado function which, if it exists, would make possible an essentially different approach to the Erdős–Rado theorem. Several combinatorial propositions, including $\omega_2 \to (\omega_1+2)^2_\omega$ and assertions about graphs on ω_2 due to Hajnal, imply the existence of Erdős–Rado functions. It is shown relatively consistent with $\mathrm{CH}+2^{\omega_1}=\omega_3$ that no Erdős–Rado functions exist. A list of open problems is included.

1 Introduction

The Erdős–Rado theorem [2], one of the cornerstones of combinatorial set theory has received several rather different-looking proofs over the years, including in particular the model-theoretic argument due to Simpson, yet, at bottom, all the proofs seem to make use of the fundamental ramification argument of Erdős and Rado. The purpose of this paper is to investigate the question whether a truly different proof is possible.

That question can be made precise as follows. Under CH, the first non-trivial version of the Erdős–Rado theorem (beyond Ramsey's theorem) says that $\omega_2 \to (\omega_1+1)^2_\omega$. In fact, the ramification argument yields rather more. For example, if $f\colon [\omega_2]^2 \to \omega$ and $g\colon [\omega_2]^2 \to \omega_1$ are given, it is possible to find a set X of order type ω_1+1 such that X is not only homogeneous for f but also end-homogeneous for g, i.e., whenever $\alpha,\beta,\gamma \in X$ and $\alpha<\beta<\gamma$ then $g\{\alpha,\beta\} = g\{\alpha,\gamma\}$. This is an example of a so-called canonical partition relation (see [1] for a

* The preparation of this paper was partially supported by National Science Foundation grant number DMS-870456.

systematic treatment). All the known proofs of the Erdős–Rado theorem yield this result; it may be regarded as characteristic of the ramification argument.

Let us say that $X \subseteq \omega_2$ is *anti-homogeneous* for g: $[\omega_2]^2 \to \omega_1$ provided that, whenever $\alpha, \beta, \gamma \in X$ and $\alpha < \beta < \gamma$ then $g\{\alpha,\beta\} \neq g\{\alpha,\gamma\}$. We call such a function an *Erdős–Rado function* (or *ER-function* for short) provided that for any f: $[\omega_2]^2 \to \omega_1$ there is an $X \subseteq \omega_2$ such that $|X| = 3$ and X is homogeneous for f and anti-homogeneous for g. (One may require $|X| > 3$ and so on, but the problems already seem non-trivial for this case.) We will be concerned with the existence of ER-functions.

The remainder of this paper is organized as follows. In Section 2 we consider some natural candidates for ER-functions, the fat and very fat functions, and we show that these functions have very large anti-homogeneous sets. In Section 3 we find several sufficient conditions for the existence of ER-functions. These include the partition relation $\omega_2 \to (\omega_1+2)^2_\omega$ and several partition relations for infinite graphs. Unfortunately we do not have consistency results for any of these sufficient conditions. Section 4 is devoted to showing how ER-functions may be destroyed by forcing and how that forcing may be iterated to find a model of the theory

$$\mathrm{ZFC} + \mathrm{CH} + 2^{\omega_1} = \omega_3 + \text{‘there are no ER-functions'}.$$

We conclude with a section on open problems.

Our set-theoretic notation, including Erdős notation for ordinary partition relations, is standard. Given a graph $G = (\kappa, E)$, where $E \subseteq [\kappa]^2$, we write $E \to (\alpha)_\mu$ if for any f: $E \to \mu$ there is an $X \subseteq \kappa$ such that $[X]^2 \subseteq E$ (i.e., X is a complete subgraph of G) and X is homogeneous for f with order type α.

This paper owes its genesis to the question of Hajnal whether there is a graph $G = (\omega_2, E)$ such that G has no uncountable complete subgraphs and yet $E \to (\omega)_\omega$. The author tenders his thanks to Hajnal for having posed this question and regrets that he has been unable to answer it.

2 Fat and very fat functions

Not every function can be an ER-function. For example, a constant function cannot be an ER-function. it is important that an ER-function be properly ‘spaced out'.

Let us call $g: [\omega_2]^2 \to \omega_1$ *fat* if, for each $\beta < \omega_2$, the function $g_\beta: \beta \to \omega_1$ is one-to-one, where $g_\beta(\alpha) = g\{\alpha, \beta\}$. We call g *very fat* if g is fat and, whenever $\alpha \neq \beta$, then g_α and g_β agree on only countably many points.

It is easy to construct a very fat function. Also, if g is an ER-function then it is easy to find a very fat ER-function $\bar{g}$. Let h be any very fat function and let $\pi: \omega_1 \times \omega_1 \to \omega_1$ be a bijection. Then $\bar{g}\{\alpha, \beta\} = \pi(g\{\alpha, \beta\}, h\{\alpha, \beta\})$. Very fat functions have large anti-homogeneous sets, as the next result shows.

Theorem 2.1 (a) *If g is fat then there is $X \subseteq \omega_2$ such that X has order type $\omega_1 + 1$ and is anti-homogeneous for g.*

(b) *If g is very fat then for all $\alpha < \omega_2$ there is $X \subseteq \omega_2$ such that X has order type α and is anti-homogeneous for g.*

Proof Suppose g is fat. Let us call a countable set $C \subseteq \omega_2$ *limited* if $\exists\, \xi < \omega_1\ \exists\, \alpha < \omega_2\ \forall\, \beta \geqslant \alpha\ \xi \cap g_\beta``C \neq 0$.

Lemma 2.2 *There is $\alpha < \omega_2$ such that, for all countable $C \subseteq \omega_2 - \alpha$, C is not limited.*

Proof Suppose not. Then for every $\alpha < \omega_2$ there is a limited $C_\alpha \subseteq \omega_2 - \alpha$. We may assume that the C_α are pairwise disjoint and that the same $\xi < \omega_1$ witnesses that each C_α is limited. But now we can find $\beta < \omega_2$ so large that, for all $\alpha < \omega_1$, $\xi \cap g_\beta``C \neq 0$, and this is impossible since g_β is one-to-one. □

For part (a) we use the lemma to find an increasing sequence $\langle \alpha_\xi : \xi < \omega_1 \rangle$ in ω_2 so that α_0 is greater or equal to the α of Lemma 2.2 and whenever $\eta < \xi$ then $g\{\alpha_\eta, \alpha_\xi\} \geqslant \xi$. Let $\alpha_{\omega_1} < \omega_2$ be arbitrary so that $\alpha_\xi < \alpha_{\omega_1}$ for all ξ. But now it is easy to find uncountable $A \subseteq \{\alpha_\xi : \xi < \omega_1\}$ so that $A \cup \{\alpha_{\omega_1}\}$ is anti-homogeneous.

Part (b) is only a little more complicated. Fix $\alpha < \omega_2$ and define a sequence $\langle A_\gamma : \gamma < \alpha \rangle$ of subsets of ω_2 so that $|A_\gamma| = \omega_1$, if $\gamma < \delta$ then $\sup A_\gamma < \inf A_\delta$ and, for each δ, if $C \subseteq \bigcup_{\gamma<\delta} A_\gamma$ is countable then $\forall\, \xi < \omega_1\ \exists\, \beta \in A_\delta\ \xi \cap g_\beta``C = 0$. Of course we use Lemma 2.2 and ensure that $\inf A_0$ is at least as large as the ordinal of the lemma. Note that CH is not required to deal with all C in ω_1 steps since there are clearly ω_1 countable subsets of $\bigcup_{\gamma<\delta} A_\gamma$ such that each countable C is contained in one of them.

Now let $\langle \alpha_\xi : \xi < \omega_1 \rangle$ enumerate α. By induction on ξ, we choose $\beta_\xi \in A_{\alpha_\xi}$ so that $\{\beta_\xi : \xi < \omega_1\}$ is anti-homogeneous (and of course has order type α). Given β_η for all $\eta < \xi$, let $B_1 = \{\beta_\eta : \alpha_\eta < \alpha_\xi\}$ and

$B_2 = \{\beta_\eta : \alpha_\xi < \alpha_\eta\}$. By construction of A_{α_ξ} there are uncountably many $\beta \in A_{\alpha_\xi}$ so that $\forall\, \beta_\eta \in B_1\ \forall\, \beta_\zeta \in B_2\ g_{\beta_\eta}(\beta) \neq g_{\beta_\zeta}(\beta)$. Choose β_ξ to be such a β. This completes the proof. □

Let us remark that Theorem 2.1 (a) is best possible, in view of the following result, an observation of Kunen.

Proposition 2.3 *There are one-to-one functions $g_\beta\colon \beta \to \omega_1$ for all $\beta < \omega_2$ such that whenever $\beta < \gamma$ then $\{\alpha < \beta : g_\beta(\alpha) \neq g_\gamma(\alpha)\}$ is countable.*

The g_β are constructed by induction on β, with the restriction that the range of g_β must be non-stationary. Details are left to the reader.

Of course the fat function given by Proposition 2.3 cannot have any anti-homogeneous set of order type ω_1+2.

3 Sufficient conditions

We begin with an observation of Galvin's.

Proposition 3.1 *If $\omega_2 \to (\omega_1+2)^2_\omega$ then any very fat function is an ER-function.*

Proof Let g be very fat and let $f\colon [\omega_2]^2 \to \omega$. Let X be homogeneous for f of order type ω_1+2 and let α and β be the last two elements of X. Since g is very fat there is $\gamma \in X$ ($\gamma < \alpha, \beta$) such that $g_\alpha(\gamma) \neq g_\beta(\gamma)$. But now $\{\gamma, \alpha, \beta\}$ is anti-homogeneous for g and homogeneous for f. □

Let P be the proposition that there is a family $\langle f_\alpha : \alpha < \omega_2\rangle$ of functions such that $f_\alpha\colon \alpha \to \omega$ and whenever $\alpha < \beta$ and $f_\beta(\alpha) = n$, say, then $\{\gamma < \alpha : f_\beta(\gamma) = f_\alpha(\gamma) = n\}$ is countable. Then it is easy to see that $\omega_2 \to (\omega_1+2)^2_\omega$ implies $\neg$P, and $\neg$P implies that every very fat function is an ER-function. □

Moreover, $\neg$P is a strong form of the almost-disjoint transversal hypothesis ADT, which we may see as follows. Recall that ADT asserts that there is no family $\langle f_\alpha : \alpha < \omega_2\rangle$ such that $f_\alpha\colon \omega_1 \to \omega$ and every two such functions agree at only countably many points. Suppose that there is in fact a family $\langle f_\alpha : \alpha < \omega_2\rangle$. Let $\langle g_\beta : \beta < \omega_2\rangle$ be as in Proposition 2.3. Then $\langle f_\alpha g_\alpha : \alpha < \omega_2\rangle$ witnesses P in a very strong way, namely, if $\alpha < \beta$ then $\{\gamma < \alpha : f_\alpha g_\alpha(\gamma) \neq f_\beta g_\beta(\gamma)\}$ is countable.

Let κ be a cardinal. Let

$$K_1(\kappa) = \{\{(\alpha,\beta),(\gamma,\delta)\} : \alpha < \gamma < \beta < \delta\};$$

$$K_2(\kappa) = \{\{(\alpha,\beta),(\gamma,\delta)\} : \gamma < \alpha < \beta < \delta\}.$$

Thus $K_1(\kappa)$ and $K_2(\kappa)$ are graphs on $\kappa\times\kappa$.

Proposition 3.2 *Assume* CH. *Then* $K_1(\omega_3) \to (\omega_1)_\omega$.

Proof It is easy to see that $K_1(\omega_3)$ has a complete subgraph of cardinality ω_2. Now we use the Erdős–Rado theorem $\omega_2 \to (\omega_1)^2_\omega$. □

Proposition 3.3 *If* $K_1(\omega_2) \to (3)_\omega$ *then there is an ER-function.*

Proof Let π: $\{(\alpha,\beta) : \alpha < \beta < \omega_2\} \to \omega_2$ be a bijection such that if $\beta_0 < \beta_1$ then $\pi(\alpha_0,\beta_0) < \pi(\alpha_1,\beta_1)$ for all $\alpha_0 < \beta_0$ and all $\alpha_1 < \beta_1$. We determine $\bar{g}$: $K_1(\omega_2) \to \omega_1$. If $\alpha < \gamma < \beta < \delta < \omega_2$, let

$$\bar{g}\{(\alpha,\beta),(\gamma,\delta)\} = k_\beta(\gamma),$$

where k_β: $\beta \to \omega_1$ is a fixed one-to-one function for each β. Define g: $[\omega_2]^2 \to \omega_1$ by letting $g\{\alpha,\beta\} = \bar{g}\{\pi(\alpha),\pi(\beta)\}$ if this is defined; let $g\{\alpha,\beta\}$ be arbitrary otherwise. Suppose now that f: $[\omega_2]^2 \to \omega$ is given. Define $\bar{f}$: $K_1(\omega_2) \to \omega$ by $\bar{f}\{x,y\} = f\{\pi(x),\pi(y)\}$. If $A = \{(\alpha_i,\beta_i) : i < 3\}$ is homogeneous for $\bar{f}$ with $\alpha_0 < \alpha_1 < \alpha_2 < \beta_0 < \beta_1 < \beta_2$, then

$$\bar{g}\{(\alpha_0,\beta_0),(\alpha_1,\beta_1)\} = k_{\beta_0}(\alpha_1) \neq k_{\beta_0}(\alpha_2) = \bar{g}\{(\alpha_0,\beta_0),(\alpha_2,\beta_2)\},$$

so π“A is homogeneous for f and anti-homogeneous for g. □

Proposition 3.4 *Assume* $2^{\omega_i} = \omega_{i+1}$ *for all* $i \leq 3$. *Then* $K_2(\omega_4) \to (n)_\omega$ *for all* $n < \omega$.

Proof This is a little more interesting since $K_2(\kappa)$ can never contain an infinite complete graph. Let $\bar{f}$: $K_2(\omega_4) \to \omega$ be given. We use the polarized partition relation

$$\begin{pmatrix}\omega_4\\ \omega_2\end{pmatrix} \to \begin{pmatrix}\omega_3\\ \omega_1\end{pmatrix}^{2,2}_{\omega},$$

which asserts that if f: $[\omega_4]^2\times[\omega_2]^2 \to \omega$ then $\exists\, A \subseteq \omega_4$ $\exists\, B \subseteq \omega_2$ $|A| = \omega_3$, $|B| = \omega_1$ and f is constant on $[A]^2\times[B]^2$. Define f on $[\omega_4]^2\times[\omega_2]^2$ so that if $\alpha_0 < \alpha_1 < \omega_2 < \beta_0 < \beta_1$ then

$$f(\{\beta_0,\beta_1\}\,\{\alpha_0,\alpha_1\}) = \bar{f}\{(\alpha_1,\beta_0)\,(\alpha_0,\beta_1)\}.$$

If A and B are as in the polarized relation, let $\langle\alpha_i : i < n\rangle$ be an increasing sequence from B and let $\langle\beta_i : i < n\rangle$ be an increasing sequence (above ω_2) from A. Then $[\{\alpha_i,\beta_{n-i-1}) : i < n\}]^2 \subseteq K_2(\omega_4)$ and is homogeneous for $\bar{f}$.

To see that the polarized relation holds, just use the Erdős–Rado theorem twice as follows. Given f: $[\omega_4]^2 \times [\omega_2]^2 \to \omega$, first define for each $x \in [\omega_4]^2$ a function f_x: $[\omega_2]^2 \to \omega$ by $f_x(y) = f(x,y)$. By $\omega_2 \to (\omega_1)^2_\omega$ there is $B_x \subseteq \omega_2$ of cardinality ω_1 homogeneous for f_x. By $2^{\omega_1} = \omega_2$ there are only ω_2 such B_x, so by $\omega_4 \to (\omega_3)^2_{\omega_2}$ may find $A \subseteq \omega_4$ of cardinality ω_3 such that $B_x = B$, say, for all $x \in [A]^2$. But now f is constant on $[A]^2 \times [B]^2$. □

Added in proof (December 20, 1989) Proposition 3.4 is contained in the proof of Theorem 7 on page 370 of P. Erdős & A. Hajnal, On decomposition of graphs, *Acta Math. (Hung.)*, **18** (1967), 359–77.

Proposition 3.5 *If* $K_2(\omega_2) \to (3)_\omega$ *then there is an ER-function.*

The proof is the same as the proof of Proposition 3.3. □

Next we attack Hajnal's question.

Theorem 3.6 *Suppose there is* $E \subseteq [\omega_2]^2$ *such that* E *contains no uncountable complete subgraph but* $E \to (3)_\omega$. *Then either* $K_1(\omega_2) \to (3)_\omega$ *or* $K_2(\omega_2) \to (3)_\omega$. *In either case there is an ER-function.*

Proof Note that CH must hold since otherwise $E \nrightarrow (3)_\omega$. For each $\alpha < \omega_2$ let N_α be the Skolem hull of $\{\alpha\}$ in the structure $(H(\omega_2), \in, <^*, E)$, where $H(\omega_2)$ is the collection of all sets hereditarily of cardinality less than ω_2 and $<^*$ is a well-ordering. Let a_α be the transitive collapse of N_α with π_α: $N_\alpha \to a_\alpha$ the unique isomorphism. If $a_\alpha = a_\beta$ then we may define an isomorphism $\pi_{\alpha\beta}$: $N_\alpha \to N_\beta$ by $\pi_{\alpha\beta} = \pi_\beta^{-1}\pi_\alpha$.

Lemma 3.7 *Suppose that* $\alpha < \beta$, $a_\alpha = a_\beta$, $\pi_{\alpha\beta}(\alpha) = \beta$ *and* $\{\alpha, \beta\} \in E$. *If* γ *is the least ordinal moved by* $\pi_{\alpha\beta}$ *then* $\gamma, \pi_{\alpha\beta}(\gamma) < \alpha$.

Proof For each $\alpha < \omega_2$ define a set $S_\alpha = \{s^\alpha_\xi : \xi < \rho_\alpha\}$ by induction on ξ as follows. Choose $s^\alpha_\xi < \alpha$ so that $s^\alpha_\xi > s^\alpha_\eta$ for all $\eta < \xi$, $[\{s^\alpha_\eta : \eta \leqslant \xi\} \cup \{\alpha\}]^2 \subseteq E$ and s^α_ξ is minimal with these properties if such an s^α_ξ exists; otherwise, set $\rho_\alpha = \xi$ and stop. Since E has no uncountable subgraph, S_α must be countable and, since $S_\alpha \in N_\alpha$, we must have $S_\alpha \subseteq N_\alpha$. If $\pi_{\alpha\beta}$ is the identity on S_α then $S_\alpha = S_\beta$, contradicting the fact that $[S_\beta \cup \{\alpha, \beta\}]^2 \subseteq E$. So some s^α_ξ must be moved by $\pi_{\alpha\beta}$. Let us look at the first one moved.

Of course $\pi_{\alpha\beta}(s^\alpha_\xi) = s^\beta_\xi$. Since α satisfies the definition of s^β_ξ we must have $s^\beta_\xi \leqslant \alpha$, that is, $\pi_{\alpha\beta}(s^\alpha_\xi) \leqslant \alpha$. Since clearly $\gamma \leqslant s^\alpha_\xi$ and $\pi_{\alpha\beta}(\gamma) \leqslant \pi_{\alpha\beta}(s^\alpha_\xi)$, it will complete the proof to show that $\alpha \notin N_\beta$.

It is easy to see that since $a_\alpha = a_\beta$ we have that $N_\alpha \cap \omega_1 = \omega_1^{a_\alpha} = N_\beta \cap \omega_1$, and it follows that $\pi_{\alpha\beta}$ is the identity on $N_\alpha \cap N_\beta$. (The proof is by induction on sets, using the fact that every element belongs to $H(\omega_2)$). But if $\alpha \in N_\beta$ then $S_\alpha \in N_\beta$; so $\pi_{\alpha\beta}$ is the identity on S_α and we know this is not true.

Now, by way of contradiction, assume that f_i $(i = 1,2)$ is a counterexample to $K_i(\omega_2) \to (3)_\omega$. We will construct a counterexample to $E \to (3)_\omega$.

Note that by CH the set $A = \{a_\alpha : \alpha < \omega_2\}$ has cardinality ω_1. For each $a \in A$ let $\sigma_a \colon a \to \omega$ be a bijection. For $\alpha < \omega_2$ let $n_\alpha = \sigma_{a_\alpha}\pi_\alpha(\alpha)$. Observe that if $a_\alpha = a_\beta$ and $n_\alpha = n_\beta$ then $\pi_{\alpha\beta}(\alpha) = \beta$. Fix $h\colon [A\times\omega]^2 \to 3$ with no homogeneous triangles.

Suppose $\{\alpha,\beta\} \in E$ and $\alpha < \beta$. Define $f\{\alpha,\beta\} = (i,j,k,l)$ as follows.

If $(a_\alpha, n_\alpha) \neq (a_\beta, n_\beta)$ let $i = 0$ and $j = h\{(a_\alpha,n_\alpha),(a_\beta,n_\beta)\}$. Let $k = l = 0$.

If $(a_\alpha, n_\alpha) = (a_\beta, n_\beta)$ let $i = 1$. Let $\gamma < \alpha$ be minimal such that γ is moved by $\pi_{\alpha\beta}$ and put $j = \sigma_{a_\alpha}(\pi_\alpha(\gamma))$. If $\gamma < \pi_{\alpha\beta}(\gamma)$ then $\{(\gamma,\alpha),(\pi_{\alpha\beta}(\gamma),\beta)\} \in K_1(\omega_2)$ by Lemma 3.7, so put $k = 1$ and $l = f_1\{(\gamma,\alpha),(\pi_{\alpha\beta}(\gamma),\beta)\}$. Otherwise $\{(\gamma,\alpha),(\pi_{\alpha\beta}(\gamma),\beta)\} \in K_2(\omega_2)$. Put $k = 2$ and $l = f_2\{(\gamma,\alpha),(\pi_{\alpha\beta}(\gamma),\beta)\}$.

We claim f witnesses $E \not\to (3)_\omega$. Suppose $\{\alpha_0,\alpha_1,\alpha_2\}$ is a complete subgraph of E homogeneous for f. Say f has constant value (i_0,j_0,k_0,l_0) on $[\{\alpha_0,\alpha_1,\alpha_2\}]^2$. If $i_0 = 0$ then h is constant on $[\{(a_{\alpha_i}, n_{\alpha_i}) : i < 3\}]^2$: contradiction. So $i_0 = 1$. Let $\gamma_i = \pi_{\alpha_i}^{-1}\sigma_{a_{\alpha_i}}^{-1}(j_0)$. If $\gamma_0 < \gamma_1$ then $k_0 = 1$ and $\{(\gamma_i,\alpha_i) : i < 3\}$ is homogeneous for f_1: contradiction. If $\gamma_0 > \gamma_1$ then $k_0 = 2$ and $\{(\gamma_i,\alpha_i) : i < 3\}$ is homogeneous for f_2: also a contradiction. This completes the proof of the theorem. □

If we had assumed $E \to (\omega)_\omega$ then we would have had a set $\{\alpha_n : n \in \omega\}$ homogeneous for f at the end of the proof above. Since γ_n cannot form a descending sequence we would have had $k_0 = 1$. Thus we draw the following stronger conclusion in this case.

Theorem 3.8 *Suppose there is $E \subseteq [\omega_2]^2$ such that E contains no uncountable complete subgraph but $E \to (\omega)_\omega$. Then $K_1(\omega_2) \to (\omega)_\omega$.*

4 There may be no Erdős–Rado functions

Next we concentrate on showing that the non-existence of ER-functions is relatively consistent with CH. Unfortunately $2^{\omega_1} = \omega_3$ in our model.

The idea is quite simple: we add functions in ω_3 steps which will destroy any potential ER-functions. The successor step of the construction is given in the following theorem. The successor step alone, however, is enough to deduce the relative consistency with GCH of $K_i(\omega_2) \nrightarrow (3)_\omega$ $(i = 1,2)$ and hence the non-existence of Hajnal's graph, as we observe in Corollary 4.2 below.

Theorem 4.1 *Assume* CH. *Let* $g\colon [\omega_2]^2 \to \omega_1$. *There is a partial ordering P which is countably closed and has the ω_2-chain condition, and is such that, in V^P, g is not an ER-function.*

Proof The partial ordering P is only one step away from the obvious ordering. Let P consist of all tripples (a,s,t), where $a \in [\omega_2]^\omega$, $s\colon [a]^2 \to \omega$, and t is a function with domain $[a]^2$ such that, for all $\alpha,\beta \in a$, $t\{\alpha,\beta\}$ is an infinite coinfinite subset of ω containing $s\{\alpha,\beta\}$, and, for all $n \in t\{\alpha,\beta\}$ and $\gamma \in a$, if $\gamma < \alpha,\beta$ and $s\{\gamma,\alpha\} = s\{\gamma,\beta\} = n$ then $g\{\gamma,\alpha\} = g\{\gamma,\beta\}$ (i.e., $g_\alpha(\gamma) = g_\beta(\gamma)$). Let $(a_1,s_1,t_1) \leqslant (a_2,s_2,t_2)$ if and only if $a_1 \supseteq a_2$, $s_1 \supseteq s_2$ and $t_1 \supseteq t_2$.

Perhaps the simplest ordering to try is the collection of all pairs (a,s) as above, where the final clause about γ is satisfied when $n = s\{\alpha,\beta\}$, and no requirement is made otherwise. Unfortunately we have not been able to prove the ω_2-chain condition for this case.

It is clear that our P is countably closed.

It is not completely trivial to see that for $\gamma \in \omega_2$ the set of all $p = (a^p,s^p,t^p)$ such that $\gamma \in a^p$ is dense in P. Suppose p is fixed with $\gamma \notin a^p$. Let A be an infinite coinfinite subset of ω. We determine $q \leqslant p$. Let $a^q = a^p \cup \{\gamma\}$. Define $s^p \supseteq s^p$ so that it carries $\{\{\beta,\gamma\} : \beta \in a^p\}$ one-to-one into $\omega - A$. For $\beta \in a^p$ let $t^p\{\beta,\gamma\} = \{s^q\{\beta,\gamma\}\} \cup A$. Clearly $q \leqslant p$ provided $q \in P$. There are only a few cases to check. Suppose $\gamma < \alpha,\beta$. Then $s^q\{\alpha,\gamma\} \neq s^q\{\beta,\gamma\}$, so there is nothing to worry about. Suppose $\delta < \alpha,\gamma$. Then $s^q\{\delta,\gamma\} \in \omega - A$, so if $s^q\{\delta,\gamma\} \in t^q\{\alpha,\gamma\}$ we must have $s^q\{\delta,\gamma\} = s^q\{\alpha,\gamma\}$, contradicting the fact that s^q is one-to-one on $\{\{\beta,\gamma\} : \beta \in a^p\}$.

We must check that P has the ω_2-chain condition. An absolutely standard Δ-system argument shows that we need only to be able to amalgamate conditions p and q when p and q are canonically order-isomorphic in the obvious sense, that is, the unique order-isomorphism $\pi\colon a^p \to a^q$ lifts to an isomorphism of s^p with s^q and t^p with t^q. We may also assume that $\Delta = a^p \cap a^q$ is an initial segment of both a^p and a^q and that π lifts to an isomorphism of $g \restriction [a^p]^2$ with $g \restriction [a^q]^2$ as well. Also assume $\sup a^p < \inf(a^q - \Delta)$.

We want to define $r \leqslant p,q$. Let $a^r = a^p \cup a^q$. It remains only to define s^r and t^r on the set $B = \{\{\alpha,\beta\} : \alpha \in a^p - \Delta, \beta \in a^q - \Delta\}$. s^r will be one-to-one on B, and if $\{\alpha,\beta\} \in B$, $\pi(\gamma) = \beta$ and $\alpha \neq \gamma$ then $s^r\{\alpha,\gamma\} \in t^p\{\alpha,\gamma\}$ and $t^p\{\alpha,\gamma\} - \text{range}(s^r \restriction B)$ will be infinite. It is easy to determine s^r by induction in ω steps, using the fact that each $t^p\{\alpha,\gamma\}$ is finite.

Now determine t^r on B as follows. If $\{\alpha,\beta\} \in B$ and $\pi(\alpha) = \beta$, let $t^r\{\alpha,\beta\} = \{s^r\{\alpha,\beta\}\} \cup (\omega - \text{range}(s^r \restriction B))$; otherwise, if $\pi(\gamma) = \beta$ and $\alpha \neq \gamma$, let $t^r\{\alpha,\beta\} = \{s^r\{\alpha,\beta\}\} \cup (t^p\{\alpha,\gamma\} - \text{range}(s^r \restriction B))$.

Let us check that in fact $r \in P$. Then it will be clear that $r \leqslant p,q$. Suppose $\gamma < \alpha < \beta$ ($\gamma,\alpha,\beta \in a^r$). We proceed by cases.

Case 1 $\gamma \in \Delta$. If $\alpha,\beta \in a^p$ or $\alpha,\beta \in a^q$ we are done since $p,q \in P$, so we may assume $\alpha \in a^p - \Delta$ and $\beta \in a^q - \Delta$.

Case 1.1 $\pi(\alpha) = \beta$. In this case, by the isomorphism $g_\alpha(\gamma) = g_\beta(\gamma)$, so no problems arise.

Case 1.2 $\pi(\alpha) \neq \beta$. Say $\pi(\delta) = \beta$. Then $t^r\{\alpha,\beta\} \subseteq t^p\{\alpha,\delta\}$, so if $n \in t^r\{\alpha,\beta\}$ and $s^r\{\alpha,\gamma\} = s^r\{\beta,\gamma\} = n$ then we must have $s^p\{\alpha,\gamma\} = s^r\{\alpha,\gamma\} = n$ and $s^q\{\beta,\gamma\} = s^r\{\beta,\gamma\} = n$, and by the isomorphism $g_\delta(\gamma) = g_\beta(\gamma)$. But now since $n \in t^p\{\alpha,\delta\}$ we have $g_\alpha(\gamma) = g_\delta(\gamma)$ and $g_\alpha(\gamma) = g_\beta(\gamma)$ and we are done.

Case 2 $\gamma \notin \Delta$.

Case 2.1 $\gamma \in a^p - \Delta$ and $\alpha,\beta \in a^q - \Delta$. Then since $s^r \restriction B$ is one-to-one we have $s^r\{\gamma,\alpha\} \neq s^r\{\gamma,\beta\}$, so this case causes no difficulties.

Case 2.2 $\gamma,\alpha \in a^p - \Delta$ and $\beta \in a^q - \Delta$. Let $n \in t^r\{\alpha,\beta\}$. If $n \in s^r\{\alpha,\beta\}$ then we are done since $s^r\{\gamma,\beta\} \neq s^r\{\alpha,\beta\}$. Otherwise, $n \notin \text{range}(s^r \restriction B)$, so $n \neq s^r\{\gamma,\beta\}$.

Case 2.3 Otherwise, that is either $\gamma,\alpha,\beta \in a^p$ or $\gamma,\alpha,\beta \in a^q$. This is trivial since $p,q \in P$.

This completes the proof that P has the ω_2-chain condition. □

Note, for Theorem 4.3 below, that in showing $r \leqslant p,q$ we made no use of the values of $g_\alpha(\beta)$ unless $\alpha,\beta \in a^p$ or $\alpha,\beta \in a^q$.

Recall that both the propositions $K_1(\omega_2) \to (3)_\omega$ and $K_2(\omega_2) \to (3)_\omega$ imply the existence of specific ER-functions. Since specific ER-functions can be destroyed by a single application of the partial ordering P of Theorem 4.1, we obtain the following.

Corollary 4.2 *If* ZF *is consistent then so is*

$$\mathrm{ZF} + \mathrm{GCH} + K_1[\omega_2] \nrightarrow (3)_\omega + K_2[\omega_2] \nrightarrow (3)_\omega .$$

Thus by Theorem 3.6 it is relatively consistent with GCH *that if*

$E \subseteq [\omega_2]^2$ and $E \to (3)_\omega$ then E contains an uncountable complete subgraph.

Of course a similar remark applies to the partition relation $\omega_2 \nrightarrow (\omega_1+2)^2_\omega$ and, more generally, to the Proposition P of Section 3, to see that these assertions are also relatively consistent with GCH. This result, however, is well known.

By iterating the orderings arising in Theorem 4.1, we arrive at the following theorem.

Theorem 4.3 *If* ZF *is consistent then so is*

$$\mathrm{ZF}+\mathrm{CH}+2^{\omega_1} = \omega_3+\text{'there are no Erdős–Rado functions'}.$$

Proof This is straightforward. Just iterate the forcing of Theorem 4.1 for ω_3 steps with countable support over a model of GCH. Countable closure is trivial and the ω_2-chain condition follows from an easy Δ-system argument together with the observation after the end of the proof of Theorem 4.1. □

5 Open questions

We conclude with some of the problems we have been unable to settle.

Question 1 *Is it consistent that* CH *holds and there is an ER-function?*

Question 2 *Is it consistent that* GCH *holds and there is no ER-function?*

Question 3 *Is there an ER-function in L?*

Question 4 *Is it consistent (relative to large cardinals) that $\omega_2 \to (\omega_1+2)^2_\omega$?*

This question is not due to the author. Note that by an application of the partition result $\forall\, \alpha < \omega_2\ \alpha \nrightarrow (\omega_1^\omega)^1_\omega$ due to Milner and Rado [3], it follows that in ZFC $\omega_2 \nrightarrow (\omega_1^\omega)^2_\omega$. (Here ω_1^ω denotes ordinal exponentiation.)

Question 5 *Is it consistent that $K_1(\omega_2) \to (3)_\omega$?*

Question 6 *Is it consistent that $K_2(\omega_2) \to (3)_\omega$?*

Question 7 *Is it provable from* GCH *that $K_2(\omega_2) \to (3)_\omega$?*

Question 8 (Hajnal) *Is it consistent that there is a graph $G = (\omega_2, E)$ which has no uncountable complete subgraphs, but is such that $E \to (\omega)_\omega$?*

References

[1] J. Baumgartner, Canonical partition relations, *J. Symbolic Logic*, **40** (1975), 541–54

[2] P. Erdős & R. Rado, A partition calculus in set theory, *Bull. Amer. Math. Soc.*, **62** (1956) 427–89

[3] E. Milner & R. Rado, The pigeonhole-principle for ordinal numbers, *Proc. London Math. Soc., Ser. 3*, **15** (1965), 750–68.

Almost collinear triples among N points on the plane

József Beck

1 Introduction

To what extent can a finite number of points in a bounded domain be in general position? Is there always a line passing close to at least three of the points? In this paper we give a partial answer to these questions.

Let $N \geqslant 1$ be an integer. Let $Q(N)$ denote the N by N square

$$\{X = (x_1, x_2) \in \mathbb{R}^2 : 0 \leqslant x_1 \leqslant N, 0 \leqslant x_2 \leqslant N\}.$$

Let $\mathcal{P} = \{P_1, P_2, \ldots, P_N\}$ be a distribution of N points in the square $Q(N)$. Let $\Omega(\mathcal{P})$ be the smallest real number $w \geqslant 0$ for which there is a strip

$$S(\theta, a, w) = \{X = (x_1, x_2) \in \mathbb{R}^2 : a - \tfrac{1}{2}w \leqslant x_2 \cos\theta - x_1 \sin\theta \leqslant a + \tfrac{1}{2}w\}$$

of width w containing at least three points of $\mathcal{P}$. It is easily seen that $\Omega(\mathcal{P})$ is also the largest real number $d \geqslant 0$ such that no straight line intersects three of the open discs with centres P_i $(i = 1, 2, \ldots, N)$ and of common diameter d.

Let

$$\Omega(N) = \max \Omega(\mathcal{P}),$$

where the maximum is taken over all point configurations $\mathcal{P}$ of cardinality N in the square $Q(N)$.

The question of studying the behaviour of $\Omega(N)$ as $N \to \infty$ was proposed, independently of each other, by T. S. Motzkin (see Problem 31 in W. Moser's summary [3] and Problem 43 in the new edition [4]) and, in the early nineteen-seventies, by W. M. Schmidt (oral communication).

A straightforward application of the pigeon-hole principle implies that if $\epsilon > 0$ and N is sufficiently large then

$$\Omega(N) \leq 2+\epsilon.$$

It was conjectured by Schmidt that

$$\lim_{N\to\infty} \Omega(N) = 0, \tag{1.1}$$

or equivalently, using the o-notation,

$$\Omega(N) = o(1).$$

Unfortunately, this conjecture remains open. In this paper we can prove only the following partial result.

Let n be a positive integer and set $N = n^2$. We call an N-element point configuration $\mathcal{P} = \{P_1,P_2,\ldots,P_N\} \subset Q(N)$ *weakly uniformly distributed* if, for every $i = 1,2,\ldots,n$ and $j = 1,2,\ldots,n$, the subsquare

$$Q_{i,j} = \{X = (x_1,x_2) \in \mathbb{R}^2 : \\ (i-1)n \leq x_1 < in,\ (j-1)n \leq x_2 < jn\} \subset Q(N)$$

contains precisely one point of $\mathcal{P}$.

We are now able to formulate our result.

Theorem 1. *Let N be a square, i.e. $N = n^2$, where n is an integer. Let $\mathcal{P} = \{P_1,P_2,\ldots,P_N\}$ be a weakly uniformly distributed set of N points in the square $Q(N)$. Then*

$$\Omega(\mathcal{P}) \ll (\log N)^{-1/9}.$$

Schmidt's conjecture is related to Heilbronn's well-known triangle problem. Let $\Delta(\mathcal{P})$ be the minimum of the areas of all the triangles P_i,P_j,P_k $(1 \leq i < j < k \leq N)$ and let $\Delta(N) = \max \Delta(\mathcal{P})$, where the maximum extends over all point distributions $\mathcal{P} \subset Q(N)$ of cardinality N. Trivially, we have

$$\Omega(N) > \frac{\Delta(N)}{N}.$$

This shows that, as far as the upper bound is concerned, the Motzkin–Schmidt triangle problem is even harder than Heilbronn's triangle problem.

The analogue of (1.1) for Heilbronn's triangle problem, namely $\lim \Delta(N)/N = 0$, is due to K. F. Roth [5], who in 1950 proved, by a combinatorial method, that

$$\frac{\Delta(N)}{N} \ll (\log\log N)^{-1/2}.$$

In 1971 Schmidt [8] improved this to

$$\frac{\Delta(N)}{N} \ll (\log N)^{-1/2}.$$

Introducing a new analytic method, in 1972 Roth succeeded in proving

$$\frac{\Delta(N)}{N} < N^{-\mu},$$

where $\mu > 0$ is an absolute constant. In fact, Roth (see [6], and somewhat later [7]) obtained the following explicit values:

$$\mu = 1-\sqrt{\tfrac{4}{5}}-\epsilon \cong 0.105-\epsilon \qquad \text{and} \qquad \mu = \tfrac{1}{8}(9-\sqrt{65})-\epsilon \cong 0.117-\epsilon.$$

By using a refinement of Roth's analytic method, in 1981 Komlós, Pintz & Szemerédi [1] improved the exponent to $\mu = \frac{1}{7}-\epsilon \cong 0.142-\epsilon$.

The first lower bound $\Delta(N) \gg 1$ was achieved by an ingenious number-theoretic construction of Erdős in 1950 (cf. Appendix of [5]). There was no further improvement until about 30 years later, when Komlós, Pintz & Szemerédi [2] established $\Delta(N) \gg \log N$.

2 Proof of Theorem 1

The proof makes use of Roth's analytic method combined with some new ideas.

Given two points P' and P'' on the plane $\mathbb{R}^2$, let $|P'-P''|$ denote the usual Euclidean distance between them. The line through P' and P'' can be written in the form

$$x_2\cos\theta - x_1\sin\theta = a \tag{2.1}$$

where $\theta \equiv \theta(P',P'')$ is the *inclination* of the line. Let $S(P',P'';w) = S(\theta,a,w)$ denote the strip

$$a-\tfrac{1}{2}w \leqslant x_2\cos\theta - x_1\sin\theta \leqslant a+\tfrac{1}{2}w$$

of width w about the line (2.1).

For notational convenience, let

$$\omega = \Omega(\mathcal{P}). \tag{2.2}$$

We shall assume throughout what follows that $N = n^2$ is sufficiently large for our various inequalities to hold. Let

$$T(P' \to P'') = \left\{ X \in \mathbb{R}^2 : \sin|\theta(P',P'') - \theta(P',X)| < \frac{\omega}{|P'-P''|} \right\}$$

and

$$T(P',P'') = T(P' \to P'') \cup T(P'' \to P').$$

Note that $T(P',P'')$ is the union of two angular domains.

The statement that no three of the points of $\mathcal{P}$ can be covered by an open strip of width $\omega = \Omega(\mathcal{P})$ is equivalent to saying that, for each pair P_i, P_j of points of $\mathcal{P}$, the set $T(P_i,P_j)$ contains no member of $\mathcal{P}$ other than the two points P_i and P_j, that is,

$$T(P_i,P_j) \cap \mathcal{P} = \{P_i,P_j\} \qquad \text{for all } 1 \leqslant i < j \leqslant N. \tag{2.3}$$

Suppose that $\omega = \Omega(\mathcal{P})$ is not 'small'. If the distance $|P_i - P_j|$ is very small, then the set $T(P_i,P_j)$ is very 'deficient' of points of $\mathcal{P}$. The basic idea is to deduce from (2.3) a result that contradicts the hypothesis that $\mathcal{P}$ is weakly uniformly distributed.

Let $D(Y,r)$ denote the open disc of centre $Y \in \mathbb{R}^2$ and of radius r. Let

$$\rho = \tfrac{1}{100}\omega n, \tag{2.4}$$

and, for every $i = 1,2,\ldots,N$, let

$$D_i = D(P_i,\rho). \tag{2.5}$$

Let χ_i denote the characteristic function of the disc D_i, i.e.

$$\chi_i(X) = \begin{cases} 1 & \text{if } X \in D_i \\ 0 & \text{if } X \notin D_i \end{cases} \tag{2.6}$$

and let

$$F = \sum_{i=1}^{N} \chi_i. \tag{2.7}$$

Following Roth, we shall construct an auxiliary function

$$G = \sum_{k=1}^{m} G_k,$$

where G_k are 'quasi-orthogonal' in the sense of the usual inner product

$$\langle G_k, G_l \rangle = \int_{\mathbb{R}^2} G_k(X)\, G_l(X)\, \mathrm{d}X$$

and $m \approx n(\log n)^{1/3}$, with the precise value of m to be given later. We shall deduce from (2.3) a contradiction, on the assumption that $\omega = \Omega(\mathcal{P})$ is not 'small', by employing the obvious inequality

$$2\langle F, G\rangle \leqslant \|F\|^2 + \|G\|^2,$$

where, as usual, $\|f\|^2 = \langle f, f\rangle$. In order to get the 'quasi-orthogonal' functions G_k ($k = 1, \ldots, m$), we shall find suitable pairs of points of $\mathcal{P}$.

In Section 3, we shall prove the following lemma.

Lemma 1. *Let $\mathcal{P}$ be any configuration of $N = n^2$ points, weakly uniformly distributed in $Q(N)$, and set $\omega = \Omega(\mathcal{P})$. There exist*

$$m \gg \min\{n\omega^3 \log n, n\omega^{3/2}(\log n)^{1/2}\}$$

disjoint pairs $\{\boldsymbol{P}_i^{(1)}, \boldsymbol{P}_i^{(2)}\}$ ($i = 1, 2, \ldots, m$) of points of $\mathcal{P}$ such that

$$\{\boldsymbol{P}_i^{(1)}, \boldsymbol{P}_i^{(2)}\} \subset [\tfrac{1}{3}N, \tfrac{2}{3}N]^2 \qquad (i = 1, 2, \ldots, m); \tag{2.8}$$

$$|\boldsymbol{P}_i^{(1)} - \boldsymbol{P}_i^{(2)}| < 3n \qquad (i = 1, 2, \ldots, m); \tag{2.9}$$

and if

$$S(\boldsymbol{P}_i^{(1)}, \boldsymbol{P}_i^{(2)}; 2n) \cap S(\boldsymbol{P}_j^{(1)}, \boldsymbol{P}_j^{(2)}; 2n) \neq \varnothing \qquad (1 \leqslant i < j \leqslant m)$$

then

$$|\theta(\boldsymbol{P}_i^{(1)}, \boldsymbol{P}_i^{(2)}) - \theta(\boldsymbol{P}_j^{(1)}, \boldsymbol{P}_j^{(2)})| > \frac{1}{n}. \tag{2.10}$$

We now use Lemma 1 to prove Theorem 1 as follows. Let $\boldsymbol{P}$, m and $\boldsymbol{P}_i^{(1)}, \boldsymbol{P}_i^{(2)}$ ($i = 1, \ldots, m$) be as in Lemma 1 and, for notational convenience, for $i = 1, 2, \ldots, m$, write

$$\begin{aligned}\theta_i &= \theta(\boldsymbol{P}_i^{(1)}, \boldsymbol{P}_i^{(2)}),\\ S_i^* &= S(\boldsymbol{P}_i^{(1)}, \boldsymbol{P}_i^{(2)}; 2n),\\ S_i^{**} &= S(\boldsymbol{P}_i^{(1)}, \boldsymbol{P}_i^{(2)}; \rho),\\ T_i &= T(\boldsymbol{P}_i^{(1)}, \boldsymbol{P}_i^{(2)}).\end{aligned}$$

Let $\tilde{Q} = [\frac{1}{6}N, \frac{5}{6}N]^2$ and let $\tilde{Q}\{\theta\}$ denote the copy of the square $\tilde{Q}$ rotated around the centre $(\frac{1}{2}N, \frac{1}{2}N) \in \mathbb{R}^2$ by an angle θ. Note that for every angle θ,

$$[\tfrac{1}{3}N, \tfrac{2}{3}N]^2 \subset \tilde{Q}\{\theta\} \subset Q(N) = [0, N]^2.$$

For every i, the intersection set

$$T_i \cap S(\boldsymbol{P}_i^{(1)}, \boldsymbol{P}_i^{(2)}; 3\rho) \cap \tilde{Q}\{\theta_i\}$$

is the disjoint union of two rectangles and a 'bow-tie' shape. Let R_i' denote the longer of the two rectangles. Let

$$R_i'' = R_i' \cap S_i^{**}.$$

Let $L''_{i(1)}$ and $L''_{i(2)}$ denote the straight lines passing through the shorter sides of R_i''. Let C_i denote the centre of the rectangle R_i'' and let $L_{i(1)}$ and $L_{i(2)}$ be the lines parallel to $L''_{i(1)}$ and $L''_{i(2)}$ such that

$$\text{dist}(C_i, L_{i(1)}) = \text{dist}(C_i, L''_{i(1)}) - \rho,$$
$$\text{dist}(C_i, L_{i(2)}) = \text{dist}(C_i, L''_{i(2)}) - \rho.$$

Here $\text{dist}(C, L)$ stands for the distance of the point C and the line L.

Let R_i^* denote the rectangle determined by the strip S_i^* and the lines $L_{i(1)}$ and $L_{i(2)}$. Let R_i^{**} denote the rectangle determined by the strip S_i^{**} and the lines $L_{i(1)}$ and $L_{i(2)}$. It follows from the construction of R_i^{**} and from (2.3) that

$$D(P_j, \rho) \cap R_i^{**} = \varnothing \qquad \text{for all } 1 \leqslant j \leqslant N;\ 1 \leqslant i \leqslant m. \tag{2.11}$$

Let V_i^* and V_i^{**} be the sets of four vertices of the rectangles R_i^* and R_i^{**}, respectively. Let $H_i(X)$ and $h_i(X)$ be the characteristic functions of the rectangles R_i^* and R_i^{**}, respectively, that is,

$$H_i(X) = \begin{cases} 1 & \text{if } X \in R_i^*, \\ 0 & \text{if } X \notin R_i^*, \end{cases}$$

and

$$h_i(X) = \begin{cases} 1 & \text{if } X \in R_i^{**}, \\ 0 & \text{if } X \notin R_i^{**}. \end{cases}$$

Let

$$g_i(X) = H_i(X) - \frac{2n}{\rho} h_i(X). \tag{2.12}$$

Observe that in (2.12) the characteristic functions H_i and h_i are weighted in the inverse proportion to their widths. This implies the truth of the following important observation of Roth.

Orthogonality property. *If* $(V_i^* \cup V_j^*) \cap S_i^* \cap S_j^* = \varnothing$, *then*

$$\int_{\mathbb{R}^2} g_i(X) g_j(X) \, dX = 0.$$

As well as the orthogonality property, we shall employ a more delicate geometric fact which we shall call the *cancellation property*.

For every real number $t \in [\frac{1}{4}, 1]$, let $L(i,1;t)$ and $L(i,2;t)$ be the lines parallel to $L_{i(1)}$ and $L_{i(2)}$ such that

$$\operatorname{dist}(C_i, L(i,1;t)) = t \operatorname{dist}(C_i, L_{i(1)}),$$
$$\operatorname{dist}(C_i, L(i,2;t)) = t \operatorname{dist}(C_i, L_{i(2)}).$$

Let $R_i^*(t)$ denote the rectangle determined by the strip S_i^* and the lines $L(i,1;t)$ and $L(i,2;t)$. Similarly, let $R_i^{**}(t)$ denote the rectangle determined by the strip S_i^{**} and the lines $L(i,1;t)$ and $L(i,2;t)$. Note that $R_i^*(1) = R_i^*$, $R_i^{**}(1) = R_i^{**}$, and

$$R_i^*(t) \subset R_i^*(t'), \qquad R_i^{**}(t) \subset R_i^{**}(t') \qquad \text{if } \tfrac{1}{4} \leqslant t < t' \leqslant 1.$$

Let $V_i^*(t)$ and $V_i^{**}(t)$ denote the sets of four vertices of the rectangles $R_i^*(t)$ and $R_i^{**}(t)$, respectively. For every real number $t \in [\frac{1}{4}, 1]$, let $H_i(t, X)$ and $h_i(t, X)$ be the characteristic functions of the rectangles $R_i^*(t)$ and $R_i^{**}(t)$, respectively. Finally, let

$$g_i(t, X) = H_i(t, X) - \frac{2n}{\rho} h_i(t, X).$$

Cancellation property. *Let* $\alpha \in [\frac{1}{2}, 1]$ *and* $\beta \in [\frac{1}{2}, 1]$. *Let* $1 \leqslant i < j \leqslant m$. *Assume that*

$$[V_i^*(\alpha) \cup V_j^*(\beta) \cup V_i^*(\tfrac{1}{2}\alpha) \cup V_j^*(\tfrac{1}{2}\beta)] \cap S_i^* \cap S_j^* = \varnothing \tag{2.13}$$

and that the two sets

$$U_{i,j}(\alpha) = \{t \in [\tfrac{1}{2}, 1] : V_i^*(\alpha t) \cap S_i^* \cap S_j^* \neq \varnothing\}, \tag{2.14a}$$
$$U_{j,i}(\beta) = \{t \in [\tfrac{1}{2}, 1] : V_j^*(\beta t) \cap S_i^* \cap S_j^* \neq \varnothing\} \tag{2.14b}$$

are disjoint. Then

$$\int_{1/2}^{1} \int_{\mathbb{R}^2} g_i(\alpha t, X) g_j(\beta t, X) \, \mathrm{d}X \, \mathrm{d}t = 0.$$

We leave the simple proof to the reader. □

If $i \neq j$ and

$$\int_{1/2}^{1} \int_{\mathbb{R}^2} g_i(\alpha t, X) g_j(\beta t, X) \, \mathrm{d}X \, \mathrm{d}t \neq 0$$

then certainly $S_i^* \cap S_j^* \neq \varnothing$. Thus by the properties of the pairs $(P_i^{(1)}, P_i^{(2)})$ given in Lemma 1, $|\theta_i - \theta_j| > 1/n$. A simple geometric consideration shows that for every $t \in [\frac{1}{2}, 1]$,

$$
\begin{aligned}
\left|\int_{\mathbb{R}^2} g_i(\alpha t, X)\, g_j(\beta t, X)\, \mathrm{d}X\right|
&\leqslant \int_{\mathbb{R}^2} \Big\{ H_i(\alpha t, X)\, H_j(\beta t, X) + \frac{2n}{\rho} h_i(\alpha t, X)\, H_j(\beta t, X) \\
&\qquad + \frac{2n}{\rho} H_i(\alpha t, X)\, h_j(\beta t, X) \\
&\qquad + \left(\frac{2n}{\rho}\right)^2 h_i(\alpha t, X)\, h_j(\beta t, X) \Big\}\, \mathrm{d}X \\
&= \operatorname{area}(R_i^*(\alpha t) \cap R_j^*(\beta t)) + \frac{2n}{\rho} \operatorname{area}(R_i^{**}(\alpha t) \cap R_j^*(\beta t)) \\
&\qquad + \frac{2n}{\rho} \operatorname{area}(R_i^*(\alpha t) \cap R_j^{**}(\beta t)) \\
&\qquad + \left(\frac{2n}{\rho}\right)^2 \operatorname{area}(R_i^{**}(\alpha t) \cap R_j^{**}(\beta t)) \\
&\ll \frac{n^2}{|\theta_i - \theta_j|}\,.
\end{aligned}
$$

Moreover, observe that for all $t \in [\frac{1}{2}, 1]$ except on a set of measure $\ll (n|\theta_i - \theta_j|)^{-1}$, the orthogonality property can be applied and yields

$$
\int_{\mathbb{R}^2} g_i(\alpha t, X)\, g_j(\beta t, X)\, \mathrm{d}X = 0.
$$

Therefore we have, for all $1 \leqslant i < j \leqslant m$,

$$
\begin{aligned}
\left|\int_{1/2}^{1} \int_{\mathbb{R}^2} g_i(\alpha t, X)\, g_j(\beta t, X)\, \mathrm{d}X\, \mathrm{d}t\right| &\ll \frac{1}{n|\theta_i - \theta_j|} \frac{n^2}{|\theta_i - \theta_j|} \\
&= \frac{n}{|\theta_i - \theta_j|^2}\,. \qquad (2.15)
\end{aligned}
$$

The desired 'quasi-orthogonal' functions G_k $(1 \leqslant k \leqslant m)$ have the form

$$
G_k(X) = \delta \int_{1/2}^{1} g_k(\alpha_k t, X)\, \mathrm{d}t,
$$

where $\delta \approx (\log n)^{-1/3}$ and the real numbers $\alpha_k \in [\frac{1}{2}, 1]$, $(1 \leqslant k \leqslant m)$ will be specified later by a probabilistic argument.

Let

$$
G = \sum_{k=1}^{m} G_k\,.
$$

As before, let F be the sum of the characteristic functions χ_i of the discs $D_i = D(\boldsymbol{P}_i, \rho)$. We shall estimate $\omega = \Omega(\mathcal{P})$ by using the trivial inequality

$$\int_{\mathbb{R}^2} [F(\boldsymbol{X}) - G(\boldsymbol{X})]^2 \, \mathrm{d}\boldsymbol{X} \geqslant 0,$$

or equivalently,

$$2\int_{\mathbb{R}^2} F(\boldsymbol{X})\, G(\boldsymbol{X}) \, \mathrm{d}\boldsymbol{X} \leqslant \int_{\mathbb{R}^2} [F(\boldsymbol{X})]^2 \, \mathrm{d}\boldsymbol{X} + \int_{\mathbb{R}^2} [G(\boldsymbol{X})]^2 \, \mathrm{d}\boldsymbol{X}. \quad (2.16)$$

From (2.4)–(2.7) and from the fact that $\mathcal{P} = \{\boldsymbol{P}_1, \dots, \boldsymbol{P}_N\}$ ($N = n^2$) is weakly uniformly distributed, we have

$$\begin{aligned}\int_{\mathbb{R}^2} [F(\boldsymbol{X})]^2 \, \mathrm{d}\boldsymbol{X} &\ll \int_{\mathbb{R}^2} F(\boldsymbol{X}) \, \mathrm{d}\boldsymbol{X} \\ &= \sum_{i=1}^{N} \int_{\mathbb{R}^2} \chi_i(\boldsymbol{X}) \, \mathrm{d}\boldsymbol{X} \\ &= N\rho^2\pi \\ &= n^4(\tfrac{1}{100}\omega)^2\pi < n^4\omega^2. \end{aligned} \quad (2.17)$$

On the other hand, by (2.11) we have, for all $\alpha_k \in [\frac{1}{2}, 1]$ and $t \in [\frac{1}{2}, 1]$,

$$\int_{\mathbb{R}^2} F(\boldsymbol{X})\, h_k(\alpha_k t, \boldsymbol{X}) \, \mathrm{d}\boldsymbol{X} = \sum_{i=1}^{N} \operatorname{area}(D(\boldsymbol{P}_i, \rho) \cap R_k^{**}(t)) = 0. \quad (2.18)$$

Moreover, since $\mathcal{P}$ is weakly uniformly distributed, every rectangle $R_k^*(t)$ ($k = 1, 2, \dots, m$; $\frac{1}{2} \leqslant t \leqslant 1$) contains $\gg n$ points of $\mathcal{P}$. Thus, we have for all $\alpha_k \in [\frac{1}{2}, 1]$,

$$\begin{aligned}\int_{1/2}^{1} \int_{\mathbb{R}^2} F(\boldsymbol{X})\, H_k(\alpha_k t, \boldsymbol{X}) \, \mathrm{d}\boldsymbol{X} \, \mathrm{d}t &= \int_{1/2}^{1} \left(\sum_{i=1}^{N} \operatorname{area}(D(\boldsymbol{P}_i, \rho) \cap R_k^*(\alpha_k t)) \right) \mathrm{d}t \\ &\gg n\rho^2\pi \gg n^3\omega^2. \end{aligned}$$

Therefore, by (2.18),

$$\begin{aligned}\int_{\mathbb{R}^2} F(\boldsymbol{X})\, G(\boldsymbol{X}) \, \mathrm{d}\boldsymbol{X} &= \delta \sum_{k=1}^{m} \int_{1/2}^{1} \int_{\mathbb{R}^2} F(\boldsymbol{X})\, H_k(\alpha_k t, \boldsymbol{X}) \, \mathrm{d}\boldsymbol{X} \, \mathrm{d}t \\ &\gg \delta m n^3 \omega^2. \end{aligned} \quad (2.19)$$

It remains to estimate the integral

$$\int_{\mathbb{R}^2} [G(\boldsymbol{X})]^2 \, \mathrm{d}\boldsymbol{X}$$

from above. By definition

$$\int_{\mathbb{R}^2} [G(\boldsymbol{X})]^2 \, \mathrm{d}\boldsymbol{X} = \delta^2 \sum_{k=1}^{m} \int_{1/2}^{1} \int_{\mathbb{R}^2} [g_k(\alpha_k t, \boldsymbol{X})]^2 \, \mathrm{d}\boldsymbol{X} \, \mathrm{d}t$$
$$+ \delta^2 \sum_{k=1}^{m} \sum_{\substack{l=1 \\ l \neq k}}^{m} \int_{1/2}^{1} \int_{\mathbb{R}^2} g_k(\alpha_k t, \boldsymbol{X}) g_l(\alpha_l t, \boldsymbol{X}) \, \mathrm{d}\boldsymbol{X} \, \mathrm{d}t. \tag{2.20}$$

From (2.4) we have, for all $k = 1, 2, \ldots, m$,

$$\int_{1/2}^{1} \int_{\mathbb{R}^2} [g_k(\alpha_k t, \boldsymbol{X})]^2 \, \mathrm{d}\boldsymbol{X} \, \mathrm{d}t$$
$$= \int_{1/2}^{1} \int_{\mathbb{R}^2} \left\{ [H_k(\alpha_k t, \boldsymbol{X})]^2 + \left(\frac{2n}{\rho}\right)^2 [h_k(\alpha_k t, \boldsymbol{X})]^2 + 2\left(\frac{2n}{\rho}\right) H_k(\alpha_k t, \boldsymbol{X}) h_k(\alpha_k t, \boldsymbol{X}) \right\} \mathrm{d}\boldsymbol{X} \, \mathrm{d}t$$
$$= \int_{1/2}^{1} \left\{ \operatorname{area}(R_k^*(\alpha_k t)) + \left(\frac{2n}{\rho}\right)^2 \operatorname{area}(R_k^{**}(\alpha_k t)) + 2\left(\frac{2n}{\rho}\right) \operatorname{area}(R_k^{**}(\alpha_k t)) \right\} \mathrm{d}t$$
$$\ll \frac{n^3}{\omega}. \tag{2.21}$$

We shall now show that there are parameters $\alpha_k \in [\frac{1}{2}, 1]$ ($k = 1, 2, \ldots, m$) such that the double sum

$$\sum_{k=1}^{m} \sum_{\substack{l=1 \\ l \neq k}}^{m} \int_{1/2}^{1} \int_{\mathbb{R}^2} g_k(\alpha_k t, \boldsymbol{X}) g_l(\alpha_l t, \boldsymbol{X}) \, \mathrm{d}\boldsymbol{X} \, \mathrm{d}t \tag{2.22}$$

is small.

For notational convenience, write

$$\int (k, \alpha; l, \beta) = \int_{1/2}^{1} \int_{\mathbb{R}^2} g_k(\alpha t, \boldsymbol{X}) g_l(\beta t, \boldsymbol{X}) \, \mathrm{d}\boldsymbol{X} \, \mathrm{d}t.$$

Let $\xi_1, \xi_2, \ldots, \xi_m$ be *mutually independent* random variables, uniformly distributed in the interval $[\frac{1}{2}, 1]$. Consider the expectation

$$E\left(\sum_{k=1}^{m}\sum_{\substack{l=1\\l\neq k}}^{m}\int(k,\xi_k;l,\xi_l)\right)=\sum_{k=1}^{m}\sum_{\substack{l=1\\l\neq k}}^{m}E\int(k,\xi_k;l,\xi_l).$$

If $k\neq l$ and $E\int(k,\xi_k;l,\xi_l)\neq 0$, then certainly $S_k^*\cap S_l^*\neq\varnothing$. Thus, by Lemma 1, $|\theta_k-\theta_l|>1/n$. For every $k=1,2,\ldots,m$ and every $b=1,2,3,\ldots$, let

$$\mathcal{L}(k;b)=\left\{l\in\{1,\ldots,m\}: l\neq k,\ S_k^*\cap S_l^*\neq\varnothing,\ \frac{b}{n}<|\theta_k-\theta_l|\leqslant\frac{b+1}{n}\right\}.$$

A simple geometric consideration and Lemma 1 show that

$$\operatorname{card}\mathcal{L}(k;b)\ll b\qquad\text{uniformly for all } k \text{ and } b. \tag{2.23}$$

Here card $\mathcal{L}$ denotes the cardinality of the set $\mathcal{L}$. Let $l\in\mathcal{L}(k;b)$. It is easily seen that

$$\operatorname{area}\{(\alpha,\beta)\in[\tfrac{1}{2},1]^2: [V_k^*(\alpha)\cup V_l^*(\beta)\cup V_k^*(\tfrac{1}{2}\alpha)\cup V_l^*(\tfrac{1}{2}\beta)]\cap S_k^*\cap S_l^*\neq\varnothing\}\ll\frac{1}{b}$$

uniformly for all k, l and b. Similarly, a simple geometric consideration shows that for every *fixed* $\alpha\in[\frac{1}{2},1]$ (see (2.14)),

$$\operatorname{length}\{\beta\in[\tfrac{1}{2},1]: U_{k,l}(\alpha)\cap U_{l,k}(\beta)\neq\varnothing\}\ll\frac{1}{b}.$$

Thus, we have

$$\operatorname{area}\{(\alpha,\beta)\in[\tfrac{1}{2},1]^2: U_{k,l}(\alpha)\cap U_{l,k}(\beta)\neq\varnothing\}\ll\frac{1}{b}$$

uniformly for all k, l and b. Since ξ_k and ξ_l are independent and uniformly distributed in the interval $[\frac{1}{2},1]$, we conclude that

$$\Pr\{[V_k^*(\xi_k)\cup V_l^*(\xi_l)\cup V_k^*(\tfrac{1}{2}\xi_k)\cup V_l^*(\tfrac{1}{2}\xi_l)]\cap S_k^*\cap S_l^*\neq\varnothing\}\ll\frac{1}{b} \tag{2.24}$$

and

$$\Pr\{U_{k,l}(\xi_k)\cap U_{l,k}(\xi_l)\neq\varnothing\}\ll\frac{1}{b}. \tag{2.25}$$

Therefore, if $l\in\mathcal{L}(k;b)$ then, by using the cancellation property, (2.15), (2.24) and (2.25),

$$E\int(k,\xi_k;l,\xi_l)\ll\frac{1}{b}\,\frac{n}{(b/n)^2}=\frac{n^3}{b^3}.$$

Hence, by (2.23),

$$\sum_{k=1}^{m} \sum_{\substack{l=1 \\ l\neq k}}^{m} E \int (k,\xi_k;l,\xi_l) = \sum_{k=1}^{m} \sum_{b\geqslant 1} \sum_{l\in\mathcal{L}(k;b)} E \int (k,\xi_k;l,\xi_l)$$

$$\ll \sum_{k=1}^{m} \sum_{b\geqslant 1} \sum_{l\in\mathcal{L}(k;b)} \frac{n^3}{b^3}$$

$$\ll \sum_{k=1}^{m} \sum_{b\geqslant 1} b\frac{n^3}{b^3}$$

$$= mn^3 \sum_{b\geqslant 1} \frac{1}{b^2} \ll mn^3. \tag{2.26}$$

Thus there is a point $(\alpha_1, \alpha_2, \ldots, \alpha_m) \in [\frac{1}{2}, 1]^m$ such that the double sum in (2.22) is $\ll mn^3$. Combining this with (2.20) and (2.21), we have

$$\int_{\mathbb{R}^2} [G(\boldsymbol{X})]^2 \, \mathrm{d}\boldsymbol{X} \ll \delta^2 m \frac{n^3}{\omega} + \delta^2 mn^3 \ll \delta^2 \frac{mn^3}{\omega}. \tag{2.27}$$

Summarizing, by (2.16), (2.17), (2.19) and (2.27),

$$\delta mn^3 \omega^2 \ll n^4 \omega^2 + \delta^2 \frac{mn^3}{\omega}. \tag{2.28}$$

Now let

$$\delta = \left(\frac{n\omega^3}{m}\right)^{1/2}.$$

Then (2.28) reduces to

$$m \ll \frac{n}{\omega^3}. \tag{2.29}$$

On the other hand, by Lemma 1,

$$m \gg \min\{n\omega^3 \log n, n\omega^{3/2} (\log n)^{1/2}\}. \tag{2.30}$$

Comparing (2.29) and (2.30), we conclude that (see also (2.2))

$$\omega = \Omega(\mathcal{P}) \ll (\log n)^{-1/9} \ll (\log N)^{-1/9},$$

as required. □

3 Proof of Lemma 1

Since $\mathcal{P} = \{\boldsymbol{P}_1, \ldots, \boldsymbol{P}_N\} \subset Q(N) = [0,N]^2$ is weakly uniformly distributed, one can easily find disjoint pairs

$$\{P_1,\tilde{P}_1\},\ \{P_2,\tilde{P}_2\},\ldots,\ \{P_M,\tilde{P}_M\} \tag{3.1}$$

of points of $\mathcal{P}$ satisfying

$$\{P_i,\tilde{P}_i\} \subset [\tfrac{1}{3}N,\tfrac{2}{3}N]^2 \qquad (1 \leqslant i \leqslant M); \tag{3.2}$$

$$|P_i-\tilde{P}_i| < 3n \qquad (1 \leqslant i \leqslant M); \tag{3.3}$$

$$|P_i-P_j| > 6n \qquad (1 \leqslant i < j \leqslant M); \tag{3.4}$$

$$M \gg N. \tag{3.5}$$

Let $S(\theta,a,n)$ be an arbitrary but fixed strip of width n. Let $q(\theta,a)$ denote the number of pairs $\{P_i,\tilde{P}_i\}$ among (3.1) such that

$$|\theta(P_i,\tilde{P}_i)-\theta| < \frac{\omega}{6n}, \tag{3.6}$$

and

$$P_i \in S(\theta,a,n). \tag{3.7}$$

In order to prove Lemma 1, it suffices to show that

$$\max q(\theta,a) = \max_{\theta}\max_{a} q(\theta,a) \ll \max\left\{\frac{n}{\omega^2\log n},\frac{n}{\sqrt{\omega\log n}}\right\}. \tag{3.8}$$

Indeed, a standard application of the greedy algorithm and (3.8) give that there are

$$m \gg \omega\frac{M}{\max q(\theta,a)} \gg \max\{n\omega^3\log n, n\omega^{3/2}(\log n)^{1/2}\}$$

suitable pairs among (3.1) satisfying (2.8)–(2.10).

Let L be a straight line perpendicular to the border lines of the strip $S(\theta,a,n)$. For notational convenience, let $q = q(\theta,a)$ and let

$$\{P_1,\tilde{P}_1\},\ \{P_2,\tilde{P}_2\},\ldots,\ \{P_q,\tilde{P}_q\}$$

be the pairs satisfying (3.6) and (3.7). For $i = 1,2,\ldots,q$, let Y_i denote the projection of P_i onto the line L. Without loss of generality, we can assume that the points $Y_1,Y_2,\ldots,Y_q$ are precisely in this linear order on the line L.

We need the following simple lemma.

Lemma 2. *Let* $0 \leqslant y_1 < y_2 < \cdots < y_q \leqslant n$. *Then there exists* $l \in \{1,2,\ldots,q\}$ *such that*

$$\sum_{\substack{i=1\\ i\neq l}}^{q} \frac{1}{\max\{|y_l-y_i|,n/q\}} \gg \frac{q}{n}\log q.$$

Proof. Consider the double sum

$$\Sigma^* = \sum_{j=1}^{q} \sum_{\substack{i=1 \\ i \neq j}}^{q} \frac{1}{\max\{|y_j - y_i|, n/q\}}. \tag{3.9}$$

Clearly

$$\Sigma^* = 2 \sum_{j=2}^{q} \sum_{i=1}^{j-1} \frac{1}{\max\{y_j - y_i, n/q\}}.$$

For every $d = 1, 2, \ldots, q-1$ and $r = 1, 2, \ldots, d$, let

$$\Sigma(d, r) = \sum_{\substack{d+1 \leq j \leq q \\ j \equiv r \pmod d}} \frac{1}{\max\{y_j - y_{j-d}, n/q\}}.$$

Note that

$$\Sigma^* = 2 \sum_{d=1}^{q-1} \sum_{r=1}^{d} \Sigma(d, r). \tag{3.10}$$

Let

$$k(q, d, r) = \left[\frac{q-r}{d}\right] \quad \text{(integral part)},$$

and

$$q(d, r) = dk(q, d, r) + r.$$

Observe that

$$\Sigma(d, r) = \sum_{k=1}^{k(q,d,r)} \frac{1}{\max\{y_{kd+r} - y_{(k-1)d+r}, n/q\}}.$$

We have

$$\sum_{k=1}^{k(q,d,r)} \max\{y_{kd+r} - y_{(k-1)d+r}, n/q\} \leq (y_{q(d,r)} - y_r) + k(q, d, r)\frac{n}{q} \leq 2n.$$

Thus, from the inequality between the arithmetic and harmonic means,

$$\frac{k(q, d, r)}{\Sigma(d, r)} \leq \frac{2n}{k(q, d, r)}.$$

Hence

$$\Sigma(d, r) \geq \frac{\{k(q, d, r)\}^2}{2n}.$$

Since

$$k(q,d,r) = \left[\frac{q-r}{d}\right] \geqslant \frac{q}{d}-2,$$

we obtain

$$\Sigma(d,r) \geqslant \frac{1}{2n}\left(\frac{q}{d}-2\right)^2.$$

Returning to (3.10), we have

$$\begin{aligned}\Sigma^* &= 2\sum_{d=1}^{q-1}\sum_{r=1}^{d}\frac{1}{2n}\left(\frac{q}{d}-2\right)^2 \\ &\geqslant \frac{1}{n}\sum_{1\leqslant d\leqslant \frac{1}{3}q}\sum_{r=1}^{d}\left(\frac{q}{d}-2\right)^2 \\ &\geqslant \frac{1}{n}\sum_{1\leqslant d\leqslant \frac{1}{3}q}\sum_{r=1}^{d}\left(\frac{q}{3d}\right)^2 = \frac{1}{n}\sum_{1\leqslant d\leqslant \frac{1}{3}q}^{d}\left(\frac{q}{3d}\right)^2 \\ &\geqslant \frac{q^2}{n}\sum_{1\leqslant d\leqslant \frac{1}{3}q}\frac{1}{d} \geqslant \frac{q^2}{n}\log q.\end{aligned}$$

Therefore, by (3.9),

$$\max_{1\leqslant j\leqslant q}\sum_{\substack{i=1\\ i\neq j}}^{q}\frac{1}{\max\{|y_j-y_i|,n/q\}} \geqslant \frac{1}{q}\Sigma^* \geqslant \frac{1}{q}\frac{q^2}{n}\log q,$$

and Lemma 2 follows. □

It follows from Lemma 2 that there is an $l \in \{1,2,\ldots,q\}$ such that

$$\sum_{\substack{i=1\\ i\neq l}}^{q}\frac{1}{\max\{|Y_l-Y_i|,n/q\}} > c_0\frac{q}{n}\log q, \tag{3.11}$$

where $c_0 > 0$ is an absolute constant independent of $\mathcal{P}$ and N.

Let $i \in \{1,\ldots,q\}\backslash\{l\}$. It follows from the definition of the set $T(P',P'')$ and from (2.3) and (3.3) that

$$|\theta(P_l,\tilde{P}_l)-\theta(P_l,P_i)| > \sin|\theta(P_l,\tilde{P}_l)-\theta(P_l,P_i)| \geqslant \frac{\omega}{|P_l-\tilde{P}_l|} > \frac{\omega}{3n}. \tag{3.12}$$

Thus, by (3.6) and (3.12) we have,

$$|\theta-\theta(P_l,P_i)| > \frac{\omega}{3n}-\frac{\omega}{6n} = \frac{\omega}{6n}. \tag{3.13}$$

For every $i \in \{1,\ldots,q\}\backslash\{l\}$, consider the angular domain

$$A_i = \left\{ \boldsymbol{X} \in \mathbb{R}^2 : |\theta(\boldsymbol{P}_l, \boldsymbol{P}_i) - \theta(\boldsymbol{P}_l, \boldsymbol{X})| < \frac{\omega}{2|\boldsymbol{P}_l - \boldsymbol{P}_i|} \right\}$$

and the angular domain

$$A_i^2 = \left\{ \boldsymbol{X} \in \mathbb{R}^2 : |\theta(\boldsymbol{P}_l, \boldsymbol{P}_i) - \theta(\boldsymbol{P}_l, \boldsymbol{X})| < \frac{\omega}{|\boldsymbol{P}_l - \boldsymbol{P}_i|} \right\}$$

with double the area of A_i. Note that $A_i^2 \subset T(\boldsymbol{P}_l \to \boldsymbol{P}_i)$. Therefore, from (2.3) we obtain that the sets A_i $(i = 1, \ldots, q;\ i \neq l)$ are already pairwise disjoint.

Consider now the wider strip $S(\theta, a, 3n)$ and consider the intersection set

$$B = S(\theta, a, 3n) \cap Q(N).$$

Since the angular domains A_i $(i = 1, \ldots, q;\ i \neq l)$ are disjoint, we have

$$\sum_{\substack{i=1 \\ i \neq l}}^{q} \operatorname{area}(A_i \cap B) \leqslant \operatorname{area}(B). \tag{3.14}$$

Clearly

$$\operatorname{area}(B) < 3n \times 2N = 6n^3. \tag{3.15}$$

In order to estimate $\operatorname{area}(A_i \cap B)$, we first note that the angle

$$\frac{\omega}{|\boldsymbol{P}_l - \boldsymbol{P}_i|}$$

of the angular domain A_i is less than $|\theta - \theta(\boldsymbol{P}_l, \boldsymbol{P}_i)|$ (see (3.4) and (3.13)). Thus, by a simple geometric consideration, we have

$$\operatorname{area}(A_i \cap B) \geqslant \left(\frac{n}{\max\{|\theta - \theta(\boldsymbol{P}_l, \boldsymbol{P}_i)|, 1/n\}} \right)^2 \frac{\omega}{|\boldsymbol{P}_l - \boldsymbol{P}_i|}. \tag{3.16}$$

For the sake of brevity, for every $i \in \{1, 2, \ldots, q\} \setminus \{l\}$, write

$$\delta_i = |\theta - \theta(\boldsymbol{P}_l, \boldsymbol{P}_i)|.$$

Note that

$$|\boldsymbol{P}_l - \boldsymbol{P}_i| = \frac{|Y_l - Y_i|}{\sin \delta_i}.$$

Applying the elementary inequality $\sin x \geqslant 2x/\pi$ if $0 \leqslant x \leqslant \frac{1}{2}\pi$, we get

$$|\boldsymbol{P}_l - \boldsymbol{P}_i| \leqslant \tfrac{1}{2}\pi \frac{|Y_l - Y_i|}{\delta_i}.$$

Returning to (3.16), we get

$$\text{area}(A_i \cap B) \geqslant \left(\frac{n}{\max\{\delta_i, 1/n\}}\right)^2 \frac{\omega}{|Y_l - Y_i|/\delta_i}$$
$$= \frac{\delta_i}{(\max\{\delta_i, 1/n\})^2} \frac{n^2\omega}{|Y_l - Y_i|}. \quad (3.17)$$

Let $K = \lceil \log q / \log 2 \rceil$ and, for every $k = 1, 2, \ldots, K$, let

$$\mathcal{J}_k = \left\{ i \in \{1, \ldots, q\} \backslash \{l\} : 2^{k-1}\frac{n}{q} \leqslant \max\{|Y_l - Y_i|, n/q\} < 2^k \frac{n}{q} \right\}. \quad (3.18)$$

Write the distances $|P_l - P_i|$, $(i \in \mathcal{J}_k)$ in increasing order. Let, say,

$$\mathcal{J}_k = \{j_1, j_2, \ldots, j_{q(k)}\}, \qquad q(k) = \text{card}\,\mathcal{J}_k$$

and

$$|P_l - P_{j_1}| \leqslant |P_l - P_{j_2}| \leqslant \cdots \leqslant |P_l - P_{j_{q(k)}}|.$$

Let

$$t(k) = \min\left\{\left[c_0 2^{k-2}\frac{\log q}{K}\right], q(k)\right\}, \quad (3.19)$$

and consider the 'truncated' set

$$\mathcal{J}_k^* = \{j_\nu : t(k) + 1 \leqslant \nu \leqslant q(k)\} \subset \mathcal{J}_k.$$

Since $\mathcal{P}$ is weakly uniformly distributed, we have

$$|P_l - P_i| \geqslant \left[c_0 2^{k-2}\frac{\log q}{K}\right] n \gg 2^k n \qquad \text{for all } i \in \mathcal{J}_k^* \ (k = 1, \ldots, K). \quad (3.20)$$

Let

$$\mathcal{J}^* = \bigcup_{k=1}^{K} \mathcal{J}_k^* \subset \{1, 2, \ldots, q\} \backslash \{l\}.$$

By (3.11), (3.18) and (3.19),

$$\sum_{i \in \mathcal{J}^*} \frac{1}{\max\{|Y_l - Y_i|, n/q\}} \geqslant \sum_{\substack{i=1 \\ i \neq l}}^{q} \frac{1}{\max\{|Y_l - Y_i|, n/q\}} - \sum_{k=1}^{K} t(k) \frac{q}{n} 2^{-k+1}$$
$$> c_0 \frac{q}{n} \log q - \sum_{k=1}^{K} \left(c_0 2^{k-2} \frac{\log q}{K}\right) \frac{q}{n} 2^{-k+1}$$
$$\geqslant \tfrac{1}{2} c_0 \frac{q}{n} \log q. \quad (3.21)$$

Let us return to (3.17). We have

$$\sin \delta_i = \frac{|Y_l - Y_i|}{|P_l - P_i|} \leqslant \frac{\max\{|Y_l - Y_i|, n/q\}}{|P_l - P_i|}. \tag{3.22}$$

Let $i \in \mathcal{I}_k^*$. By (3.20),

$$|P_l - P_i| \geqslant 2^k n,$$

and so by (3.18) and (3.22) we have

$$\delta_i \leqslant \tfrac{1}{2}\pi \sin \delta_i \ll \frac{2^k n/q}{2^k n} = \frac{1}{q}. \tag{3.23}$$

Summarizing, by (3.13) and (3.23), for every $i \in \mathcal{I}^*$,

$$\frac{\omega}{6n} < \delta_i \ll \frac{1}{q}.$$

It follows that, for every $i \in \mathcal{I}^*$,

$$\frac{\delta_i}{(\max\{\delta_i, 1/n\})^2} \gg \min\{\omega n, q\}. \tag{3.24}$$

Now by (3.17), (3.21) and (3.24),

$$\begin{aligned}\sum_{i \in \mathcal{I}^*} \operatorname{area}(A_i \cap B) &\gg \sum_{i \in \mathcal{I}^*} \min\{\omega n, q\} n^2 \omega \frac{1}{|Y_l - Y_i|} \\ &\geqslant \min\{\omega n, q\} n^2 \omega \sum_{i \in \mathcal{I}^*} \frac{1}{\max\{|Y_l - Y_i|, n/q\}} \\ &\gg \min\{\omega n, q\} n^2 \omega \frac{q}{n} \log q \\ &= \min\{\omega n, q\} nq\omega \log q. \end{aligned} \tag{3.25}$$

On the other hand, by (3.14), (3.15) and (3.25),

$$\min\{\omega n, q\} nq\omega \log q \ll n^3. \tag{3.26}$$

It follows from (3.26) that

$$q \ll \max\left\{\frac{n}{\omega^2 \log n}, \frac{n}{\sqrt{\omega \log n}}\right\}.$$

This proves (3.8), and the proof of Lemma 1 is complete. □

References

[1] J. Komlós, J. Pintz, & E. Szemerédi, On Heilbronn's triangle problem, *J. London Math. Soc.* (2), **24** (1981), 385–96

[2] J. Komlós, J. Pintz, & E. Szemerédi, A lower bound for Heilbronn's problem, *J. London Math. Soc.* (2), **25** (1982), 13–24
[3] W. Moser, *Research Problems in Discrete Geometry*, mimeograph notes, 1981
[4] W. Moser, *Research Problems in Discrete Geometry*, mimeograph notes, 1984
[5] K. F. Roth, On a problem of Heilbronn, *J. London Math. Soc.*, **26** (1951), 198–204
[6] K. F. Roth, On a problem of Heilbronn. II, *Proc. London Math. Soc.* (3), **25** (1972), 193–212
[7] K. F. Roth, On a problem of Heilbronn. III, *Proc. London Math. Soc.*, (3), **25** (1972), 543–49
[8] W. M. Schmidt, On a problem of Heilbronn, *J. London Math. Soc.* (2), **4** (1971), 545–50

Hamilton cycles in random graphs of minimal degree at least k

B. Bollobás*, T. I. Fenner and A. M. Frieze†

0 Introduction

This paper is a contribution to the theory of random graphs (see Bollobás [4]). Thanks to the efforts of a great many people, including Pósa [13], Korshunov [12], Komlós and Szemerédi [11] and Bollobás [2], the threshold function of a Hamilton cycle is known to be $\frac{1}{2}n(\log n + \log\log n + \omega(n))$, where $\omega(n) \to \infty$.

In fact, Bollobás [3] showed that almost every random graph process is such that the hitting time of minimal degree 2 is equal to the hitting time of a Hamilton cycle. Thus if at time 0 we start with the empty graph with vertex set $[n] = \{1, 2, \ldots, n\}$ and at time t, $1 \leqslant t \leqslant N = \binom{n}{2}$, we add the t-th edge at random, then in almost every case it is true that if we stop as soon as the minimal degree becomes 2, the graph at hand is Hamiltonian. Since, trivially, a graph of minimal degree less than 2 is not Hamiltonian, this means that the primary obstruction to a Hamilton cycle is the existence of a vertex of degree less than 2. It is a classical result of Erdős and Rényi [7] that the threshold function of the minimal degree being at least 2 is $\frac{1}{2}n(\log n + \log\log n + \omega(n))$, where $\omega(n) \to \infty$, so, in particular, the threshold function of a Hamilton cycle is also $\frac{1}{2}n(\log n + \log\log n + \omega(n))$.

A simpler result in the vein of the results above, proved by Erdős and Rényi [8], is that the primary obstruction to a matching (assuming that n, the number of vertices, is even) is the existence of an isolated vertex. In particular, having a matching has the same threshold function as having minimal degree 1, namely $\frac{1}{2}n(\log n + \omega(n))$, where $\omega(n) \to \infty$.

* Partially supported by NFS grant DMS 8806097.
† Partially supported by NFS grant CCR 8900112.

Bollobás & Frieze [6] studied the secondary obstructions to a matching in a graph of even order. If we have a condition on the minimal degree being at least 1 then the threshold function goes down to about half the original one: $\frac{1}{2}n(\frac{1}{2}\log n + \log\log n + \omega(n))$, where $\omega(n) \to \infty$, the main secondary obstruction being the existence of two vertices of degree 1 having a common neighbour.

Our main aim in this paper is the study of secondary obstructions to a Hamilton cycle. If we condition on the minimal degree being at least 2 then what is the threshold function of a Hamilton cycle and what is the crucial obstruction to a Hamilton cycle? We shall show that the secondary obstruction is a 2-*spider*: three vertices of degree 2 having a common neighbour; the new threshold function is

$$\tfrac{1}{2}n(\tfrac{1}{3}\log n + 2\log\log n + \omega(n)), \qquad \text{where } \omega(n) \to \infty.$$

In fact, we shall study the secondary obstructions to $\lfloor\frac{1}{2}k\rfloor$ edge-disjoint Hamilton cycles and $k-2\lfloor\frac{1}{2}k\rfloor$ matchings, the obvious primary obstruction being the existence of a vertex of degree less than k. These turn out to be k-*spiders*: $k+1$ vertices of degree k having a common neighbour.

1 Generating a random graph of minimal degree at least k

For a detailed study of the standard models of random graphs we refer the reader to [4]. Here we shall concern ourselves with a natural model which has rarely been studied because of the technical difficulties involved. Given natural numbers n, m and k, let $\mathcal{G}(n,m;\delta \geqslant k)$ be the set of all graphs with vertex set $[n] = \{1,2,\ldots,n\}$ having m edges and minimal degree at least k. Let us turn $\mathcal{G}(n,m;\delta \geqslant k)$ into a probability space by giving all members of it the same probability. In this paper we shall study this probability space, but as the space itself is not amenable to direct investigation, we shall generate the members of $\mathcal{G}(n,m;\delta \geqslant k)$ in a rather roundabout way.

We shall need a probability space rather close to $\mathcal{G}(n,m;\delta \geqslant k)$: the space $\mathcal{MG}(n,m;\delta \geqslant k)$ consisting of all multigraphs on $[n]$ with m edges and loops, having minimal degree at least k, with all members of this set equiprobable.

For natural numbers s and t, let $[s]^t$ be the set of all s^t sequences of length t with the terms taken from the set $[s] = \{1,2,\ldots,s\}$. Consider $[s]^t$ as a probability space in which any two points (i.e., sequences) have the same probability, namely s^{-t}. The space $[s]^t$ has the following intuitive interpretation which we shall use in the sequel. Put t distinguishable balls, say $b_1, b_2, \ldots, b_t$ into s boxes, with probability $1/s$ of putting a ball into any of the boxes. Every arrangement corresponds to a sequence of length t: if b_j goes into the i-th box then set $x_j = i$. Then the sequence $(x_1, x_2, \ldots, x_t)$ is a random element of the space $[s]^t$.

The *degree* of a number i in a sequence $X = (x_1, x_2, \ldots, x_t) \in [s]^t$, denoted by $d_X(i)$, is the number of times i occurs in the sequence: $d_X(i) = |\{j : x_j = i\}|$. Thus $d_X(i)$ is the number of balls in the i-th box. The *minimal degree* of X is $\delta(X) = \min\{d_X(i) : i \in [s]\}$. Similarly the *maximal degree* of X is $\Delta(X) = \max\{d_X(i) : i \in [s]\}$.

Let $[n \mid \delta \geqslant k]^l = \{X \in [n]^l : \delta(X) \geqslant k\}$ and consider this set as a probability space consisting of equiprobable elements. This space is much less pleasant than $[s]^t$ but it is not very far from $\mathcal{G}(n, m; \delta \geqslant k)$, the probability space we intend to study. Indeed for $Y = (y_1, y_2, \ldots, y_l) \in [n \mid \delta \geqslant k]^l$ (l even) we take the multigraph with vertex set $[n]$ and edge set $\{y_1y_2, y_3y_4, \ldots, y_{l-1}y_l\}$. Ignoring the loops and replacing multiple edges by simple edges, we obtain a graph with vertex set $[n]$. Conditional on this graph having precisely m edges and minimal degree at least k, as we shall see, we obtain exactly a random element of $\mathcal{G}(n, m; \delta \geqslant k)$.

Let us see then how we can pass from $[s]^t$ to $[n \mid \delta \geqslant k]^l$. For $X \in [s]^t$, let

$$U(X) = U(X, k) = \{i \in [s] : d_X(i) \geqslant k\}$$
$$= \{i_1, i_2, \ldots, i_n\},$$

where $i_1 < i_2 < \cdots < i_n$. Omit the terms of X not belonging to U and replace i_r by r. Let $\rho(X)$ be the sequence obtained in this way; call $\rho(X)$ the *reduced sequence*. By construction, $\rho(X) \in [n \mid \delta \geqslant k]^l$ for some l. We call n the *order* and l the *length* of the reduced sequence. For example, if s=7, $t = 16$ and $k = 2$ then from

$$X = (4, 7, 1, 5, 6, 7, 2, 7, 1, 4, 7, 2, 3, 5, 7, 4)$$

we first obtain

$$(4, 7, 1, 5, 7, 2, 7, 1, 4, 7, 2, 5, 7, 4),$$

as $U(X) = \{1, 2, 4, 5, 7\}$, and then

$$\rho(X) = (3, 5, 1, 4, 5, 2, 5, 1, 3, 5, 2, 4, 5, 3) \in [5 \mid \delta \geqslant 2]^{14}.$$

Let us use the reduced sequence $\rho(X) = (y_1, y_2, \ldots, y_l) \in [n \mid \delta \geqslant k]^l$ to construct a multigraph $MG(X, k)$ as follows: the vertex set is $[n]$ and the edge set is $\{y_1y_2, y_3y_4, \ldots, y_{l'-1}y_{l'}\}$, where $l' = 2\lfloor \frac{1}{2}l \rfloor$. Finally, let $G(X,k)$ be the graph obtained from $MG(X,k)$ by deleting the loops and replacing the multiple edges by simple edges. Let us write $v(X)$ and $l(X)$ for the order and length of $\rho(X)$, let $e_m(X) = \lfloor \frac{1}{2}l(X) \rfloor$ be the number of edges and loops of $MG(X,k)$ and let $e(X)$ be the number of edges of $G(X,k)$. Note that the number of vertices of $G(X,k)$ (and $MG(X,k)$) is also $v(X)$.

For $Y \in [n \mid \delta \geqslant k]^l$ write $P(\rho(X) = Y) = P(X \in [s]^t : \rho(X) = Y)$, with the probability taken in $[s]^t$.

Lemma 1 *Let $Y_1, Y_2 \in [n \mid \delta \geqslant k]^l$. Then*

$$P(\rho(X) = Y_1) = P(\rho(X) = Y_2).$$

Proof Let $Y \in [n \mid \delta \geqslant k]^l$. In how many ways can we choose a sequence $X \in [s]^t$ satisfying $\rho(X) = Y$? The set $U(X)$ can be chosen in $\binom{s}{n}$ ways and the set $W(X) = \{j : x_j \in U\}$ can be chosen in $\binom{t}{l}$ ways. Having chosen $U(X)$ and $W(X)$, we have fixed l terms of the sequence X. The remaining $t-l$ terms come from a set of $s-n$ elements, with no element occuring more than $k-1$ times. Hence $P(X \in [s]^t : \rho(X) = Y)$ does not depend on Y, provided $Y \in [n \mid \delta \geqslant k]^l$. □

Lemma 1 states that conditional on $v(X) = n$ and $l(X) = l$, all sequences in $[n \mid \delta \geqslant k]^l$ are equally likely to arise as $\rho(X)$. A similar assertion holds for $G(X,k)$ and $\mathcal{G}(n,m;\delta \geqslant k)$.

Lemma 2 *If $G_1, G_2 \in \mathcal{G}(n,m;\delta \geqslant k)$ then*

$$P(G(X,k) = G_1) = P(G(X,k) = G_2),$$

with the probabilities taken in $[s]^t$.

Proof Fix an ordering of $[n]^{(2)}$, the set of possible edges, say take the lex order: 12, 13, ..., $1n$, 23, 24, ..., $2n$, 34, ..., $(n-1)n$.

Pick two sequences of non-negative integers $(\mu_j)_1^m$ and $(\nu_i)_1^n$, such that $\mu_j \geqslant 1$ for every j and $m_1 = \mu + \nu \leqslant \frac{1}{2}t$, where $\mu = \sum \mu_j$ and $\nu = \sum \nu_i$.

Given a graph $G \in \mathcal{G}(n,m;\delta \geqslant k)$, let $\varphi(G)$ be the multigraph obtained from G by taking the j-th edge with multiplicity μ_j and adding ν_i loops at vertex i. All we have to check is that the number of sequences X for which $MG(X,k) = \varphi(G)$ is independent of G.

Suppose that $MG(X,k) = \varphi(G)$. The multigraph $\varphi(G)$ has m_1 edges and loops, so the set of edges and loops can be ordered in

$$\binom{m_1}{\mu_1, \ldots, \mu_m, \nu_1, \ldots, \nu_n}$$

ways; furthermore, the edges can be oriented in 2^μ ways. The list of oriented edges and unoriented loops determines the first $2m_1$ terms of the reduced sequence $\rho(X)$. If $\rho(X)$ has more than $2m_1$ terms then it has precisely $2m_1+1$ terms, giving us n choices for the last term. By Lemma 1, this shows that the number of sequences $X \in [s]^t$ for which $MG(X,k) = \varphi(G)$ is independent of G. □

It is easily seen that Lemma 2 does not extend to $MG(X,k)$ and $\mathcal{MG}(n,m;\delta \geqslant k)$. However, the proof of Lemma 2 implies a variant of it for multigraphs with given sequences $(\mu_j)_1^m$ and $(\nu_i)_1^n$.

Our main problem in investigating the space $\mathcal{G}(n,m;\delta \geqslant k)$ is that, although for a given space $[s]^t$ there is a pair (n,m) such that $P(v(X) = n$ and $e(X) = m)$ is rather large (in the sense that it is much larger than the average though still close to 0), we cannot pinpoint the exact dependence of a pair (n,m) on (s,t). As we cannot come close to determining this exact dependence, we cannot find suitable pairs $(n,m) = (n,m(n))$ $(n = 1,2,\ldots,)$ which are reached with not too small a probability from a pair $(s,t) = (s(n),t(n))$. To overcome this difficulty, we shall nest the original models $[s]^t$ in such a way that a small change in the parameters (s,t) yields only a small change in the distribution of $(v(X),e(X))$. Then we shall locate rather crudely the peak of this distribution and shall show that if (s,t) is varied suitably then for some choice of (s,t) we must hit our preselected pair (n,m) with a reasonable probability.

Let Σ_s be the set of all $s!$ sequences of natural numbers $\boldsymbol{a} = (a_1,a_2,\ldots,a_s)$ of length s such that $a_i \leqslant i$ for every i. Consider Σ_s as a probability space by giving all sequences the same probability, $1/s!$. Given $\boldsymbol{a} \in \Sigma_s$ define $\beta(\boldsymbol{a}) = (b_1,\ldots,b_s)$ by setting $b_1 = a_1 = 1$ and, having defined $b_1,\ldots,b_i$ for $i < s$,

$$b_{i+1} = \begin{cases} b_i & \text{if } a_{i+1} \leqslant i, \\ i+1 & \text{if } a_{i+1} = i+1. \end{cases}$$

It is easily checked that if $1 \leqslant h \leqslant i$ then

$$P(b_i = h) = P(b_i = h \mid b_{i+1} = i+1, b_{i+2} = c_{i+2},\ldots, b_s = c_s) = \frac{1}{i} \quad (1)$$

and

$$P(b_i = b_{i-1} \mid b_i \leqslant i-1) = 1. \quad (2)$$

To see this, note that $P(b_i = i) = P(a_i = i) = 1/i$ and, by induction on

i, for $h \leqslant i-1$ we have

$$P(b_i = h) = P(a_i \leqslant i-1)\, P(b_{i-1} = h) = \frac{i-1}{i}\,\frac{1}{i-1} = \frac{1}{i}\,.$$

The probability space $\Sigma_s^t = \Sigma_s \times \cdots \times \Sigma_s$ allows us to pick a random sequence from each member of a family of probability spaces $[i]^j$; this will enable us to pass from a random sequence in a space $[i]^j$ to a random sequence in a space $[i_0]^{j_0}$. Indeed, for

$$A = (a^{(1)}, a^{(2)}, \ldots, a^{(t)}) \in \Sigma_s^t, \qquad 1 \leqslant i \leqslant s, \qquad 1 \leqslant j \leqslant t,$$

define

$$X(i,j) = X_A(i,j) = (x(i,1), x(i,2), \ldots, x(i,j)),$$

where $x(i,l)$ is defined by

$$\beta(a^{(l)}) = (x(1,l), x(2,l), \ldots, x(s,l)).$$

Since, by relation (1), for fixed i and l, the random variable $x(i,l)$ is uniformly distributed on $[i]$, we have the following lemma.

Lemma 3 *Let $1 \leqslant i \leqslant s$ and $1 \leqslant j \leqslant t$. Then the map $\Sigma_s^t \to [i]^j$, given by $A \mapsto X_A(i,j)$, is a measure-preserving map onto $[i]^j$.*

There is another, more intuitive, way of describing a random array $(X_A(i,j))$. This time we start with the only element of $[1]^t$, namely $(1,1,\ldots,1)$, we change it randomly into an element of $[2]^t$, then apply another random transformation to obtain an element of $[3]^t$, and so on. What are these random transformations we apply? Having got a random element $X(i,t)$ of $[i]^t$, say $(x(i,1), x(i,2), \ldots, x(i,t))$, let

$$x(i+1,j) = \begin{cases} x(i,j) & \text{with probability } \dfrac{i}{i+1}, \\[2ex] i+1 & \text{with probability } \dfrac{1}{i+1}. \end{cases}$$

Then $X(i+1,t) = (x(i+1,1), x(i+1,2), \ldots, x(i+1,t))$ is our random element of $[i+1]^t$. For $1 \leqslant j \leqslant t$ we take

$$X(i,j) = (x(i,1), x(i,2), \ldots, x(i,j)).$$

Relations (1) and (2) imply that the random array $\{X(i,j) : 1 \leqslant i \leqslant s, 1 \leqslant j \leqslant t\}$ obtained in this way has the same distribution as the array $\{X_A(i,j) : 1 \leqslant i \leqslant s, 1 \leqslant j \leqslant t\}$, with A chosen from Σ_s^t.

Lemma 4 *Let $A = (a^{(1)}, a^{(2)}, \ldots, a^{(t)}) \in \Sigma_s^t$. Then the random variables $\beta(a^{(j)}) = (x(1,j), x(2,j), \ldots, x(s,j))$, $1 \leqslant j \leqslant t$, are independent and,* *conditional on* $x(1,j), x(2,j), \ldots, x(i-1,j)$,

$$x(i,j) = \begin{cases} i & \textit{with probability } \dfrac{1}{i}, \\ x(i-1,j) & \textit{with probability } 1-\dfrac{1}{i}. \end{cases}$$

Let now $1 \leqslant m \leqslant N$ be a natural number satisfying

$$m = m(n) = \tfrac{1}{2}n\left(\frac{1}{k+1}\log n + k\log\log n + c\right) + O(1),$$

where k is a fixed natural number and c is a real constant. Furthermore, set

$$s_0 = n, \qquad s = n + n^{k/(k+1)},$$
$$t_0 = 2m, \qquad t = 2m + n^{k/(k+1)}.$$

Strictly speaking, we should set

$$s = n + \lfloor n^{k/(k+1)} \rfloor \quad \text{and} \quad t = 2m + \lfloor n^{k/(k+1)} \rfloor$$

but, in order to avoid inessential, purely formal complications, throughout this paper we shall dispense with the integer signs. In addition, all our inequalities are claimed to hold if n is *sufficiently large*.

Given a sequence $X = (x_1, x_2, \ldots, x_j) \in [i]^j$, set

$$D_l(X) = |\{h : d_X(h) = l\}|,$$

that is, denote by $D_l(X)$ the number of boxes containing precisely l balls.

Lemma 5 *Set*

$$D_l = \frac{1}{\min\{k!, l!\}(k+1)^l} \mathrm{e}^{-c} (\log n)^{l-k} n^{k/(k+1)} \left(1 + \frac{1}{\sqrt{\log n}}\right), \tag{3}$$

and let $E_0 \subset \Sigma_s^t$ *be the event*

$$\{A \in \Sigma_s^t : D_l(X_A(i,j)) \leqslant D_l \text{ for all } 0 \leqslant l \leqslant t,\ s_0 \leqslant i \leqslant s \text{ and } t_0 \leqslant j \leqslant t\}.$$

Then

$$P(\bar{E}_0) = o(n^{-\log n}).$$

Proof The bound above is tighter than needed immediately, but is important later. Fix i and j ($s_0 \leqslant i \leqslant s$, $t_0 \leqslant j \leqslant t$) and, for simplicity, write $d = D_l$ and let $\lambda = (\log n)^3$. The (generalized) Markov inequality implies

$$
\begin{aligned}
&P[D_l(X_A(i,j)) > d] \\
&\quad \leqslant \binom{d}{\lambda}^{-1} E\left[\binom{D_l(X_A(i,j))}{\lambda}\right] \\
&\quad \leqslant \binom{d}{\lambda}^{-1} \binom{i}{\lambda} \frac{j^{\lambda l}}{(l!)^{\lambda}} \left(\frac{1}{i}\right)^{\lambda l} \left(1 - \frac{\lambda}{i}\right)^{j-\lambda l} \\
&\quad \leqslant \left[\frac{i-\lambda}{d-\lambda}\left(\frac{j}{i}\right)^l \frac{1}{l!} \exp\left(-\frac{j}{i} + \frac{\lambda l}{i}\right)\right]^{\lambda} \\
&\quad = \left[\frac{1}{d}\frac{\mathrm{e}^{-c}}{(k+1)^l}(\log n)^{l-k} n^{k/(k+1)} \frac{1}{l!}\left\{1 + O\left(\frac{\log\log n}{\log n}\right)\right\}^l\right]^{\lambda} \\
&\quad \leqslant \left(1 + \frac{1}{2\sqrt{\log n}}\right)^{-\lambda}.
\end{aligned}
$$

Since we have fewer than n^4 choices for the triple (l, i, j), the assertion follows. □

Lemma 6 *Almost every $A \in \Sigma_s^t$ is such that*

$$e_m(X_A(i,j)) - e(X_A(i,j)) \leqslant 8(\log n)^4$$

for all i, j, $s_0 \leqslant i \leqslant s$ and $t_0 \leqslant j \leqslant t$.

The proof of Lemma 6 is deferred to the next section as it relies upon concepts developed in that section.

Lemmas 5 and 6 enable us to show that as we vary i and j, the functions $v(X_A(i,j))$, $e(X_A(i,j))$ and $e_m(X_A(i,j))$ vary in a more or less predictable way.

Lemma 7 *Let the maps $\Sigma_s^t \to [i]^j$, given by $A \mapsto X(i,j) = X_A(i,j)$, be as before. Then almost every A is such that if $s_0 \leqslant i \leqslant s$ and $t_0 \leqslant j \leqslant t$ then*

$$v(X(s_0, j)) \leqslant n \leqslant v(X(s,j)), \tag{4}$$

$$e_m(X(i, t_0)) \leqslant m \leqslant e(X(i,t)), \tag{5}$$

$$|f(X(i, j+1)) - f(X(i,j))| < 9(\log n)^4, \tag{6}$$

$$|f(X(i+1, j)) - f(X(i,j))| < 9(\log n)^4, \tag{7}$$

for $f = v$, e and e_m.

Proof Let $E_1 \subset \Sigma_s^t$ be the event that E_0 (see Lemma 5) holds and so does the conclusion of Lemma 6. We have to show that the probability of $A \in E_1$ not satisfying all the inequalities (4), (5), (6) and (7) tends to 0 as $n \to \infty$.

Note that

$$v(X(i,j)) = i - \sum_{l=0}^{k-1} D_l(X(i,j)) \tag{8}$$

and

$$e_m(X(i,j)) = \left\lfloor \frac{1}{2}\left(j - \sum_{l=1}^{k-1} lD_l(X(i,j))\right)\right\rfloor. \tag{9}$$

Since for $A \in E_1$ we have

$$0 \leqslant D_l(X(i,j)) \leqslant D_l \leqslant \frac{2e^{-c}}{(k+1)^l}(\log n)^{l-k} n^{k/(k+1)} \tag{10}$$

and

$$0 \leqslant e_m(X(i,j)) - e(X(i,j)) \leqslant 8(\log n)^4 \tag{11}$$

for all i and j ($s_0 \leqslant i \leqslant s$, $t_0 \leqslant j \leqslant t$), equations (8) and (9) imply (4) and (5).

Let us turn to the proof of (6) and (7). When we add the $(j+1)$-th ball, there is precisely one box whose contents changes, namely it has one more ball. Hence

$$0 \leqslant v(X(i,j+1)) - v(X(i,j)) \leqslant 1 \tag{12}$$

and

$$0 \leqslant e_m(X(i,j+1)) - e_m(X(i,j)) \leqslant \tfrac{1}{2}(k+1). \tag{13}$$

Inequalities (11), (12) and (13) imply (6).

In order to prove some analogous inequalities about $X(i+1,j)$ and $X(i,j)$, we shall make use of Lemma 4. Given

$$\sum_{l=0}^{k+1} lD_l = O((\log n) n^{k/(k+1)})$$

elements of $X(i,j)$, the probability that when passing to $X(i+1,j)$ we change at least $2k+1$ of them (into $i+1$) is

$$O((n^{k/(k+1)} \log n)^{2k+1}) n^{-2k-1} = o(n^{-2k/(k+1)}).$$

Since for $s_0 \leqslant i \leqslant s$ and $t_0 \leqslant j \leqslant t$ we have about $n^{2k/(k+1)}$ choices for (i,j), almost every $A \in E_1$ is such that at most $2k$ of the balls belonging to boxes with at most $k+1$ balls are changed at each stage $X_A(i,j) \to X_A(i+1,j)$.

Similarly, the probability that, given $D_{k+l} = O((\log n)^l n^{k/(k+1)})$ groups of $k+l \geqslant k+2$ balls each, we change at least $l+1$ in some group is at most

$$D_{k+l}\binom{k+l}{l+1}n^{-l-1} = O((\log n)^l n^{k/(k+1)} n^{-l-1})$$
$$= O\left(n^{-1/(k+1)}\left\{\frac{\log n}{n}\right\}^l\right).$$

Hence, summing these inequalities for $l \geq 2$, we see that almost every $A \in E_1$ is such that, when changing $X_A(i, j)$ into $X_A(i+1, j)$, no box with at least $k+2$ balls is reduced to a box with fewer than k balls.

Consequently, almost every $A \in E_1$ is such that

$$0 \leq \sum_{l=0}^{k-1} D_l(X(i+1, j)) - \sum_{l=0}^{k-1} D_l(X(i, j)) \leq 2k+1$$

and

$$-2k \leq \sum_{l=0}^{k-1} lD_l(X(i+1, j)) - \sum_{l=0}^{k-1} lD_l(X(i, j)) \leq 2k(k-1)+k-1 \leq 2k^2.$$

The additional terms 1 and $k-1$ arise because it may happen that the $(i+1)$-th box contains at most $k-1$ balls. Hence

$$-2k \leq v(X(i+1, j)) - v(X(i, j)) \leq 1 \tag{14}$$

and

$$-k^2 \leq e_m(X(i+1, j)) - e_m(X(i, j)) \leq k. \tag{15}$$

Inequalities (11), (14) and (15) imply inequality (7). □

Lemma 8 *Let n, m, s_0, s, t_0 and t be as before and set $L = L(n) = 9(\log n)^4$.*

(a) *There exist i, j, n' and m' with $|n-n'| \leq L$, $|m-m'| \leq L$, $s_0 \leq i \leq s$ and $t_0 \leq j \leq t$ such that*

$$P[v(X(i, j)) = n' \text{ and } e_m(X(i, j)) = m'] > 10^{-3}(\log n)^{-8} n^{-2k/(k+1)}.$$

(b) *There exist i, j, n'' and m'' with $|n-n''| \leq L$, $|m-m''| \leq L$, $s_0 \leq i \leq s$ and $t_0 \leq j \leq t$ such that*

$$P[v(X(i, j)) = n'' \text{ and } e(X(i, j)) = m''] > 10^{-3}(\log n)^{-8} n^{-2k/(k+1)}.$$

Proof (a) Let $E_2 \subset \Sigma_s^t$ be the set of A's satisfying the conclusions of Lemma 7 so that, by Lemma 7, $P(E_2) = 1 - o(1)$. Let

$$\Lambda = \{(i, j) \in \mathbb{Z}^2 : s_0 \leq i \leq s, t_0 \leq j \leq t\} \subset Q,$$

where

$$Q = \{(i, j) \in \mathbb{R}^2 : s_0 \leq i \leq s, t_0 \leq j \leq t\} \subset \mathbb{R}^2.$$

Define $F: \Lambda \to \mathbb{R}^2$ by

$$F(i,j) = \big(v(X(i,j)), e_m(X(i,j))\big).$$

By inequalities (6) and (7), this function F can be extended to a function $\tilde{F}: Q \to \mathbb{R}^2$ such that

$$\|\tilde{F}(x',y') - \tilde{F}(x,y)\|_\infty \leqslant 2L\max\{|x-x'|, |y-y'|\}$$

and $\tilde{F}$ is linear on the segments $[(i,j),(i+1,j)]$ and $[(i,j),(i,j+1)]$.

We claim that $\tilde{F}$ maps some point $(x,y) \in Q$ into (n,m). Indeed, by inequalities (4) and (5), the function $\tilde{F}$ maps the side $x = s_0$ of Q into the half-plane $\{(x,y) : x \leqslant n\}$, the side $y = t$ into the half-plane $\{(x,y) : y \geqslant m\}$, $x = s$ into $\{(x,y) : x \geqslant n\}$, and $y = t_0$ into $\{(x,y) : y \leqslant m\}$. Hence the image of the boundary of Q has winding number 1 about (n,m), unless (n,m) is in the image of the boundary. But this implies that the continuous function $\tilde{F}$ maps some point of Q into (n,m).

Having found $(x,y) \in Q$ with $\tilde{F}(x,y) = (n,m)$, let $(i,j) \in \Lambda$ be a lattice point satisfying $\max\{|i-x|, |j-y|\} \leqslant \frac{1}{2}$. Then $\tilde{F}(i,j) = F(i,j)$ is close enough to (n,m):

$$|v(X(i,j))-n| \leqslant L \quad \text{and} \quad |e_m(X(i,j))-m| \leqslant L. \tag{16}$$

Since for every $A \in E_2$ there is a point $(i,j) \in \Lambda$ satisfying (16) and we have $|\Lambda| \sim n^{2k/(k+1)}$ choices for (i,j), some point $(i,j) \in \Lambda$ will do for at least $\frac{2}{3}n^{-2k/(k+1)}$ portion of E_2. As, moreover, there are $(2L+1)^2$ pairs (v, e_m) satisfying (16), the assertion (a) follows.

The proof of (b) is analogous. □

We are ready to prove the main theorem of the section about generating random graphs and multigraphs of minimal degree at least k.

Theorem 9 *Let*

$$m = m(n) = \tfrac{1}{2}n\left(\frac{1}{k+1}\log n + k\log\log n + c\right),$$

where k is a natural number and $c \in \mathbb{R}$.

(a) *There exist i' and j' such that $n \leqslant i' \leqslant n+n^{k/(k+1)}$, $2m \leqslant j' \leqslant 2m+n^{k/(k+1)}$ and*

$$P[v(X(i',j')) = n \text{ and } e_m(X(i',j')) = m] \geqslant 10^{-4}(\log n)^{-8}n^{-2k/(k+1)}.$$

(b) *There exist i'' and j'' such that $n \leqslant i'' \leqslant n+n^{k/(k+1)}$, $2m \leqslant j'' \leqslant 2m+n^{k/(k+1)}$ and*

$$P[v(X(i'',j'')) = n \text{ and } e(X(i'',j'')) = m] \geqslant 10^{-4}(\log n)^{-8}n^{-2k/(k+1)}.$$

Proof (a) Let i, j, n' and m' be as in Lemma 8 (a): with probability at least $10^{-3}(\log n)^{-8} n^{-2k/(k+1)}$ the pair $(v(X(i,j)), e_m(X(i,j))) = (n', m')$ is close to (n, m), so let $a = n - n'$ and $b = m - m'$.

We claim that $i' = i + a$ and $j' = j + 2b$ will do for (a). To see this, start with $MG(X_A(i,j), k)$ and add or remove $|a|$ boxes and $2|b|$ balls to $X_A(i,j)$, as appropriate, that is, consider the multigraph $MG(X_A(i+a, j+2b), k)$. Let E_0 be the event defined in Lemma 5. Recalling the calculations in the proof of Lemma 7, we see that for $A \in E_0$ the probability that $MG(X_A(i+1,j),k)$ has one more vertex than $MG(X_A(i,j),k)$ and these multigraphs have the same number of edges is at least

$$1 - O(n^{-1/(k+1)}) - O(n^{-(k+2)/(k+1)} \log n) = 1 - O(n^{-1/(k+1)}), \qquad (17)$$

the second term being the probability that we change a ball in a box with at most k balls, or that the box $i+1$ contains fewer than k balls, and the third is the probability that we change at least $l+1$ balls in a box with $k+l$ balls ($l \geqslant 1$).

Similarly, for $A \in E_0$ the probability that $MG(X(i, j+2), k)$ has precisely one more edge than $MG(X(i,j),k)$ and they have the same number of vertices is at least

$$(1 - n^{k/(k+1)} n^{-1})^2 \geqslant 1 - 2n^{-1/(k+1)}, \qquad (18)$$

where $n^{k/(k+1)} n^{-1}$ is an upper bound for the probability that a ball is added to a box containing at most $k-1$ balls.

Since $a + b = o(n^{1/(k+1)})$, part (a) of the theorem follows, recalling that by (4) and (5) the numbers i' and j' satisfy the required inequalities, and noting that by changing i to $i-1$ and j to $j-2$, if necessary, the result holds whatever the signs of a and b.

(b) This is proved analogously. For the case of adding the $(i+1)$-th box, we have to subtract a term of order $(\log n)^2/n$ from the left-hand side of (17) to account for the probability that either a loop or multiple edge has at least one of its end vertices changed to vertex $i+1$, or that $MG(X(i+1,j),k)$ has some loop or multiple edge incident with vertex $i+1$.

For the case of adding balls $j+1$ and $j+2$, we have to subtract a term of order $\log n/n$ from the left-hand side of (18) to account for the probability that the 2 new balls yield a loop or a multiple edge.

We also need $A \in E_1$ rather than E_0 for (5) to hold (see Lemma 7).
□

2 Configurations

We describe a useful way of partitioning $[n \mid \delta \geqslant k]^l$ according to the degrees of the numbers in the sequence.

Let

$$DX(k,\nu,\mu) = \left\{ \boldsymbol{d} \in [\mu]^\nu : \sum_{i=1}^{\nu} d_i = \mu \text{ and } d_i \geqslant k \text{ for } i = 1,2,\ldots,\nu \right\},$$

where μ is even. For $\boldsymbol{d} \in DX(k,\nu,\mu)$ let

$$[\nu|\boldsymbol{d}]^\mu = \{X \in [\nu]^\mu : d_X(i) = d_i \text{ for } i = 1,2,\ldots,\nu\}.$$

It turns out that several properties can most easily be proved by conditioning on $\rho(X) \in [\nu|\boldsymbol{d}]^\mu$ for fixed $\boldsymbol{d} \in DX(k,\nu,\mu)$ for some ν and μ.

We work with $MG(X,k)$ conditioned on $\rho(X) \in [\nu|\boldsymbol{d}]^\mu$. This has the same distribution as the multigraph produced in the configuration model of Bollobás [4]. Thus let $W_1, W_2, \ldots, W_\nu$ be disjoint sets, where $|W_i| = d_i$ for $i = 1,2,\ldots,\nu$, and let $W = \bigcup_{i=1}^{\nu} W_i$. For $S \subset [\nu]$ let $W_S = \bigcup_{i\in S} W_i$ and for $w \in W$ we define $\psi(w)$ by $w \in W_{\psi(w)}$. A configuration F is a partition of W into $\frac{1}{2}\mu$ pairs. Φ is the set of configurations and for $F \in \Phi$ we let $\varphi(F)$ be the multigraph with vertices $[\nu]$ and an edge $\{\psi(x), \psi(y)\}$ for each pair $\{x,y\} \in F$.

We claim that if each $F \in \Phi$ is equally likely then $\varphi(F)$ is distributed as $MG(X,k)$ conditional on $\rho(X) \in [\nu|\boldsymbol{d}]^\mu$. To see this, note the following.

We can generate F at random by taking a random permutation $w_1, w_2, \ldots, w_\mu$ of W and then taking pairs $\{w_1, w_2\}$, $\{w_3, w_4\}$, ... (Note that each F appears $(\frac{1}{2}\mu)!\,2^{\frac{1}{2}\mu}$ times). (19a)

If we replace each w_i by $\psi(w_i)$ then we obtain a member of $[\nu|\boldsymbol{d}]^\mu$ and each such member appears $\prod_{i=1}^{\nu} d_i!$ times. (19b)

The next lemma gives a list of the likely properties of the sequence $\boldsymbol{d}_{\rho(X)}$.

Lemma 10 *Let s_0, s, t_0 and t be as in the previous section and suppose that i and j satisfy $s_0 \leqslant i \leqslant s$ and $t_0 \leqslant j \leqslant t$.*

Let $DX_0 = \{\boldsymbol{d} \in \bigcup_{\nu,\mu} DX(k,\nu,\mu) : (20) \text{ below holds}\}$. Then

$$P(\boldsymbol{d}_{\rho(X)} \notin DX_0) = o(n^{-\log n}),$$

where $X = X(i,j)$.

$$d_r \leqslant (\log n)^2 \qquad (r = 1,2,\ldots,\nu). \tag{20a}$$

$$\left| |\{r : d_r = k\}| - \frac{e^{-c}}{k!\,(k+1)^k} n^{k/(k+1)} \right| = o(n^{k/(k+1)}). \quad (20b)$$

$$\left|\left\{r : d_r < (1-\epsilon_n)\frac{\log n}{k+1} \text{ or } d_r > (1+\epsilon_n)\frac{\log n}{k+1}\right\}\right| < 2n^{1-\epsilon_n^2/(4k+4)}, \quad (20c)$$

where $\epsilon_n = 1/\log\log n$.

$$|\text{TINY}| \leqslant n^{(2k+1)/(2k+3)}, \quad (20d)$$

where

$$\text{TINY} = \left\{r : d_r \leqslant \frac{\alpha_k \log n}{k+1} + 2\right\}$$

and where α_k *satisfies*

$$\frac{\alpha_k}{k+1} \log \frac{e}{\alpha_k} = \frac{1}{2(k+1)(k+2)}.$$

(The exact value for α_k is unimportant; we only require that α_k is positive and sufficiently small.)

Proof (a) Now $\Delta(\rho(X)) = \Delta(X)$ and

$$\begin{aligned} P(\Delta(X) \geqslant (\log n)^2) &\leqslant iP(d_X(1) \geqslant (\log n)^2) \\ &\leqslant i\binom{j}{(\log n)^2}\left(\frac{1}{i}\right)^{(\log n)^2} \\ &= o(n^{-\log n}). \end{aligned} \quad (21)$$

(b) Lemma 5 gives an upper bound to the number of r such that $d_X(r) = k$ which is tight enough and is satisfied with the required probability. Now let $Z_k = |\{r : d_X(r) \geqslant k+1\}|$, $\lambda = n^{1/(k+1)}(\log n)^3$ and

$$K = \frac{e^{-c}}{k!\,(k+1)^k} n^{-1/(k+1)}.$$

Then, using a simple monotonicity argument for the third inequality,

$$\begin{aligned} P\left(Z_k \geqslant \hat{\imath} = i - \left[1 - \frac{1}{\sqrt{\log n}}\right] \frac{e^{-c}}{k!\,(k+1)^k} n^{k/(k+1)}\right) &\leqslant \binom{\hat{\imath}}{\lambda}^{-1} E\left[\binom{Z_k}{\lambda}\right] \\ &\leqslant \binom{\hat{\imath}}{\lambda}^{-1} \binom{i}{\lambda} P(d_X(r) \geqslant k+1,\ r = 1, 2, \ldots, \lambda) \\ &\leqslant \binom{\hat{\imath}}{\lambda}^{-1} \binom{i}{\lambda} P(d_X(1) \geqslant k+1)^\lambda \end{aligned}$$

$$\leqslant \left(\frac{i-\lambda}{\hat{\imath}-\lambda}\left[1-K\left\{1-O\left(\frac{\log\log n}{\log n}\right)\right\}\right]\right)^{\lambda}$$
$$\leqslant \left(\left[1+\left(1-\frac{1}{2\sqrt{\log n}}\right)K\right]\left[1-K\left\{1-O\left(\frac{\log\log n}{\log n}\right)\right\}\right]\right)^{\lambda}$$
$$\leqslant \left(1-\frac{K}{3\sqrt{\log n}}\right)^{\lambda}$$
$$= o(n^{-\log n}).$$

(c) In the proof of (c) we need an inequality for the binomial random variable $B(n,p)$ (see [4, p.13, Theorem 7 (i)]): if $0 < p \leqslant \frac{1}{2}$, $0 < \delta \leqslant \frac{1}{12}$ and $\delta^2 pqn \geqslant 1$ then

$$P(|B(n,p)-pn| \geqslant \delta pn) \leqslant e^{-\delta^2 np/3}. \tag{22}$$

Let now

$$Z_> = \left|\left\{r : d_X(r) \geqslant \frac{1+\epsilon_n}{k+1}\log n\right\}\right|, \qquad \lambda = (\log n)^2.$$

Then

$$P(Z_> > \alpha = n^{1-\frac{1}{4}\epsilon_n^2/(k+1)}) \leqslant \binom{\alpha}{\lambda}^{-1} E\left[\binom{Z_>}{\lambda}\right]$$
$$\leqslant \binom{\alpha}{\lambda}^{-1}\binom{i}{\lambda}P\left(d_X(1) \geqslant (1+\epsilon_n)\frac{\log n}{k+1}\right)^{\lambda}$$
$$\leqslant \left(\frac{i}{\alpha}e^{-\frac{1}{3}\epsilon_n^2\log n/(k+1)}(1+o(1))\right)^{\lambda}$$
$$\leqslant [e^{-\frac{1}{12}\epsilon_n^2\log n/(k+1)}(1+o(1))]^{\lambda}$$
$$= o(n^{-\log n}),$$

where in the third inequality we made use of (22). The other half of (c) is proved in an identical way.

(d) We can be cruder with our estimates as we do not need a tight bound. Let $n_0 = n^{(2k+1)/(2k+3)}$. Then

$$P(|\text{TINY}| \geqslant n_0) \leqslant E\left[\binom{|\text{TINY}|}{n_0}\right]$$
$$\leqslant \binom{i}{n_0}P\left(d_X(1) \leqslant d = \frac{\alpha_k \log n}{k+1}+2\right)^{n_0}$$

$$\leqslant \binom{i}{n_0}\left\{2\binom{j}{d}\left(\frac{1}{i}\right)^d\left(1-\frac{1}{i}\right)^{j-d}\right\}^{n_0}$$

$$\leqslant \left\{\frac{2ie}{n_0}\left(\frac{je}{di}\right)^d e^{-(j-d)/i}\right\}^{n_0}$$

$$\leqslant \{n^{2/(2k+3)}n^{\{1+o(1)\}/2(k+1)(k+2)}n^{-\{1-o(1)\}/(k+1)}\}^{n_0}$$

$$= o(n^{-\log n}). \qquad \square$$

Armed with these results we can now prove Lemma 6.

Proof of Lemma 6 Fix i and j and consider $X = X_A(i,j)$. Condition on $\rho(X) \in [\nu|\boldsymbol{d}]^{\mu}$, where, by Lemmas 5 and 10 we have, with probability $1-o(n^{-\log n})$,

$$s \geqslant \nu \geqslant n - o(n^{k/(k+1)}), \tag{23a}$$

$$t \geqslant \mu \geqslant 2m - o(n^{k/(k+1)}), \tag{23b}$$

$$\boldsymbol{d} \in DX_0. \tag{23c}$$

Let us call DX_1 the set of $\boldsymbol{d}$ where (23a, b, c) are satisfied.

We work with the configuration model. We first consider the number of loops in $MG(X,k)$. Let $a = \lceil 10(\log n)^2 \rceil$ and $J \subset \text{EVENS} = \{2,4,6,\dots,\mu\}$, where $|J| = a$, be given. Then (see (19a))

$$P(\psi(w_{i-1}) = \psi(w_i),\ i \in J) \leqslant \left(\frac{(\log n)^2}{\mu-2a}\right)^a.$$

Hence

$$P(MG(X,k) \text{ has} \geqslant a \text{ loops}) \leqslant \binom{\frac{1}{2}\mu}{a}\left(\frac{(\log n)^2}{\mu-2a}\right)^a \tag{23d}$$

$$= o(n^{-\log n}).$$

Next let $b = \lceil 7.5(\log n)^4 \rceil$. If $MG(X,k)$ contains at least b edges that are removed in the reduction to $G(X,k)$ then there exist disjoint sets $J_1, J_2 \subset \text{EVENS}$ with $|J_1| = |J_2| = b$ such that

$$\{\{\psi(w_{l-1}),\psi(w_l)\} : l \in J_2\} \subset \{\{\psi(w_{l-1}),\psi(w_l)\} : l \in J_1\}.$$

Thus

$$P(e_m(X) - e(X) \geqslant a+b) \leqslant o(n^{-\log n}) + \binom{\frac{1}{2}\mu}{b}^2\left\{2b\left(\frac{(\log n)^2}{\mu-4b}\right)^2\right\}^b$$

$$= o(n^{-\log n}). \tag{23e}$$

The result follows as there are $o(n^2)$ values for i and j. $\square$

The reader may have noticed that in order to apply Theorem 9 we need to show $\delta(G(X,k)) \geqslant k$ with high probability.

Lemma 11 (a) *Let i' and j' be as in Theorem* 9(a). *Then*

$$\Pi_1 = P[\delta(G(X,k)) < k \mid v(X) = n,\ e_m(X) = m] = o(1),$$

where $X = X(i', j')$.

(b) *Let i'' and j'' be as in Theorem* 9(b). *Then*

$$\Pi_2 = P[\delta(G(X,k)) < k \mid v(X) = n,\ e(X) = m] = o(1),$$

where $X = X(i'', j'')$.

Proof (a)

$$\Pi_1 = \sum_{d \in DX(k,n,2m)} P[\delta(G(X,k)) < k \mid \rho(X) \in [n|\boldsymbol{d}]^{2m}] \times P[\rho(X) \in [n|\boldsymbol{d}]^{2m} \mid v(X) = n,\ e_m(X) = m].$$

Now (a) will follow from

$$P[\delta(G(X,k)) < k \mid \rho(X) \in [n|\boldsymbol{d}]^{2m}] = o(1) \qquad \text{if } \boldsymbol{d} \in DX_0, \quad (24)$$

since, using Theorem 9 and Lemma 10, we have

$$\sum_{d \notin DX_0} P(\rho(X) \in [n|\boldsymbol{d}]^{2m} \mid v(X) = n,\ e_m(X) = m) \leqslant \frac{P(\rho(X) \in [n|\boldsymbol{d}]^{2m}, \text{ where } \boldsymbol{d} \notin DX_0)}{P(v(X) = n,\ e_m(X) = m)} = o(1).$$

We prove a set of results (25) that imply (24) and will be useful later. Assume that $\boldsymbol{d} \in DX_0$.

$$P(\text{there exists a loop or repeated edge within distance } 10k \text{ of TINY in } MG(X,k) \mid \rho(X) \in [n|\boldsymbol{d}]^{2m}) = o(n^{-1/(k+2)}). \quad (25a)$$

$$P(\text{there exists a vertex incident with 3 edges of } MG(X,k) \text{ which are not in } G(X,k) \mid \rho(X) \in [n|\boldsymbol{d}]^{2m}) = O(n^{-2}\{\log n\}^{18}). \quad (25b)$$

Proof of (25a)

$$P(\text{there exists a loop}\ldots) \leqslant n^{(2k+1)/(2k+3)} \sum_{r=0}^{10k} n^r \left(\frac{(\log n)^4}{2m-2r-1}\right)^{r+1} = O(\{\log n\}^{30k+3} n^{-2/(2k+3)}).$$

$$P(\text{there exists a repeated edge}\ldots) \leqslant n^{(2k+1)/(2k+3)} \sum_{r=0}^{10k} n^{r+1}\left(\frac{(\log n)^4}{2m-2r-3}\right)^{r+2} = O(\{\log n\}^{30k+6} n^{-2/(2k+3)}).$$

Proof of (25b)

$$P(\text{there exists a vertex}\ldots) \leqslant P(\text{there exist } t+1 \text{ vertices inducing } t+3 \text{ edges in } MG(X,k) \text{ for some } 0 \leqslant t \leqslant 3) \leqslant \sum_{t=0}^{3} n^{t+1}\left(\frac{(\log n)^4}{2m-2t-5}\right)^{t+3} = O(n^{-2}\{\log n\}^{18}).$$

This completes the proof of (24) and thus of (a).

(b) For future reference, let E_2 denote the event specified by (25b). Let

$$DX_1(n) = \{(d_1, d_2, \ldots, d_n) \in DX_1\};$$

$$DX_2 = \left\{\boldsymbol{d} \in DX_1(n) : m \leqslant \sum_{i=1}^{n} d_i \leqslant m + 8(\log n)^4\right\}.$$

Define an equivalence relation $\approx$ on DX_2 by $\boldsymbol{d} \approx \boldsymbol{d}'$ if and only if $d_i' = d_{\sigma(i)}$ for some permutation σ of $[n]$. Let Ω be the set of equivalence classes of $\approx$. Then, letting $D_\omega = \bigcup_{d\in\omega} [n|\boldsymbol{d}]^{2m'}$ for $\omega \in \Omega$ and $2m' = \sum_{i=1}^{n} d_i$, we find that

$$\Pi_2 \leqslant \sum_{\omega\in\Omega} P[\delta(G(X,k)) < k \text{ and } \bar{E}_2 \mid \rho(X) \in D_\omega, e(X) = m] \times P[\rho(X) \in D_\omega \mid v(X) = n,\ e(X) = m]$$
$$+ \frac{P\left(\rho(X) \in \bigcup_{d\notin DX_1} [n|\boldsymbol{d}]^{2m'}\right) + P(e_m(X) - e(X) > 8(\log n)^4) + P(E_2)}{P(v(X) = n,\ e(X) = m)} \tag{26}$$

The calculations for (25b) are clearly valid for any i and j, so $P(E_2) = O(n^{-2}\{\log n\}^{18})$. Hence, by Theorem 9 and the proof of Lemma 6, the second term of (26) is $o(1)$.

Let us now fix $\omega \in \Omega$. We have the following inequalities:

$$
\begin{aligned}
&P[\delta(G(X,k)) < k \text{ and } \bar{E}_2 \mid \rho(X) \in D_\omega,\ e(X) = m] \\
&\quad \leqslant nP(1 \in \text{TINY},\ 1 \text{ is incident with a loop} \\
&\qquad\qquad \text{or multiple edge in } MG(X,k) \mid \rho(X) \in D_\omega,\ e(X) = m) \\
&\quad \leqslant n^{(2k+1)/(2k+3)} \max\{P(1 \text{ is incident with a loop} \\
&\qquad\qquad\qquad \text{or multiple edge in } MG(X,k) \\
&\qquad \mid \rho(X) \in [n|\boldsymbol{d}]^{2m'},\ e(X) = m) : \boldsymbol{d} \in \omega \text{ such that } 1 \in \text{TINY}\}.
\end{aligned}
$$

If TINY $= \varnothing$ then $\delta(G(X,k)) < k$ and $\bar{E}_2$ is impossible, so we may assume that TINY $\neq \varnothing$. Fix $\boldsymbol{d} \in \omega$ for which $1 \in$ TINY.

Referring to the configuration model, let

$$
\begin{aligned}
\Phi_1 = \{F \in \Phi :\ &1 \text{ is incident with a loop} \\
&\text{or multiple edge, } \bar{E}_2 \text{ holds and } e(\hat{\varphi}(F)) = m\},
\end{aligned}
$$

where $\hat{\varphi}(F)$ is the graph underlying the multigraph $\varphi(F)$, and

$$
\begin{aligned}
\Phi_2 = \{F \in \Phi :\ &1 \text{ is not incident with a loop} \\
&\text{or multiple edge, } \bar{E}_2 \text{ holds and } e(\hat{\varphi}(F)) = m\},
\end{aligned}
$$

The result will follow from the following relation:

$$\frac{|\Phi_1|}{|\Phi_2|} = O\left(\frac{(\log n)^4}{n}\right). \tag{27}$$

To prove (27) we consider the bipartite graph $BG_1 = (\Phi_1, \Phi_2, \Lambda)$, where if $F_1 \in \Phi_1$ and $F_2 \in \Phi_2$ then $(F_1, F_2) \in \Lambda$ if there exists a vertex $v \neq 1$ such that

(i) v is not incident with any loops or multiple edges in $\varphi(F_1)$;
(ii) the distance from 1 to v in $\varphi(F_1)$ is at least 3;
(iii) $v \notin$ TINY;
(iv) F_2 is obtained from F_1 as follows: suppose that

$$W_1 = \{x_1, x_2, \ldots, x_p\} \qquad \text{and} \qquad W_v = \{y_1, y_2, \ldots, y_q\},$$

where $q \geqslant p$. If there is a loop (x_a, x_b) and $(y_1, w), (y_2, w') \in F_1$, then these pairs are replaced by $(x_a, w), (x_b, w'), (y_1, y_2) \in F_2$. If F_1 contains at most three pairs (x_a, z_a), (x_b, z_b), $\ldots$ $(a < b < \cdots)$, where $\{z_a, z_b, \ldots\} \subset W_t$ for some $t \neq 1$ and pairs (y_1, u_1), (y_2, u_2), $\ldots$, then we replace these pairs in F_2 by (x_a, u_1), (x_b, u_2), $\ldots$ and (y_1, z_a), (y_2, z_b), $\ldots$. All other pairs are in $F_1 \cap F_2$. (We interchange the loop/multiple edge pairings for W_1 and the beginning of W_v.)

It is straightforward to check that this is a proper definition. Furthermore (27) follows since $F_1 \in \Phi_1$ has degree at least

$$n - n^{(2k+1)/(2k+3)} - (\log n)^4 - 16(\log n)^4$$

in BG_1 and $F_2 \in \Phi_2$ has degree at most $16(\log n)^4$ in BG_1 $\square$

3 Basic properties of random graphs of minimal degree at least k

We can now prove some properties of the random graph $G = G_{n,m}^{(k)} \in \mathscr{G}(n, m; \delta \geqslant k)$. Let

$$\text{SMALL} = \left\{ v \in [n] : d_G(v) \leqslant \frac{\alpha_k \log n}{k+1} \right\}; \qquad \text{LARGE} = [n] - \text{SMALL}.$$

Let

$$\text{PETIT} = \left\{ v \in [n] : d_G(v) \leqslant \frac{\log n}{3k+3} \right\};$$

$$N_G(S) = \{w \in V(G) - S : \exists\, v \in S \text{ with } \{v, w\} \in E(G)\}.$$

Lemma 12 *$G = G_{n,m}^{(k)}$ satisfies* (a) *to* (d) *below with probability $1 - o(n^{-A \log n})$ and* (e) *and* (f) *with probability $1 - o(1)$, where A is some positive constant.*

(a) $\Delta(G) \leqslant (\log n)^2$.

(b) $|\text{SMALL}| \leqslant 2n^{(2k+1)/(2k+3)}$ *and* $|\text{PETIT}| \leqslant n^{(4k+3)/(4k+4)}$.

(c) $S \subset \text{LARGE}, |S| \leqslant \dfrac{n}{2 \log n} \Rightarrow |N_G(S)| \geqslant \dfrac{\alpha_k \log n}{2k+2} |S|$.

(d) $S, T \subset [n]$, $|S| = |T| = n/\log \log n$ *and* $S \cap T = \varnothing$

$$\Rightarrow |\{(v, w) \in E(G) : v \in S,\, w \in T\}| \geqslant \frac{n \log n}{9(k+1)(\log \log n)^2}.$$

(e) *No connected subgraph of order at most $2k+5$ contains $k+2$ small vertices, i.e., vertices in* SMALL.

(f) *No small vertex is on a cycle of length at most $2k+2$.*

Proof Our proofs of (a)–(d) are on the following lines: let i'' and j'' be as in Theorem 9 (b) and let $X = X(i'', j'')$.

Now let Π and $\hat{\Pi}$ be properties such that

$$G(X, k) \in \Pi \Rightarrow \rho(X) \in \hat{\Pi} \text{ or } e_m(X) - e(X) > 8(\log n)^4. \tag{28}$$

Then, on using Lemma 2, we have

$$\begin{aligned}
&P(G_{n,m}^{(k)} \in \Pi) \\
&\quad = P[G(X,k) \in \Pi \mid v(X) = n,\ e(X) = m,\ \delta(G(X,k)) \geqslant k] \\
&\quad \leqslant P[\rho(X) \in \hat{\Pi} \mid v(X) = n,\ e(X) = m,\ \delta(G(X,k)) \geqslant k] \\
&\qquad + \frac{P(e_m(X) - e(X) > 8(\log n)^4)}{P[\delta(G(X,k)) \geqslant k \mid v(X) = n,\ e(X) = m]P[v(X) = n,\ e(X) = m]}.
\end{aligned} \tag{29}$$

Note that the second term on the right-hand side of (29) is $o(n^{-\frac{1}{2}\log n})$ by (23e) in the proof of Lemma 6, Lemma 11 and Theorem 9 (b). Continuing,

$$\begin{aligned}
P[\rho(X) \in \hat{\Pi} \mid v(X) = n,\ e(X) = m,\ \delta(G(X,k)) \geqslant k] \\
\leqslant \frac{P(\rho(X) \in \hat{\Pi} \mid v(X) = n,\ e(X) = m)}{P[\delta(G(X,k)) \geqslant k \mid v(X) = n,\ e(X) = m]} \\
= (1+o(1))P(\rho(X) \in \hat{\Pi} \mid v(X) = n,\ e(X) = m)
\end{aligned} \tag{30}$$

by Lemma 11. Now

$$\begin{aligned}
&P(\rho(X) \in \hat{\Pi} \mid v(X) = n,\ e(X) = m) \\
&\quad \leqslant \sum_{d \in DX_1(n)} P(\rho(X) \in \hat{\Pi} \text{ and } \rho(X) \in [n|d]^{\mu} \mid v(X) = n,\ e(X) = m) \\
&\qquad + \frac{P(\rho(X) \notin \bigcup_{d \in DX_1} [\nu|d]^{\mu})}{P(v(X) = n,\ e(X) = m)}.
\end{aligned} \tag{31}$$

Now, by the remark preceding (23) the second term on the right-hand side of (31) is $o(n^{-\frac{1}{2}\log n})$.

Furthermore,

$$\begin{aligned}
P(\rho(X) \in \hat{\Pi} \text{ and } \rho(X) \in [n|d]^{\mu} \mid v(X) = n,\ e(X) = m) \\
\leqslant \frac{P(\rho(X) \in \hat{\Pi} \mid \rho(X) \in [n|d]^{\mu})P(\rho(X) \in [n|d]^{\mu})}{P(v(X) = n,\ e(X) = m)}.
\end{aligned} \tag{32}$$

Hence (29) to (32) imply

$$\begin{aligned}
&P(G_{n,m}^{(k)} \in \Pi) \\
&\quad \leqslant \frac{(1+o(1))\max_{d \in DX_1(n)}\{P(\rho(X) \in \hat{\Pi} \mid \rho(X) \in [n|d]^{\mu})\}}{P(v(X) = n,\ e(X) = m)} + o(n^{-\frac{1}{2}\log n}).
\end{aligned} \tag{33}$$

(a) Here we take $\hat{\Pi} = \{\rho(X) : \Delta(\rho(X) > (\log n)^2\}$ and use (33). It is easy to see that the first term is zero.

(b) For |SMALL| we take $\hat{\Pi} = \{\rho(X) : |\text{TINY}| > n^{(2k+1)/(2k+3)}\}$ and again use (33). (Again the first term is zero.) For |PETIT| we would need a similar result for a set like TINY, but this is no problem.

(c) Here we take

$$\hat{\Pi} = \{\rho(X) : \exists\, S, T \subset [n] \text{ such that}$$

$$\text{(i)}\quad |S| \leqslant \frac{n}{2\log n},\ d_X(r) \geqslant \frac{\alpha_k \log n}{k+1} \text{ for } r \in S;$$

$$\text{(ii)}\quad T \subset [n] - S,\ |T| \leqslant \frac{\alpha_k \log n}{2k+2}|S|;$$

$$\text{(iii)}\quad N_{MG(X,k)}(S) \subset T\}.$$

Fix $\boldsymbol{d} \in DX_1(n)$ and let S and T satisfy (i) and (ii) above.

Let E^* be the event that each element of W_S is paired with an element of $W_{S\cup T}$ in the configuration model. Assume first that

$$|S| \leqslant s_0 = \frac{n}{4\alpha_k \mathrm{e}(k+1)(\log n)^3}.$$

Then

$$P(E^*) \leqslant \left\{\frac{1}{\mu-1}(\log n)^2\left(\frac{\alpha_k \log n}{2(k+1)}+1\right)|S|\right\}^{\{\alpha_k \log n/(k+1)\}|S|}$$

$$\leqslant \left(\frac{\alpha_k(\log n)^2}{n}|S|\right)^{\{\alpha_k \log n/(k+1)\}|S|}.$$

When $|S| > s_0$ we consider $\{t \in T : d_X(t) > (1+\epsilon_n)(\log n)/(k+1)\}$ and its complement in T. Then we find, letting $n_0 = n^{1-\epsilon_n^2/(4k+4)}$, that

$$P(E^*)$$

$$\leqslant \left\{\frac{1}{\mu-1}\left[(\log n)^2 n_0 + (1+\epsilon_n)\frac{\log n}{k+1}\left(\frac{\alpha_k \log n}{2(k+1)}+1\right)|S|\right]\right\}^{\{\alpha_k \log n/(k+1)\}|S|}$$

$$\leqslant \left(\frac{\alpha_k \log n}{(k+1)n}|S|\right)^{\{\alpha_k \log n/(k+1)\}|S|}.$$

Hence

$$P(\rho(X) \in \hat{\Pi} \mid \rho(X) \in [n|\boldsymbol{d}]^\mu)$$

$$\leqslant \sum_{\sigma=1}^{s_0}\binom{n}{\sigma}\binom{n}{(\sigma\alpha_k \log n)/(2k+2)}\left(\frac{\alpha_k\sigma(\log n)^2}{n}\right)^{(\alpha_k\sigma\log n)/(k+1)}$$

$$+\sum_{\sigma=s_0+1}^{n/2\log n}\binom{n}{\sigma}\binom{n}{(\sigma\alpha_k \log n)/(2k+2)}\left(\frac{\alpha_k\sigma\log n}{(k+1)n}\right)^{(\alpha_k\sigma\log n)/(k+1)}$$

$$\leqslant \sum_{\sigma=1}^{s_0} \left\{ \left(\frac{ne}{\sigma}\right)^{(k+1)/\alpha_k \log n} \left(\frac{2(k+1)en}{\alpha_k \sigma \log n}\right)^{1/2} \left(\frac{\alpha_k \sigma (\log n)^2}{n}\right)\right\}^{(\alpha_k \sigma \log n)/(k+1)}$$

$$+ \sum_{\sigma=s_0+1}^{n/2\log n} \left\{ \left(\frac{ne}{\sigma}\right)^{(k+1)/\alpha_k \log n} \left(\frac{2(k+1)en}{\alpha_k \sigma \log n}\right)^{1/2} \times \left(\frac{\alpha_k \sigma \log n}{(k+1)n}\right)\right\}^{(\alpha_k \sigma \log n)/(k+1)}$$

$$= o(n^{-(\alpha_k \log n)/(3k+3)}).$$

We can now apply (33).

(d) Here we take

$$\hat{\Pi} = \Big\{\rho(X) : \exists\, S, T \subset [n] \text{ such that}$$

$$\text{(i) } |S| = |T| = \frac{n}{\log\log n} \text{ and } S \cap T = \varnothing;$$

$$\text{(ii) } MG(X,k) \text{ contains fewer than } \frac{n \log n}{8(k+1)(\log\log n)^2}\ S\text{–}T \text{ edges}\Big\}.$$

Fix $d \in DX_1(n)$ and let S and T satisfy (i) of $\hat{\Pi}$. Consider the configuration model. Now

$$|W_S|, |W_T| \geqslant \frac{\{1-o(1)\}n\log n}{(k+1)\log\log n}.$$

Thus let $U \subset W_S$ be of size $(n\log n)/2(k+1)\log\log n$ and suppose that in the construction of F we first choose the pairs containing elements of U. Now, at any stage, if $x \in U$ then the probability it is paired with something in W_T is at least $(|W_T| - |U|)/\mu \geqslant 1/3\log\log n$. Hence the number of pairs stochastically dominates the binomial distribution $B(|U|, 1/3\log\log n)$. But

$$P\left[B\left(|U|, \frac{1}{3\log\log n}\right) \leqslant \frac{|U|}{4\log\log n}\right] \leqslant e^{-\frac{1}{32}|U|/3\log\log n}$$

$$\leqslant e^{-(n\log n)/200(k+1)(\log\log n)^2}.$$

Now

$$\frac{|U|}{4\log\log n} = \frac{n\log n}{8(k+1)(\log\log n)^2},$$

so

$$P(\rho(X) \in \hat{\Pi} \mid \rho(X) \in [n|d]^{\mu}) \leq \binom{n}{n/\log\log n}^2 e^{-(n\log n)/200(k+1)(\log\log n)^2}$$
$$= o(n^{-\log n})$$

and thus (d) follows by (33).

(e) The proof of this (and (f)) is more complex than (a)–(d) as the failure probability in $G(X,k)$ is *not* $o(n^{-2})$ *in the unconditioned case.*

So let i' and j' be as in Theorem 9 (a) and let $X = X(i', j')$. Let E_3 denote the event 'there is a vertex incident with 3 edges of $MG(X,k)$ which are not in $G(X,k)$, or there exists a loop or multiple edge incident with a vertex of degree k, $k+1$, $k+2$ or $k+3$ in $MG(X,k)$'.

Suppose that Π and $\hat{\Pi}$ are properties such that

$$MG(X,k) \in \hat{\Pi} \text{ and } \bar{E}_3 \Rightarrow G(X,k) \in \Pi. \tag{34}$$

Then

$$P[G(X,k) \in \Pi \mid v(X) = n,\ e_m(X) = m,\ \delta(G(X,k)) \geq k]$$
$$\geq P[MG(X,k) \in \hat{\Pi} \text{ and } \bar{E}_3 \mid v(X) = n,\ e_m(X) = m,\ \delta(G(X,k)) \geq k]$$
$$= P[MG(X,k) \in \hat{\Pi} \text{ and } \bar{E}_3 \mid v(X) = n,\ e_m(X) = m] - o(1)$$

(by applying Lemma 11)

$$= \sum_{d \in DX_1(n)} P(MG(X,k) \in \hat{\Pi} \text{ and } \bar{E}_3 \mid \rho(X) \in [n|d]^{2m})$$
$$\times P(\rho(X) \in [n|d]^{2m} \mid v(X) = n, e_m(X) = m) - o(n^{2-\log n}) - o(1)$$

(by applying Lemma 10)

$$= \sum_{d \in DX_1(n)} P(MG(X,k) \in \hat{\Pi} \mid \rho(X) \in [n|d]^{2m})$$
$$\times P(\rho(X) \in [n|d]^{2m} \mid v(X) = n,\ e_m(X) = m)$$
$$-o(n^{-1/(k+2)}) - O(n^{-2}\{\log n\}^{18}) - o(1)$$

(by applying (25a) and (25b))

$$\geq \min_{d \in DX_1(n)} \{P(MG(X,k) \in \hat{\Pi} \mid \rho(X) \in [n|d]^{2m})\} - o(n^{2-\log n}) - o(1). \tag{35}$$

On the other hand,

$$P(G(X,k) \in \Pi \mid v(X) = n,\ e_m(X) = m,\ \delta(G(X,k)) \geq k)$$
$$= P(G(X,k) \in \Pi \text{ and } e_m(X) - e(X) \leq 8(\log n)^4$$
$$\mid v(X) = n,\ e_m(X) = m,\ \delta(G(X,k)) \geq k) + o(1)$$

$$= \sum_{m'=m-8(\log n)^4}^{m} P(e(X) = m' \mid v(X) = n, e_m(X) = m, \delta(G(X,k)) \geqslant k)$$

$$\times P(G_{n,m'}^{(k)} \in \Pi) + o(1) \quad (36)$$

on using Lemma 2. What we deduce from (35) and (36) is that if (34) holds then there exists an m' $(m-8(\log n)^4 \leqslant m' \leqslant m)$ such that

$$P(G_{n,m'}^{(k)} \in \Pi) \geqslant \min_{d \in DX_1(n)} \{P(MG(X,k) \in \hat{\Pi} \mid \rho(X) \in [n|d]^{2m})\} - o(1). \quad (37)$$

To prove (e) we let Π denote the property described in (e) and $\hat{\Pi}$ the equivalent property in $MG(X,k)$, where TINY takes the place of SMALL; it is easy to see that (34) holds. For a fixed $\boldsymbol{d} \in DX_1(n)$ we have

$$P(MG(X,k) \notin \hat{\Pi} \mid \rho(X) \in [n|\boldsymbol{d}]^{2m})$$

$$\leqslant \sum_{h=0}^{k+3} \binom{n^{(2k+1)/(2k+3)}}{k+2} \binom{n}{h} (k+2+h)^{h+k} \left(\frac{(\log n)^4}{2m-2(2k+5)} \right)^{k+h+1}$$

$$= o(1).$$

Applying (37) we see that

$$P(G_{n,m'}^{(k)} \in \Pi) = 1 - o(1) \quad (38)$$

for some m' $(m-8(\log n)^4 \leqslant m' \leqslant m)$.

Now let $\tilde{\mathscr{G}}(n,m';\delta \geqslant k) = \mathscr{G}(n,m';\delta \geqslant k) \cap \Pi$. Set

$$g_{m'} = |\mathscr{G}(n,m';\delta \geqslant k)|; \qquad \tilde{g}_{m'} = |\tilde{\mathscr{G}}(n,m';\delta \geqslant k)|.$$

We can derive our result from (38) by showing

$$\frac{\tilde{g}_{m'+1}}{g_{m'+1}} \geqslant \frac{\tilde{g}_{m'}}{g_{m'}} \left\{ 1 - O\left(\frac{1}{n^{1/(k+2)}} \right) \right\} \quad (39)$$

for $m-8(\log n)^4 \leqslant m' \leqslant m$.

To prove (39) we consider the bipartite graph

$$H \equiv (\mathscr{G}(n,m';\delta \geqslant k), \mathscr{G}(n,m'+1;\delta \geqslant k), E(H)L),$$

where if

$$G_1 \in \mathscr{G}(n,m';\delta \geqslant k) \qquad \text{and} \qquad G_2 \in \mathscr{G}(n,m'+1;\delta \geqslant k)$$

then $(G_1, G_2) \in E(H)$ if and only if $E(G_1) \subset E(G_2)$. Also let $\tilde{H}$ denote the subgraph of H induced by $\tilde{\mathscr{G}}(n,m';\delta \geqslant k)$ and $\tilde{\mathscr{G}}(n,m'+1;\delta \geqslant k)$. Furthermore, let d and $\tilde{d}$ refer to degrees in H and $\tilde{H}$, respectively, so that, for example,

$$\sum_{G\in\mathcal{G}(n,m';\delta\geq k)} d(G) = \sum_{G\in\mathcal{G}(n,m'+1;\delta\geq k)} d(G). \tag{40}$$

Now if $G \in \mathcal{G}(n,m';\delta \geq k)$ then $d(G) = N-m'$ (where $N = \binom{n}{2}$) and if $G \in \mathcal{G}(n,m'+1;\delta \geq k)$ then $d(G) \geq m'+1-kn_k(G)$, where $n_k(G)$ is the number of vertices of degree k in G. Hence (40) implies

$$(N-m')g_{m'} \geq \{m'+1-kE_{m'+1}(n_k(G))\}g_{m'+1}, \tag{41}$$

where $E_{m'+1}(n_k(G))$ is the expectation over $\mathcal{G}(n,m'+1;\delta \geq k)$ of $n_k(G)$.

Applying the same ideas to $\tilde{H}$ we find that if $G \in \tilde{\mathcal{G}}(n,m';\delta \geq k)$ then $\tilde{d}(G) \geq N-m'-|\mathrm{SMALL}(G)|\Delta(G)^{10k}$ and if $G \in \tilde{\mathcal{G}}(n,m'+1;\delta \geq k)$ then $\tilde{d}(G) \leq m'+1$. We deduce then that

$$\{N-m'-E_{m'}(|\mathrm{SMALL}(G)|\Delta(G)^{10k})\}\tilde{g}_{m'} \leq (m'+1)\tilde{g}_{m'+1}. \tag{42}$$

Inequality (39) will follow from (41) and (42) once we prove

$$E_{m'}(|\mathrm{SMALL}(G)|\Delta(G)^{10k}) \leq 3n^{(2k+1)/(2k+3)}(\log n)^{20k} \tag{43}$$

for $m-8(\log n)^4 \leq m' \leq m$. But (43) follows since

$$P(|\mathrm{SMALL}(G^{(k)}_{n,m'})| \geq 2n^{(2k+1)/(2k+3)}) = o(n^{-A\log n}), \tag{44}$$

$$P(\Delta(G^{(k)}_{n,m'}) \geq (\log n)^2) = o(n^{-A\log n}), \tag{45}$$

where (44) and (45) are immediate from Lemma 12 (a) and (b).

(f) We proceed as in (e), taking Π to be the property described in (f) and obtain $\hat{\Pi}$ from Π by replacing SMALL by TINY as in (e). Then, for a fixed $\boldsymbol{d} \in DX_1(n)$,

$$\begin{aligned} P(MG(X,k) \notin \hat{\Pi} \mid \rho(X) \in [n|\boldsymbol{d}]^{2m}) \\ &\leq \sum_{h=3}^{2k+2} \binom{n}{h-1} n^{(2k+1)/(2k+3)} \frac{(h-1)!}{2}\left(\frac{(\log n)^4}{2m-2h}\right)^h \\ &= o(1) \end{aligned}$$

The proof then continues in a manner analogous to (e). □

Let $\mathcal{G}_0 = \mathcal{G}_0(n,m;\delta \geq k)$ denote the members of $\mathcal{G}(n,m;\delta \geq k)$ having no k-spiders and satisfying conditions (e) and (f) of Lemma 12 and the following conditions (a)–(d), which are somewhat weaker than those of Lemma 12:

(a) $\Delta(G) \leq 2(\log n)^2$;

(b) $|\mathrm{SMALL}| \leq 3n^{(2k+1)/(2k+3)}$ *and* $|\mathrm{PETIT}| \leq 2n^{(4k+3)/(4k+4)}$;

(c) $S \subset \mathrm{LARGE}$; $|S| \leq \dfrac{n}{2\log n} \Rightarrow |N_G(S)| \geq \dfrac{\alpha_k \log n}{3k+3}|S|$;

(d) $S, T \subset [n]$, $|S| = |T| = n/\log\log n$ *and* $S \cap T = \emptyset$

$$\Rightarrow |\{(v,w) \in E(G) : v \in S, w \in T\}| \geqslant \frac{n \log n}{10(k+1)(\log\log n)^2}.$$

Let $\mathcal{G}_0' = \mathcal{G}_0'(n,m;\delta \geqslant k)$ be the set of graphs in $\mathcal{G}_0$ which satisfy the more stringent conditions of Lemma 12. From Lemma 12 and Lemma 13 below, it is easy to see that

$$\frac{|\mathcal{G}_0'|}{|\mathcal{G}_0|} = 1 - o(n^{-A\log n}). \tag{46}$$

4 k-spiders in random graphs

We now investigate the existence of k-spiders in $G_{n,m}^{(k)}$.

Lemma 13 *Let*

$$\theta_k = \frac{e^{-(k+1)c}}{(k+1)!\{(k-1)!\}^{k+1}(k+1)^{k(k+1)}}.$$

Then $\lim_{n\to\infty} P(G_{n,m}^{(k)}$ has a k-spider$) = 1 - e^{-\theta_k}$.

Proof We will not give all the details as (1) the case $k = 1$ is given in detail in [6] and (2) the most important ideas are already explained in Lemma 12 (e).

Call a k-spider *isolated* if it does not share any of its $k+2$ vertices with any other k-spider. By Lemma 12 (e) and (f) we have

$$\lim_{n\to\infty} P(G_{n,m}^{(k)} \text{ has a non-isolated } k\text{-spider}) = o(1).$$

We can therefore restrict our attention to isolated k-spiders. Let i' and j' be as in Theorem 9 (a) and let $X = X(i', j')$. The first task is to show that

$$\lim_{n\to\infty} P(MG(X,k) \text{ has an isolated } k\text{-spider} \mid \rho(X) \in [n|d]^{2m}) = 1 - e^{-\theta_k} \tag{47}$$

for $d \in DX_1(n)$. Thus let $d \in DX_1(n)$ be fixed and let ζ denote the number of isolated k-spiders in $MG(X,k)$. If $p > 0$ is a fixed integer and $(\zeta)_p = \zeta(\zeta-1)\ldots(\zeta-p+1)$, we show that

$$\lim_{n\to\infty} E((\zeta)_p) = \theta_k^p \tag{48}$$

and then applying a basic result from probability theory (see, for example, [4], Theorem 1.20), we obtain (47).

Let

$$n_1 = \frac{1}{k!\,(k+1)^k} e^{-c_n k/(k+1)}.$$

It then follows, using (20), that

$$\begin{aligned} E((\zeta)_p) &\leqslant \left(\frac{\{1+o(1)\}n_1}{k+1}\right)^p \sum_{r=0}^{p} \binom{n}{r}\binom{o(n)}{p-r} p! \\ &\quad \times \left(\frac{k(1+\epsilon_n)\log n}{(k+1)\{2m-2(k+1)p\}}\right)^{r(k+1)} \left(\frac{k(\log n)^2}{2m-2(k+1)p}\right)^{(p-r)(k+1)} \\ &= \{1+o(1)\} \frac{n_1^{p(k+1)}}{(k+1)!^p} \binom{n}{p} p! \left(\frac{k\log n}{2(k+1)m}\right)^{(k+1)p} \\ &= \{1+o(1)\}\theta_k^p. \end{aligned} \tag{49}$$

A similar calculation shows that $E((\zeta)_p) \geqslant \{1-o(1)\}\theta_k^p$, so (48) follows.

We can now use (37) with Π and $\hat{\Pi}$ being the property of having an isolated k-spider to show that there exists m' such that

$$P(G_{n,m'}^{(k)} \text{ has an isolated } k\text{-spider}) \geqslant 1 - e^{-\theta_k} - o(1), \tag{50}$$

where $m - 8(\log n)^4 \leqslant m' \leqslant m$. Using a similar argument to the final argument of Lemma 12 (e) we can replace m' by m in the probability inequality of (50).

To get the lower bound we again use (37) but this time with Π and $\hat{\Pi}$ being the property of *not* having a k-spider and then proceed as in Lemma 12 (e). This yields

$$P(G_{n,m}^{(k)} \text{ has no isolated } k\text{-spider}) \geqslant e^{-\theta_k} - o(1)$$

and the result follows. □

Let S be a set of vertices of degree at least $k \geqslant 1$ is a graph G. For k even, an *S-system* is a set of $\frac{1}{2}k$ edge-disjoint systems of vertex-disjoint paths in G such that, for each system, every vertex in S is an internal vertex of some path of that system. For k odd, an *S-system* is a set of $\frac{1}{2}(k-1)$ edge-disjoint systems of paths, as before, together with an independent set of edges incident with each vertex in S, which is edge-disjoint from the path systems.

We shall need the following lemma.

Lemma 14 *Let G be a graph and let $S \subset V(G)$ consist of vertices of degree at least k. Denote by S_0 the set of vertices of S having degree k in G. Suppose that*

(a) *no vertex of* G *is adjacent to* $k+1$ *vertices in* S_0*;*
(b) *no connected subgraph of order at most* $2k+3$ *contains* $k+2$ *vertices of* S*;*
(c) *no vertex in* S *is on a cycle of length at most* $2k+2$.

Then G *contains an* S*-system.*

Proof Consider the subgraph induced by the edges incident with the vertices of S. Without loss of generality, we may take G to be this graph and, moreover, we can assume that G is connected. It is easy to show that $|S| \leqslant k+1$; otherwise we could construct a subtree of G with $k+2$ vertices of S and at most $2k+3$ vertices, contradicting (b). It follows that no cycle has length greater than $2k+2$ and thus, by (c), G must be a tree. Let T be the subtree obtained by deleting all leaves of G which are not in S. All leaves of T are in S, so $|V(T)| \leqslant 2k+1$. If $|S| = k+1$ and some vertex u is adjacent to every vertex in S, then some vertex w in S has degree at least $k+1$ in G, by (a). In this case consider $G-uw$; the component containing w clearly has an S-system, so we may assume that (a) holds with S_0 replaced by S. This, together with the observation that, for any vertex $v \in V(T)$, every component of $T-v$ contains a vertex of S, implies that $\Delta(T) \leqslant k$. (We are assuming that T is not trivial, otherwise the result is immediate.)

If $k = 1$ or 2 the result is immediate. If $k = 3$ then $|V(T)| \leqslant 7$ and it is easy to show that any maximal path in G containing a maximum number of vertices of S can be extended to an S-system.

Suppose then that $k \geqslant 4$ and that the assertion holds for smaller values of k. Let P_0 be a maximal path in T containing two vertices of highest degree in T. We claim that $\Delta(T-E(P_0)) \leqslant k-2$. Otherwise, T would contain three vertices of degree at least $k-1$ which were not on a path. But then T would have at least

$$3(k-2) > k+1 \geqslant |S|$$

leaves, which is impossible.

Now extend P_0 to a maximal path P in G. Every vertex w of S not on P is adjacent to at least two leaves of G; this gives a path P_w of length two through w. We will show that the path system $P^* = \{P\} \cup \{P_w : w \in S - V(P)\}$ can be extended to an S-system. Let $G_0 = G-E(P^*)$. Then G_0 is a forest, two components of which contain just one vertex of S, and no vertex not in S is of degree greater than $\Delta(T-E(P_0)) \leqslant k-2$. Thus G_0 trivially satisfies the conditions of the lemma with k replaced by $k-2$. So, by our assumption, G_0 contains an S-system, which, together with P^*, yields an S-system for G. □

5 The colouring argument

It follows from Lemma 14 that if $G \in \mathcal{G}_0$ then SMALL is *k-coverable* in the following sense: there exist sets of vertex-disjoint paths $\mathcal{P}_1, \mathcal{P}_2, \ldots, \mathcal{P}_{\lfloor k/2 \rfloor}$ plus a matching $\mathcal{P}_{\lceil k/2 \rceil}$ if k is odd, where

(i) both endpoints of every path in $\mathcal{P}_1, \mathcal{P}_2, \ldots, \mathcal{P}_{\lfloor k/2 \rfloor}$ are in LARGE;
(ii) each element of SMALL is an internal vertex of one path in each of $\mathcal{P}_1, \mathcal{P}_2, \ldots, \mathcal{P}_{\lfloor k/2 \rfloor}$;
(iii) if k is odd then each element of SMALL is incident with an edge of $\mathcal{P}_{\lceil k/2 \rceil}$;
(iv) (minimality) if EP_i denotes the edges of $\mathcal{P}_i$ $(i = 1, 2, \ldots, \lceil \frac{1}{2}k \rceil)$ then $e \in \bigcup_{i=1}^{\lceil k/2 \rceil} EP_i$ implies $e \cap \text{SMALL} \neq \emptyset$;
(v) the edge sets $EP_1, EP_2, \ldots, E_{\lceil k/2 \rceil}$ are pairwise disjoint.

For each $G \in \mathcal{G}_0$ choose fixed sets of paths $\mathcal{P}_1, \mathcal{P}_2, \ldots,$ together with a matching if k is odd. Let $EP_i(G)$ refer to this fixed choice.

Suppose $H_1, H_2, \ldots, H_r$ is a sequence of edge-disjoint Hamilton cycles of G. They are said to be *compatible* if $H_i \supset EP_i$ for $i = 1, 2, \ldots, r$, where $r \leq \lfloor \frac{1}{2}k \rfloor$.

We are now close to proving our main theorem. The main tools will be Pósa's theorem [13] plus the colouring argument of Fenner and Frieze [9]. To use the colouring argument we need to consider the deletion of a small set of edges. So if $G = G_{n,m}^{(k)}$ and $X \subseteq E(G)$ we say that X is *deletable* if

(i) X is a matching;
(ii) X is not incident with any vertex of PETIT.

Let G_X be the graph $G - X$ obtained by deleting X from G.

We shall need to consider the state of G after deleting some edge-disjoint Hamilton cycles. Thus we prove the following result.

Lemma 15 *Let $G \in \mathcal{G}_0$; let X be deletable and $H_1, H_2, \ldots, H_r$, $r < \lceil \frac{1}{2}k \rceil$, be compatible edge-disjoint Hamilton cycles. Let*

$$K = G - \left(X \cup \bigcup_{i=1}^{r} E(H_i) \right)$$

be the graph obtained by deleting X and the edges of these cycles, and let $\Sigma = \text{SMALL} \cup N_G(\text{SMALL})$.

(a) *If $\emptyset \neq S \subseteq \text{LARGE}$ and $|S| \leq \beta_k n$ then*

$$|N_K(S) - \Sigma| \geq 3|S|,$$

where $\beta_k = \alpha_k / 7(k+1)(2k+5)$.

(b) *If $r < \lfloor \frac{1}{2}k \rfloor$ then K is connected.*

Proof (a) By Lemma 12 (e) and (f) we have

$$|N_G(S) \cap \Sigma| \leq (k+1)|S|. \tag{51}$$

Let $S_1 \subseteq S$ be of size $\min\{|S|, \lfloor n/2\log n \rfloor\}$. Then

$$\begin{aligned}
|N_K(S)| &\geq |N_G(S)| - (2r+1)|S| \\
&\geq |N_G(S_1)| - (2r+2)|S| \\
&\geq \frac{\alpha_k \log n}{3(k+1)}|S_1| - (2r+2)|S| \\
&\geq (2k+5)|S| - (2r+2)|S| \\
&\geq (k+4)|S|. \qquad (52)
\end{aligned}$$

The result now follows after using (51).

(b) Suppose S is a component of K ($|S| \leq \frac{1}{2}n$) and suppose first that $|S| \leq \beta_k n$. Since $E(K) \supseteq EP_{\lfloor k/2 \rfloor}$ we have $S_1 = S \cap \text{LARGE} \neq \emptyset$. Thus

$$\begin{aligned}
|N_K(S)| &\geq |N_K(S_1)| - |N_K(S_1) \cap \text{SMALL}| \\
&\geq (k+4)|S_1| - (k+1)|S_1| > 0,
\end{aligned}$$

where we have used (51) and (52) to establish the second inequality. Thus $|S| \leq \beta_k n$ is not possible, and condition (d) of the definition of $\mathscr{G}_0$ shows that we have not deleted enough edges to create a component of size greater than $\beta_k n$. □

We now proceed as in the proof of Theorem 1.2 of [6].

Let $\mathscr{A}_k$ be the graph property of having $\lfloor \frac{1}{2}k \rfloor$ edge-disjoint Hamilton cycles plus a further edge-disjoint matching of size $\lfloor \frac{1}{2}n \rfloor$ if k is odd. Let r be a non-negative integer and $G \in \mathscr{G}_0$. Let $H_1, H_2, \ldots, H_r$ be a compatible sequence of edge-disjoint Hamilton cycles. $G - \bigcup_{i=1}^{r} E(H_i)$ is called an *r-subgraph* of G.

Let now $\phi(G) = (r, s)$, where r is the maximal length of a compatible sequence of edge-disjoint Hamilton cycles and

$$s = \begin{cases} 0 & \text{if } k = 2r, \\ \text{maximal cardinality of a matching containing} & \\ EP_{\lceil k/2 \rceil} \text{ in any } r\text{-subgraph of } G & \text{if } k = 2r+1, \\ \text{maximal length of a path } P \text{ in any } r\text{-subgraph} & \\ \text{of } G \text{ such that } Q \in \mathscr{P}_{r+1} \text{ implies } Q \text{ is a} & \\ \text{subpath of } P \text{ or } Q \text{ is disjoint from } P & \text{if } k \geq 2r+2. \end{cases}$$

Thus if $\phi(G) = \theta(k, n) = (\lfloor \frac{1}{2}k \rfloor, \lfloor \frac{1}{2}n \rfloor (k - 2\lfloor \frac{1}{2}k \rfloor))$ then $G \in \mathscr{A}_k$.

If $\phi(G) = (r,s)$ we define a *ϕ-subgraph* of G to be any r-subgraph containing either a matching of size s or a path of length s as the case may be.

Lemma 16 *Suppose $G \in \mathcal{G}_0 - \mathcal{A}_k$ and X is deletable. Let $u = \lceil \beta_k n \rceil$ and $\phi(G) = (r,s)$. Then for n large there exist a ϕ-subgraph H of G_X, $A = \{a_1, a_2, \ldots, a_t\}$, $A_1, A_2, \ldots, A_t \subseteq [n]$ $(t > u)$ such that, for $i = 1, 2, \ldots, t$, $|A_i| \geqslant u$, $a_i \notin A_i$ and, if $a \in A_i$, then $e = \{a, a_i\} \notin E(H)$ and $\phi(H+e) \neq \phi(H)$.*

Proof Let, in fact, H be any ϕ-subgraph of G_X. Suppose first that $k = 2r+1$. Let

$$\mathcal{M} = \{M : M \supseteq EP_{\lceil k/2 \rceil} \text{ and } M \text{ is a matching of size } s \text{ in } H\};$$

$$A = \{a : a \text{ is left exposed by some matching in } \mathcal{M}\} = \{a_1, a_2, \ldots, a_t\};$$

$$A_i = \{a : a \text{ and } a_i \text{ are left exposed by some matching in } \mathcal{M}\} \subseteq A \quad (i = 1, 2, \ldots, t).$$

Clearly $a \in A_i$ implies $e = \{a, a_i\} \notin E(G_X)$ and $\phi(G_X + e) \neq \phi(G_X)$. We must show that $t \geqslant u$. Consider, for example, A_1. Let $A_1' = A_1 \cap \text{LARGE}$. $A_1' \neq \emptyset$, else a_1 is the only vertex left exposed by any matching in $\mathcal{M}$, contradicting $G \notin \mathcal{A}_k$. We will show that

$$|N_H(A_1') - \Sigma| < |A_1'| \tag{53}$$

and then Lemma 15 (a) implies that $|A_1'| \geqslant u$.

To prove (53), let $\{x, y\} \in E(H)$, $x \in A_1'$ and $y \notin \Sigma$, and let $M \in \mathcal{M}$ leave x and a_1 exposed. y is not exposed, so let $\{y, z\} \in M$. Then $z \in \text{LARGE}$ (otherwise $y \in \Sigma$) and, because $M' = M + \{x, y\} - \{y, z\}$ leaves a_1 and z exposed, we have that $z \in A_1'$. Thus $y \in N_H(A_1') - \Sigma$ implies that y is adjacent to A_1' via an edge of M and (53) follows.

Assume next that $k > 2r+1$. If $P = (v_1, v_2, \ldots, v_k)$ is a path of H and $e \equiv \{v_k, v_i\} \in E(H)$ then, as has often been done before, we consider the path

$$\text{ROTATE}(P, e) = (v_1, v_2, \ldots, v_i, v_k, v_{k-1}, \ldots, v_{i+1}) \tag{54}$$

and let v_1 be called the *fixed endpoint* in this construction.

Now let P be a path of length s in H either wholly containing or wholly disjoint from any path of $\mathcal{P}_{r+1}$. Let a_1 be one endpoint of P. Consider all paths that may be obtained from P by a sequence of rotations with a_1 as a fixed endpoint, subject to the restriction that no rotation is allowed in which the deleted edge ($\{v_i, v_{i+1}\}$ in (54)) is in EP_{r+1}.

Let A_1 be the set of endpoints, other than a_1, of the paths produced by this procedure and let $A_1' = A_1 \cap \text{LARGE}$. $A_1' \neq \emptyset$ as vertices in SMALL can only be internal vertices of any of the paths produced. Following Pósa [13] we show that

$$|N_H(A_1') - \Sigma| < 2|A_1'| \tag{55}$$

and then Lemma 15 (a) shows that $|A_1'| \geqslant u$. Note also that there is no edge $\{a, a_1\} \in E(H)$ where $a \in A_1$, since, by Lemma 15 (b), H is connected and such an edge could be used to give a longer path.

So let $\{x, y\} \in E(H)$, $x \in A_1'$ and $y \notin \Sigma$ (y is an internal vertex of P as P is maximal). We observe as in [6] that y has at least one neighbour in A_1' on P and (55) follows.

Finally, take $A = A_1 \cup \{a_1\}$ and repeat the argument for $a \in A_1$ using any path of length s with a as fixed endpoint. □

We now give a final lemma which essentially proves our main theorem.

Lemma 17

$$\lim_{n \to \infty} P(G_{n,m}^{(k)} \in \mathscr{A}_k \mid G_{n,m}^{(k)} \text{ has no } k\text{-spider}) = 1.$$

Proof Let

$$\mathscr{G}_2 = \mathscr{G}_2(n, m; \delta \geqslant k) = \mathscr{G}_0 - \mathscr{A}_k.$$

If $G \in \mathscr{G}(n, m; \delta \geqslant k)$ and $X \subseteq E(G)$, let X be *strongly deletable* if X is deletable and $\phi(G_X) = \phi(G)$. Let $\omega = \lceil \sqrt{\log n} \rceil$ and for $X \subseteq E(G)$ ($|X| = \omega$) let

$$a(X, G) = \begin{cases} 1 & \text{if } G \in \mathscr{G}_2 \text{ and } X \text{ is strongly deletable,} \\ 0 & \text{otherwise.} \end{cases}$$

Let $G \in \mathscr{G}_2$. We show first that

$$\sum_{\substack{X \subset E(G) \\ |X| = \omega}} a(X, G) > \binom{m}{\omega} \left(1 - \frac{2k^2}{\log n}\right)^{\omega}.$$

To see this let $\phi(G) = (r, s)$. Choose a random ω-subset X of $E(G)$. It is easy to see that the right-hand side of (55) is a lower bound for the expected number of X satisfying (1) X is a matching (use Lemma 12 (a)), (2) X is not incident with a vertex of PETIT (use Lemma 12 (b)) and (3) X contains no edge of some fixed compatible sequence of r Hamilton cycles and s further edges.

If (1), (2) and (3) hold then $a(X,G) = 1$. Hence

$$\sum_{G\in\mathscr{G}_2}\sum_{\substack{X\subseteq E(G)\\|X|=\omega}} a(X,G) \geqslant \binom{m}{\omega}\left(1-\frac{2k^2}{\log n}\right)^{\omega}|\mathscr{G}_2|. \tag{56}$$

We now bound the sum on the left-hand side of (56) from above. Let

$$\Omega = \{H : \exists G \in \mathscr{G}_0,\ X \subseteq E(G),\ |X|=\omega,\ X \text{ is not incident with any vertex of PETIT}(G), \text{ and } H = G_X\}.$$

For $H \in \Omega$ let $S_H = |\{X : G = H+X \in \mathscr{G}_2 \text{ and } a(X,G) = 1\}|$. We show

$$S_H \leqslant \binom{\binom{n}{2}-\gamma n^2}{\omega}, \tag{57}$$

where $\gamma = \frac{1}{2}\beta_k^2$ and β_k is as in Lemma 15. To see this note that if $S_H > 0$ then H contains sets $A, A_1, A_2, \ldots, A_t$ as in Lemma 16 and $X \in S_H$ implies (through $a(X,G) = 1$) that X contains no edge of the form $\{a, a_i\}$, where $a_i \in A$ and $a \in A_i$ $(i = 1,2,\ldots,t)$. This implies (57).

But

$$\sum_{G\in\mathscr{G}_2}\sum_{\substack{X\subseteq E(G)\\|X|=\omega}} a(X,G) = \sum_{H\in\Omega} S_H.$$

Thus (56) and (57) imply

$$|\mathscr{G}_2| \leqslant \binom{\binom{n}{2}-\gamma n^2}{\omega}\binom{m}{\omega}^{-1}\left(1-\frac{2k^2}{\log n}\right)^{-\omega}|\Omega|. \tag{58}$$

Clearly we must produce an upper bound for $|\Omega|$. To do this we count pairs (H,X), where (1) H is a graph with $V(H) = [n]$, (2) $|E(H)| = m-\omega$, (3) $X \subset [n]^{(2)} - E(H)$ and $|X| = \omega$, (4) $G = H+X \in \mathscr{G}_0$ and (5) X is not incident with any vertex in PETIT(G); in this case we call (H,X) a *proper* pair. Let ζ be the number of such pairs. Clearly

$$\zeta = \{1-o(1)\}\binom{m}{\omega}|\mathscr{G}_0| \tag{59}$$

as $(G-X, X)$ is almost always a proper pair when $G \in \mathscr{G}_0$, $|X| = \omega$ and $X \subseteq E(G)$. On the other hand

$$\zeta = \sum_{H\in\Omega} \epsilon_H, \tag{60}$$

where $\epsilon_H = |\{X : X \text{ is a proper partner for } H\}|$ for $H \in \Omega$. Let

$$\Omega' = \{H \in \Omega : \exists \text{ a proper partner } X \text{ for which } H+X \in \mathscr{G}_0'\}|.$$

We observe that

$$H \in \Omega' \Rightarrow \epsilon_H \geqslant \{1-o(1)\}\binom{\binom{n}{2}-m+\omega}{\omega}. \tag{61}$$

To see this let $H \in \Omega'$ and (H,X) be proper with $H+X \in \mathscr{G}_0'$. Let $Y \subseteq [n]^{(2)} - E(H)$ and suppose that the edges in Y are not incident with any vertex within $10k$ of PETIT$(H+X)$. Observe that

$$\text{SMALL}(H+X) = \text{SMALL}(H+Y).$$

It is easy then to see that $H+Y \in \mathscr{G}_0$ (but not necessarily $\mathscr{G}_0'$). A simple calculation yields (61). It follows from (59) to (61) that

$$|\mathscr{G}_0| \geqslant \{1-o(1)\}|\Omega'|\binom{\binom{n}{2}-m+\omega}{\omega}\bigg/\binom{m}{\omega}. \tag{62}$$

We will show that

$$|\Omega'| = \{1-o(1)\}|\Omega|, \tag{63}$$

to obtain from (58) and (62) that

$$\frac{|\mathscr{G}_2|}{|\mathscr{G}_0|} \leqslant \{1+o(1)\}\binom{N-\gamma n^2}{\omega}\bigg/\binom{N-m+\omega}{\omega} = o(1).$$

which proves the lemma.

If $H \in \Omega - \Omega'$ then there exist $G \in \mathscr{G}_0 - \mathscr{G}_0'$ and $X \subseteq E(G)$ ($|X| = \omega$) such that $H = G - X$. Hence

$$|\Omega| - |\Omega'| \leqslant \binom{m}{\omega}(|\mathscr{G}_0| - |\mathscr{G}_0'|). \tag{64}$$

On the other hand,

$$|\Omega'| \geqslant |\mathscr{G}_0'|\bigg/\binom{N-m+\omega}{\omega} \tag{65}$$

as, given $G \in \mathscr{G}_0'$, one can always find a proper pair (H,X) such that $H+X = G$.

From (64) and (65) we obtain

$$\frac{|\Omega|}{|\Omega'|} - 1 \leqslant \binom{m}{\omega}\binom{N-m+\omega}{\omega}\left(\frac{|\mathscr{G}_0|}{|\mathscr{G}_0'|} - 1\right)$$
$$= o(n^{4\omega} n^{-A\log n}) = o(1),$$

which is equivalent to (63), with the last relation following from (46). □

6 The main theorem

We now come finally to the main result of our paper.

Theorem 18 *Let*

$$m = \tfrac{1}{2}n\left(\frac{\log n}{k+1} + k\log\log n + c_n\right).$$

Then

$$\lim_{n\to\infty} P(G_{n,m}^{(k)} \in \mathscr{A}_k) = \begin{cases} 0 & \text{if } c_n \to -\infty \text{ sufficiently slowly,} \\ e^{-\theta_k} & \text{if } c_n \to c, \\ 1 & \text{if } c_n \to +\infty. \end{cases}$$

Proof When $c_n \to c$ we need only consult Lemmas 13 and 17. For $c_n \to \pm\infty$ we have to rework all the calculations. For $c_n \to +\infty$ things get easier, but we can only allow $c_n \to -\infty$ so that

$$\frac{1}{k+1} + \frac{c_n}{\log n} \geqslant \epsilon$$

for some fixed $\epsilon > 0$. In this case our methods work. □

The above theorem describes rather neatly the secondary obstruction to membership of $\mathscr{A}_k$. In a related vein one could consider $P(G_{n,m}^{(k+1)} \in \mathscr{A}_k)$. We strongly believe and seem close to proving that there exist constants $c_1, c_2, \ldots$ such that if $c \geqslant c_k$ then

$$\lim_{n\to\infty} P(G_{n,cn}^{(k+1)} \in \mathscr{A}_k) = 1.$$

Finally we mention that we could have based our proof on an analysis of HAM in [5]. This would show that the Hamilton cycles and matchings are almost always constructable in polynomial time. The proof, however, would have been even longer.

References

[1] B. Bollobás, A probabilistic proof of an assymptotic formula for the number of labelled regular graphs, *Europ. J. Combinatorics*, **1** (1980), 311–16

[2] B. Bollobás, Almost all regular graphs are Hamiltonian, *Europ. J. Combinatorics*, **4** (1983), 97–106

[3] B. Bollobás, The evolution of sparse graphs, in *Graph Theory and Combinatorics*, Proc. Cambridge Combinatorial Conference in Honour of Paul Erdős (ed. B. Bollobás), Academic Press, London, 1984, 35–57

[4] B. Bollobás, *Random Graphs*, Academic Press, London, 1985

[5] B. Bollobás, T. I. Fenner & A. M. Frieze, An algorithm for finding Hamiltonian paths and cycles in random graphs, *Combinatorica*, **7** (1987), 327–42

[6] B. Bollobás & A. M. Frieze, On matchings and Hamilton cycles in random graphs, to appear

[7] P. Erdős & A. Rényi, On random graphs I, *Publ. Math. Debrecen*, **6** (1959), 290–7

[8] P. Erdős & A. Rényi, On the existence of a factor of degree one of a connected random graph, *Acta Math. Acad. Sci. Hungar.*, **17** (1966), 359–68

[9] T. I. Fenner & A. M. Frieze, On the existence of Hamiltonian cycles in a class of random graphs, *Discrete Math.*, **45** (1983), 301–5

[10] T. I. Fenner & A. M. Frieze, Hamiltonian cycles in random regular graphs, *J. Combinatorial Theory* (B), **37** (1984), 103–12

[11] J. Komlós & E. Szemerédi, Limit distributions for the existence of Hamilton cycles in a random graph, *Discrete Math.*, **43** (1983), 55–63

[12] A. D. Korshunov, Solution of a problem of Erdős and Rényi on Hamilton cycles in non-oriented graphs, *Soviet Mat. Dokl.*, **17** (1976), 760–4

[13] L. Pósa, Hamiltonian circuits in random graphs, *Discrete Math.*, **14** (1976), 359–64

The circumference of a graph with a given minimal degree

Béla Bollobás* and Roland Häggkvist

Abstract

Extending a theorem of Alon, we prove a conjecture of Katchalski that every graph of order n and minimal degree at least $n/k > 1$ contains a cycle of length at least $n/(k-1)$. The result is best possible for all values of n and k ($2 \leqslant k < n$).

Introduction

A well-known result of Erdös and Gallai [4] states that, for $n \geqslant k \geqslant 3$, a graph of order n and size greater than $\frac{1}{2}(k-1)(n-1)$ has circumference at least k, that is, it contains a cycle of length at least k. According to Dirac's [3] classical theorem, every graph of order $n \geqslant 3$ and minimal degree at least $\frac{1}{2}n$ is Hamiltonian. What can one say about the circumference of a graph of order n and minimal degree at least $d \geqslant 2$? Recently Alon [1] came close to giving a complete answer to this question when he proved that for $2 \leqslant k < n$ every graph of order n and minimal degree at least n/k has circumference at least $\lfloor n/(k-1) \rfloor$. Our aim here is to improve on this slightly, namely to show that the assertion holds without the integer sign, as conjectured by Katchalski. Although this seems to need a surprising amount of work, we feel it is worth it since the new result is best possible for all values of n and k ($2 \leqslant k < n$) and implies a complete answer to the question above concerning the minimal circumference of a graph of order n and minimal degree $d \geqslant 2$.

* The research in this paper was conducted during the first author's visit to Stockholm in July 1987.

Our notation is standard (see [2]). In particular, given a graph G, we write $\delta(G)$ for the *minimal degree* of G and $c(G)$ for the *circumference*, i.e., the length of the longest cycle.

We shall need four lemmas. The first two are essentially due to Dirac [3], although he did not state them precisely in this form. The first is very simple; for the second, which is somewhat more involved, see Pósa [5]. The third is also rather simple but, for the sake of completeness, we prove it here.

Lemma 1 *Let G be a graph of order n. If $d(x)+d(y) \geqslant n$ then $c(G+xy) = c(G)$ for any two non-adjacent vertices. In particular, if G is a maximal graph with $c(G) = c$ for some c, and $x, y \in G$ $(x \neq y)$ then $d(x)+d(y) \geqslant n$ implies $xy \in E(G)$.*

Lemma 2 *Let G be a 2-connected graph and let $P = x_1 x_2 \ldots x_p$ be a path in G. Then $c(G) \geqslant \min\{p, d_P(x_1)+d_P(x_2)\}$, where $d_P(x)$ is the number of vertices of P joined to x.*

Lemma 3 *Let G be a 2-connected graph of order n with degree sequence $d_1 \leqslant d_2 \leqslant \cdots \leqslant d_n$. Then for $1 \leqslant u \leqslant n-2$ we have*

$$c(G) \geqslant \min\{n-u+1, d_{u+1}+d_{u+2}-u\}.$$

Proof Let $V(G) = \{x_1, \ldots, x_n\}$ with $d(x_i) = d_i$ and set $U = \{x_1, \ldots, x_u\}$ and $W = \{x_{u+1}, \ldots, x_n\}$. Thus U is the set of vertices of small degrees and W is the set of vertices having large degrees. Put $p = \min\{n-u+1, d_{u+1}+d_{u+2}-u\}$. Let us assume that $c(G) \leqslant p-1$.

Note that if we add to G all the edges joining vertices of W then the new graph has circumference at least p. Hence there is a *minimal* set of edges whose addition to G results in a graph H with $c(H) \geqslant p$. Let xy be one of the edges added to G such that $x, y \in W$ and $xy \in E(H) \backslash E(G)$. Then in $H' = H - xy$ there is an x–y path of order at least p. Extend this path to a maximal path P joining vertices of W: say P is a path from w_1 to w_2 $(w_1, w_2 \in W)$. Then since $c(H') \leqslant p-1$, by Lemma 2 we have

$$d_P(w_1)+d_P(w_2) \leqslant p-1, \tag{1}$$

where $d_P(w_i)$ is the number of vertices of P joined to w_i in H'. Now the maximality of P implies that every neighbour of w_i not on P belongs to U; furthermore, since $c(H') \leqslant p$ (in fact, $c(H')$ is at most $p-1$), the vertices w_1 and w_2 have no common neighbour outside P. Hence

$$p-1 \geqslant d_P(w_1)+d_P(w_2) \geqslant d_G(w_1)+d_G(w_2)-u \geqslant d_{u+1}+d_{u+2}-u,$$

contradicting (1) and proving the lemma. □

If $\delta(G) \geqslant d \geqslant 2$ then, rather trivially, $c(G) \geqslant d+1$. Indeed, if $x_1 x_2 \ldots x_p$ is a longest path in G then every neighbour of x_1 belongs to this path. Hence there is an x_l $(3 \leqslant d+1 \leqslant l \leqslant p)$ adjacent to x_1. Then $x_1 x_2 \ldots x_l$ is a cycle of length $l \geqslant d+1$. This argument also shows that every endblock of G contains a cycle of length at least $d+1$.

Our final lemma concerns graphs satisfying $c(G) = d+1$.

Lemma 4 *Let G be a graph of order n such that $\delta(G) \geqslant d \geqslant 2$ and $c(G) = d+1$. Then $n = bd+c$ for some integers b and c with $1 \leqslant c \leqslant b$.*

Proof If $n \geqslant d^2+1$ or $n = d+1$ then n is of the form above. Hence we may assume that $d+2 \leqslant n \leqslant d^2$. Furthermore, we may and shall assume that G is a *maximal* graph satisfying $c(G) = d+1$. In particular, G is connected.

A block of G is said to be *proper* if it has at least $d+1$ vertices; otherwise it is an *improper block*. A cutvertex is a *proper cutvertex* if it belongs to (and so separates) at least two proper blocks; otherwise it is *improper*. Note that every endblock of G is proper. Furthermore, every vertex of G belongs to a proper block. Indeed, if $x \in G$ belongs to no proper block then every neighbour of x is a cutvertex, so G has at least d disjoint endblocks. Hence $n \geqslant d(d+1)+1$, contradicting our assumption on n.

Let B be a proper block of G, with a set S of s proper cutvertices, a set $T = \{x_1, \ldots, x_t\}$ of improper cutvertices and a set W of $w \geqslant d+1-s-t$ other vertices.

Claim *Either $w+t \leqslant d$ or $w+t = d+1$ and $s = 0$.*

Most of the work in the proof of our lemma goes into justifying this claim. If $s = t = 0$ then $B = G$ is a 2-connected graph with minimal degree at least d and so, by Lemma 2, $c(G) \geqslant \min\{n, 2d\} \geqslant d+2$, since if G is non-Hamiltonian then it has a path of order $c(G)+1$. Hence $s+t \geqslant 1$.

Let x_i be adjacent to l_i vertices not in B and so in other proper blocks. Then there are at least $s + \sum_{i=1}^{t} l_i$ other proper blocks disjoint from each other and from $B \backslash S$, so

$$d+1-s+(d+1)\left(s + \sum_{i=1}^{t} l_i\right) \leqslant d^2.$$

Hence

$$ds+d+1+(d+1)\sum_{i=1}^{t} l_i \leqslant d^2,$$

and so

$$s+t \leqslant s + \sum_{i=1}^{t} l_i \leqslant d-2. \qquad (2)$$

Inequality (2) implies that for every $x_i \in T$ we have

$$d_B(x_i) \geqslant d-l_i \geqslant d-l_i - \sum_{j \neq i} l_j + (t-1) \geqslant s+t+1. \qquad (3)$$

Since B has at least w vertices having degree at least d in B, by applying Lemma 3 to B with $u = s+t \geqslant 1$ we see that

$$\min\{w+1, 2d-s-t\} \leqslant c(B) \leqslant c(G) \leqslant d+1.$$

Therefore, as $s+t \leqslant d-2$, we see that

$$w \leqslant d, \qquad d+1 \leqslant |B| = w+s+t \leqslant 2d-2, \qquad w \geqslant 3.$$

As $|B| \leqslant 2d$ and G is a maximal graph satisfying $c(G) \leqslant d+1$, Lemma 1 implies that W spans a complete graph in B.

Suppose first that $s = 0$. As for $x \in W$ and $x_i \in T$ we have

$$d_B(x) + d_B(x_i) \geqslant d+t+1 \geqslant w+t = |B|,$$

by Lemma 1 the vertex x is joined to x_i, so every vertex of W is joined to every vertex of T. If $t \leqslant w$ then this implies that $w+t = c(B) \leqslant d+1$. If $t > w$ then $d_B(x_i) \geqslant t+1 > \frac{1}{2}(t+w)$, so, by Lemma 1 again, T also spans a complete graph and therefore $w+t = c(B) \leqslant d+1$. As $|B| = w+t \geqslant d+1$, we have, in fact, $w+t = d+1$.

Suppose now that $s \geqslant 1$. Since B is 2-connected, there is a path P in S whose initial vertex, say z_1, is joined to a vertex x in $W \cup T$ and whose terminal vertex, say z_2, is joined to a vertex $y \neq x$ in $W \cup T$. Let H' be the graph constructed from the subgraph $B' = B[W \cup T]$ spanned by $W \cup T$, by adding to it a vertex z and joining z to all the neighbours of z_1 and z_2 in $W \cup T$. Let H be a maximal graph containing H', with vertex set $V(H')$ and $c(H) = c(H')$. Then

$$c(H) = c(H') \leqslant c(B) \leqslant c(G) = d+1.$$

Furthermore, by (3) we see that $d_H(x_i) \geqslant d_B(x_i) - (s-1) \geqslant t+2$, so if $x \in W$ and $x_i \in T$ then

$$d_H(x) + d_H(x_i) \geqslant w-1+t+2 = w+t+1 = |H|.$$

Therefore, by Lemma 1, the graph H contains all the $W-T$ edges. But then $d_H(x) \geqslant |H|-2$ for all $x \in W$, so z is joined to every vertex in W. Also $d_H(x_i) \geqslant \max\{t+2, w\} \geqslant \frac{1}{2}|H|$, so T spans a complete graph in H.

Hence so does the set $W \cup T$ and consequently $c(H) = |H| = w+t+1$. Therefore $w+t \leqslant d$, completing the proof of our claim.

The rest of the proof is straightforward. As $n \geqslant d = 2$, our claim implies that G has at least two blocks. Let c_0 be a cutvertex. For a proper block B, if $c_0 \in B$ then $c_B = c_0$. Otherwise let c_B be the cutvertex belonging to B and separating it from c_0. By our claim, there is a set V_B of $d+1$ vertices of $V(B)$ such that $c_B \in V_B$ and $V(B) \backslash V_B$ consists of proper cutvertices. For each proper block B, let K_B be the complete graph with vertex set V_B, let $\tilde{G}$ be the union of these complete graphs K_B. Then $\tilde{G}$ has no isolated vertices and all its blocks are complete graphs of order $d+1$. Hence if $\tilde{G}$ has c components and b blocks then $n = bd+c$. □

It is clear that Lemma 4 is best possible. Given $n = bd+c$, where $1 \leqslant c \leqslant b$ and $d \geqslant 2$, let $\tilde{G}$ be a graph with c non-trivial components and b blocks, each of which is a K_{d+1}. Then $|\tilde{G}| = n$, $\delta(\tilde{G}) = d$ and $c(\tilde{G}) = d+1$.

We are now ready to prove Katchalski's conjecture.

Theorem 5 *Let G be a graph of order n and minimal degree at least $\lceil n/k \rceil \geqslant 2$, where $k \geqslant 2$. Then G contains a cycle of length at least $\lceil n/(k-1) \rceil$.*

Proof For $k = 2$ the result is just Dirac's theorem (equivalently, an immediate consequence of Lemma 1), so we may assume that $k \geqslant 3$. Set $d = \lceil n/k \rceil$ and $l = \lceil n/(k-1) \rceil$. We know that $c(G) \geqslant d+1$. Suppose first that $c(G) = d+1$. Then, by Lemma 4, $n = bd+c$, where $1 \leqslant c \leqslant b$. Since $n \leqslant kd$, we have that $b \leqslant k-1$. Therefore $l = \lceil n/(k-1) \rceil \leqslant \lceil n/b \rceil = d+1$ and so the assertion holds in this case. Thus we may assume that $c(G) \geqslant d+2$.

Therefore we may also assume that $l \geqslant d+3$. This implies that

$$\frac{n+(k-2)}{k-1} \geqslant d+3$$

and so

$$d \geqslant 2k+1. \tag{4}$$

Let us suppose that the assertion of the theorm is false, i.e., $c(G) \leqslant l-1$. We may and shall also assume that G is a maximal graph satisfying $c(G) \leqslant l-1$.

As in the proof of Lemma 4, call a block of G *proper* if it has at least $d+1$ vertices; otherwise it is *improper*. A cutvertex is *proper* if it belongs to at least two proper blocks; otherwise it is *improper*. Let $B_1, \ldots, B_p$ be proper blocks of G with vertex sets $V_1, \ldots, V_p$ of cardinalities $n_1, \ldots, n_p$. The set U_i obtained from V_i by omitting the proper cutvertices is the *interior* of B_i; similarly, the *kernel* W_i of B_i is obtained from V_i by omitting all cutvertices. Set $u_i = |U_i|$ and $w_i = |W_i|$.

Note that every endblock of G is a proper block. Also, every vertex of G belongs to a proper block. Indeed, otherwise there is a cutvertex x which is in no proper block. But then every neighbour of x is a cutvertex, so G has at least d disjoint endblocks, implying that $d(d+1) \leqslant n \leqslant kd \leqslant d^2$.

As $n \leqslant kd$, there are at most $k-1$ proper blocks, so a proper block contains at most $k-2$ cutvertices. Moreover, if the improper blocks containing a vertex have t vertices altogether then G has at least t endblocks, so it has at least t proper blocks. Hence $t \leqslant k-1$. Trivial calculations show that

$$l \leqslant 2d-k+2. \tag{5}$$

Note that, by (5),

$$2d-(n_i-w_i) \geqslant 2d-(k-2) \geqslant l.$$

Therefore, applying Lemma 3 to a proper block B_i with $u = n_i - w_i \geqslant 1$, we see that

$$l-1 \geqslant c(G) \geqslant c(B_i) \geqslant \min\{w_i+1, 2d-(n_i-w_i)\} \geqslant \min\{w_i+1, l\},$$

and so

$$w_i \leqslant l-2 \qquad (i = 1, \ldots, p).$$

Let us investigate the subgraph B_i for a moment. Note that, by (5),

$$n_i \leqslant w_i+k-2 \leqslant l+k-4 < 2d.$$

As G is a maximal graph with $c(G) \leqslant l-1$, Lemma 1 implies that W_i, the kernel of B_i, spans a complete graph. Hence in the graph B_i every vertex in the kernel W_i has degree at least $w_i - 1 \geqslant n_i-(k-2)-1$ and every vertex in the interior U_i has degree at least $d-(k-2)$. By inequality (4),

$$\{n_i-(k-2)-1\} + \{d-(k-2)\} \geqslant n_i$$

and so, by Lemma 1, every vertex in W_i is joined to every vertex in U_i.

In particular, in B_i every vertex in $U_i \backslash W_i$ has degree at least w_i. By inequality (4) we have

$$n_i \geqslant d+1 \geqslant 2(k-2)$$

and so

$$2w_i \geqslant 2\{n_i-(k-2)\} \geqslant n_i.$$

Therefore, by Lemma 1, the set $U_i \backslash W_i$ spans a complete subgraph and hence so does U_i. This shows that

$$\text{if } u_i < n_i, \qquad \text{then} \qquad c(G) \geqslant u_i+1. \tag{6}$$

The proof is almost complete. Let G have h blocks of order at least $l-1$ that contain no proper cutvertices. Then each of these blocks has precisely $l-1$ vertices. Since $(l-1)h \leqslant (l-1)(k-1) < n$, the graph G has some other proper blocks, say $B_1,\ldots,B_j$. Note that $j \leqslant k-1-h$ and G has at most $k-2-h$ proper cutvertices.

We claim that some u_{i_0} $(1 \leqslant i_0 \leqslant j)$ is at least $l-1$. Indeed, otherwise

$$n-(l-1)h-(k-2-h) \leqslant \sum_{i=1}^{j} u_i \leqslant (l-2)j \leqslant (l-2)(k-1-h),$$

which implies the contradiction

$$n \leqslant kl-k-l < (k-1)(l-1).$$

Finally, as $n_{i_0} > u_{i_0} \geqslant l-1$, relation (6) implies that $c(G) \geqslant l$, contradicting our assumption on G and so completing the proof of the theorem. □

Theorem 5 is all we need to determine the minimal circumference of a graph of order n and minimal degree d. For $2 \leqslant d \leqslant n-1$ set

$$f(n,d) = \min\{c(G) : |G| = n \text{ and } \delta(G) \geqslant d\}.$$

Theorem 6 *Let* $2 \leqslant d \leqslant n-1$ *and set* $k = \lceil n/d \rceil$. *Then*

$$f(n,d) = \lceil n/(k-1) \rceil.$$

Proof By the definition of k, we have $k \geqslant 2$ and

$$2 \leqslant \lceil n/k \rceil \leqslant d.$$

Hence, by Theorem 5, $f(n,d) \geqslant \lceil n/(k-1) \rceil$.

On the other hand, set $l = \lceil n/(k-1) \rceil$ and note that

$$d \leqslant \lceil n/(k-1) \rceil - 1 = l-1.$$

Hence $n = (k-1)(l-1)+c$ for some $1 \leqslant c \leqslant k-1$, so there is a graph

G of order n without isolated vertices, with c components made up of complete blocks, each of which is a K_l. Then $\delta(G) = l-1 \geqslant d$ and $c(G) = l$, showing that $f(n,d) \leqslant f(n,l-1) \leqslant l$. □

Added in May 1988 We have just learned that, by a different method, Y. Egawa and T. Miyamoto have also proved the main result of this note. The paper by Egawa and Miyamoto, entitled 'The longest cycle in a graph of minimum degree at least $|G|/k$', will appear in the *Journal of Combinatorial Theory, Series B*.

References

[1] N. Alon, A longest cycle of a graph with a large minimal degree, *J. Graph Theorey*, **10** (1986), 123–7

[2] B. Bollobás, *Extremal Graph Theory*, Academic Press, London, 1978

[3] G. A. Dirac, Some theorems in abstract graphs, *Proc. London Math. Soc.* (3), **2** (1952), 69–81

[4] P. Erdös & T. Gallai, On maximal paths and circuits of graphs, *Acta Math. Acad. Sci. Hungar.*, **10** (1959), 337–56

[5] L. Pósa, On the circuits of finite graphs, *Publ. Math. Inst. Hungar. Acad. Sci.*, **8** (1963), 355–61

On arithmetic progressions in sums of sets of integers

J. Bourgain

0 Statement of results

Let A and A' be sets of integers contained in the interval $[0, N]$ with densities $\delta = |A|/N$ and $\delta' = |A'|/N$. Then, provided only that N is large enough, $A+A' = \{n+n' : n \in A, n' \in A'\}$ contains an arithmetic progression of length L whenever

$$L < \exp[c(\delta\delta' \log N)^{1/3} - \log\log N]$$

for some constant c. This investigation was suggested by [1] and [2].

1 Reduction to a problem in harmonic analysis

Assume for the sake of convenience that A and A' are contained in $[0, \frac{1}{2}N)$ and denote by $G = \mathbb{Z}/N\mathbb{Z}$ the cyclic group of residue classes modulo N. The harmonic analysis will be performed on this group.

Define a function f on G by setting

$$f(k) = \frac{1}{N}|A \cap (k - A')| \qquad (k = 0, 1, \ldots, N-1),$$

where $k - A'$ is taken in G. Note that if $f(k) \neq 0$ then $k \in A + A'$ (in G) and hence, since $0 \leqslant k < N$, $\max A < \frac{1}{2}N$ and $\max A' < \frac{1}{2}N$, also $k \in A + A'$ in $\mathbb{Z}$. Thus we are led to construct an arithmetic progression in $[0, N)$ on which f does not vanish.

Suppose $y \in G$ $(y \neq 0)$ is such that

$$\int_G \max_{1 \leqslant k < K} |f(x + ky) - f(x)| \, dx < \int_G f(x) \, dx. \tag{1}$$

Then there is a point $x \in G$ such that

$$\max_{1\leq k<K} |f(x+ky)-f(x)| < f(x)$$

and so

$$f(x), f(x+y), \ldots, f(x+Ky) \neq 0.$$

If we now consider the orbit $x, x+y, \ldots, x+Ky$ in G, standard arguments permit us to find an arithmetic progression in $[0, N)$ of length greater than $[\sqrt{K}]$ contained in the given orbit. This completes the proof, provided that we may take $L = \sqrt{K}$ of the size given above.

Defining

$$\hat{f}(j) = \int_G f(x)\mathrm{e}^{-2\pi ijx}\,\mathrm{d}x = \frac{1}{N}\sum_{k=0}^{N-1} f(k)\mathrm{e}^{-2\pi ijk/N},$$

we have

$$\hat{f}(0) = \int_G f(x)\,\mathrm{d}x = \frac{1}{N^2}\sum_{k=0}^{N-1} |A\cap(k-A')|$$

$$= \frac{1}{N^2}\sum_{j\in A} |(A'+j)\cap[0,N)| = \delta\delta'$$

and also

$$\|f\|_{A(G)} = \sum_{j=0}^{N-1} |\hat{f}(j)| = \frac{1}{N^2}\sum_{j=0}^{N-1}\left|\sum_{k=0}^{N-1} \mathrm{e}^{-2\pi ijk/N}|A\cap(k-A')|\right|$$

$$= \frac{1}{N^2}\sum_{j=0}^{N-1}\left|\sum_{k\in A} \mathrm{e}^{-2\pi ijk/N}\right|\left|\sum_{k\in A'} \mathrm{e}^{-2\pi ijk/N}\right|,$$

which, by Hölder's inequality and Parseval's theorem, in bounded by

$$\sum_{j=0}^{N-1} |\hat{\chi}_A(j)||\hat{\chi}_{A'}(j)| \leq \sqrt{\int_G \chi_A(x)\,\mathrm{d}x}\sqrt{\int_G \chi_{A'}(x)\,\mathrm{d}x} = \sqrt{\delta\delta'}.$$

Thus, letting $\epsilon = \sqrt{\delta\delta'}$ and replacing f by $\epsilon^{-1}f$, we need only to prove (1) for some function f on G fulfilling the conditions

$$\hat{f}(0) = \epsilon \qquad \text{and} \qquad \|f\|_A \leq 1, \tag{2}$$

taking K with

$$K < \exp[c(\epsilon^2 \log N)^{1/3} - \log\log N]. \tag{3}$$

2 Cutting of the Fourier transform

Recall that a subset A of the dual group Γ of a compact Abelian group G is said to be *dissociated* if

$$\sum_{\gamma\in\Gamma}{}' \epsilon_\gamma \gamma = 0, \qquad \epsilon_\gamma = 0, 1, -1 \Rightarrow \epsilon_\gamma = 0 \qquad \text{for } \gamma \neq 0.$$

(The latter sum refers to the group operation in Γ.)

Let us take $1 \leqslant p < \infty$ and write $L^p(G, \mathrm{d}x)$ for the approriate Lebesgue space with norm $\| \ \|_p$. It follows from a result of W. Rudin that if $\operatorname{supp}\hat{f}$ is dissociated then, for all p, we have

$$\|f\|_p \leqslant 10\sqrt{p}\|f\|_2. \tag{4}$$

Thus as a corollary to (4) we have the following lemma.

Lemma *Assume that Λ is a dissociated set and $f \in L^2(G)$ with $\operatorname{supp}\hat{f} \subset \Lambda$. Denote by f_x $(x \in G)$ the translates of f. Then, for any finite subset I of G, we have*

$$\|\max_{x\in I}|f_x|\|_2 \leqslant 10\sqrt{\log|I|}\|f\|_2. \tag{5}$$

Let f be the function on $\mathbb{Z}/N\mathbb{Z}$ considered in (2). Fix an integer $\nu > 0$ and define for $s = 0, \ldots, \nu$ the sets

$$E_s = \{0 \leqslant n < N : 2^{-s-1} < |\hat{f}(n)| \leqslant 2^{-s}\},$$
$$E = \{0 \leqslant n < N : |\hat{f}(n)| \leqslant 2^{-\nu}\}.$$

Fix an integer l. For each s, we can write

$$E_s = \bigcup_t D_s^t \cup R_s,$$

where the D_s^t are disjoint dissociated sets of size $|D_s^t| = l$ and the elements of R_s can be written in the form

$$n = \sum_{k\in H_s} \epsilon_k k \qquad (\epsilon_k = 0, 1, -1),$$

where $|H_s| < l$. The details of this inductive construction are straight forward and are left to the reader.

We now estimate the left-hand side of (1) using the triangle inequality to obtain

$$\int_G \max_{1\leq k<K} |f(x+ky)-f(x)| \, \mathrm{d}x$$

$$\leq 2\sum_{s=0}^{\nu}\sum_t \int \max_{0\leq k<K}\left|\sum_{n\in D_s'} \hat{f}(n)\mathrm{e}^{2\pi i n(x+ky)}\right| \mathrm{d}x \tag{6}$$

$$+ \int \max_{0\leq k<K}\left|\sum_{n\in R} \hat{f}(n)[1-\mathrm{e}^{2\pi inky}]\mathrm{e}^{2\pi inx}\right| \mathrm{d}x \tag{7}$$

$$+ 2\int \max_{0\leq k<K}\left|\sum_{n\in E} \hat{f}(n)\mathrm{e}^{2\pi in(x+ky)}\right| \mathrm{d}x, \tag{8}$$

where $R = \bigcup_{s=0}^{\nu} R_s$.

3 Estimates

In view of the lemma, (6) is dominaed by

$$20\sum_s\sum_t \sqrt{\log K}\sqrt{\sum_{n\in D_s'} |\hat{f}(n)|^2}$$

$$\leq 40\sqrt{\frac{\log K}{l}}\sum_{s=0}^{\nu}\sum_t\sum_{n\in D_s'} |\hat{f}(n)| \leq 40\sqrt{\frac{\log K}{l}}. \tag{9}$$

Since $\|f\|_A \leq 1$, (7) is bounded by

$$K\sup_{n\in R}|1-\mathrm{e}^{2\pi iny}|. \tag{10}$$

Write $H = \bigcup_{s=0}^{\nu} H_s$. Since each element $n \in R$ is of the form

$$n = \sum_{k\in H} \epsilon_k k \quad \text{with } \epsilon_k = 0, 1, -1 \quad \text{and} \quad \sum|\epsilon_k| < l,$$

(10) is bounded by

$$Kl\sup_{n\in H}|1-\mathrm{e}^{2\pi iny}|. \tag{11}$$

Note that

$$|H| < \nu l \tag{12}$$

The last term (8) is estimated as follows:

$$\sqrt{2\sum_{0\leq k<K}\left\|\sum_{n\in E}\hat{f}(n)\mathrm{e}^{2\pi in(x+nky)}\right\|^2_{L^2(\mathrm{d}x)}} \leq 2\sqrt{K}\sqrt{\sum_{n\in E}|\hat{f}(n)|^2} < 2\sqrt{2^{-\nu}K}. \tag{13}$$

The inequalities (9), (11) and (13) show that, in order to satisfy (1), we need only find a $y \in G$ with $y \neq 0$ such that

$$40\sqrt{\frac{\log N}{l}} + Kl \sup_{n\in H} |1 - e^{2\pi i n y}| + 2\sqrt{2^{-\nu}K} < \epsilon = \hat{f}(0). \tag{14}$$

If we consider the homomorphism $G \to \Pi^{|H|}$, where Π is the torus, mapping y to $(e^{2\pi i n y})_{n\in H}$, a simple volume argument shows that there exists a non-trivial y satisfying

$$|1 - e^{2\pi i n y}| < x \qquad \text{for } n \in H,$$

provided that

$$\left(\frac{2\pi}{x}\right)^{|H|} < N.$$

Thus if

$$\left(\frac{4\pi}{\epsilon} Kl\right)^{\nu l} < N \tag{15}$$

then by (12) we get

$$Kl \sup_{n\in H} |1 - e^{2\pi i n y}| < \tfrac{1}{2}\epsilon$$

for some $y \neq 0$. In order to obtain (14), let

$$K < \exp \frac{\epsilon^2}{160^2}\, l, \qquad \nu \sim \log \frac{K}{\epsilon}$$

and let l satisfy (15). This will be possible provided (3) holds.

References

[1] H. Halberstam, private communication
[2] G. Freiman & H. Halberstam, preprint

On graphs not containing prescribed induced subgraphs

F. R. K. Chung and R. L. Graham

1 Introduction

Given a fixed graph H on t vertices, a typical graph G on n vertices contains many induced subgraphs isomorphic to H as n becomes large. Indeed, for the usual model of a random graph G^* on n vertices (see [4]), in which potential edges are independently included or not each with probability $\frac{1}{2}$, almost all such G^* contain $\{1+o(1)\}n^t 2^{-\binom{t}{2}}$ induced copies of H as $n \to \infty$. Thus, if a large graph G contains *no* induced copy of H, it deviates from being 'typical' in a rather strong way. In this case, we would expect it to behave quite differently from random graphs in many other ways as well. That this in fact must happen follows from recent work of several authors, e.g., see Chung, Graham & Wilson [5] and Thomason [7], [8]. In this paper we initiate a quantitative study of how various deviations of randomness are related. The particular property we investigate ('uniform edge density for half sets' – see Section 3) is just one of many which might have been selected and for which the same kind of analysis could be carried out.

This work also shares a common philosophy with several recent papers of Alon & Bollobás [1] and Erdős & Hajnal [6], which investigate the structure of graphs which have an unusually small number of non-isomorphic induced subgraphs. This is a strong restriction and such graphs must have very large subgraphs which are (nearly) complete or independent.

2 Preliminaries

By a graph G we will mean a finite set $V(G)$ called *vertices*, together with a set $E(G)$ of unordered pairs of vertices called *edges*. We often denote the fact that G has n vertices by writing G as $G(n)$. If $X \subseteq V(G)$, we let $e(X)$ denote the number of edges $\{v, v'\} \in E(G)$ with $v, v' \in X$. The adjacency matrix $A(G)$ is the matrix $(a(v, v'))$ indexed by (a fixed ordering of) the vertices of G with

$$a(v, v') = \begin{cases} 1 & \text{if } \{v, v'\} \in E(G), \\ 0 & \text{otherwise.} \end{cases}$$

For $v, v' \in V(G)$, we define

$$s(v, v') = |\{v'' \in V(G) : a(v, v'') = a(v', v'')\}|.$$

In other words, $s(v, v')$ is the number of vertices of G which are either joined to both v and v' or joined to neither of them.

We say that H is an *induced subgraph* of G (written $H < G$) if there is a 1–1 mapping λ: $V(H) \to V(G)$ such that

$$\{v, v'\} \in E(H) \Leftrightarrow \{\lambda(v), \lambda(v')\} \in E(G).$$

We let $|\{\lambda : H < G\}|$ denote the number of such mappings.

Other terminology will be introduced as it is needed. The reader can consult [3] for standard graph-theory terminology.

3 The main result

Theorem *Let $H(t)$ be an arbitrary fixed graph on t vertices and suppose $H(t)$ is not an induced subgraph of $G(n)$. Then there exists $S \subset V(G(n))$ with $|S| = \lfloor \frac{1}{2}n \rfloor$ and*

$$|e(S) - \tfrac{1}{16}n^2| > 2^{-2(t^2+27)}n^2, \tag{1}$$

provided n is sufficiently large.

Comment With probability tending to 1 as $n \to \infty$, a random graph $G^*(n)$ has the property that *every* subset S of vertices of size $\lfloor \frac{1}{2}n \rfloor$ spans $\frac{1}{16}\{1 + o(1)\}n^2$ edges. This we call the 'uniform edge density for half sets' property. The inequality (1) asserts that this property must fail in a strong way whenever any fixed t-vertex graph fails to occur as an induced subgraph.

The following outline gives the main ideas of the proof. More detailed calculations for some of the steps can be found in [5].

Sketch of the proof We first show that

$$|\{\lambda : H(t) < G(n)\}| < n_{(t)}(2^{-\binom{t}{2}} - \sqrt{t}\,2^{-t^2/2}) \tag{2}$$

implies

$$\sum_{v,v' \in V(G(n))} |s(v,v') - \tfrac{1}{2}n| \geq 2^{-(t^2+3)} n^3, \tag{3}$$

where $n_{(k)} := n(n-1)\cdots(n-k+1)$ and n is large.

Assume the contrary, i.e. that (2) holds but (3) does *not* hold. Write $V(H(t)) = \{v_1, v_2, \ldots, v_t\}$. For $1 \leq r \leq t$, define $H(r)$ to be the subgraph of $H(t)$ induced by $\{v_1, \ldots, v_r\}$. Let N_r denote $|\{\lambda : H(r) < G(n)\}|$. We prove by induction on r that

$$N_r = n_{(r)}(2^{-\binom{r}{2}} - \sqrt{r}\,2^{-t^2/2}). \tag{4}$$

This is immediate for $r = 1$ since $N_1 = n$. Assume for some r $(1 \leq r < t)$ that (4) holds. Define $\boldsymbol{\alpha} := (\alpha_1, \ldots, \alpha_r)$, where the α_j are distinct elements of $[n] := \{1, 2, \ldots, n\}$, which we can take for $V(G(n))$ without loss of generality. Also define $\boldsymbol{\epsilon} := (\epsilon_1, \ldots, \epsilon_r)$ $(\epsilon_j = 0$ or $1)$ and

$$f_r(\boldsymbol{\alpha}, \boldsymbol{\epsilon}) := |\{i \in [n] : i \neq \alpha_1, \ldots, \alpha_r \text{ and } a(i, \alpha_j) = \epsilon_j \ (1 \leq j \leq r)\}|.$$

Observe that N_{r-1} is the sum of exactly N_r quantities $f_r(\boldsymbol{\alpha}, \boldsymbol{\epsilon})$. Namely, for each embedding of $H(r)$ in $G(n)$, say $\lambda(v_j) = \alpha_j$ $(1 \leq j \leq r)$, $f_r(\boldsymbol{\alpha}, \boldsymbol{\epsilon})$ counts the number of ways of choosing $i \in [n]$ so that if we extend λ to $\{v_1, v_2, \ldots, v_{r+1}\}$ by defining $\lambda(v_{r+1}) = i$ and we take $\epsilon_j = a(v_{r+1}, v_j)$, then λ becomes an embedding of $H(r+1)$ into $G(n)$. Also, there are just $n_{(r)}2^r$ quantities $f_r(\boldsymbol{\alpha}, \boldsymbol{\epsilon})$, since there are $n_{(r)}$ choices of $\boldsymbol{\alpha}$ and 2^r choices for $\boldsymbol{\epsilon}$. Simple counting arguments now show that

$$\bar{f}_r = \frac{1}{n_{(r)}2^r} \sum_{\boldsymbol{\alpha}, \boldsymbol{\epsilon}} f_r(\boldsymbol{\alpha}, \boldsymbol{\epsilon}) = \frac{n-r}{2^r}$$

and

$$S_r := \sum_{\boldsymbol{\alpha}, \boldsymbol{\epsilon}} f_r(\boldsymbol{\alpha}, \boldsymbol{\epsilon})\{f_r(\boldsymbol{\alpha}, \boldsymbol{\epsilon}) - 1\} = \sum_{i \neq j} s(i,j)_{(r)}. \tag{5}$$

Define $\epsilon_{ij} := s(i,j) - \frac{1}{2}n$. Thus

$$\begin{aligned}
\sum_{i \neq j} s(i,j)_{(r)} &= \sum_{i \neq j} (\tfrac{1}{2}n + \epsilon_{ij})_{(r)} \\
&\leq \sum_{i \neq j} (\tfrac{1}{2}n + \epsilon_{ij})^r \\
&\leq (\tfrac{1}{2}n)^r n_{(2)} + \sum_{i \neq j} \sum_{k=0}^{r-1} \binom{r}{k} (\tfrac{1}{2}n)^k |\epsilon_{ij}|^{r-k} \\
&\leq (\tfrac{1}{2}n)^r n_{(2)} + 2^r \sum_{k=0}^{r-1} (\tfrac{1}{2}n)^k \sum_{i \neq j} |\epsilon_{ij}|^{r-k} \\
&\leq \frac{n^{r+2}}{2^r} + 2^r \sum_{k=0}^{r-1} (\tfrac{1}{2}n)^k (\tfrac{1}{2}n)^{r-k-1} \sum_{i \neq j} |\epsilon_{ij}|
\end{aligned}$$

$$\leqslant \frac{n^{r+2}}{2^r} + 2rn^{r-1} \sum_{i \neq j} |\epsilon_{ij}|$$
$$= \frac{n^{r+2}}{2^r} + 2rn^{r-1} \sum_{v, v' \in V(G(n))} |s(v, v') - \tfrac{1}{2}n|$$
$$\leqslant n^{r+2}2^{-r} + 2rn^{r+2}2^{-(t^2+3)} \qquad (6)$$

by hypothesis. Thus,

$$S_r \leqslant n^{r+2}(2^{-r} + r2^{-(t^2+2)})$$

so that

$$\sum_{\alpha, \epsilon} \{f_r(\alpha, \epsilon) - \bar{f}_r\}^2 = \sum_{\alpha, \epsilon} f_r^2(\alpha, \epsilon) - \sum_{\alpha, \epsilon} \bar{f}_r^2$$
$$= \sum_{\alpha, \epsilon} \{f_r^2(\alpha, \epsilon) - f_n(\alpha, \epsilon)\} + \sum_{\alpha, \epsilon} f_r(\alpha, \epsilon)$$
$$- n_{(r)}2^r(n-r)^2 2^{-2r}$$
$$= S_r + n_{(r+1)} - n_{(r)}(n-r)^2 2^{-r}$$
$$\leqslant n^{r+2}(2^{-r} + r2^{-(t^2+2)}) + n_{(r+1)} - n_{(r)}(n-r)^2 2^{-r}$$
$$\leqslant 2^{-r}\{n^{r+2} - (n-r)^{r+2}\} + rn^{r+2}2^{-(t^2+2)} + n_{(r+1)}$$
$$\leqslant 2^{-r}r(r+2)n^{r+1} + n_{(r+1)} + rn^{r+2}2^{-(t^2+2)}$$
$$\leqslant n^{r+2}\left(r2^{-(t^2+2)} + \frac{3}{n}\right) \qquad (7)$$

for n large. Since we have noted that

$$N_{r+1} = \sum_{N_r \text{ choices of } (\alpha, \epsilon)} f_r(\alpha, \epsilon),$$

then

$$|N_{r+1} - N_r\bar{f}_r|^2 = \left| \sum_{N_r \text{ terms}} \{f_r(\alpha, \epsilon) - \bar{f}_r\} \right|^2$$
$$\leqslant N_r \sum_{N_r \text{ terms}} \{f_r(\alpha, \epsilon) - \bar{f}_r\}^2,$$

by the Cauchy–Schwartz inequality

$$\leqslant N_r \sum_{\alpha, \epsilon} \{f_r(\alpha, \epsilon) - \bar{f}_r\}^2$$
$$\leqslant n^{2r+2}\left(r2^{-(t^2+2)} + \frac{3}{n}\right), \qquad (9)$$

since $N_r \leqslant n^r$. Thus,

$$N_{r+1} \geqslant N_r \bar{f}_r - n^{r+1}\sqrt{r2^{-(t^2+2)}+\frac{3}{n}}$$

$$= N_r(n-r)2^{-r} - n^{r+1}\sqrt{r2^{-(t^2+2)}+\frac{3}{n}}$$

$$\geqslant n_{(r)}(2^{-\binom{r}{2}} - \sqrt{r}2^{-t^2/2})(n-2)2^{-r} - n^{r+1}\sqrt{r2^{-(t^2+2)}+\frac{3}{n}}$$

by induction

$$\geqslant n_{(r+1)}(2^{-\binom{r+1}{2}} - \sqrt{r}2^{-r}2^{-t^2/2}) - n^{r+1}\sqrt{r2^{-(t^2+2)}+\frac{3}{n}}$$

$$\geqslant n_{(r+1)}(2^{-\binom{r+1}{2}} - \sqrt{r+1}2^{-t^2/2}) \tag{10}$$

for n sufficiently large, where the last step follows by a simple calculation. This completes the inductive step and proves (4), which for $r = t$ asserts

$$N_t \geqslant n_{(t)}(2^{-\binom{t}{2}} - \sqrt{t}2^{-t^2/2})$$

contradicting (2). Therefore (2) ⇒ (3) as claimed.

We next claim that (3) implies: for some $S \subset V(G(n))$,

$$\left|e(S) - \tfrac{1}{4}|S|^2\right| \geqslant \tfrac{1}{3}\epsilon^2 n^2, \tag{11}$$

where $\epsilon = \frac{1}{10}2^{-(t^2+3)}$.

To see this, suppose (11) does *not* hold, that is, for all $S \subset V(G(n))$, we have

$$\left|e(S) - \tfrac{1}{4}|S|^2\right| < \tfrac{1}{3}\epsilon^2 n^2. \tag{12}$$

A standard argument now shows that all vertices of $G(n)$ except for a set Y of size at most $2\epsilon n$ have degrees between $(\frac{1}{2}-\epsilon)n$ and $(\frac{1}{2}+\epsilon)n$. For vertices $v, v' \in V(G(n))$, define

$$f_{ij}(v,v') := |\{w \in V(G(n)) : a(v,w) = i,\ a(v',w) = j\}|$$

for $0 \leqslant i, j \leqslant 1$. Thus,

$$|f_{ij}(v,v') + f_{i'j'}(v,v') - \tfrac{1}{2}n| < \epsilon n$$

if $v, v' \in V(G(n))\backslash Y := V'$ and $(i,j) = (0,0)$ or $(1,1)$ and $(i',j') = (1,0)$ or $(0,1)$. Therefore, in this case,

$$|f_{11}(v,v') - f_{00}(v,v')| \leqslant 2\epsilon n, \qquad |f_{10}(v,v') - f_{01}(v,v')| \leqslant 2\epsilon n.$$

For a fixed $v \in V'$, let $X(v)$ denote the set

$$\{v' \in V' : |s(v,v') - \tfrac{1}{2}n| > 8\epsilon n\}.$$

We consider two cases.

Case (a) For all $v \in V'$, $|X(v)| \leqslant 2\epsilon n$.

We then have

$$\begin{aligned}\sum_{v,v' \in V(G(n))} |s(v,v') - \tfrac{1}{2}n| &\leqslant \sum_{v \in V'} \sum_{v' \in V(G(n))} |s(v,v') - \tfrac{1}{2}n| \\ &\quad + \sum_{v \notin V'} \sum_{v' \in V(G(n))} |s(v,v') - \tfrac{1}{2}n| \\ &< n(2\epsilon n \times \tfrac{1}{2}n + n \times 8\epsilon n) + 2\epsilon n \times \tfrac{1}{2}n^2 \\ &= 10\epsilon n^3.\end{aligned}$$

This contradicts (3).

Case (b) For some $v_0 \in V'$, $|X(v_0)| > 2\epsilon n$.

Either there are ϵn vertices in the set

$$X_1(v_0) = \{u : s(v_0,u) > \tfrac{1}{2}n + 8\epsilon n\},$$

or there are ϵn vertices u' with $s(v_0,u') < \frac{1}{2}n - 8\epsilon n$. We will examine the case that there are ϵn vertices u with $s(v_0,u) > \frac{1}{2}n + 8\epsilon n$ and omit the (similar) proof for the other case. Now, $s(v_0,u) > \frac{1}{2}n + 8\epsilon n$ implies

$$f_{11}(v_0,v) \geqslant \tfrac{1}{2}\{s(v_0,u) - 2\epsilon n\} \geqslant \tfrac{1}{4}n + 3\epsilon n.$$

The number of ordered pairs (u,v) with $u \in X_1$ and $v \in \mathrm{nd}(v_0)$, denoted by $e(X_1, \mathrm{nd}(v_0))$, is at least $|X_1|(\frac{1}{4}n + 3\epsilon n)$ (where $\mathrm{nd}(v_0)$ denotes the neighbourhood of v_0, i.e., the set of vertices adjacent to v_0). From (12) we have:

$$\begin{aligned}e(X_1) &\geqslant \tfrac{1}{4}|X_1|^2 - \tfrac{1}{3}\epsilon^2 n^2, \\ e(\mathrm{nd}(v_0)) &\geqslant \tfrac{1}{4}|\mathrm{nd}(v_0|^2 - \tfrac{1}{3}\epsilon^2 n^2, \\ e(X_1 \cap \mathrm{nd}(v_0)) &\leqslant \tfrac{1}{4}|X_1 \cap \mathrm{nd}(v_0|^2 + \tfrac{1}{3}\epsilon^2 n^2, \\ e(X_1 \cup \mathrm{nd}(v_0)) &\leqslant \tfrac{1}{4}|X_1 \cup \mathrm{nd}(v_0|^2 + \tfrac{1}{3}\epsilon^2 n^2.\end{aligned}$$

However, we now reach a contradiction since a simple counting argument shows we must always have

$$e(X_1 \cup \mathrm{nd}(v_0)) \geqslant e(X_1) + e(\mathrm{nd}(v_0)) + |X_1|(\tfrac{1}{4}n + 3\epsilon n) - 3e(X_1 \cap \mathrm{nd}(v_0)).$$

Therefore we conclude that (12) does not hold and, so, (11) follows.

The last step is to show that (11) implies the following: for some $S \subset V(G(n))$ with $|S| = \lfloor \frac{1}{2}n \rfloor$ we have

$$|e(S) - \tfrac{1}{16}n^2| \geqslant 2^{-(2t^2+27)}n^2. \tag{13}$$

Suppose (13) does not hold, i.e., for all $S \subset V(G(n))$ with $|S| = \lfloor \frac{1}{2}n \rfloor$ we have

$$|e(S) - \tfrac{1}{16}n^2| < \tfrac{1}{24}\epsilon^2 n^2. \tag{14}$$

From (11) we know that there is a set $S' \subset V(G(n))$ such that

$$\left|e(S') - \tfrac{1}{4}|S'|^2\right| \geqslant \tfrac{1}{3}\epsilon^2 n^2.$$

There are two possibilities.

Case (a') $|S'| \geqslant \frac{1}{2}n$.

By averaging over all subsets S'' of S' of size $\lfloor \frac{1}{2}n \rfloor$, we get

$$e(S') \leqslant \sum_{S'' \subseteq S'} \frac{e(S'')}{\binom{|S'|-2}{\lfloor \frac{1}{2}n \rfloor - 2}}$$
$$< \binom{|S'|}{2}(\tfrac{1}{2} + \tfrac{1}{3}\epsilon^2).$$

Similarly, we can also show that

$$e(S') > \binom{|S'|}{2}(\tfrac{1}{2} - \tfrac{1}{3}\epsilon^2).$$

This implies $\left|e(S') - \frac{1}{4}|S'|^2\right| \leqslant \frac{1}{6}\epsilon^2 n^2$, contradicting (11).

Case (b') $|S'| < \frac{1}{2}n$.

Let $\bar{S}'$ denote $V(G(n))\backslash S'$. From the proof of Case (a') we have

$$\left|e(\bar{S}') - \tfrac{1}{4}|\bar{S}'|^2\right| \leqslant \tfrac{1}{3}\epsilon^2\binom{|S'|}{2}$$

and

$$|e(G(n)) - \tfrac{1}{4}n^2| \leqslant \tfrac{1}{3}\epsilon^2\binom{n}{2}.$$

First we note that

$$e(S', \bar{S}') = e(G) - e(S') - e(\bar{S}').$$

Now consider the average value of $e(S' \cup S'')$, where S'' ranges over all subsets of $\bar{S}'$ with $\lfloor \frac{1}{2}n \rfloor - |S'|$ elements. This average is

$$\sum_{S''\subseteq \bar{S}'} \frac{e(S'\cup S'')}{\binom{n-|S'|}{\lfloor \frac{1}{2}n\rfloor - |S'|}}$$

$$= e(S') + \frac{(\lfloor \frac{1}{2}n\rfloor - |S'|)(\lfloor \frac{1}{2}n\rfloor - |S'| - 1)}{(n-|S'|)(n-|S'|-1)} e(|\bar{S}'|) + \frac{\lfloor \frac{1}{2}n\rfloor - |S'|}{n-|S'|} e(S, |\bar{S}'|)$$

$$= \frac{\lceil \frac{1}{2}n\rceil}{n-|S'|} e(S') - \frac{(\lfloor \frac{1}{2}n\rfloor - |S'|)\lceil \frac{1}{2}n\rceil}{(n-|S'|)(n-|S'|-1)} e(\bar{S}') + \frac{\lfloor \frac{1}{2}n\rfloor - |S'|}{n-|S'|} e(G)$$

$$> \frac{\lceil \frac{1}{2}n\rceil}{n-|S'|}(\tfrac{1}{4}|S'|^2 + \tfrac{1}{3}\epsilon^2 n^2) - \frac{(\lfloor \frac{1}{2}n\rfloor - |S'|)\lceil \frac{1}{2}n\rceil}{(n-|S'|)(n-|S'|-1)} \binom{n-|S'|}{2}(\tfrac{1}{2}+\tfrac{1}{3}\epsilon^2) + \frac{\lfloor \frac{1}{2}n\rfloor - |S'|}{n-|S'|}\binom{n}{2}(\tfrac{1}{2}-\tfrac{1}{3}\epsilon^2)$$

$$\geqslant \tfrac{1}{16}n^2 + \tfrac{1}{24}\epsilon^2 n^2.$$

This contradicts (14) asserting that any set with $\lfloor \frac{1}{2}n\rfloor$ elements spans at most $\frac{1}{16}n^2 + \frac{1}{24}\epsilon^2 n^2$ edges.

This completes the proof of the main theorem since $H(t) \not< G(n)$ certainly implies (2). □

4 Concluding remarks

As we remarked earlier, the assumption $H(t) \not< G(n)$ must also be reflected in the failure of *all* so-called quasi-random properties for $G(n)$ (see [5]). For example, it follows that, for some $\delta(t) > 0$, either $e(G(n)) < \{\frac{1}{4} - \delta(t)\}n^2$, or $|\lambda_1(G(n)) - \frac{1}{2}n| > \delta(t)n$, or $\lambda_2(G(n)) > \delta(t)n$ for n sufficiently large, where $\lambda_k(G(n))$ denotes the kth largest eigenvalue of the adjacency matrix of $G(n)$. However, we leave the quantitative interrelationships between these various properties for a later paper.

With respect to the condition studied here, namely $H(t) \not< G(n)$, it would be of interest to know what the 'correct' values of the constants are. In particular, can the factor $2^{-(2t^2+27)}$ be replaced by a substantially larger quantity, such as c^{-t}, for a constant $c > 1$? On the other hand, we have no interesting upper bounds here. We have not tried to see how different graphs $H(t)$ on t vertices affect the estimates. Clearly some graphs have a stronger effect than others. We have no idea which graphs are the most influential from this point of view.

The best value of the constants are not even known for the small cases of $H(t)$. For example, when $H(t) = K_3$, the complete graph on three vertices, an old conjecture of Erdős asserts the following.

Conjecture *If* $e(S) > 2n^2$ *for every* $S \subseteq G(10n)$ *with* $|S| = 5n$ *then* $K_3 \subset G(10n)$.

The graph $G'(10n)$ consisting of 5 independent sets $I_i(2n)$ of size $2n$, with complete bipartite graphs between $I_i(2n)$ and $I_{i+1}(2n)$ $(1 \leqslant i \leqslant 5)$, with $I_6(2n) := I_1(2n)$, shows that, if true, this result would be best possible. For K_t $(t \geqslant 4)$ the corresponding conjecture is the following. Let $T_t(n)$ denote the Turán graph for K_t (see [3]), i.e., the (unique) graph on n vertices having the maximum possible number of edges which contains no K_t. Let $b_t(n)$ denote the minimum number of edges spanned by any set of $\frac{1}{2}n$ vertices of $T_t(n)$.

Conjecture *If every set of* $\frac{1}{2}n$ *vertices of* $G(n)$ *spans more than* $b_t(n)$ *edges then* $K_t \subset G(n)$.

Finally, let us call a set E of edges of $G(n)$ a *bisector* if E is the set of edges joining a set $S \subseteq V$ with $|S| = \lfloor \frac{1}{2}n \rfloor$ to $\bar{S} := V \backslash S$ (see [2]). In almost all random graphs on n vertices, all bisectors have size $\frac{1}{8}\{1+o(1)\}n^2$. One might easily guess that the analogue of the theorem holds for bisectors, i.e., if $H(t) \not< G(n)$ then, for some $\delta(t) > 0$, there is a bisector E with $||E| - \frac{1}{8}n^2| > \delta(t)n^2$. This is *not* the case, however, as the following graph $B(n)$ shows. $B(n)$ will consist of disjoint vertex sets V_1 and V_2 with $|V_1| = \lfloor \frac{1}{2}n \rfloor$ and $|V_2| = \lceil \frac{1}{2}n \rceil$. V_1 spans a complete graph and V_2 spans an empty graph (i.e., with no edges). Between V_1 and V_2 we choose a random (bipartite) graph with edge probability $\frac{1}{2}$. A simple computation shows that every bisector of $B(n)$ has size $\frac{1}{8}n^2 + O(n)$. However, $B(n)$ has no induced 4-cycle C_4. This cannot happen for graphs which also have all but $o(n)$ vertices with degrees $\frac{1}{2}\{1+o(1)\}n$ (i.e., 'almost regular'). In this case, it is not hard to show that 'almost regular' together with 'all bisectors have size $\frac{1}{8}\{1+o(1)\}n^2$' is a quasi-random property (see [5]).

References

[1] N. Alon & B. Bollobás, Graphs with a small number of distinct induced subgraphs, *Discrete Mathematics*, **75** (1989), 23–30

[2] N. Alon & V. D. Milman, λ_1, isoperimetric inequalities on graphs, and superconcentrators, *J. Combin. Th. (B)*, **38** (1985), 73–88

[3] B. Bollobás, *Graph Theory – An Introductory Course*, GTM Springer Verlag, New York, 1979
[4] B. Bollobás, *Random Graphs*, Academic Press, New York.
[5] F. R. K. Chung, R. L. Graham & R. M. Wilson, Quasi-random graphs, *Combinatorica*, **9** (1989), 345–62
[6] P. Erdős & A. Hajnal, On the number of induced subgraphs of a graph, *Discrete Mathematics*, **75** (1989), 145–54
[7] A. T. Thomason, Pseudo-random graphs, *Proceedings of Random Graphs, Poznán 1985* (ed. M. Karonski), *Annals of Discrete Math.*, **33** (1987), 307–31
[8] A. T. Thomason, Random graphs, strongly regular graphs and pseudo-random graphs, *Survey in Combinatorics 1987* (ed. C. Whitehead), L.M.S. Lecture Notes **123**, Cambridge University Press, Cambridge, 1987, 173–96

Partitions sans petits sommants

Jacques Dixmier et Jean-Louis Nicolas*

Introduction

Soient m et n des entiers $\geqslant 0$. On note $p(n)$ le nombre de partitions de n et $r(n,m)$ le nombre de partitions de n dont toutes les parts sont $\geqslant m$. Depuis Hardy–Ramanujan, le comportement asymptotique de $p(n)$ quand $n \to \infty$ est très bien connu. D'autre part, quand $n \to \infty$, on a

$$r(n,m) = p(n)\left(\frac{\pi}{\sqrt{6n}}\right)^{m-1}(m-1)!\left\{1+O\left(\frac{m^2}{\sqrt{n}}\right)\right\}$$

uniformément pour $1 \leqslant m \leqslant n^{1/4}$ ([3], th. 1); et il existe une constante $\alpha > 0$ telle que, quand $n \to \infty$, on a

$$\begin{aligned} p(n)\left(\frac{\pi}{\sqrt{6n}}\right)^{m-1}(m-1)!\exp O\left(\frac{m^2}{\sqrt{n}}\right) &\leqslant r(n,m) \\ &\leqslant p(n)\left(\frac{\pi}{\sqrt{6n}}\right)^{m-1}(m-1)!\left\{1+O\left(\frac{1}{\sqrt{n}}\right)\right\} \end{aligned}$$

uniformément pour $1 \leqslant m \leqslant \alpha\sqrt{n}$ ([4], §2, proposition). Le but du présent mémoire est de compléter ces résultats: (1) par une évaluation de $\log r(n,m)$ pour $m = \lambda\sqrt{n}$ (λ, constante > 0); (2) par une évaluation de $r(n,m)$ pour $m \leqslant n^{\frac{1}{3}-\epsilon}$.

Conformément à l'usage, on note $p(n,m)$ le nombre de partitions de n dont toutes les parts sont $\leqslant m$. Il sera commode de poser $r(n,x) = r(n,\lceil x\rceil)$ et $p(n,x) = p(n,\lfloor x\rfloor)$ pour x réel > 0.

Les calculs de développements asymptotiques ont été effectués par les systèmes de calcul formel MAPLE et MACSYMA.

* Recherche partiellement financée par le CNRS, Greco 'Calcul formel, et P. R. C. Mathématiques-Informatique'.

1 Rappel de résultats de G. Szekeres

Dans cette partie, nous ne faisons qu'expliciter des résultats de G. Szekeres ([9], [10]), en ajoutant quelques détails utiles pour la suite.

1.1 Pour $x \geqslant 0$, posons

$$F(x) = \int_0^x \frac{x}{e^x - 1}\,dx.$$

Le fonction F est analytique et strictement croissante de 0 à $\frac{1}{6}\pi^2$. On a

$$F(x) \sim x \qquad \text{quand } x \to 0. \tag{1}$$

D'autre part,

$$F(x) = \tfrac{1}{6}\pi^2 - \int_x^\infty x\left(\sum_{n=1}^{\infty} e^{-nx}\right) dx = \tfrac{1}{6}\pi^2 - \sum_{n=1}^{\infty} \frac{nx+1}{n^2} e^{-nx},$$

d'où

$$F(x) = \tfrac{1}{6}\pi^2 - x e^{-x} + O(e^{-x}) \qquad \text{quand } x \to \infty. \tag{2}$$

1.2 Pour $x > 0$, posons

$$G(x) = xF(x)^{-1/2}.$$

La fonction G est analytique. On a $G(x)^{-2} = x^{-2}F(x)$, d'où

$$\begin{aligned} -2G(x)^{-3}\frac{dG}{dx} &= -2x^{-3}F(x) + x^{-2}\frac{x}{e^x-1} \\ &= -x^{-3}\left(2F(x) - \frac{x^2}{e^x-1}\right). \end{aligned}$$

Or

$$\frac{d}{dx}\left(2F(x) - \frac{x^2}{e^x-1}\right) = \frac{2x}{e^x-1} - \frac{2x}{e^x-1} + \frac{x^2 e^x}{(e^x-1)^2} > 0.$$

Pour $x > 0$, on a donc

$$2F(x) - \frac{x^2}{e^x-1} > \lim_{x\to 0}\left(2F(x) - \frac{x^2}{e^x-1}\right) = 0,$$

de sorte que $dG/dx > 0$. On a, d'après (1),

$$G(x) \sim \sqrt{x} \qquad \text{quand } x \to 0 \tag{3}$$

et $x^{-1/2}G(x)$ est holomorphe au voisinage de 0. D'après (2),

$$G(x) = x\left\{\tfrac{1}{6}\pi^2\left(1-\frac{6}{\pi^2}xe^{-x}+O(e^{-x})\right)\right\}^{-1/2} \quad \text{quand } x \to \infty,$$

d'où

$$\frac{\pi}{\sqrt{6}}G(x) = x+\frac{3}{\pi^2}x^2e^{-x}+O(xe^{-x}) \qquad \text{quand } x \to \infty. \tag{4}$$

1.3 L'application G: $]0, +\infty[\to]0, +\infty[$ admet, d'après 1.2, une application réciproque H: $]0, +\infty[\to]0, +\infty[$, qui est une fonction analytique de dérivée > 0, holomorphe au voisinage de 0. La relation définissant H est

$$x^{-2}H(x)^2 = \int_0^{H(x)} \frac{x}{e^x-1}\,dx. \tag{5}$$

On peut sans difficulté pousser plus loin les dévelopement de $F(x)$ et $G(x)$ quand $x \to 0$ et $x \to \infty$. Un procédé d'inversion classique (cf. par exemple [1], p. V. 45, prop. 5) fournit alors

$$H(x) = x^2-\tfrac{1}{4}x^4+\tfrac{13}{144}x^6-\tfrac{7}{192}x^8+\tfrac{8081}{518\,400}x^{10}+O(x^{12}) \quad \text{quand } x \to 0 \tag{6}$$

et, quand $x \to \infty$,

$$\begin{aligned} H(x) = {} & \frac{\pi}{\sqrt{6}}x-\left(\tfrac{1}{2}x^2+\frac{\sqrt{6}}{2\pi}x\right)\exp\left\{-\frac{\pi}{\sqrt{6}}x\right\} \\ & -\left\{\tfrac{1}{4}x^4+\frac{3\sqrt{6}}{8\pi}x^3+\left(\frac{1}{4}+\frac{3}{2\pi^2}\right)x^2+\left(\frac{\sqrt{6}}{8\pi}+\frac{3\sqrt{6}}{4\pi^3}\right)x\right\}\exp\left\{-\frac{2\pi}{\sqrt{6}}x\right\} \\ & +O\left(x^6\exp\left(-\frac{3\pi}{\sqrt{6}}x\right)\right). \end{aligned} \tag{7}$$

De (5), on déduit

$$2x^{-1}H(x)[x^{-1}H(x)]' = \frac{H(x)\,H'(x)}{e^{H(x)}-1},$$

d'où

$$[x^{-1}H(x)]' = \frac{xH'(x)}{2(e^{H(x)}-1)} \tag{8}$$

et une équation différentielle vérifiée par H:

$$2xH'-2H = \frac{x^3H'}{e^H-1}. \tag{9}$$

1.4 Pour $\lambda > 0$, posons

$$f(\lambda) = 2\frac{H(\lambda)}{\lambda}-\lambda\log(1-e^{-H(\lambda)}). \tag{10}$$

La fontion f est analytique. On a, d'après (8),

$$f'(\lambda) = \frac{\lambda H'(\lambda)}{e^{H(\lambda)}-1} - \log(1-e^{-H(\lambda)}) - \lambda\frac{H'(\lambda)e^{-H(\lambda)}}{1-e^{-H(\lambda)}},$$

ou

$$f'(\lambda) = -\log(1-e^{-H(\lambda)}) > 0. \tag{11}$$

Les relations (10) et (11) entraînent

$$\lambda f'(\lambda) = f(\lambda) - \frac{2H(\lambda)}{\lambda},$$

d'où

$$2H(\lambda) = \lambda f(\lambda) - \lambda^2 f'(\lambda), \tag{12}$$

$$2H'(\lambda) = f(\lambda) - \lambda f'(\lambda) - \lambda^2 f''(\lambda). \tag{13}$$

En portant dans (9), on obtient l'équation différentielle suivante vérifiée par f:

$$\lambda^2 f'' + \lambda f' - f = 2f''[\exp(\tfrac{1}{2}\lambda f - \tfrac{1}{2}\lambda^2 f'') - 1]. \tag{14}$$

Par ailleurs, (11) entraîne

$$f''(\lambda) = -\frac{H'(\lambda)}{e^{H(\lambda)}-1} < 0. \tag{15}$$

Du développement (6), on déduit, quand $\lambda \to 0$,

$$f(\lambda) = -2\lambda\log\lambda + 2\lambda + \tfrac{1}{4}\lambda^3 - \tfrac{13}{288}\lambda^5 + \tfrac{5}{576}\lambda^7 - \tfrac{8081}{2073\,600}\lambda^9 + O(\lambda^{11}), \tag{16}$$

la fonction $f(\lambda) + 2\lambda\log\lambda$ étant holomorphe au voisinage de 0.

La ressemblance avec les coefficients de (6) vient de (12), qui peut s'écrire

$$\frac{d}{d\lambda}\left(\frac{f(\lambda)}{\lambda}\right) = -\frac{2H(\lambda)}{\lambda^3}.$$

D'après (7),

$$f(\lambda) = \pi\sqrt{\tfrac{2}{3}} - \frac{\sqrt{6}}{\pi}\exp\left\{-\frac{\pi}{\sqrt{6}}\lambda\right\} - \left(\frac{\sqrt{6}}{4\pi}\lambda^2 + \frac{3}{\pi^2}\lambda + \frac{\sqrt{6}}{4\pi} + \frac{3\sqrt{6}}{2\pi^3}\right)\exp\left\{-\frac{2\pi}{\sqrt{6}}\lambda\right\} + O\left(\lambda^4\exp\left\{-\frac{3\pi}{\sqrt{6}}\lambda\right\}\right) \quad \text{quand } \lambda \to \infty. \tag{17}$$

1.5 Théoreme (G. Szekeres) *Soit* $\lambda > 0$. *On a, quand* $n \to \infty$,

$$\frac{1}{\sqrt{n}}\log p(n, \lambda\sqrt{n}) \to f(\lambda)$$

uniformément, pourvu que λ reste minoré par une constante positive aussi petite qu'on veut.

En fait, cela est un cas très particulier des résultats de [9] et [10].

2 Etude de $\log r(n,m)$ pour $m = \lambda\sqrt{n}$.

2.1 Soit $x \in]0,1[$. La fonction $x \mapsto x/\sqrt{1-x}$ est strictement croissante de 0 à $+\infty$. Posons, pour $\lambda > 0$ et $0 < x < 1$,

$$J(x,\lambda) = H\left(\frac{x}{\lambda\sqrt{1-x}}\right)$$
$$K(x,\lambda) = 2\lambda\frac{1-x}{x}J(x,\lambda) - \frac{x}{\lambda}\log(1-\mathrm{e}^{-J(x,\lambda)}).$$

Les fonctions J et K sont analytiques. Pour λ fixé, J est strictement croissante de 0 à ∞. D'après (6) et (7),

$$J(x,\lambda) \sim \begin{cases} \dfrac{x^2}{\lambda^2} & (x \to 0,\ \lambda \text{ fixé}), \\ \dfrac{\pi}{\lambda\sqrt{6}\sqrt{1-x}} & (x \to 1,\ \lambda \text{ fixé}), \end{cases}$$

d'où

$$K(x,\lambda) \sim \begin{cases} -2\dfrac{x\log x}{\lambda} & (x \to 0,\ \lambda \text{ fixé}), \\ \pi\sqrt{\tfrac{2}{3}}\sqrt{1-x} & (x \to 1,\ \lambda \text{ fixé}). \end{cases}$$

On déduit de là que, pour λ fixé, $\partial K/\partial x$ prends dans $]0,1[$ des valeurs > 0 et des valeurs < 0.

2.2 Lemme *On a* $\dfrac{\partial}{\partial x}\{x^{-1}J(x,\lambda)\} > 0$ *pour* $\lambda > 0$ *et* $0 < x < 1$.

D'après (5), on a

$$\lambda^2\frac{1-x}{x^2}J(x,\lambda)^2 = \int_0^{J(x,\lambda)} \frac{x}{\mathrm{e}^x-1}\,\mathrm{d}x.$$

En dérivant par rapport à x, on obtient

$$2\lambda^2\frac{1-x}{x^2}J\frac{\partial J}{\partial x} + \lambda^2\left(-\frac{2}{x^3}+\frac{1}{x^2}\right)J^2 = \frac{J}{\mathrm{e}^J-1}\frac{\partial J}{\partial x},$$

d'où

$$\frac{\partial J}{\partial x}\left(2\lambda^2\frac{1-x}{x^2} - \frac{1}{\mathrm{e}^J-1}\right) = \lambda^2 J\frac{2-x}{x^3} \tag{18}$$

$$x\{2\lambda^2(1-x)(e^J-1)-x^2\}\frac{\partial J}{\partial x} = \lambda^2(2-x)J(e^J-1). \tag{19}$$

Il résulte de (19) que

$$2\lambda^2(1-x)(e^J-1)-x^2 > 0 \tag{20}$$

et

$$\left(x\frac{\partial J}{\partial x}-J\right)\{2\lambda^2(1-x)(e^J-1)-x^2\}$$
$$= \lambda^2(2-x)J(e^J-1)-2\lambda^2(1-x)J(e^J-1)+x^2J$$
$$= \lambda^2xJ(e^J-1)+x^2J > 0,$$

d'où, compte tenu de (20), $x\,\partial J/\partial x - J > 0$, ce qui prouve le lemme. □

2.3 Lemme (a) *Pour* λ *fixé et* x *parcourant* $]0,1[$, *la fonction* $K(x,\lambda)$ *a un maximum* $M(\lambda)$ *et un seul.*

(b) $M(\lambda)$ *est l'unique nombre* $x \in\,]0,1[$ *tel que*

$$\exp\{-J(x,\lambda)\}+\exp\left\{-\frac{\lambda^2}{x}J(x,\lambda)\right\} = 1.$$

(c) *On a*

$$K(M(\lambda),\lambda) = \lambda\frac{2-M(\lambda)}{M(\lambda)}J(M(\lambda),\lambda).$$

On a

$$\begin{aligned}\frac{\partial K}{\partial x} &= 2\lambda\frac{1-x}{x}\frac{\partial J}{\partial x}-2\frac{\lambda}{x^2}J-\frac{1}{\lambda}(1-e^{-J})-\frac{x}{\lambda}\frac{e^{-J}}{1-e^{-J}}\frac{\partial J}{\partial x}\\ &= \left(2\lambda\frac{1-x}{x}-\frac{x}{\lambda(e^J-1)}\right)\frac{\partial J}{\partial x}-2\frac{\lambda}{x^2}J-\frac{1}{\lambda}\log(1-e^{-J})\\ &= \lambda J\frac{2-x}{x^2}-\frac{2\lambda}{x^2}J-\frac{1}{\lambda}\log(1-e^{-J}) \qquad \text{d'après(18)}\\ &= -\frac{\lambda}{x}J-\frac{1}{\lambda}\log(1-e^{-J}).\end{aligned}$$

Les zéros éventuels de $\partial K/\partial x$ sont donc fournis par l'équation

$$-\frac{\lambda^2}{x}J = \log(1-e^{-J}) \tag{21}$$

ou

$$\exp\left(-\frac{\lambda^2}{x}J\right)+\exp(-J) = 1. \tag{22}$$

Compte tenu du lemme 2.2, la fonction

$$x \mapsto \exp\left(-\frac{\lambda^2}{x}J\right)+\exp(-J)$$

a une dérivée < 0, donc l'égalité (22) ne peut être vérifiée en plus d'un point de $]0,1[$. Compte tenu des remarques du §2.1, on voit que $\partial K/\partial x$ a exactement une racine $M(\lambda)$ et que $x \mapsto K(x,\lambda)$ est maximum pour $x = M(\lambda)$. On a ainsi prouvé (a) et (b). Compte tenu de (21), on a

$$\begin{aligned} K(M(\lambda),\lambda) &= 2\lambda\frac{1-M(\lambda)}{M(\lambda)}J(M(\lambda),\lambda)+\frac{M(\lambda)}{\lambda}\,\frac{\lambda^2}{M(\lambda)}J(M(\lambda),\lambda) \\ &= \frac{\lambda}{M(\lambda)}\{2-2M(\lambda)+M(\lambda)\}J(M(\lambda),\lambda), \end{aligned}$$

d'où (c). □

2.4 Posons, pour $\lambda > 0$,

$$g(\lambda) = K(M(\lambda),\lambda) = \lambda\frac{2-M(\lambda)}{M(\lambda)}J(M(\lambda),\lambda). \tag{23}$$

Théorème *Soit* $\lambda > 0$. *On a*

$$\log r(n,\lambda\sqrt{n}) \sim g(\lambda)\sqrt{n} \qquad \textit{quand } n \to \infty.$$

(a) Soit $m_n = \lceil\lambda\sqrt{n}\rceil$. Soit $\mathcal{P}$ l'ensemble des partitions de n dont toutes les parts sont $\geqslant \lambda\sqrt{n}$, ou, ce qui revient au même, $\geqslant m_n$. On a $r(n,\lambda\sqrt{n}) = r(n,m_n) = \operatorname{Card}\mathcal{P}$. Soit $\mathcal{P}_i$ l'ensemble des $\pi \in \mathcal{P}$ ayant exactement i parts. Si $i > \sqrt{n}/\lambda$, on a $\mathcal{P} = \varnothing$; on prendra donc $i \leqslant \sqrt{n}/\lambda$. L'ensemble $\mathcal{P}$ est réunion disjointe des $\mathcal{P}_i$, donc

$$\operatorname{Card}\mathcal{P} = \sum_{i\geqslant 1}\operatorname{Card}\mathcal{P}_i.$$

Si $\pi \in \mathcal{P}_i$, on a $\pi = (m_n+a_1, m_n+a_2,\ldots,m_n+a_i)$ où $a_1 \geqslant a_2 \geqslant \cdots \geqslant a_i \geqslant 0$ et $a_1+a_2+\cdots a_i = n-im_n$. Donc

$$\operatorname{Card}\mathcal{P}_i = p(n-im_n, i). \tag{24}$$

(b) Posons $i\lambda/\sqrt{n} = \rho \in]0,1]$. Choisissons une petite constante $\omega \in]0,\frac{1}{2}]$ (nous préciserons ce choix ultérieurement) et considérons seulement dans la partie (b) les i tels que $\omega \leqslant \rho \leqslant 1-\omega$.

Alors,

$$n-im_n \geqslant n-\sqrt{n}(1-\omega)\frac{m_n}{\lambda} \to \infty \qquad \text{quand } n \to \infty.$$

D'autre part,

$$\frac{i}{\sqrt{n-im_n}} = \frac{\rho\sqrt{n}}{\lambda\sqrt{n-\rho\sqrt{n}m_n/\lambda}} \to \frac{\rho}{\lambda\sqrt{1-\rho}} \geqslant \frac{\omega}{\lambda\sqrt{1-\omega}},$$

donc

$$\frac{i}{\sqrt{n-im_n}} \geqslant \frac{\omega}{2\lambda\sqrt{1-\omega}} \qquad \text{pour } n \gg 1.$$

D'après 1.5, on a, uniformément en i,

$$\begin{aligned}
&\frac{1}{\sqrt{n}}\log p(n-im_n, i) \\
&\quad = \sqrt{1-\frac{im_n}{n}}\,\frac{i}{\sqrt{n-im_n}}\log p(n-im_n, i) \\
&\quad \to \sqrt{1-\rho}\left(\frac{2H\left(\frac{\rho}{\lambda\sqrt{1-\rho}}\right)}{\frac{\rho}{\lambda\sqrt{1-\rho}}} - \frac{\rho}{\lambda\sqrt{1-\rho}}\log\left[1-\exp\left\{-H\left(\frac{\rho}{\lambda\sqrt{1-\rho}}\right)\right\}\right]\right) \\
&\quad = 2\lambda\frac{1-\rho}{\rho}J(\rho,\lambda) - \frac{\rho}{\lambda}\log(1-\exp\{-J(\rho,\lambda)\}) = K(\rho,\lambda).
\end{aligned}$$

Soit $\epsilon > 0$. Compte tenu de (24), on a donc, pour $n \geqslant N_1(\epsilon)$,

$$|\log \operatorname{Card}\mathcal{P}_i - \sqrt{n}K(\rho,\lambda)| \leqslant \epsilon\sqrt{n}$$

et par suite, pour $n \geqslant N_2(\epsilon)$,

$$\operatorname{Card}\mathcal{P}_i = \exp\{[K(\rho,\lambda)+\epsilon_1]\sqrt{n}\} \tag{25}$$

avec $|\epsilon_1| \leqslant \epsilon$; et ceci, pour tous les i considérés. Par suite,

$$\sum_i \operatorname{Card}\mathcal{P}_i \leqslant \frac{\sqrt{n}}{\lambda}\exp\{[g(\lambda)+\epsilon]\sqrt{n}\}.$$

On a

$$\frac{(i+1)\lambda}{\sqrt{n}} - \frac{i\lambda}{\sqrt{n}} = \frac{\lambda}{\sqrt{n}} \to 0.$$

Imposons à ω d'être tel que $\omega < M(\lambda) < 1-\omega$. Alors, pour $n \geqslant N_3(\epsilon)$, il existe i tel que $|\rho - M(\lambda)| \leqslant \epsilon$. Donc, pour $n \geqslant N_4(\epsilon)$, il existe i tel que $K(\rho,\lambda) \geqslant g(\lambda)-\epsilon$. Compte tenu de (25), on a, pour un tel i, et pourvu que $n \geqslant N_5(\epsilon)$,

$$\operatorname{Card}\mathcal{P}_i \geqslant \exp\{[g(\lambda)-\epsilon]\sqrt{n}\}.$$

Donc

$$\exp\{[g(\lambda)-\epsilon]\sqrt{n}\} \leqslant \sum_i \operatorname{Card}\mathcal{P}_i \leqslant \frac{\sqrt{n}}{\lambda}\exp\{[g(\lambda)+\epsilon]\sqrt{n}\}.$$

(c) Envisageons dans (c) les i tels que $\rho < \omega$, c'est-à-dire $i < \omega\sqrt{n}/\lambda$. Imposons à ω d'être tel que $f(\omega/\lambda) < \frac{1}{2}g(\lambda)$. Alors, pour $n \geqslant N_6(\epsilon)$,

$$\sum_i \operatorname{Card} \mathscr{P}_i < \exp\{\tfrac{1}{2}g(\lambda)\sqrt{n}\}.$$

(d) Envisageons dans (d) les i tel que $\rho > 1-\omega$, c'est-à-dire $i > (1-\omega)\sqrt{n}/\lambda$. Alors

$$n - im_n \leqslant \frac{1-\omega}{\lambda}\sqrt{n}m_n \leqslant n-(1-\omega)n = \omega n.$$

Donc, pour $n \geqslant N_7(\epsilon)$,

$$\sum_i \operatorname{Card} \mathscr{P}_i \leqslant \sum_{j=1}^{\lfloor \omega n \rfloor} p(j) \leqslant \exp \omega'\sqrt{n},$$

où ω' peut être rendu aussi petit qu'on veut en choisant ω assez petit; en particulier, on peut faire en sorte que $\omega' \leqslant \frac{1}{2}g(\lambda)$.

(e) Finalement

$$\exp\{[g(\lambda)-\epsilon]\sqrt{n}\} < \operatorname{Card} \mathscr{P} \leqslant \frac{\sqrt{n}}{\lambda}\exp\{[g(\lambda)+\epsilon]\sqrt{n}\} + 2\exp\{\tfrac{1}{2}g(\lambda)\sqrt{n}\}.$$

Donc, pour $n \geqslant N_8(\epsilon)$,

$$[g(\lambda)-\epsilon]\sqrt{n} < \log \operatorname{Card} \mathscr{P} \leqslant [g(\lambda)+\epsilon]\sqrt{n},$$

d'où le théorème. □

2.5 Théorème (a) *La fonction* $\lambda \mapsto g(\lambda)$ *est analytique pour* $\lambda > 0$.

(b) *Posant* $J(M(\lambda),\lambda) = J_\lambda$, *on a*

$$g'(\lambda) = -J_\lambda < 0, \qquad g''(\lambda) = \frac{2\lambda J_\lambda(e^{J_\lambda}-1)}{M(\lambda)\{2+\lambda^2(e^{J_\lambda}-1)\}} > 0.$$

(c) *On a*

$$\lambda^2 g''(\lambda) + \lambda g'(\lambda) - g(\lambda) = \frac{2g''(\lambda)}{1-e^{-g'(\lambda)}}.$$

Rappelons que $M(\lambda)$ s'obtient en résolvant l'équation en x

$$\exp\{-J(x,\lambda)\} + \exp\left\{-\frac{\lambda^2}{x}J(x,\lambda)\right\} = 1$$

et que la dérivée par rapport à x du 1er membre est < 0 (cf. notamment le lemme 2.2). Donc $M(\lambda)$ est fonction analytique de λ. Compte tenu de (23), g est analytique.

On a, d'après (23),

$$g(\lambda) = -\lambda J_\lambda + 2\lambda \frac{J_\lambda}{M(\lambda)},$$

d'où

$$g'(\lambda) = -J_\lambda - \lambda J'_\lambda + 2\frac{J_\lambda}{M(\lambda)} + 2\lambda\left(\frac{J_\lambda}{M(\lambda)}\right)'. \tag{26}$$

D'après le lemme 2.3,

$$\exp\{-J_\lambda\} + \exp\left\{-\lambda^2 \frac{J_\lambda}{M(\lambda)}\right\} = 1, \tag{27}$$

d'où, en dérivant,

$$-\exp\{-J_\lambda\}J'_\lambda + \exp\left\{-\lambda^2\frac{J_\lambda}{M(\lambda)}\right\}\left\{-2\lambda\frac{J_\lambda}{M(\lambda)} - \lambda^2\left(\frac{J_\lambda}{M(\lambda)}\right)'\right\} = 0$$

ou, compte tenu de (27),

$$\exp\{-J_\lambda\}J'_\lambda + (1-\exp\{-J_\lambda\})\left\{2\lambda\frac{J_\lambda}{M(\lambda)} + \lambda^2\left(\frac{J_\lambda}{M(\lambda)}\right)'\right\} = 0$$

ou, encore,

$$\lambda^2\left(\frac{J_\lambda}{M(\lambda)}\right)' = -2\lambda\frac{J_\lambda}{M(\lambda)} - \frac{J'_\lambda}{e^{J_\lambda}-1}. \tag{28}$$

Reportons dans (26). On obtient

$$g'(\lambda) = -J_\lambda - \lambda J'_\lambda + 2\frac{J_\lambda}{M(\lambda)} - 4\frac{J_\lambda}{M(\lambda)} - \frac{2J'_\lambda}{\lambda(e^{J_\lambda}-1)},$$

ou

$$g'(\lambda) = -J_\lambda + 2\frac{J_\lambda}{M(\lambda)} + \frac{\lambda^2 - 2 - \lambda^2 e^{J_\lambda}}{\lambda(e^{J_\lambda}-1)}J'_\lambda. \tag{29}$$

On a

$$\frac{\partial}{\partial x}\left(\frac{x}{\lambda\sqrt{1-x}}\right) = \frac{2-x}{2\lambda}(1-x)^{-3/2}, \qquad \frac{\partial}{\partial\lambda}\left(\frac{x}{\lambda\sqrt{1-x}}\right) = -\frac{x}{\lambda^2}(1-x)^{-1/2},$$

donc

$$\left(2x(1-x)\frac{\partial}{\partial x} + \lambda(2-x)\frac{\partial}{\partial\lambda}\right)\frac{x}{\lambda\sqrt{1-x}} = 0,$$

donc, compte tenu de la définition de $J(x,\lambda)$,

$$\left(2x(1-x)\frac{\partial}{\partial x} + \lambda(2-x)\frac{\partial}{\partial\lambda}\right)J(x,\lambda) = 0.$$

Alors, d'après (19),

$$\{2\lambda^2(1-x)(e^J-1)-x^2\}\frac{\partial J}{\partial\lambda} = -\frac{2x(1-x)}{\lambda(2-x)}\{2\lambda^2(1-x)(e^J-1)-x^2\}\frac{\partial J}{\partial x}$$
$$= -2\lambda(1-x)J(e^J-1). \quad (30)$$

Or

$$J'_\lambda = \frac{\partial J}{\partial x}M'(\lambda)+\frac{\partial J}{\partial\lambda}.$$

Compte tenu de (19) et (30):

$$M(\lambda)[2\lambda^2\{1-M(\lambda)\}(e^{J_\lambda}-1)-M(\lambda)^2]J'_\lambda$$
$$= \lambda^2\{2-M(\lambda)\}J_\lambda(e^{J_\lambda}-1)M'(\lambda)-2\lambda\{1-M(\lambda)\}M(\lambda)J_\lambda(e^{J_\lambda}-1). \quad (31)$$

L'équation (28) s'écrit

$$\lambda^2\{J'_\lambda M(\lambda)-J_\lambda M(\lambda)'\}(e^{J_\lambda}-1)+2\lambda M(\lambda)J_\lambda(e^{J_\lambda}-1)+M(\lambda)^2J'_\lambda = 0,$$

ou

$$\lambda^2J_\lambda(e^{J_\lambda}-1)M(\lambda)^{-1}M(\lambda)' = \{\lambda^2(e^{J_\lambda}-1)+M(\lambda)\}J'_\lambda+2\lambda J_\lambda(e^{J_\lambda}-1).$$

Reportant dans (31) on obtient:

$$[2\lambda^2\{1-M(\lambda)\}(e^{J_\lambda}-1)-M(\lambda)^2]J'_\lambda$$
$$= \{2-M(\lambda)\}\{\lambda^2(e^{J_\lambda}-1)+M(\lambda)\}J'_\lambda$$
$$+\{2-M(\lambda)\}2\lambda J_\lambda(e^{J_\lambda}-1)$$
$$-2\lambda\{1-M(\lambda)\}M(\lambda)J_\lambda(e^{J_\lambda}-1),$$

soit, après calculs,

$$M(\lambda)(\lambda^2-2-\lambda^2e^{J_\lambda})J'_\lambda = 2\lambda J_\lambda(e^{J_\lambda}-1).$$

Reportant dans (29):

$$g'(\lambda) = -J_\lambda-2\frac{J_\lambda}{M(\lambda)}+\frac{2J_\lambda}{M(\lambda)} = -J_\lambda,$$

d'où

$$g''(\lambda) = -J'_\lambda = \frac{2\lambda J_\lambda(e^{J_\lambda}-1)}{M(\lambda)\{2+\lambda^2(e^{J_\lambda}-1)\}}.$$

D'après (23) et (b), on a

$$\frac{2\lambda J_\lambda}{M(\lambda)} = g(\lambda)+\lambda J_\lambda = g(\lambda)-\lambda g'(\lambda),$$

d'où, en portant dans l'expression de g'' donnée par (b),

$$g''(\lambda) = \{g(\lambda)-\lambda g'(\lambda)\}\frac{e^{-g'(\lambda)}-1}{2+\lambda^2(e^{-g'(\lambda)}-1)},$$

d'où facilement (c). □

2.6 Théorème

$$g(\lambda) \sim 2\frac{\log\lambda}{\lambda} \qquad \textit{quand } \lambda \to \infty.$$

Soit $\epsilon \in]0, \frac{1}{2}]$. D'après (6), quand $\lambda \to \infty$, on a

$$J(x,\lambda) = H\left(\frac{x}{\lambda\sqrt{1-x}}\right) \to 0 \qquad \text{uniformément dans }]0,1-\epsilon]. \tag{32}$$

D'autre part, pour $0 < x \leqslant 1-\epsilon$, on a

$$\begin{aligned}\frac{\lambda^2}{x}J(x,\lambda) = \frac{\lambda^2}{x}H\left(\frac{x}{\lambda\sqrt{1-x}}\right) &\leqslant \frac{\lambda^2}{x^2}2H\left(\frac{x}{\lambda\sqrt{\epsilon}}\right)\\ &\leqslant \frac{\lambda^2}{x^2}2\frac{x^2}{\lambda^2\epsilon} \qquad \text{pour } \lambda \geqslant \Lambda_1(\epsilon), \text{ d'après (6)}\\ &= \frac{2}{\epsilon}.\end{aligned}$$

Alors, pour $0 < x \leqslant 1-\epsilon$ et $\lambda \geqslant \Lambda_1(\epsilon)$,

$$\exp\{-J(x,\lambda)\}+\exp\left\{-\frac{\lambda^2}{x}J(x,\lambda)\right\} \geqslant \exp\{-J(x,\lambda)\}+\exp\left\{-\frac{2}{\epsilon}\right\},$$

d'où, d'après (32),

$$\exp\{-J(x,\lambda)\}+\exp\left\{-\frac{\lambda^2}{x}J(x,\lambda)\right\} > 1 \qquad \text{pour } \lambda \geqslant \Lambda_2(\epsilon).$$

Il résulte de là et du lemme 2.3 que

$$M(\lambda) > 1-\epsilon \qquad \text{pour } \lambda \geqslant \Lambda_2(\epsilon). \tag{33}$$

Ainsi

$$M(\lambda) \to 1. \tag{34}$$

Dans toute la fin de la preuve, on posera

$$\frac{M(\lambda)}{\lambda\sqrt{1-M(\lambda)}} = M_\lambda.$$

On a

$$\frac{\lambda^2}{M(\lambda)}J_\lambda \geqslant \frac{\lambda^2}{1-\epsilon}J(1-\epsilon,\lambda) \qquad \text{pour } \lambda \geqslant \Lambda_2(\epsilon) \text{ ((33) et lemme 2.2)}$$

$$= \frac{\lambda^2}{1-\epsilon} H\left(\frac{1-\epsilon}{\lambda\sqrt{\epsilon}}\right)$$
$$\geqslant \frac{\lambda^2}{1-\epsilon}\,\frac{1}{2}\,\frac{(1-\epsilon)^2}{\lambda^2\epsilon} \qquad \text{pour } \lambda \geqslant \Lambda_3(\epsilon) \text{ (d'après (6))}$$
$$= \frac{1-\epsilon}{2\epsilon} \geqslant \frac{1}{4\epsilon}.$$

Donc, pour $\lambda \geqslant \Lambda_4(\epsilon)$, on a, compte tenu du lemme 2.3,

$$\exp(-J_\lambda) = 1-\exp\left(-\frac{\lambda^2}{M(\lambda)}J_\lambda\right) \geqslant 1-\exp\left(-\frac{1}{4\epsilon}\right).$$

Comme

$$1-\exp\left(-\frac{1}{4\epsilon}\right) \to 1 \qquad \text{quand } \epsilon \to 0,$$

on voit que

$$J_\lambda \to 0, \tag{35}$$

donc $M_\lambda \to 0$ et alors, d'après (6),

$$J_\lambda = H(M_\lambda) \sim M_\lambda^2 = \frac{M(\lambda)^2}{\lambda^2\{1-M(\lambda)\}}. \tag{36}$$

Alors

$$\frac{1}{1-M(\lambda)} \sim \frac{\lambda^2}{M(\lambda)}J_\lambda \qquad \text{d'après (34) et (36)}$$
$$= -\log(1-e^{-J_\lambda}) \qquad \text{(lemme 2.3)}$$
$$\sim -\log J_\lambda \qquad \text{d'après (35)}$$
$$\sim 2\log\lambda + \log\{1-M(\lambda)\} \qquad \text{d'après (34) et (36),}$$

d'où

$$\frac{1}{1-M(\lambda)} \sim 2\log\lambda. \tag{37}$$

Enfin

$$g(\lambda) \sim \lambda J_\lambda \qquad \text{d'après (23) et (34)}$$
$$\sim \frac{1}{\lambda\{1-M(\lambda)\}} \qquad \text{d'après (34) et (36)}$$
$$\sim 2\frac{\log\lambda}{\lambda} \qquad \text{d'après (37).}$$

2.7 Le théorème 2.6 peut être précisé. Il résulte de (37) que:

$$M(\lambda) = 1 - \frac{1+o(1)}{2\log\lambda}.$$

Par ailleurs, l'équation du lemme (2.3) (b) s'écrit:

$$\log(1-\exp\{-J(x,\lambda)\}) = -\frac{\lambda^2}{x}J(x,\lambda).$$

D'après (35), $J_\lambda = J(M(\lambda),\lambda) \to 0$, et d'après (6) et (36),

$$J_\lambda = M_\lambda^2 + O(M_\lambda^4) = M_\lambda^2[1+O(\lambda^{-2}\log\lambda)].$$

On a donc

$$\begin{aligned}\log(1-\exp\{-J_\lambda\}) &= \log J_\lambda + O(J_\lambda)\\ &= 2\log M(\lambda) - 2\log\lambda - \log(1-M(\lambda))\\ &\qquad + O(\lambda^{-2}\log\lambda).\end{aligned}$$

$M(\lambda)$ vérifie donc l'équation

$$\begin{aligned}2\log M(\lambda) - 2\log\lambda - \log(1-M(\lambda)) + O(\lambda^{-2}\log\lambda)&\\ = -\frac{M(\lambda)}{1-M(\lambda)} + O(\lambda^{-2}\log^2\lambda)&,\end{aligned}$$

ou encore,

$$M(\lambda) = 1 + \frac{M(\lambda)}{-2\log\lambda - \log(1-M(\lambda)) + 2\log M(\lambda) + O(\lambda^{-2}\log^2\lambda)}.$$

Si l'on reporte

$$M(\lambda) = 1 - \frac{1+o(1)}{2\log\lambda}$$

dans le membre de droite de la relation ci-dessus, on obtient

$$M(\lambda) = 1 - \frac{1}{2\log\lambda} + \frac{1-\log(2\log\lambda)}{(2\log\lambda)^2}\{1+o(1)\},$$

et l'on peut continuer par la méthode d'itération mentionnée en 1.3. On obtient, en posant $u = 2\log\lambda$ et $L = \log u$,

$$M(\lambda) = 1 - \frac{1}{u} + \frac{1-L}{u^2} + \frac{3L-L^2}{u^3} + \frac{-2L^3+11L^2-6L-5}{2u^4} + O\left(\frac{L^4}{u^5}\right).$$

De (23), on déduit:

$$\begin{aligned}\lambda g(\lambda) &= \frac{2-M(\lambda)}{1-M(\lambda)}M(\lambda) + O(\lambda^{-2}\log^2\lambda)\\ &= u + 1 - L + \frac{L}{u} + \frac{L^2-2L-1}{2u^2} + \frac{2L^3-9L^2+5}{6u^3} + O\left(\frac{L^4}{u^4}\right).\end{aligned}$$

2.8 Lemme *Pour $n \geqslant 2$, on a:*

$$\sum_{k=1}^{n-1} \frac{1}{k^2(n-k)^2} \leqslant \frac{41}{9n^2}.$$

Pour $n \geqslant 4$, on a:

$$\sum_{k=2}^{n-2} \frac{1}{k(n-k)^2} = \sum_{k=2}^{n-2} \frac{1}{k^2(n-k)} \leqslant \frac{0.87231}{n}.$$

La décomposition en éléments simples donne:

$$\frac{1}{k^2(n-k)^2} = \frac{1}{n^2}\left(\frac{1}{k^2} + \frac{1}{(n-k)^2} + \frac{2/n}{k} + \frac{2/n}{n-k}\right)$$

et comme

$$\sum_{k=1}^{n-1} \frac{1}{k} \leqslant \gamma + \log n \leqslant 0.578 + \log n,$$

on obtient:

$$\sum_{k=1}^{n-1} \frac{1}{k^2(n-k)^2} \leqslant \frac{1}{n^2}\left(\tfrac{1}{3}\pi^2 + \frac{4}{n}(0.578 + \log n)\right) \leqslant \frac{41}{9n^2}$$

pour $n \geqslant 9$. On achève ensuite pour $2 \leqslant n \leqslant 8$ par un calcul numérique. Il y a égalité pour $n = 4$.

Similairement,

$$\frac{1}{k^2(n-k)} = \frac{1}{n}\left(\frac{1}{k^2} + \frac{1/n}{k} + \frac{1/n}{n-k}\right)$$

et

$$\sum_{k=2}^{n-2} \frac{1}{k^2(n-k)} \leqslant \frac{1}{n}\left(\tfrac{1}{6}\pi^2 - 1 + \frac{2}{n}\{\gamma + \log(n-1) - 1\}\right).$$

Le crochet ci-dessus est $\leqslant 0.872$ pour $n \geqslant 25$. On calcule ensuite

$$n\sum_{k=2}^{n-2} \frac{1}{k^2(n-k)}$$

pour $4 \leqslant n \leqslant 24$. Le maximum est obtenu pour $n = 11$. □

2.9 Lemme *Soient deux séries entières à coefficients complexes*

$$\sum_{m=1}^{\infty} u_m z^m \qquad \text{et} \qquad \sum_{m=1}^{\infty} v_m z^m;$$

on pose

$$\sum_{m=1}^{\infty} w_m z^m = \left(\sum_{m=1}^{\infty} u_m z^m\right)\left(\sum_{m=1}^{\infty} v_m z^m\right).$$

Soit $n \geqslant 2$. On suppose qu'il existe 3 constantes $U, V, \beta \geqslant 0$ telles que

$$|u_m| \leqslant \frac{U\beta^m}{m^2}, \qquad |v_m| \leqslant \frac{V\beta^m}{m^2}$$

pour $1 \leqslant m \leqslant n-1$. Alors on a, pour $2 \leqslant m \leqslant n$,

$$|w_m| \leqslant \frac{41}{9}\,\frac{UV\beta^m}{m^2}.$$

On a

$$w_m = \sum_{k=1}^{m-1} u_k v_{m-k}$$

et

$$|w_m| \leqslant UV\beta^m \sum_{k=1}^{m-1} \frac{1}{k^2(m-k)^2}.$$

On conclut en appliquant le lemme 2.8. □

2.10 Lemme *Soit $\sum_{m=1}^{\infty} u_m z^m$ une série entière à coefficients complexes. On définit la fonction φ par*

$$\varphi(z) = \mathrm{e}^z - 1 - z$$

et l'on définit les coefficients t_m par

$$\varphi\left(\sum_{m=1}^{\infty} u_m z^m\right) = \sum_{m=2}^{\infty} t_m z^m.$$

Soit $n \geqslant 2$. On suppose qu'il existe deux constantes $U, \beta \geqslant 0$ telles que

$$|u_m| \leqslant \frac{U\beta^m}{m^2}$$

pour $1 \leqslant m \leqslant n-1$. Alors on a, pour $2 \leqslant m \leqslant n$,

$$|t_m| \leqslant \frac{T\beta^m}{m^2},$$

où

$$T = \tfrac{9}{41}\varphi(\tfrac{41}{9}U).$$

On pose d'abord pour $k \geqslant 2$,

$$\left(\sum_{m=1}^{\infty} u_m z^m\right)^k = \sum_{m=k}^{\infty} u_{m,k} z^m.$$

D'après le lemme 2.9, on a

$$|u_{m,2}| \leqslant \tfrac{41}{9} U^2 \beta^m m^{-2}$$

pour $2 \leqslant m \leqslant n$ et, par récurrence sur k,

$$|u_{m,k}| \leqslant (\tfrac{41}{9})^{k-1} U^k \beta^m m^{-2}$$

pour $k \leqslant m \leqslant n$. On a ensuite

$$t_m = \sum_{k=2}^{m} \frac{u_{m,k}}{k!}$$

et

$$|t_m| \leqslant \frac{9}{41} \frac{\beta^m}{m^2} \sum_{k=2}^{m} \frac{(\frac{41}{9}U)^k}{k!} = \tfrac{9}{41}\varphi(\tfrac{41}{9}U)\beta^m m^{-2}.$$

2.11 Lemme *Soient a_0 un nombre réel positif, a_1 un nombre réel et $a_2 = -\frac{1}{8}a_0 - \frac{1}{2}a_0^{-1}$. On définit la suite (a_n) par récurrence. Soit $n \geqslant 2$, on suppose connus $a_0, a_1, \ldots, a_n$. On définit $b_1 = 0$, $b_2, \ldots, b_n$ par*

$$-\varphi\left(-\sum_{i=1}^{\infty}(i+1)a_{i+1}z^i\right) = \sum_{i=2}^{\infty} b_i z^i,$$

où $\varphi(t) = \mathrm{e}^t - 1 - t$. (On observe que $b_2, \ldots, b_n$ ne dépendent que de $a_2, \ldots, a_n$.) On définit ensuite a_{n+1} par

$$\begin{aligned} 2(n+1)^2 a_{n+1} = {} & -2b_n + \frac{4}{a_0}(na_n + b_{n-1}) \\ & + \sum_{k=2}^{n-2} \frac{2}{a_0}(k^2-1)a_k\{(n-k+1)a_{n-k+1} + b_{n-k}\} \\ & + \{(n-1)^2-1\}a_{n-1}\left(1+\frac{4a_2}{a_0}\right) - \frac{2}{a_0}(n^2-1)a_n. \end{aligned}$$

(Par convention la somme en k est vide pour $n \leqslant 3$.) Alors il existe $A, \alpha \geqslant 0$ dépendant de a_0 tels que, pour tout $n \geqslant 2$, on ait

$$|a_n| \leqslant A\alpha^n n^{-3}.$$

On fixe d'abord $A = 1/4\alpha$ et l'on choisit α assez grand pour que $|a_2| \leqslant \frac{1}{8}A\alpha^2$. On raisonne ensuite par récurrence: on fixe $n \geqslant 2$, on suppose que

$$|a_m| \leqslant A\alpha^m m^{-3}$$

pour $2 \leqslant m \leqslant n$. On applique le lemme 2.10 avec $u_m = -(m+1)a_{m+1}$.

On a, pour $1 \leqslant m \leqslant n-1$,

$$|u_m| \leqslant \frac{A\alpha^{m+1}}{(m+1)^2} \leqslant A\alpha\frac{\alpha^m}{m^2} = \frac{\alpha^m}{4m^2}.$$

On en déduit que

$$|b_m| \leqslant \frac{B\alpha^m}{m^2}$$

pour $2 \leqslant m \leqslant n$, avec $B = \frac{9}{41}\varphi(\frac{41}{36}) \leqslant 0.22$. Il vient ensuite

$$\begin{aligned}
&\left|\frac{2}{a_0}\sum_{k=2}^{n-2}(k^2-1)a_k\{(n-k+1)a_{n-k+1}+b_{n-k}\}\right| \\
&\qquad \leqslant \frac{2}{a_0}\sum_{k=2}^{n-2}\frac{A\alpha^k}{k}\left(\frac{A\alpha^{n-k+1}}{(n-k+1)^2}+\frac{B\alpha^{n-k}}{(n-k)^2}\right) \\
&\qquad \leqslant \frac{2A}{a_0}(A\alpha+B)\alpha^n\sum_{k=2}^{n-2}\frac{1}{k(n-k)^2} \\
&\qquad \leqslant \frac{0.83}{a_0}\frac{A\alpha^n}{n} \leqslant \frac{1.3}{a_0}\frac{A\alpha^n}{n+1}
\end{aligned}$$

en utilisant le lemme 2.8. Puis on a successivement:

$$\begin{aligned}
|2b_n| &\leqslant \frac{2B\alpha^n}{n^2} = \frac{A\alpha^n}{n+1}\left(\frac{2B}{A}\frac{n+1}{n^2}\right) \\
&= \frac{A\alpha^n}{n+1}\frac{8B\alpha(n+1)}{n^2} \leqslant 1.32\alpha\frac{A\alpha^n}{n+1};
\end{aligned}$$

$$\left|\frac{4}{a_0}na_n\right| \leqslant \frac{4}{a_0}A\frac{\alpha^n}{n+1}\frac{n+1}{n^2} \leqslant \frac{3}{a_0}A\frac{\alpha^n}{n+1};$$

$$\begin{aligned}
\left|\frac{4}{a_0}b_{n-1}\right| &\leqslant \frac{4}{a_0}B\frac{\alpha^{n-1}}{(n-1)^2} = \frac{A\alpha^n}{n+1}\left(\frac{4}{a_0}\frac{B}{A\alpha}\frac{n+1}{(n-1)^2}\right) \\
&\leqslant \frac{48B}{a_0}\frac{A\alpha^n}{n+1} \leqslant \frac{11}{a_0}\frac{A\alpha^n}{n+1};
\end{aligned}$$

$$|\{(n-1)^2-1\}a_{n-1}| \leqslant \frac{A\alpha^{n-1}}{n-1} \leqslant \frac{A\alpha^n}{n+1}\frac{n+1}{(n-1)\alpha} \leqslant \frac{3}{\alpha}\frac{A\alpha^n}{n+1};$$

$$\left|\{(n-1)^2-1\}a_{n-1}\frac{4a_2}{a_0}\right| \leqslant \frac{3}{\alpha}\frac{4A\alpha^2}{8a_0}\frac{A\alpha^n}{n+1} \leqslant \frac{3}{8a_0}\frac{A\alpha^n}{n+1};$$

$$\left|\frac{2}{a_0}(n^2-1)a_n\right| \leqslant \frac{2}{a_0}A\frac{\alpha^n}{n} \leqslant \frac{3}{a_0}A\frac{\alpha^n}{n+1}.$$

La formule de définition de a_{n+1} nous donne alors

$$|2(n+1)^2 a_{n+1}| \leqslant \frac{A\alpha^n}{n+1}\left(1.32\alpha + \frac{19}{a_0} + \frac{3}{\alpha}\right).$$

Pour α sufisamment grand, le crochet sera $\leqslant 2\alpha$ et l'on obtiendra bien

$$|a_{n+1}| \leqslant \frac{A\alpha^{n+1}}{(n+1)^3}. \qquad \square$$

2.12 Avec les notations de 2.11, on trouve

$$a_3 = \frac{a_0^2+36}{288},$$

$$a_4 = -\frac{17a_0^4+72a_0^2-48}{1152a_0^3},$$

$$a_5 = -\frac{a_0^4-200a_0^2-5200}{230\,400},$$

$$a_6 = -\frac{14a_0^8+7425a_0^6+31\,500a_0^4-32\,400a_0^2+25\,920}{2073\,600a_0^5},$$

$$a_7 = \frac{5a_0^6-882a_0^4+113\,288a_0^2+2469\,600}{406\,425\,600}.$$

2.13 Dans les sections 2.13 à 2.15 nous prendons

$$a_0 = \pi\sqrt{\tfrac{2}{3}}, \qquad a_1 = \log\tfrac{1}{2}a_0 - 1,$$

d'où une suite $(a_0, a_1, a_2, \ldots)$ bien déterminée. La calcul donne

$a_0 = 2.565\,10$	$a_5 = 0.028\,09$	$a_{10} = -0.001\,53$
$a_1 = -0.751\,15$	$a_6 = -0.014\,41$	$a_{11} = 0.000\,93$
$a_2 = -0.515\,56$	$a_7 = 0.007\,82$	$a_{12} = -0.000\,57$
$a_3 = 0.147\,85$	$a_8 = -0.004\,42$	
$a_4 = -0.059\,75$	$a_9 = 0.002\,57$	

En reprenant la preuve de 2.11, on voit que la majoration de $|a_n|$ est valable en prenant $\alpha = 17$ et $A = \frac{1}{68}$. (Des majorations plus fines pourraient être obtenues au prix de calculs plus techniques.) Soit $R > 0$ le rayon de convergence de la série $\sum_{n=0}^{\infty} a_n\lambda^n$. On a $R \geqslant \frac{1}{17}$. Pour $0 < \lambda < R$, posons

$$g_1(\lambda) = \lambda\log\lambda + \sum_{n=0}^{\infty} a_n\lambda^n. \qquad (38)$$

2.14 Lemme *On a*

$$\lambda^2 g_1'' + \lambda g_1' - g_1 = \frac{2g_1''}{1-\exp(-g_1')}.$$

On a

$$g_1'(\lambda) = \log\lambda + 1 + \sum_{n=1}^{\infty} n a_n \lambda^{n-1},$$

$$g_1''(\lambda) = \lambda^{-1} + \sum_{n=2}^{\infty} n(n-1) a_n \lambda^{n-2};$$

$$\lambda^2 + \lambda g_1' - g_1 = 2\lambda + \sum_{n=0}^{\infty} (n^2-1) a_n \lambda^n$$

$$= -a_0 + 2\lambda + \sum_{n=2}^{\infty} (n^2-1) a_n \lambda^n;$$

$$\exp\{-g_1'(\lambda)\} = \exp\{-1-\log\lambda - a_1\}\exp\left\{-\sum_{n=2}^{\infty} n a_n \lambda^{n-1}\right\}$$

$$= \frac{2}{a_0\lambda}\exp\{-u\},$$

avec

$$u = \sum_{n=2}^{\infty} n a_n \lambda^{n-1},$$

d'où

$$\exp(-u) = \varphi(-u) + 1 - u = -\sum_{n=2}^{\infty} b_n \lambda^n + 1 - \sum_{n=2}^{\infty} n a_n \lambda^{n-1}.$$

Ensuite

$$\lambda\{1-\exp(-g_1')\} = -\frac{2}{a_0} + \left(1+\frac{4a_2}{a_0}\right)\lambda + \frac{2}{a_0}\sum_{n=2}^{\infty} \{(n+1)a_{n+1} + b_n\}\lambda^n,$$

$$2\lambda g_1'' = 2 + \sum_{n=1}^{\infty} 2n(n+1) a_{n+1} \lambda^n.$$

Nous devons vérifier que

$$\lambda\{1-\exp(-g_1')\}(\lambda^2 g_1'' + \lambda g_1' - g_1) = 2\lambda g_1''.$$

L'identification des termes constants donne $(-2/a_0)(-a_0) = 2$. Celle des termes de degré 1 donne $(-2/a_0)2 + (1+4a_2/a_0)(-a_0) = 4a_2$, ce qui est vrai compte tenu de la valeur de a_2. En degré $n \geqslant 2$, on obtient le formule de définition récurrente de a_{n+1}.

2.15 Théorème *Soit (a_n) la suite définie en 2.13. Soit R le rayon de convergence de la série $\sum_{n=0}^{\infty} a_n\lambda^n$ (rappelons que $R \geqslant \frac{1}{17}$). Pour $0 < \lambda < R$, on a*

$$g(\lambda) = \lambda \log \lambda + \sum_{n=0}^{\infty} a_n\lambda^n.$$

On reprend la notation g_1 de 2.13 et 2.14. On a, quand $\lambda \to 0$,

$$g_1(\lambda) \to a_0 > 0, \qquad g_1'(\lambda) \to -\infty, \qquad \lambda g_1'(\lambda) \to 0.$$

Montrons que

$$g_1^2(\lambda) - \lambda^2 g_1'^2(\lambda) = 4F(-g_1'(\lambda)). \tag{39}$$

Les dérivées des 2 membres sont

$$2g_1g_1' - 2\lambda g_1'^2 - 2\lambda^2 g_1'g_1'', \quad 4\frac{g_1'g_1''}{e^{-g_1'}-1}$$

et elles sont égales d'après 2.14. D'autre part, les 2 membres de (39) tendent vers $\frac{2}{3}\pi^2$ quand $\lambda \to 0$. Cela prouve (39).

Posons maintenant

$$\mu(\lambda) = \frac{2\lambda g_1'(\lambda)}{\lambda g_1'(\lambda) - g_1(\lambda)} \tag{40}$$

donc

$$1 - \mu(\lambda) = \frac{g_1(\lambda) + \lambda g_1'(\lambda)}{g_1(\lambda) - \lambda g_1'(\lambda)}.$$

Pour λ assez petit, on a $\mu(\lambda) > 0$ et $1 - \mu(\lambda) > 0$. Les relations (39) et (40) donnent

$$\frac{\mu(\lambda)^2}{\lambda^2\{1-\mu(\lambda)\}}F(-g_1'(\lambda)) = g_1'(\lambda)^2,$$

ce qui, avec la définition des fonctions H et J, donne

$$-g_1'(\lambda) = H\left(\frac{\mu(\lambda)}{\lambda\sqrt{1-\mu(\lambda)}}\right) = J(\mu(\lambda), \lambda). \tag{41}$$

Montrons maintenant que

$$\exp\{g_1'(\lambda)\} + \exp\left\{\frac{\lambda^2}{\mu(\lambda)}g_1'(\lambda)\right\} = 1. \tag{42}$$

Cette relation est équivalente à

$$\tfrac{1}{2}\lambda^2 g_1'(\lambda) - \tfrac{1}{2}\lambda g_1(\lambda) = \log(1 - e^{g_1'(\lambda)}). \tag{43}$$

Les dérivées des 2 membres sont

$$\tfrac{1}{2}\lambda^2 g_1'' + \tfrac{1}{2}\lambda g_1' - \tfrac{1}{2}g_1, \qquad -\frac{e^{g_1'}g_1''}{1-e^{g_1'}}$$

et elles sont égales d'après 2.14. Les 2 membres de (43) tendent vers 0 quand $\lambda \to 0$. Cela prouve (42). On a donc

$$\exp\{-J(\mu(\lambda),\lambda)\} + \exp\left\{-\frac{\lambda^2}{\mu(\lambda)}J(\mu(\lambda),\lambda)\right\} = 1$$

et le lemme 6 (b) montre que

$$M(\lambda) = \mu(\lambda). \tag{44}$$

Finalement, on a, d'après (23), (44), (40) et (41),

$$g(\lambda) = \lambda\frac{2-\mu(\lambda)}{\mu(\lambda)}J(\mu(\lambda),\lambda) = -\lambda\frac{2g_1'(\lambda)}{2\lambda g_1'(\lambda)}\{-g_1'(\lambda)\} = g_1(\lambda).$$

Cette relation a été en fait démontrée pour λ assez petit. Comme g et g_1 sont analytiques sur $]0,R[$, on en déduit le théorème. □

2.16 Remarque Etant donné le théorème de Hardy–Ramanujan, il n'est pas surprenant que $g(\lambda) \to \pi\sqrt{\frac{2}{3}}$ quand $\lambda \to 0$. Toutefois, cela ne semble pas évident a priori.

2.17 Remarque Soient $\lambda > 0$, n un entier tendant vers $+\infty$ et (m_n) une suite d'entiers tels que $m_n/\sqrt{n} \to \lambda$. Alors, grâce à la continuité de g, le théorème 2.4 entraîne facilement que $\log r(n,m_n) \sim g(\lambda)\sqrt{n}$.

2.18 Comme dans [4], notons $R(n,a)$ le nombre de partitions de n dont aucune sous-somme n'est égale à a. Le résultat suivant est beaucoup plus spécial que ceux de [4] et [5], mais ne semble pas une conséquence de ces articles.

Théorème *Soit m un nombre impair tel que $m = \sqrt{n}\{1+o(1)\}$. On a $\log R(n,m) \geqslant 2.0138\sqrt{n}$ pour $n \geqslant 1$.*

Nous utiliserons la fonction $\lambda \mapsto s(\lambda)$ suivante, définie pour $0 < \lambda < 1$:

$$s(\lambda) = \sqrt{\tfrac{1}{2}\lambda}\,f\left(\frac{1}{\sqrt{2\lambda}}\right) + \sqrt{1-\lambda}\,g\left(\frac{1}{\sqrt{1-\lambda}}\right).$$

Soit i un entier pair tel que $0 < i < n$. Posons $n/i = \lambda$. Soit $\mathcal{P}_1$ l'ensemble des partitions de i dont toutes les parts sont paires et $< m$. Soit $\mathcal{P}_2$ l'ensemble des partitions de $n-i$ dont toutes les parts sont $> m$. La 'somme directe' d'un élément de $\mathcal{P}_1$ et d'un élément de $\mathcal{P}_2$ est une partition de n dont aucune sous-somme n'est égale à m. D'autre part, 2 couples distincts de $\mathcal{P}_1 \times \mathcal{P}_2$ ont des sommes directes distinctes. Donc

$$R(n,m) \geqslant \mathrm{Card}(\mathcal{P}_1 \times \mathcal{P}_2);$$

$$\begin{aligned}
\log R(n,m) &\geqslant \log \mathrm{Card}\,\mathcal{P}_1 + \log \mathrm{Card}\,\mathcal{P}_2 \\
&= \log p\big(\tfrac{1}{2}i, \tfrac{1}{2}(m-1)\big) + \log r(n-i, m+1) \\
&= \log p\big(\tfrac{1}{2}\lambda n, \tfrac{1}{2}\sqrt{n}\{1+o(1)\}\big) + \log r\big((1-\lambda)n, \sqrt{n}\{1+o(1)\}\big) \\
&= \log p\left(\tfrac{1}{2}\lambda n, \frac{1}{\sqrt{2\lambda}}\sqrt{\tfrac{1}{2}\lambda n}\{1+o(1)\}\right) \\
&\quad + \log r\left((1-\lambda)n, \frac{1}{\sqrt{1-\lambda}}\sqrt{(1-\lambda)n}\{1+o(1)\}\right).
\end{aligned}$$

Maintenant, supposons que $n \to \infty$ et que i est choisi de telle sorte que $\lambda \to \lambda_0 \in \,]0,1[$. On a, d'après les théorèmes 1.5 et 2.4,

$$\log p\left(\tfrac{1}{2}\lambda n, \frac{1+o(1)}{\sqrt{2\lambda}}\sqrt{\tfrac{1}{2}\lambda n}\right) \sim f\left(\frac{1}{\sqrt{2\lambda_0}}\right)\sqrt{\tfrac{1}{2}\lambda_0 n},$$

$$\log r\left((1-\lambda)n, \frac{1+o(1)}{\sqrt{1-\lambda}}\sqrt{(1-\lambda)n}\right) \sim g\left(\frac{1}{\sqrt{1-\lambda_0}}\right)\sqrt{(1-\lambda_0)n}.$$

Soit $\epsilon > 0$. Pour $n \gg 1$, on a donc

$$\begin{aligned}
\log R(n,m) &\geqslant \left\{\sqrt{\tfrac{1}{2}\lambda_0}\, f\left(\frac{1}{\sqrt{2\lambda_0}}\right) + \sqrt{1-\lambda_0}\, g\left(\frac{1}{\sqrt{1-\lambda_0}}\right) - \epsilon\right\}\sqrt{n} \\
&= \{s(\lambda_0) - \epsilon\}\sqrt{n}.
\end{aligned}$$

Il semble probable que $s(\lambda)$ est croissant pour $0 < \lambda < \lambda_1$ et décroissant pour $\lambda_1 < \lambda < 1$, avec un nombre λ_1 voisin de 0.34. En tous cas, un calcul numérique fournit $s(0.34) = 2.013\,844$, d'où le théorème. □

3 Etude de $r(n,m)$ pour $m \leqslant n^{\frac{1}{3}-\epsilon}$.

Dans cette partie, nous utilisons des méthodes probabilistes en usage depuis longtemps en théorie additive de nombres. Pour un emploi récent de ces méthodes dans un sujet voisin, cf. [6].

3.1 Lemme *Soient $u_1, u_2, \ldots, u_k$ des nombres réels > 0. On définit, pour $1 \leqslant i \leqslant k$, la fonction $\chi_i(t)$ qui vaut $1/u_i$ pour $0 \leqslant t \leqslant u_i$ et 0 ailleurs. On pose*

$$\varphi_k = \chi_1 * \chi_2 * \cdots * \chi_k,$$

la convolée des fonctions $\chi_1, \ldots, \chi_k$.

Soit maintenant f une fonction définie sur $\mathbb{R}$. On définit par récurrence l'opérateur $D^{(j)}(u_1, \ldots, u_j; f, x)$ par

$$D^{(1)}(u_1;f,x) = f(x)-f(x-u_1),$$
$$D^{(j)}(u_1,\dots,u_j;f,x) = D^{(j-1)}(u_1,\dots,u_{j-1};f,x) - D^{(j-1)}(u_1,\dots,u_{j-1};f,x-u_j).$$

Alors, si f est de classe C^k sur $[\frac{1}{2},x]$ et si $x \geqslant u_1+u_2+\cdots+u_k+\frac{1}{2}$, on a

$$D^{(k)}(u_1,\dots,u_k;f,x) = \left(\prod_{i=1}^{k} u_i\right)\int_{\frac{1}{2}}^{x} f^{(k)}(t)\varphi_k(x-t)\,\mathrm{d}t.$$

La démonstration se fait par récurrence sur k. Lorsque les u_i sont égaux à 1, la démonstration est donnée dans [7]. Nous remercions A. Odlyzko pour nous avoir signalé ce lemme. □

3.2 Rappelons quelques formules qui se trouvent dans [3].

$$p(n) = \frac{C^3}{2\pi\sqrt{2}}\varphi'(C^2[n-\tfrac{1}{24}])+f_1(n),$$

où

$$C = \pi\sqrt{\tfrac{2}{3}}, \qquad \varphi(x) = \frac{\exp\sqrt{x}}{\sqrt{x}}, \qquad f_1(n) = O\left(\frac{1}{n}\exp\tfrac{1}{2}C\sqrt{n}\right)$$

$$p(n) \sim \frac{1}{4\sqrt{3}n}\exp C\sqrt{n}, \qquad r(n,m) = D^{(m-1)}(1,2,\dots,m-1;p,n). \tag{45}$$

Enfin, la dérivée $\varphi^{(m)}(x)$ s'écrit:

$$\varphi^{(m)}(x) = \frac{\exp\sqrt{x}}{2^m x^{(m+1)/2}}y_m\left(-\frac{1}{\sqrt{x}}\right),$$

où y_m est le m-ieme polynôme de Bessel. En utilisant le lemme précédent avec $k = m-1$ et $u_i = i$, on obtient

$$r(n,m) = \frac{(m-1)!\,C^m}{2^{m+1}\pi\sqrt{2}}\int_{n-\frac{1}{2}m(m-1)}^{n}\varphi(n-t)\frac{\exp C\sqrt{t-\frac{1}{24}}}{(t-\frac{1}{24})^{(m+1)/2}}y_m\left(-\frac{1}{C\sqrt{t-\frac{1}{24}}}\right)\mathrm{d}t + O\left(\frac{2^{m-1}}{n}\exp\tfrac{1}{2}C\sqrt{n}\right) \tag{46}$$

où l'on a noté φ au lieu de φ_{m-1}.

3.3 On fixe comme précédemment $u_i = i$ et l'on désigne par X_i une variable aléatoire réelle dont la fonction de répartition est χ_i. On choisit les X_i de telle manière que $X_1,\dots,X_{m-1}$ soient independantes. On appellera Y la somme $X_1+\cdots+X_{m-1}$, et la fonction de répartition de Y est φ. On a

$$E(X_i)=\int x\chi_i(x)\,\mathrm{d}x = \tfrac{1}{2}i, \qquad V(X_i)=\int x^2\chi_i(x)\,\mathrm{d}x-E(X_i)^2 = \tfrac{1}{12}i^2,$$

d'où l'on déduit

$$E(Y) = \tfrac{1}{4}m(m-1), \qquad V(Y) = \tfrac{1}{72}m(m-1)(2m-1).$$

On note $\sigma = \sqrt{V(Y)}$ l'écart type de Y et l'on applique l'inégalité de Bernstein (cf. [8], p. 365): soit μ un nombre vérifiant $0 < \mu \leqslant \sigma/m$, alors

$$\text{Proba}(|Y-\tfrac{1}{4}m(m-1)| \geqslant \mu\sigma) \leqslant 2\exp\left(-\frac{\mu^2}{2(1+\mu m/2\sigma)^2}\right). \tag{47}$$

3.4 Théorème *Soit* $0 < \epsilon < \frac{1}{3}$ *et* $m \leqslant n^{\frac{1}{3}-\epsilon}$; *alors on a, quand* $n \to \infty$,

$$r(n,m) \sim p(n)(m-1)!\left(\frac{C}{2\sqrt{n}}\right)^{m-1} \exp\left\{-\left(\tfrac{1}{8}C+\frac{1}{2C}\right)\frac{m^2}{\sqrt{n}}\right\}.$$

Nous allons partir de la formule (46). Rappelons d'abord un résultat de M. Chellali (cf. [2]); si $m^{3/2}x \to 0$, alors

$$y_m(x) = \exp(\tfrac{1}{2}m^2x)\{1+O(mx)\}.$$

Soit maintenant $t \in [n-\frac{1}{2}m(m-1), n]$, on a

$$t = n+O(n^{2/3}),$$

$$x := -\frac{1}{C\sqrt{t-\frac{1}{24}}} = -\frac{1}{C\sqrt{n}}+O(n^{-5/6}),$$

$$y_m(x) = \exp\left(-\frac{m^2}{2C\sqrt{n}}\right)\{1+O(n^{-1/6})\}.$$

Le formule (46) devient

$$r(n,m) = \frac{(m-1)!\,C^m}{2^{m+1}\pi\sqrt{2}}\exp\left(-\frac{m^2}{2C\sqrt{n}}\right)\{1+O(n^{-1/6})\}I + O\left(\frac{2^{m-1}}{n}\exp\tfrac{1}{2}C\sqrt{n}\right) \tag{48},$$

avec

$$I = \int_{n-\frac{1}{2}m(m-1)}^{n} \varphi(n-t)\frac{\exp C\sqrt{t-\frac{1}{24}}}{(t-\frac{1}{24})^{(m+1)/2}}\,dt.$$

Pour $t \in [n-\frac{1}{2}m(m-1), n]$, on a

$$t = n+O(m^2)$$

et

$$(t-\tfrac{1}{24})^{(m+1)/2} = \exp\left\{\tfrac{1}{2}(m+1)\left[\log n+O\left(\frac{m^2}{n}\right)\right]\right\}$$

$$= n^{(m+1)/2}\left\{1+O\left(\frac{m^3}{n}\right)\right\}.$$

On a ainsi

$$I = n^{-(m+1)/2}\{1+O(n^{-3\epsilon})\}J, \tag{49}$$

avec

$$J = \int_{n-\frac{1}{2}m(m-1)}^{n} \varphi(n-t)\exp C\sqrt{t-\tfrac{1}{24}}\ \mathrm{d}t.$$

Lorsque $m = o(n^{1/4})$ le même raisonnement montre que

$$J \sim \exp C\sqrt{n} \int_{n-\frac{1}{2}m(m-1)}^{n} \varphi(n-t)\ \mathrm{d}t = \exp C\sqrt{n}$$

et le théorème se réduit à

$$r(n,m) \sim p(n)(m-1)!\left(\frac{C}{2\sqrt{n}}\right)^{m-1},$$

qui a déjà été démontré dans [3].

On peut donc supposer $m \geqslant n^{1/5}$.

On coupe alors l'intervalle d'intégration de J de la façon suivante. Rappelons que

$$\sigma = \sqrt{\tfrac{1}{72}m(m-1)(2m-1)} = \sqrt{V(Y)}.$$

On pose

$$\tau = \min(\epsilon, \tfrac{1}{20}), \qquad \mu = n^{\tau},$$

$$a_0 = n-\tfrac{1}{2}m(m-1), \qquad a_1 = n-\tfrac{1}{4}m(m-1)-\mu\sigma,$$

$$a_2 = n-\tfrac{1}{4}m(m-1)+\mu\sigma, \qquad a_3 = n$$

et, pour $0 \leqslant i \leqslant 2$,

$$J_i = \int_{a_i}^{a_{i+1}} \varphi(n-t)\exp(C\sqrt{t-\tfrac{1}{24}})\ \mathrm{d}t.$$

On observe ensuite que, comme m est supposé $\geqslant n^{1/5}$, on a $\mu \leqslant \sigma/m$ pour n assez grand, et que, pour x vérifiant $\mu \leqslant x \leqslant \sigma/m$, on a

$$0 \leqslant \frac{xm}{4\sigma} \leqslant \tfrac{1}{4}.$$

L'inégalité de Bernstein (47) implique alors:

$$\mathrm{Proba}(|Y-\tfrac{1}{4}m(m-1)| \geqslant x\sigma) \leqslant 2\exp(-\tfrac{2}{9}\mu^2). \tag{50}$$

Par application de la formule de Taylor, si $t \to +\infty$ et si $h = o(t^{3/4})$, alors on a

$$\exp(C\sqrt{t+h}) = \exp(C\sqrt{t})\exp\left(\frac{Ch}{2\sqrt{t}}\right)\left\{1+O\left(\frac{h^2}{t^{3/2}}\right)\right\}. \tag{51}$$

Evaluons maintenant J_1. Il est commode de poser $t_0 = n-\frac{1}{4}m(m-1)$. Pour $t \in [a_1, a_2]$, on a

$$t = t_0 + O(\mu\sigma),$$

et, par (50),

$$\begin{aligned}\exp(C\sqrt{t-\tfrac{1}{24}}) &= \exp(C\sqrt{t_0})\{1+O(\mu\sigma n^{-1/2})\}\\ &= \exp(C\sqrt{t_0})\{1+O(n^{-\tau/2})\}.\end{aligned}$$

On obtient alors

$$J_1 = \exp(C\sqrt{t_0})\{1+O(n^{-\tau/2})\}\int_{a_1}^{a_2} \varphi(n-t)\, \mathrm{d}t.$$

Mais

$$\int_{a_1}^{a_2} \varphi(n-t)\, \mathrm{d}t = 1-\mathrm{Proba}(|Y-\tfrac{1}{4}m(m-1)| \geqslant \mu\sigma)$$

et, d'après (50),

$$J_1 = \exp(C\sqrt{t_0})\{1+O(n^{-\tau/2})\}$$

et, par (51),

$$J_1 \sim \exp(C\sqrt{n})\exp\left(-\frac{Cm^2}{8\sqrt{n}}\right). \tag{52}$$

Il reste à montrer que

$$J_0 + J_2 = o(J_1). \tag{53}$$

On a

$$J_0 \leqslant \exp(C\sqrt{t_0})\int_{a_0}^{a_1} \varphi(n-t)\, \mathrm{d}t,$$

et comme

$$\int_{a_0}^{a_1} \varphi(n-t)\, \mathrm{d}t = \mathrm{Proba}(Y-\tfrac{1}{4}m(m-1) \leqslant -\mu\sigma) \leqslant 2\exp(-\tfrac{2}{5}n^{2\tau}),$$

on a bien $J_0 = o(J_1)$.

Pour évaluer J_2, on fait une intégration par parties, en posant $u = -\exp(C\sqrt{t-\frac{1}{24}})$ et $dv = -\varphi(n-t)\,dt$,

$$v(t) = \int_0^{n-t} \varphi(w)\,dw.$$

On obtient

$$J_2 = \exp(C\sqrt{t_0+\mu\sigma-\tfrac{1}{24}}) \int_0^{\frac{1}{4}m(m-1)-\mu\sigma} \varphi(w)\,dw + K,$$

avec

$$K = \int_{t_0+\mu\sigma}^{n} v(t)\frac{C}{2}\frac{\exp(C\sqrt{t-\frac{1}{24}})}{\sqrt{t-\frac{1}{24}}}\,dt.$$

Mais

$$\int_0^{\frac{1}{4}m(m-1)-\mu\sigma} \varphi(w)\,dw = \text{Proba}(Y-\tfrac{1}{4}m(m-1) \leqslant -\mu\sigma)$$

$$\leqslant 2\exp(-\tfrac{2}{9}n^{2\tau}),$$

et il rester à montrer que $K = o(J_1)$. On écrit:

$$K = K_1+K_2 = \int_{t_0+\sigma\mu}^{t_0+\sigma^2/m} + \int_{t_0+\sigma^2/m}^{n}.$$

On a

$$K_2 \leqslant \tfrac{1}{2}C\exp(C\sqrt{n})\left(n-t_0-\frac{\sigma^2}{m}\right)v\left(t_0+\frac{\sigma^2}{m}\right).$$

De plus

$$v\left(t_0+\frac{\sigma^2}{m}\right) = \text{Proba}\left(Y-\tfrac{1}{4}m(m-1) \leqslant -\frac{\sigma^2}{m}\right) \leqslant 2\exp\left(-\frac{2}{9}\frac{\sigma^2}{m^2}\right).$$

Comme

$$\frac{\sigma^2}{m^2} \sim \tfrac{1}{72}m \geqslant \tfrac{1}{72}n^{1/5} \qquad \text{et} \qquad \frac{m^2}{\sqrt{n}} = O(n^{1/6}),$$

on voit avec (52) que $K_2 = o(J_1)$.

Dans K_1, on fait le changement de variable $t = t_0+x\sigma$. On obtient, pour n assez grand,

$$K_1 \leqslant \sigma \int_{\mu}^{\sigma/m} v(t_0+x\sigma)\exp(C\sqrt{t_0+x\sigma})\,dx$$

$$\leqslant 2\sigma \int_{\mu}^{\sigma/m} 2\exp(-\tfrac{2}{9}x^2)\exp(C\sqrt{t_0})\exp\left(C\frac{x\sigma}{2\sqrt{t_0}}\right) \mathrm{d}x.$$

On observe que

$$\frac{\sigma}{\sqrt{t_0}} = O\left(\frac{m^{3/2}}{\sqrt{n}}\right)$$

et donc tend vers 0. On aura donc

$$K_1 \ll \exp(C\sqrt{t_0}) \int_{\mu}^{\infty} \exp(-\tfrac{1}{9}x) \,\mathrm{d}x \ll \exp(C\sqrt{t_0})\exp(-\tfrac{1}{9}n^{\tau}).$$

On a ainsi montré que $K_1 = o(J_1)$, doù (53).

Dans (48), le terme de reste est négligeable devant le terme principal. En effet, le logarithme du terme principal est équivalent à $C\sqrt{n}$, tandis que

$$\log\left(\frac{2^{m-1}}{2} \exp \frac{C\sqrt{n}}{2}\right) \sim \frac{C\sqrt{n}}{2}.$$

Finalement, en utilisant (48), (49), (53), (52) et (45), on a

$$\begin{aligned}
r(n,m) &\sim \frac{(m-1)!\,C^m}{2^{m+1}\pi\sqrt{2}} \exp\left(-\frac{m^2}{2C\sqrt{n}}\right) I \\
&\sim \frac{(m-1)!\,C^m}{2^{m+1}\pi\sqrt{2}} \exp\left(-\frac{m^2}{2C\sqrt{n}}\right) n^{-\frac{1}{2}(m+1)} J \\
&\sim \frac{(m-1)!\,C^m}{2^{m+1}\pi\sqrt{2}} \exp\left(-\frac{m^2}{2C\sqrt{n}}\right) n^{-\frac{1}{2}(m+1)} J_1 \\
&\sim \frac{(m-1)!\,C^m}{2^{m+1}\pi\sqrt{2}} \exp\left(-\frac{m^2}{2C\sqrt{n}}\right) n^{-\frac{1}{2}(m+1)} \exp(C\sqrt{n})\exp\left(-\frac{Cm^2}{8\sqrt{n}}\right) \\
&= \frac{1}{4\sqrt{3}n} \exp(C\sqrt{n})(m-1)!\left(\frac{C}{2\sqrt{n}}\right)^{m-1} \exp\left\{-\left(\tfrac{1}{8}C + \frac{1}{2C}\right)\frac{m^2}{\sqrt{n}}\right\} \\
&\sim p(n)(m-1)!\left(\frac{C}{2\sqrt{n}}\right)^{m-1} \exp\left\{-\left(\tfrac{1}{8}C + \frac{1}{2C}\right)\frac{m^2}{\sqrt{n}}\right\}. \qquad \square
\end{aligned}$$

3.5 Remarque On reconnaît dans le théorème $-(\frac{1}{8}C + 1/2C)$ qui est le coefficient de λ^2 dans le développement de $g(\lambda)$. On espérait que, en posant $\lambda = m/\sqrt{n}$ ou $\lambda = (m-1)/\sqrt{n}$, on aurait obtenu une formule du genre

$$\log r(n,m) - g(\lambda)\sqrt{n} \to 0$$

ou

$$\log r(n,m) - \frac{g(\lambda)}{C} \log p(n) \to 0.$$

Jusqu'à présent, nous n'avons pas trouvé une telle relation.

4 Tables numériques

On construit des procédures permettant calculer les différentes fonctions intervenant dans la partie théorique.

4.1 Calcul de $F(x)$. Pour $x > 1$, on utilise la formule:

$$F(x) = \tfrac{1}{6}\pi^2 - \sum_{n \geqslant 1} \frac{nx+1}{n^2} \mathrm{e}^{-nx}.$$

Pour $x \leqslant 1$, on utilise le développement en série entière

$$F(x) = x - \tfrac{1}{4}x^2 + \tfrac{1}{36}x^3 - \tfrac{1}{3600}x^5 + \cdots.$$

4.2 Calcul de $H(y)$. $x = H(y)$ est solution de l'équation

$$x^2 = y^2 F(x).$$

On part de $x_0 = y$, puis par la méthode de Newton, on calcule

$$x_{n+1} = x_n - \frac{x_n^2 - y^2 F(x_n)}{2x_n - \dfrac{y^2 x_n}{\mathrm{e}^{x_n} - 1}}.$$

4.3 Calcul de $f(\lambda)$. On a

$$f(\lambda) = \frac{2H(\lambda)}{\lambda} - \lambda \log(1 - \mathrm{e}^{-H(\lambda)}).$$

4.4 Calcul de $M(\lambda)$. On résoud l'équation

$$\Phi(x) = \exp\{-J(x,\lambda)\} + \exp\left\{-\frac{\lambda^2}{x} J(x,\lambda)\right\} - 1 = 0,$$

où

$$J(x,\lambda) = H\left(\frac{x}{\lambda\sqrt{1-x}}\right).$$

Là encore on utilise la méthode de Newton. Comme valeur de départ x_0 on choisit

$$x_0 = \begin{cases} -0.78\lambda \log 0.78\lambda & \text{si } \lambda \leqslant 0.25, \\ 0.4 & \text{si } 0.25 < \lambda < 3, \\ 1 - \dfrac{1}{2\log\lambda} & \text{si } \lambda \geqslant 3. \end{cases}$$

Puis on construit la suite

$$x_{n+1} = x_n - \frac{\Phi(x_n)}{\Phi'(x_n)},$$

qui tend vers $M(\lambda)$. Le calcul de $\Phi'(x)$ n'est pas très simple: on utilise (19).

4.5 Calcul de $g(\lambda)$. On a

$$g(\lambda) = \lambda \frac{2-M(\lambda)}{M(\lambda)} J(M(\lambda), \lambda).$$

λ	$f(\lambda)$	$g(\lambda)$	λ	$f(\lambda)$	$g(\lambda)$
0.01	0.112 10	2.511 48	2	2.490 79	1.064 95
0.02	0.196 48	2.471 63	3	2.546 89	0.876 90
0.03	0.270 40	2.436 91	4	2.560 31	0.753 78
0.04	0.337 53	2.405 48	5	2.563 80	0.665 55
0.05	0.399 60	2.376 48	6	2.564 74	0.598 58
0.06	0.457 66	2.349 40	7	2.565 00	0.545 67
0.07	0.512 38	2.323 89	8	2.565 07	0.502 62
0.08	0.564 24	2.299 72	9	2.565 09	0.466 79
0.09	0.613 61	2.276 71	10	2.565 10	0.436 41
0.1	0.660 77	2.254 71	20		0.273 72
0.2	1.045 76	2.073 45	30		0.204 95
0.3	1.329 03	1.935 73	40		0.165 89
0.4	1.548 59	1.823 81	50		0.140 35
0.5	1.723 07	1.729 51	60		0.122 21
0.6	1.863 79	1.648 18	70		0.108 57
0.7	1.978 38	1.576 85	80		0.097 91
0.8	2.072 36	1.513 49	90		0.089 32
0.9	2.149 88	1.456 63	100		0.082 24
1.0	2.214 12	1.405 19	200		0.047 28
1.1	2.267 58	1.358 31	300		0.033 95
1.2	2.312 23	1.315 35	400		0.026 76
1.3	2.349 65	1.275 77	500		0.022 22
1.4	2.381 11	1.239 14	600		0.019 07
1.5	2.407 63	1.205 11	700		0.016 75
1.6	2.430 07	1.173 37	800		0.014 96
1.7	2.449 09	1.143 69	900		0.013 54
1.8	2.465 25	1.115 84	1000		0.012 38
1.9	2.479 03	1.089 64			

Bibliographie

[1] N. Bourbaki, *Fonctions d'une variable réelle*, Hermann, Paris, 1976

[2] M. Chellali, Sur les zéros des polynômes de Bessel, III, *CRAS*, **307**, (I) (1988), 651–4, et Thèse de l'Université de Grenoble, 1989

[3] J. Dixmier & J.-L. Nicolas, Partitions without small parts, *Proceedings of the Number Theory Conference of Budapest, 1987* (to appear)

[4] P. Erdős, J.-L. Nicolas & A. Sárközy, On the number of partitions of *n* without a given subsum, I, *Discrete Math.*, **75** (1989), 155–66

[5] P. Erdős, J.-L. Nicolas & A. Sárközy, On the number of partitions of *n* without a given subsum, II (to appear in *Number Theory at Allerton Park* (eds. B. Berndt, H. Diamond, H. Halberstam & A. Hildebrand), Birkhaüser, 1990)

[6] G. A. Freiman, On extremal additive problems of Paul Erdős (to appear)

[7] A. Odlyzko, Differences of the partition function, *Acta Arithmetica*, **49** (1988), 237–54

[8] A. Rényi, *Calcul des probabilités*, Dunod, Paris, 1966

[9] G. Szekeres, An asymptotic formula in the theory of partitions, *Quart. J. Math. Oxford*, **2** (1951), 85–108

[10] G. Szekeres, Some asymptotic formulae in the theory of partitions, II, *Quart. J. Math. Oxford*, **4** (1953), 96–111

A compact sequential space

Alan Dow*

Abstract

We give an Ostaszewski-type inductive construction of a locally countable locally compact space which is not α-realcompact but whose one-point compactification is sequential. This answers a question of Nyikos. The essential ingredient is the use of the Balcar–Vojtas almost-disjoint refinement technique to guide the induction through continuum-many steps.

1 Introduction

A subset Y of a space X is *sequentially closed* if no sequence which is a subset of Y converges to a point outside of Y. A space is *sequential* if each sequentially closed subset is closed. There are not many absolute examples of 'complicated' compact sequential spaces in the literature. Furthermore, several important recent results of Balogh, Fremlin and Nyikos, which use Todorčević's 'forcing positive partition relations' techniques, show that such spaces cannot be too complicated. For example, they must contain points of first countability and no subspace can be mapped by a closed map onto ω_1. The technique, roughly speaking, is to take a countably complete maximal filter of closed sets of a subspace and diagonalize through it with an ω_1 sequence that is homogeneous with respect to a certain partition. The homogeneity with respect to the partition guarantees that the sequence ends up being a free sequence in the sense of Arhangel'skii (see [1] or [6]). The upshot is that there cannot be too many countably complete maximal filters on subspaces.

* Research supported by NSERC of Canada.

A space is said to be *α-realcompact* if every countably complete maximal closed filter is fixed. Gardner & Pfeffer [5, 8.2] show that each Radon space is hereditarily α-realcompact and Nyikos subsequently showed that each compact bisequential space is hereditarily α-realcompact. Nyikos asks [10, C54—problem section] if every compact sequential space is hereditarily α-realcompact. The example in this paper will not be α-realcompact and its one-point compactification will be sequential. Let us point out the following result of Fremlin and Nyikos which adds interest to the question. PFA is the proper forcing axiom (see [4]).

Proposition 1 [PFA] *A space which is not α-realcompact contains a free ω_1-sequence.*

I conjecture that, in addition, compact sequential spaces of cardinality $\aleph_1$ should be hereditarily α-realcompact under PFA. Let us note another property which compact sequential spaces which are not hereditarily α-realcompact must have. First a trivial lemma. A point is said to be a complete accumulation point of a set if every neighbourhood of the point hits the set in full size.

Lemma 2 *If a compact space X of character $\aleph_1$ is not hereditarily α-realcompact then there is a set $\{x_\alpha : \alpha < \omega_1\}$ with a unique complete accumulation point.*

Proof Suppose X is a compact space of character $\aleph_1$ and $Y \subset X$ is not α-realcompact. Choose $\mathcal{F}$, a free countably complete maximal filter of closed subsets of Y and let x be the unique point in $\bigcap\{\bar{F} : F \in \mathcal{F}\}$. Clearly, $\{N_\alpha : \alpha < \omega_1\} \subset \mathcal{F}$, where $\{N_\alpha : \alpha < \omega_1\}$ is any neighbourhood base for x. So choose $x_\alpha \in \bigcap\{Y \cap N_\beta : \beta < \alpha\}$. Now if U is any neighbourhood of x, $\{\beta \in \omega_1 : x_\beta \notin U\}$ is countable. □

Lemma 3 [MA(ω_1)] *If X is a compact sequential space of character at most $\aleph_1$ and x is the unique complete accumulation point of $\{x_\alpha : \alpha < \omega_1\}$, then there exists an $I \in [\omega_1]^{\omega_1}$ such that $\{x\} \cup \{x_\alpha : \alpha \in I\}$ is compact.*

Proof Identify $\{x_\alpha : \alpha \in \omega_1\}$ with ω_1. Define a poset P so that $p \in P$ iff $p = (F_p, H_p) \in [\omega_1]^{<\omega} \times [\omega_1]^{<\omega}$ and fix a neighbourhood base $\{U_\alpha : \alpha \in \omega_1\}$ for x. Define $p < q$ iff $F_p - F_q \subset \bigcap_{\beta \in H_q} U_\beta$. Let us check that P is c.c.c. Indeed, let $\{(F_\alpha, H_\alpha) : \alpha \in \omega_1\}$ be a subset of P such that $\{F_\alpha : \alpha \in \omega_1\}$ and $\{H_\alpha : \alpha \in \omega_1\}$ form Δ-systems with roots F and H, respectively, and such that $|F_\alpha - F| = n$ for all $\alpha \in \omega_1$. Furthermore we may assume that for each $\alpha < \beta < \omega_1$ we have that

$F_\beta - F \subset W_\alpha = \bigcap_{\gamma \in H_\alpha} U_\gamma$ since, for each α there is a γ such that $(\gamma, \omega_1) \subset W_\alpha$. Therefore we need only find an $\alpha < \beta$ so that $F_\alpha - F \subset W_\beta$. But now consider $F_\alpha - F$ as an element of X^n and observe that $\vec{x} = (x, \ldots, x) \in X^n \in \overline{\{F_\alpha - F : \alpha \in \omega_1\}}$. By [7], X^n has countable tightness since it is compact. Therefore there is a $\lambda < \omega_1$ such that $\vec{x} \in \overline{\{F_\alpha - F : \alpha \in \lambda\}}$. Choose any $\beta > \lambda$ and find an $\alpha < \lambda$ such that $F_\alpha - F \in W_\beta \times \cdots \times W_\beta$. Therefore (F_α, H_α) and (F_β, H_β) are compatible which shows that P is c.c.c. Furthermore, for each $\alpha \in \omega_1$, $D_\alpha = \{(F, H) : \alpha \in H \text{ and } F - \alpha \neq \emptyset\}$ is obviously a dense subset of P. Let $G \subset P$ be generic over $\{D_\alpha : \alpha \in \omega_1\}$ and let

$$I = \bigcup \{F : (\exists\, p \in G)\ F \subset F_p\}.$$

The usual density argument shows that, for each $\alpha \in \omega_1$, $I \setminus U_\alpha$ is finite. □

Proposition 4 [PFA] *If $X = \{x\} \cup \{x_\alpha : \alpha < \omega_1\}$ is a sequential space with character at most $\aleph_1$ and x is a limit point of every uncountable set, then $\{x\} \cup \{x_\alpha : \alpha \in I\}$ is compact for some uncountable $I \subset \omega_1$.*

Proof Todorčević indicates in [9] that [8, Theorem 6] can be modified so as to show that X^2 has countable tightness under PFA. By induction, it follows that X^n has countable tightness. The proof then proceeds exactly as in Lemma 3. □

Remark There is no example in the literature (to my knowledge) to show that the character restriction in Lemma 3 is necessary. The example in this paper will contain an uncountable discrete set with a unique complete accumulation point and for which the conclusion of Lemma 3 fails. Indeed this was the author's original goal in constructing this example.

2 The compact sequential space

Let $X = \omega_1 \times \mathfrak{c}$ and let π be the first coordinate projection of X onto ω_1. We shall inductively construct a locally countable, locally compact topology on X so that any filter of closed sets extending $\{\pi^{-1}(\alpha, \omega_1) : \alpha < \omega_1\}$ will be countably complete and such that the one-point compactification, X^*, will be sequential. The idea is simple. Let S be a stationary–costationary set of countable limit ordinals. We shall ensure that for any limit $\lambda \in \omega_1$ and any subset Y of X such that $\pi[Y]$ is a cofinal subset of λ, λ will be in $\pi[\overline{Y}]$ if and only if $\lambda \in S$. This is like saying that π is 'S-closed'. Recall that if π were 'C-closed' for some cub C, then PFA would imply that X contained a copy of the ordinal space ω_1 and so X^* would not be sequential.

For each $\lambda \in S$, choose a set $\{\alpha_{\lambda,n} : n \in \omega\}$ which is strictly increasing and cofinal in λ. Next, by [2], we may choose a dense tree $T \subset \mathcal{P}(\omega)$ (mod finite), ordered by reverse inclusion, which has height at most $\mathfrak{b}$. Recall that $\mathfrak{b}$ is the least cardinal of a family of functions from ω to ω which cannot be dominated mod finite. When we say that $t \in T$ for some $t \in [\omega]^\omega$, we mean of course that the equivalence class of t is a member of T.

By induction on $\alpha \in \omega_1$, we shall define a topology, τ_α, on $\alpha \times \mathfrak{c}$ so that

(a) for each $\beta < \alpha$, $\tau_\beta \subset \tau_\alpha$;
(b) τ_β equals the subspace topology induced on $\beta \times \mathfrak{c}$ by τ_α;
(c) for each $\beta < \alpha$, $\{\beta\} \times \mathfrak{c}$ is closed and discrete;
(d) τ_α is locally compact, locally countable, 0-dimensional and regular.

For α a limit let $\tau_\alpha = \bigcup_{\beta<\alpha} \tau_\beta$ and for $\alpha = \lambda+1$ with $\lambda \notin S$ let τ_α be the topology generated by $\tau_\lambda \cup \{(\lambda,\xi) : \xi < \mathfrak{c}\}$. For the remaining case, namely $\alpha = \lambda+1$ with $\lambda \in S$, we perform another induction of length $\mathfrak{c}$. Recall that we have fixed $\{\alpha_{\lambda,n} : n \in \omega\}$ cofinal in λ: so let $\alpha_n = \alpha_{\lambda,n}$ for $n \in \omega$. Let $\{X_\xi^\lambda : \xi < \mathfrak{c}\}$ list all countable subsets of $\lambda \times \mathfrak{c}$ such that $\pi[X_\xi^\lambda]$ is cofinal in λ for each $\xi \in \mathfrak{c}$. When the context is clear we shall drop the superscript in X_ξ^λ. We shall inductively define neighbourhood bases $\{W_n^\xi : n \in \omega\}$ for the points of the form (λ,ξ) so as to ensure that each X_ξ picks up a limit point in the set $\{\lambda\} \times \xi+1$. To this end let τ_λ^ξ denote the topology on $\lambda \times \xi$ generated by

$$\tau_\lambda \cup \bigcup \{W_n^\eta : n \in \omega,\, \eta < \xi\}.$$

Our inductive hypotheses for this construction are:

(e) for all $\eta < \xi$ there exist a $t_\eta \in T$ and a sequence $\{U_n^\eta : n \in t_\eta\} \subset \tau_\lambda$ of compact clopen countable sets such that

$$W_0^\eta = \{(\lambda,\xi)\} \cup \bigcup_{n \in t_\eta} U_n^\eta \quad \text{and} \quad U_n^\eta \subset (\alpha_n, \alpha_{n+1}] \times \mathfrak{c};$$

(f) t_ξ does not contain any t_η for $\eta < \xi$;
(g) for each $\eta < \xi$, $\tau_\lambda^{\eta+1}$ is a locally compact, locally countable, 0-dimensional Hausdorff topology .

So, by induction, we may assume that τ_λ^ξ is a locally compact, locally countable Hausdorff topology. Let us now choose t_ξ etc. If for some $\eta < \xi$ we already have that $(\lambda,\eta) \in \overline{X_\xi}$ then (λ,ξ) can be isolated. Otherwise, let $I_\xi = \{n : X_\xi \cap (\alpha_n, \alpha_{n+1}] \times \mathfrak{c} \neq \emptyset\}$. One of the key ideas to the proof in [3] is that we may now choose $t_\xi \in T-\{t_\eta : \eta < \xi\}$ such that $t_\xi \subset I_\xi$ and such that induction hypothesis (f) holds. The proof is simple. Choose any $\tilde{t} \in T$ such that $\tilde{t} \subset I_\xi$, which we may do since T is dense in $\mathcal{P}(\omega)$ (mod finite). Next note that since there are antichains of size $\mathfrak{c}$ below $\tilde{t}$ there is some member t_ξ of such an antichain which does not contain any member of $\{t_\eta : \eta < \xi\}$.

Now let $\{y_n^\xi : n \in t_\xi\}$ be chosen so that $y_n^\xi \in X_\xi \cap (\alpha_n, \alpha_{n+1}] \times \mathfrak{c}$ for each $n \in t_\xi$. Note that $\{y_n^\xi : n \in t_\xi\}$ does not have any limit points in τ_λ^ξ since it is a subset of X_ξ. For each $n \in t_\xi$ let $\{U(n,m) : m \in \omega\}$ be a neighbourhood base for y_n^ξ consisting of compact open countable subsets of $(\alpha_n, \alpha_{n+1}] \times \mathfrak{c}$. We claim that

(h) $\{\eta < \xi : t_\eta \,{}^*{\supset}\, t_\xi\} \supset \{\eta < \xi : |\{n \in t_\xi : W_0^\eta \cap U(n,0) \neq \emptyset\}| = \omega\}$;

(i) for all $\eta < \xi$ there exist m_η and $f_\eta \in {}^{t_\xi}\omega$ so that

$$W_0^\eta \cap U(n, f_\eta(n)) = \emptyset$$

for all $n \in t_\xi \backslash m_\eta$.

Indeed, for (h), note that

$$W_0^\eta \subset \bigcup_{n \in t_\eta} (\alpha_n, \alpha_{n+1}] \times \mathfrak{c} \cup \{\lambda\} \times \mathfrak{c}.$$

Also, for (i), if $t_\eta \cap t_\xi$ is finite then choose m_η so that $t_\eta \cap t_\xi \subset m_\eta$ and simply choose f_η to be identically 0. Now, if $t_\eta \cap t_\xi$ is infinite, first choose m_η so that $W_0^\eta \cap \{y_{n_\eta} : n \geqslant m_\eta\} = \emptyset$. Then for $n \in t_\xi \backslash m_\eta$, choose $f_\eta(n)$ so that $U(n, f_\eta(n)) \cap W_n^\eta = \emptyset$.

Now choose an $f \in {}^{t_\xi}\omega$ so that

$$\forall\, \eta < \xi\ \exists\, k_\eta \geqslant m_\eta \quad \text{such that} \quad f_\eta(n) < f(n) \text{ for all } n \in t_\xi - k_\eta.$$

The reason that we can do this is that

$$\{\eta < \xi : \exists\, n \text{ such that } f_\eta(n) > 0\} \subset \{\eta < \xi : t_\eta \,{}^*{\supset}\, t_\xi\}$$

and these sets have cardinality less than the height of T.

Now we define, for $n \in \omega$,

$$W_n^\xi = \{(\lambda, \xi)\} \cup \bigcup \{U(m, f(m)) : m \in t_\xi \backslash n\}.$$

We obtain that

$$W_{k_\eta}^\xi \cap W_0^\eta = \emptyset \qquad \text{for all } \eta < \xi.$$

It should be clear that inductive hypotheses (e)–(g) are satisfied.

It is similarly clear that if

$$\tau_{\lambda+1} = \bigcup_{\xi < \mathfrak{c}} \tau_\lambda^\xi$$

then inductive hypotheses (a)–(d) are satisfied. Therefore we may let

$$\tau_{\omega_1} = \bigcup_{\lambda\in\omega_1} \tau_\lambda$$

and we obtain that $X = \langle \omega_1 \times \mathfrak{c}, \tau_{\omega_1}\rangle$ is locally compact, locally countable, 0-dimensional and Hausdorff. Let X^* denote the one-point compactification of X.

As we indicated before the construction, the main idea is the following.

Fact 1 *For each* $A \subset X$

$$\pi[A'] = S \cap [\pi[A]]' = \{\lambda \in S : \pi[A] \cap \lambda \textit{ is cofinal in } \lambda\},$$

where $A' = \{x \in X\,;\, x \in \overline{A\backslash\{x\}}\}$.

Proof of Fact 1 Let A be a subset of X and suppose first that $\lambda \in [\pi[A]]' \cap S$. Since λ is a limit point of $\pi[A]$ and $\lambda \in S$, there is some $\xi < \mathfrak{c}$ such that $X_\xi^\lambda \subset A$. Our construction ensured that X_ξ^λ has a limit point in $\{\lambda\}\times\xi+1$. Therefore $\lambda \in \pi[A']$. Now suppose that $\lambda \in \pi[A']$ and choose $\xi < \mathfrak{c}$ so that $(\lambda, \xi) \in A'$. Clearly $\lambda \in S$ since these are the only non-isolated points. Furthermore, for each $n \in \omega$, the neighbourhood which we denoted W_n^ξ of (λ, ξ) is contained in $(\alpha_n, \lambda+1]\times\mathfrak{c}$: hence $\pi[A]$ is cofinal in λ. □

Fact 2 X^* *is sequential.*

Proof of Fact 2 It suffices to show that every closed non-compact subset of X contains an infinite closed discrete subset. Assume that $A \subset X$ is a closed non-compact subset. If $\pi[A]$ is uncountable then we may choose $\lambda \in [\pi[A]]' - S \neq \emptyset$, since S is costationary. Choose any subset Y of A such that π is one-to-one on Y and $\pi[Y]$ is an ω-sequence converging to λ. By Fact 1 and the fact that π is continuous, it follows that Y is closed and discrete in X. Now assume that $\lambda < \omega_1$ is minimal such that $A \cap (\lambda+1 \times \mathfrak{c})$ is not compact. If $\lambda \notin S$ then just as in the previous case it is clear that A contains an infinite set which is closed and discrete in X. Finally let us assume that $\lambda \in S$. It suffices to show that $A' \cap \pi^{\leftarrow}(\lambda)$ is infinite since $\pi^{\leftarrow}(\lambda)$ is a closed discrete subset of X. To see that it is, suppose there is a compact open set $U \supset A' \cap \pi^{\leftarrow}(\lambda)$ and note that by the minimality of λ, $(A \cap \pi^{\leftarrow}[\lambda+1]) - U$ is still cofinal in λ. Therefore, by Fact 1, A has other limit points outside of U in $\pi^{\leftarrow}(\lambda)$. □

Finally, to see that X is not α-realcompact we prove the following.

Fact 3 *If* $\mathcal{F}$ *is any filter of closed sets such that* $\pi^{\leftarrow}(\lambda, \omega_1) \in \mathcal{F}$ *for all* $\lambda \in \omega_1$ *then* $\mathcal{F}$ *has the countable-intersection property.*

Proof of Fact 3 Let $\{F_n : n \in \omega\} \subset \mathcal{F}$ and assume that $F_{n+1} \subset F_n$ for each $n \in \omega$. Let $\lambda \in S \cap \bigcap_{n\in\omega} [\pi[F_n]]'$ and choose $x_n \in F_n$ so that $\{\pi(x_n) : n \in \omega\}$ is strictly increasing and cofinal in λ. By Fact 1, $\{x_n : n \in \omega\}$ has a limit point. Clearly this limit point is in $\bigcap_{n\in\omega} F_n$. □

Remark Besides the conjecture immediately following Proposition 1, one other obvious question remains. Is there a compact sequential space X such that

(a) X contains an uncountable set with a unique complete accumulation point, and yet

(b) no uncountable subset of X has a unique accumulation point?

References

[1] A. Arhangel'skii, On bicompacta hereditarily satisfying Suslin's condition. Tightness and free sequences, *Soviet Math. Dokl.* **12** (1971), 1253–7

[2] B. Balcar, J. Pelant & P.Simon, The Space of Ultrafilters on N covered by nowhere dense sets, *Fundamenta Mathematica*, **110** (1980), 11–24

[3] B. Balcar & P. Vojtas, Almost disjoint refinements of subsets of N, *Proc. AMS.*, **79** (1980), 465–70

[4] J. Baumgartner, Applications of the proper forcing axiom, in *Handbook of Set-Theoretic Topology* (eds. K. Kunen & J. E. Vaughn) North-Holland, Amsterdam, (1984), 913–60

[5] R. J. Gardner & W. F. Pfeffer, Borel measures, in *Handbook of Set-Theoretic Topology* (eds. K. Kunen & J. E. Vaughn) North-Holland, Amsterdam, (1984), 961–1043

[6] I. Juhász, *Cardinal Functions in Topology—Ten Years Later*, Mathematical Centre Tracts, **123**, 1980

[7] V. Malyhin, On tightness and Suslin number in exp$X$ and in a product of spaces, *Soviet Math. Dokl.*, **13** (1972), 496–9

[8] S. Todorčević, Forcing positive partition relations, *Transactions Amer. Math. Soc.*, **280** (1983), 703–20

[9] S. Todorčević, *Partition Problems in Topology*, Contemporary Mathematics **84**, American Mathematical Society, 1989

[10] P. Nyikos, Problems, *Topology Proceedings*, **11**, 2 (1986), 415–26

The critical parameter for connectedness of some random graphs

R. Durrett and H. Kesten

Abstract

We consider $P\{\mathcal{G}$ is connected$\}$ when $\mathcal{G}$ is a graph with vertex set $\mathcal{V} = \{1, 2, \ldots\}$ and the edge between i and j present with probability $\lambda p(i, j)$. For certain functions $p(i, j)$, homogeneous of degree -1, we extend a calculation of Shepp. This identifies the critical value λ_c of λ, such that $P\{\mathcal{G}$ is connected$\} = 0$ or 1 according to $\lambda < \lambda_c$ or $\lambda > \lambda_c$. In many cases we can even show that this probability is 0 when $\lambda = \lambda_c$.

1 Introduction

In 1959 Erdős & Rényi wrote two seminal papers on random graphs [3], [4]. Since then the subject has grown enormously, as witnessed by the recent books, surveys and conference proceedings [2], [13], [5], [9], [10] (see also the references cited therein). In general the setup is as follows: One fixes a vertex set $\mathcal{V}$ and constructs a random graph $\mathcal{G}$ by choosing which of the pairs v_1, v_2 (with $v_i \in \mathcal{V}$) are connected by an edge. Erdős & Rényi took all subgraphs of the complete graph on $\mathcal{V}$ (for a finite $\mathcal{V}$) with a fixed number N of edges equally likely (some variants of this model are discussed in [2] and [13]). They investigated $P\{\mathcal{G}$ has property $A\}$ for various choices of A. Of particular interest was how this probability behaves as the cardinality of $\mathcal{V}$, $|\mathcal{V}|$, and N tend to infinity in a suitable way. Examples of properties A considered in Erdős & Rényi are $\{\mathcal{G}$ contains a specified subgraph$\}$, $\{\mathcal{G}$ contains j subgraphs of size $k\}$, $\{\mathcal{G}$ is planar$\}$, $\{\mathcal{G}$ has exactly k components$\}$ and, for $k = 1$, $\{\mathcal{G}$ is connected$\}$. In many cases there is a threshold $f(|\mathcal{V}|)$ such that, as $|\mathcal{V}| \to \infty$, $P\{\mathcal{G}$ has property $A\}$ tends to 0 or 1, according as $\lim N/|f(|\mathcal{V}|)$ is less than or greater than 1.

In this note we also consider a threshold, or critical phenomenon for $P\{\mathcal{G}$ is connected$\}$. Our model will be somewhat different from the Erdős–Rényi one, in that we take $\mathcal{V}$ infinite and choose all edges independently. In fact we always take $\mathcal{V} = \mathbb{N} = \{1,2,\dots\}$. The probability that the edge between i and j is included in $\mathcal{G}$ is denoted by $p(i,j)$. For the time being we consider only undirected graphs, so that we have to take $p(i,j) = p(j,i)$ and only make one choice for each pair $\{i,j\}$. (Some directed graphs will be considered in Section 3.) We consider the situation with

$$p(i,j) = \min\{\lambda h(i,j), 1\} \tag{1.1}$$

for some fixed function h. The choice of $p(i,i)$ has no influence on the probability of connectedness of $\mathcal{G}$, but it simplifies the formulation of our results to assume (1.1) also for $j = i$, even if $h(i,i) > 0$ (thus allowing loops in $\mathcal{G}$). In the case (1.1) we use the notation

$$C(h,\lambda) = P\{\mathcal{G} \text{ is connected}\}$$

to bring out the dependence on h and λ. Clearly $C(h,\lambda)$ is increasing in λ and, in our case, there will be a critical value, λ_c, of λ such that $C(h,\lambda) = 0$ for $\lambda < \lambda_c$ and $C(h,\lambda) = 1$ for $\lambda > \lambda_c$. Our main result is the following theorem which identifies λ_c in some specific cases.

Theorem *Assume that (1.1) holds for some function* $h\colon [0,\infty)^2\backslash\{0\} \to (0,\infty)$ *which satisfies*

$$h(x,y) = h(y,x) \qquad (\textit{symmetry}), \tag{1.2}$$

h is homogeneous of degree -1, i.e.,

$$h(tx,ty) = t^{-1}h(x,y) \qquad (t > 0), \tag{1.3}$$

and

$$h \text{ is continuous and strictly positive on } \{(x,y) : x+y = 1\}. \tag{1.4}$$

Then the critical value for λ is

$$\lambda_c = \left(\int_0^1 \frac{h(u,1-u)}{\sqrt{u(1-u)}}\, du\right)^{-1} = \left(\int_0^\infty \frac{h(1,y)}{\sqrt{y}}\, dy\right)^{-1} =: \theta^{-1} \tag{1.5}$$

in the sense that

$$C(h,\lambda) = \begin{cases} 0 & \text{if } \lambda < \lambda_c, \\ 1 & \text{if } \lambda < \lambda_c. \end{cases} \tag{1.6}$$

If, in addition to (1.1)–(1.4), h satisfies

$$h \text{ is decreasing in each of its arguments,} \tag{1.7}$$

then also

$$C(h,\lambda) = 0 \qquad \text{if } \lambda = \lambda_c. \tag{1.8}$$

Examples (a) If, for some $C \geqslant 0$,

$$p(i,j) = \lambda(i+j+C|i-j|)^{-1} = \lambda\{(1+C)\max(i,j)+(1-C)\min(i,j)\}^{-1},$$

then

$$\lambda_c = \lambda_c(C) = \frac{1}{2}\left(\int_0^{\frac{1}{2}} \frac{1}{\sqrt{u(1-u)}}(1+C-2Cu)^{-1}\,du\right)^{-1}.$$

In this example (1.7), and hence (1.8), will hold as long as $0 \leqslant C \leqslant 1$. For $C = 1$ we have

$$p(i,j) = \lambda[2\max(i,j)]^{-1}, \tag{1.9}$$

which (except for the factor 2) is the special case treated by Shepp [15]. With our normalization we obtain $\lambda_c = \frac{1}{2}$ for this case, which is equivalent to Shepp's result.

(b) If for some $p > 0$

$$p(i,j) = \lambda(i^p + j^p)^{-1/p}, \tag{1.10}$$

then

$$\lambda_c = \lambda_c(p) = p\Gamma\left(\frac{1}{p}\right)\left[\Gamma\left(\frac{1}{2p}\right)\right]^{-2}.$$

Note that as $p \to \infty$, $\lambda_c(p) \to \frac{1}{4}$, since $\Gamma(z) = \pi[(\sin \pi z)\Gamma(1-z)]^{-1} \sim 1/z$ as $z \to 0$, [17, Section 12.14]. Thus we again recover Shepp's case (1.9) as a limiting case of (1.10).

Some possible extensions of the theorem will be discussed in Section 3.

We conclude this introduction by listing some of the previous work on this problem. Grimmett, Keane and Marstrand [6] considered the homogeneous case where $\mathscr{V} = \mathbb{Z}^d$ and

$$P\{\mathscr{G} \text{ contains an edge from } v_1 \text{ to } v_2\} = p(v_1,v_2) = p(v_2,v_1) = f(v_1 - v_2).$$

For this case they showed (under some obvious irreducibility conditions) that

$$P\{\mathscr{G} \text{ is connected}\} = \begin{cases} 0 & \text{if } \sum_{w\in\mathbb{Z}^d} f(w) < \infty, \\ 1 & \text{if } \sum_{w\in\mathbb{Z}^d} f(w) = \infty. \end{cases} \tag{1.11}$$

An alternative proof of (1.11) was given in [1]. Kalikow & Weiss [11] proved (1.11) when $\mathbb{Z}^d$ is replaced by $\mathbb{N}$ (i.e., 'half' of $\mathbb{Z}$). Kalikow & Weiss also proved a general zero–one law for any countable $\mathcal{V}$: if $p(v_1, v_2)$ satisfies

$$\sum_{\substack{v_1 \in A \\ v_2 \in A^c}} p(v_1, v_2) = \infty \tag{1.12}$$

for any subset A of $\mathcal{V}$ with $A \neq \emptyset$ and $A^c = \mathcal{V} \setminus A \neq \emptyset$, then

$$P\{\mathcal{G} \text{ is connected}\} \in \{0, 1\}. \tag{1.13}$$

The inhomogeneous example with $\mathcal{V} = \mathbb{N}$ and p given by (1.9) seems to have been suggested first by Dubins in 1984, who conjectured $C(h, 1) = 1$ for this case. [11] investigated this example and showed $C(h, \lambda) = 0$ if $\lambda < \frac{1}{4}$ and $C(h, \lambda) = 1$ for $\lambda > 1$. Shepp improved the last result to $C(h, \lambda) = 0$ for $\lambda \leqslant \frac{1}{4}$ and $C(h, \lambda) = 1$ if $\lambda > \frac{1}{4}$. The $\frac{1}{4}$ in the lower bound comes from a combinatorial argument, the one in the upper bound from an eigenvalue. Because of the special nature of the combinatorial argument, Shepp could not decide what happened for simple variants such as (1.10).

Shepp's upper bound comes from using a supercritical branching process to get a lower bound on the set of vertices that are connected to each other. The reader will see below that this part of the argument generalizes easily to the p's of our theorem. Our main contribution is the simple observation that certain results of Vere-Jones [16] imply that Shepp's upper bound is the critical value for λ. The lower bound comes from showing that the expected number of self-avoiding paths between two points in $\mathcal{G}$ is finite. We find it surprising that this works up to, and sometimes at, the critical value.

2 Proofs

The first step is to relate the right-hand side of (1.5) to the largest eigenvalue of submatrices of $(h(i, j))_{i, j \geqslant 1}$ and to give some properties of these eigenvalues. We write $H_{K,M}$ for the submatrix

$$(h(i, j))_{K \leqslant i, j < M}.$$

If S is a positive finite square matrix we set

$$\rho(S) = \text{Perron–Frobenius eigenvalue of } S.$$

$\rho(S) \geqslant 0$ and $\rho(S)$ is at least as large as the absolute value of any other eigenvalue of S (see for instance [14, Chapter 1] or [8, Appendix]).

Lemma 1 *If (1.2)–(1.4) hold, then*

$$\lim_{K\to\infty}\lim_{M\to\infty}\rho(H_{K,M}) = \int_0^1 \frac{h(u,1-u)}{\sqrt{u(1-u}}\,du = \int_0^1 \frac{h(u,y)}{\sqrt{y}}\,dy. \tag{2.1}$$

Moreover, for any $\epsilon > 0$ we can find an N such that

$$\liminf_{K\to\infty}\rho(H_{K,NK}) \geqslant \int_0^1 \frac{h(u,1-u)}{\sqrt{u(1-u}}\,du - \epsilon. \tag{2.2}$$

If in addition h satisfies (1.7), then

$$\lim_{M\to\infty}\rho(H_{1,M}) = \int_0^1 \frac{h(u,1-u)}{\sqrt{u(1-u}}\,du. \tag{2.3}$$

Remark It seems quite possible that (2.3) holds even without (1.7). If one can prove this, then one can prove (1.8) also without (1.7).

Proof It is standard that the Perron–Frobenius eigenvalue, $\rho(S)$, of a matrix S is an increasing function of each of the entries of S [14, formula (1.1)] or [8, Appendix, Corollary 2.3]. Since, for $K \leqslant K' < M' \leqslant M$, $H_{K',M'}$ can be viewed as obtained from $H_{K,M}$ by setting all the $h(i,j)$ with $K \leqslant i < K'$ or $M' \leqslant j < M$ equal to zero, it follows that $\rho(H_{K,M})$ increases with M and decreases with K. Thus the limits in the left-hand side of (2.1) exist.

We now prove (2.1) by exhibiting, at least approximately, the eigenvector of $H_{K,M}$ corresponding to $\rho(H_{K,M})$. This approximate eigenvector is suggested by [15, Section 2]. Let $\xi(l) = l^{-1/2}$. Then for $K \leqslant i < M$

$$H_{K,M}\xi(i) = \sum_{j=K}^{M-1} h(i,j)\frac{1}{\sqrt{j}} = \frac{1}{\sqrt{i}}\sum_{j=K}^{M-1} h\left(1,\frac{j}{i}\right)\frac{1}{\sqrt{j/i}}\,\frac{1}{i}.$$

For large K and $K \leqslant i < M$ the right-hand side lies between $(1-\epsilon)$ and $(1+\epsilon)$ times

$$\frac{1}{\sqrt{i}}\int_{K/i}^{M/i}\frac{h(1,y)}{\sqrt{y}}\,dy \leqslant \frac{1}{\sqrt{i}}\int_0^\infty \frac{h(1,y)}{\sqrt{y}}\,dy = \frac{1}{\sqrt{i}}\int_0^1 \frac{h(u,1-u)}{\sqrt{u(1-u)}}\,du.$$

(set $u = (1+y)^{-1}$). Let us use the abbreviation

$$\theta = \int_0^1 \frac{h(u,1-u)}{\sqrt{u(1-u}}\,du.$$

Then the above computation shows that for large K

$$H_{K,M}\xi \leqslant (1+\epsilon)\,\theta\xi$$

holds componentwise. Thus, also

$$(H_{K,M})^n \xi \leqslant [(1+\epsilon)\theta]^n \xi.$$

It is easy to deduce from this (compare [8, Corollary 2.2 in Appendix] or [14, Exercise 1.6]) that

$$\rho(H_{K,M}) \leqslant (1+\epsilon)\theta$$

and therefore

$$\lim_{K\to\infty} \lim_{M\to\infty} \rho(H_{K,M}) \leqslant \theta. \tag{2.4}$$

So far we have not used the symmetry of h. However, for symmetric h we can use the Courant–Weyl lemma, as in [15], to conclude for large K that

$$\begin{aligned}\rho(H_{K,M}) &\geqslant \left(\sum_K^{M-1} \xi^2(j)\right)^{-1} (\xi, H_{K,M}\xi)\\ &\geqslant \left(\sum_K^{M-1} \frac{1}{j}\right)^{-1} (1-\epsilon) \sum_K^{M-1} \frac{1}{i} \int_{K/i}^{M/i} \frac{h(1,y)}{\sqrt{y}}\, dy.\end{aligned}$$

Therefore

$$\lim_{M\to\infty} \rho(H_{K,M}) \geqslant (1-\epsilon) \int_{K^{-1}}^{\infty} \frac{h(1,y)}{\sqrt{y}}\, dy,$$

and letting $K \to \infty$ gives (2.1).

To obtain (2.2) we redo the last estimate slightly more carefully. Taking $M = NK$ we have, for large K and N,

$$\begin{aligned}\rho(H_{K,NK}) &\geqslant (1-2\epsilon)(\log N)^{-1} \sum_K^{NK-1} \frac{1}{i} \int_{K/i}^{N/Ki} \frac{h(1,y)}{\sqrt{y}}\, dy\\ &\geqslant (1-2\epsilon)\left((\log N)^{-1} \sum_{K\log N}^{NK/\log N} \frac{1}{i}\right) \int_{(\log N)^{-1}}^{\log N} \frac{h(1,y)}{\sqrt{y}}\, dy.\end{aligned}$$

We obtain (2.2) by picking N so large that

$$\log \frac{N}{(\log N)^2} \geqslant (1-\epsilon)\log N$$

and

$$\int_{(\log N)^{-1}}^{\log N} \frac{h(1,y)}{\sqrt{y}}\, dy \geqslant \int_0^{\infty} \frac{h(1,y)}{\sqrt{y}}\, dy - \epsilon.$$

Finally, the proof of (2.3) is implicit in [15]. Indeed, if we define

$$\sigma(r,s) = \sigma_{K,L}(r,s) = h(K+rL, K+sL) \qquad (r,s \geqslant 0), \tag{2.5}$$

then, under (1.7)

$$h(i,j) \geqslant \sigma(r,s)$$

for

$$K+(r-1)L \leqslant i < K+rL, \qquad K+(s-1)L \leqslant j < K+sL. \tag{2.6}$$

Consequently, if $\tilde{H}_{K,K+NL}(i,j) = \sigma(r,s)$ for i and j satisfying (2.6) and $1 \leqslant r,s \leqslant N$, then

$$\rho(H_{K,K+NL}) \geqslant \rho(\tilde{H}_{K,K+NL}).$$

As $L \to \infty$ (with K fixed)

$$\sigma_{K,L}(r,s) = \frac{1}{L}h\left(\frac{K}{L}+r, \frac{K}{L}+s\right) \sim \frac{1}{L}h(r,s).$$

Thus for fixed K and N, $\tilde{H}_{K,K+NL}$ is asymptotically (as $L \to \infty$) equal to the Kronecker product of the matrix $H_{1,N} = (h(r,s))_{1 \leqslant r,s<N}$ and the $L \times L$ matrix all of whose entries equal $1/L$. The latter matrix has largest eigenvalue 1, and the eigenvalues of the Kronecker product of two matrices are the products of the eigenvalues of the two matrices (e.g. [12, Section 12.2]). Therefore

$$\liminf_{L\to\infty} \rho(H_{K,K+NL}) \geqslant \rho(H_{1,N}).$$

Consequently

$$\lim_{M\to\infty} \rho(H_{1,M}) \leqslant \lim_{K\to\infty} \lim_{M\to\infty} \rho(H_{K,M}) = \theta.$$

The converse inequality follows from the monotonicity property

$$\rho(H_{1,M}) \geqslant \rho(H_{K,M}) \qquad (1 \leqslant K < M)$$

mentioned at the beginning of this proof. □

We need another characterization of θ. This will come from an application of the following proposition of Vere-Jones [16]; see also [14, Theorems 6.1, 6.8].

Proposition *Let $T = (T(i,j))_{i,j\in\mathbb{N}}$ be a matrix with strictly positive entries, T^n be the n-th power of T, and T_{N+1} the $N \times N$ matrix $(T(i,j))_{1\leqslant i,j\leqslant N}$. Also let*

$$F^n(i,i) = \sum^*_{i_1,i_2,\ldots,i_{n-1}} T(i,i_1)\,T(i_1,i_2)\cdots T(i_{n-2},i_{n-1})\,T(i_{n-1},i), \tag{2.7}$$

where $\sum^*$ *runs over all sequences,* $i_1,\ldots,i_{n-1}$ *with* $i_k \neq i$. *Then all the power series*

$$\sum_{n=0}^{\infty} T^n(i,i)\,z^n \qquad (i \in \mathbb{N})$$

have a common radius of convergence $R = R(T)$:

$$R = \lim_{N\to\infty} \frac{1}{\rho(T_N)}. \tag{2.8}$$

Moreover,

$$\sum_{n=1}^{\infty} F^n(i,i)\,R^n \leqslant 1 \qquad (i \in \mathbb{N}). \tag{2.9}$$

When this proposition is applied to the matrices $T = H_{K,\infty} = (h(i,j))_{i,j\geqslant K}$ then we see from (2.1) that

$$\theta = \lim_{K\to\infty} \frac{1}{R(H_{K,\infty})}.$$

We are ready to prove the lower bound $\lambda_c \geqslant \theta^{-1}$.

Lemma 2 *If (1.1)–(1.4) hold and* $\lambda < \theta^{-1}$, *then* $C(h,\lambda) = 0$. *If in addition (1.7) is assumed, then even* $C(h,\theta^{-1}) = 0$.

Proof As in Lemma 2 of [15] we have that $C(h,\lambda) < 1$ whenever

$$\sum_{n=1}^{\infty} G^n(i,j)\,\lambda^n < \infty \qquad \text{for some } i \neq j, \tag{2.10}$$

where

$$G^n(i,j) = \sum^{**}_{i_1,\ldots,i_{n-1}} h(i,i_1)\,h(i_1,i_2)\cdots h(i_{n-1},j) \tag{2.11}$$

and $\sum^{**}$ is the sum over all sequences $i_1,\ldots,i_{n-1}$ such that $i,i_1,\ldots,i_{n-1},j$ are all distinct. (2.10) is simply the expected number of self-avoiding paths in $\mathcal{G}$ between i and j; it is the same as Shepp's $\nu_{i,j}$. Once we have $C(h,\lambda) < 1$, it follows from the zero–one law (1.13) that actually $C(h,\lambda) = 0$.

Now fix $i \neq j$ and $K \geqslant i,j$. Any sequence $i_1,\ldots,i_{n-1}$ entering into the sum $\sum^{**}$ can contain at most $K-2$ terms in $[1,K]$. We shall replace all of these, including the i and j at the beginning and end, by K. The sequence $i,i_1,\ldots,i_{n-1},j$ then goes over into a sequence $K,i_1',i_2',\ldots,i_{n-1}',K$ which has at most K terms equal to K and no terms less than K. Let $0 < \nu_1 < \cdots < \nu_r < n$, with $r \leq K-2$, be the indices ν for which $i'_\nu = K$ ($i'_0 = 0$, $i'_n = K$). It is easy to see from (1.3) and (1.4) that there exists an $1 \leq A_K < \infty$ such that

$$h(i,j) \leq A_K h(K,j) \qquad \text{for all } i \leq K \text{ and } j \geq 1,$$

and hence also

$$h(i,j) \leq A_K h(i,K) \qquad \text{for all } i \geq 1 \text{ and } j \leq K.$$

For such an A_K we have

$$\begin{aligned} &\sum_{n=1}^{\infty} G^n(i,j)\lambda^n \\ &\quad = \sum_{n=1}^{\infty} \lambda^n \sum_{i_1,\ldots,i_{n-1}}^{**} h(i,i_1)\cdots h(i_{n-1},j) \\ &\quad \leq A_K^{2k} \sum_{r=0}^{K-2} \sum_{n=1}^{\infty} \lambda^n \sum_{0<\nu_1<\cdots<\nu_r<n} \sum_{i'_1,\ldots,i'_{n-1}}^{(K)} h(K,i'_1)h(i'_1,i'_2)\cdots h(i'_{n-1},K), \end{aligned} \tag{2.12}$$

where the sum $\sum^{(K)}$ runs over all sequences $i'_1,\ldots,i'_{n-1}$ with $i'_\nu = K$ for $\nu \in \{\nu_1,\ldots,\nu_r\}$, $i'_j > K$ for $j \notin \{\nu_1,\ldots,\nu_r\}$ and all $i'_1,\ldots,i'_{n-1}$ distinct. Simple rearranging of the sums shows that the right-hand side of (2.12) is at most equal to

$$A_K^{2K} \sum_{r=0}^{K-2} \left(\sum_{n=1}^{\infty} F_K^n(K,K)\lambda^n \right)^{r+1}, \tag{2.13}$$

where F_K is defined as in (2.7) for $T = H_{K,\infty}$. In particular, by (2.8) and (2.9),

$$\sum_{n=1}^{\infty} G^n(i,j)\lambda^n < \infty \tag{2.14}$$

if

$$\lambda \leq [\lim_{M\to\infty} \rho(H_{K,M})]^{-1} \qquad \text{for some } K.$$

Thus, if

$$\lambda < \theta^{-1} = \lim_{K\to\infty} \lim_{M\to\infty} \rho(H_{K,M}),$$

then we can choose K so large that (2.14) holds and, as observed before, this implies $C(h,\lambda) = 0$. This proves the first statement of the lemma.

If also (1.7) holds then, by (2.3), $\lambda \leqslant \theta^{-1}$ gives (2.14) (take $K = 1$), so that even $C(h, \theta^{-1}) = 0$ under (1.7). □

We turn to an upper bound for λ_c. Lemmas 2 and 3 together will prove our theorem.

Lemma 3 *Under (1.1)–(1.4), if* $\lambda > \theta^{-1}$, *then* $C(h, \lambda) = 1$.

Proof Since the proof so closely follows Section 2 of [15] we merely give an outline. Because of (2.2) it is enough to prove

$$\lambda\rho(H_{K,NK}) \geqslant 1+4\eta > 1 \text{ for all large } K \Rightarrow C(h,\lambda) = 1. \quad (2.15)$$

This is done by showing that there exist $\gamma > 0$ and $\delta > 0$ such that for all sufficiently large K the random subgraph $\mathcal{G}_K$ with vertex set $[K, NK-1]$ and edge probabilities $\lambda h(i,j)$ $(K \leqslant i, j < NK)$ has the property

$$P\{i \text{ belongs to a component of } \mathcal{G}_K \text{ with at least } \gamma NK \text{ vertices}\} \geqslant \delta \quad (K \leqslant i < NK). \quad (2.16)$$

(2.16) will imply $\mathscr{C}(h, \lambda) = 1$ by the following argument. Let u and v be two fixed vertices. Pick $K_1 < NK_1 < K_2 < NK_2 < \cdots$ such that (2.16) holds for each K_r (with the same N). u will be connected to v if, for some r, $\mathcal{G}_{K_r}$ has a component of γNK_r vertices in $[K_r, NK_{r-1}]$ and if u and v are each connected by at least one edge to this component. For a set C write $|C|$ for the number of vertices in C. Now, for any component $\mathscr{C} \in [K_r, NK_{r-1}]$ with $|\mathscr{C}| \geqslant \gamma NK_r$, one has

$$\begin{aligned}
&P\{u \text{ and } v \text{ are connected to } \mathscr{C}\} \\
&\quad \geqslant (1-P\{\text{no edge from } u \text{ to } \mathscr{C} \text{ is present}\}) \\
&\qquad \times(1-P\{\text{no edge from } v \text{ to } \mathscr{C} \text{ is present}\}) \\
&\quad \geqslant \left[1-\left\{1-K_r^{-1} \min_{1\leqslant y\leqslant N} h\left(\frac{u}{K_r}, y\right)\right\}^{\gamma NK_r}\right] \\
&\qquad \times\left[1-\left\{1-K_r^{-1} \min_{1\leqslant y\leqslant N} h\left(\frac{v}{K_r}, y\right)\right\}^{\gamma NK_r}\right] \qquad (2.17)
\end{aligned}$$

by (1.3) and (1.4). Since the last member of (2.17) is bounded away from 0 and since, for each r,

$$P\{\exists \text{ component } \mathscr{C} \text{ in } \mathcal{G}_{K_r} \text{ with } |\mathscr{C}| \geqslant \gamma NK_r\} \geqslant \delta$$

(once we have (2.16)) and since the $\mathcal{G}_{K_r}$ as well as the edges from u and

v to $\mathscr{G}_{K_r}$ are independent for different r, one easily concludes that

$$P\{u \text{ is connected to } v\} = 1.$$

We have therefore reduced the proof to showing that $\lambda > \theta^{-1}$ implies (2.16). This is done by comparing the number of vertices in $[K, NK-1]$ to which i is connected to a supercritical multitype branching process. For the remainder of the proof we fix N such that for some $\eta > 0$

$$\lambda\rho(H_{K,NK}) \geqslant 1+4\eta \qquad \text{for all large } K \tag{2.18}$$

(see (2.2)). Next, we bound the matrix $H_{K,NK}$ from below in a manner analogous to (2.5) and (2.6). Define

$$\begin{aligned}\boldsymbol{\sigma}(r,s) &= \boldsymbol{\sigma}_{K,L}(r,s)\\ &= \min\{h(x,y): K+(r-1)L \leqslant x < K+rL,\ K+(s-1)L \leqslant y < K+sL\}\end{aligned}$$

and take $\boldsymbol{H}(i,j) = \boldsymbol{\sigma}(r,s)$ for i,j in the block (2.6). Again $h(i,j) \geqslant \boldsymbol{H}(i,j)$. This time we take L small with respect to K (in contrast to the lines following (2.6)). For simplicity pick K and L such that L divides K, say $K = ML$. Then

$$\boldsymbol{\sigma}(r,s) = \frac{1}{K}\min\left\{h(x,y): 1+\frac{r-1}{M} \leqslant x < 1+\frac{r}{M},\ 1+\frac{s-1}{M} \leqslant y < 1+\frac{s}{M}\right\}.$$

Thus, for our fixed N, we can choose M so large that, uniformly in $1 \leqslant r,s \leqslant MN$ and i,j in the block (2.6),

$$h(i,j) \geqslant \boldsymbol{\sigma}(r,s) \geqslant \frac{1}{K}\frac{1+3\eta}{1+4\eta}h\left(\frac{i}{K},\frac{j}{K}\right) = \frac{1+3\eta}{1+4\eta}h(i,j). \tag{2.19}$$

Consequently, if $\boldsymbol{H}_{K,NK}$ denotes the submatrix of $\boldsymbol{H}$ with i,j restricted to $[K, NK-1]$, then

$$\lambda\rho(\boldsymbol{H}_{K,NK}) \geqslant \lambda\frac{1+3\eta}{1+4\eta}\rho(H_{K,NK}) \geqslant 1+3\eta.$$

Now we construct our branching process. Part of this construction is very reminiscent of [6, Section 3]. Let $i \in [K, NK-1]$ and let $Z_0 = i$ be the zero-th generation of our branching process. Assume that $Z_0, Z_1, \ldots, Z_\alpha$ have already been determined, and let

$$Z_\alpha = \{z_{\alpha,1}, \ldots, z_{\alpha,\nu(\alpha)}\},$$

with the $z_{\alpha,j}$ distinct vertices in $[K, NK-1]$. Assume further that also the set of children of $z_{\alpha,j}$ has been determined for $0 \leqslant j \leqslant \tau < \nu(\alpha)$. Denote these sets of children by $W_{\alpha,j}$. The $W_{\alpha,j}$ are also sets of vertices in $[K, NK-1]$. Assume that these sets have been chosen such that

$Z_0,\ldots,Z_\alpha, W_{\alpha,j}$ $(0 \leqslant j \leqslant \tau)$ are pairwise disjoint. If

$$|Z_0|+\cdots+|Z_\alpha|+|W_{\alpha,1}|+\cdots+|W_{\alpha,\tau}| \geqslant \eta L = \frac{\eta K}{M} \tag{2.20}$$

then we stop. Otherwise, a vertex

$$y \in [K, NK-1] \Bigg\backslash \left(\bigcup_{i=1}^{\alpha} Z_i \cup \bigcup_{j=1}^{\tau} W_{\alpha,j}\right) \tag{2.21}$$

is chosen as a child of $z_{\alpha,\tau+1}$ with probability

$$\lambda H_{K,NK}(z_{\alpha,\tau+1}, y), \tag{2.22}$$

and all the y satisfying (2.21) are chosen independently. (Here we assume that K is large enough so that $\lambda h(i,j) \leqslant 1$ for $i,j \geqslant K$ so that (2.22) is permissible as a probability.) The y's chosen in this way make up $W_{\alpha,\tau+1}$, and we can now continue with the choice of $W_{\alpha,\tau+2}$ or $W_{\alpha+1,1}$ as long as $\tau < \nu(\alpha)-1$ or $|Z_{\alpha+1}| > 0$. Let us call a vertex y of type s if $K+(s-1)L \leqslant y < K+sL$. Then, as long as (2.20) *fails*, there are at least $(1-\eta)L$ vertices y of type s which satisfy (2.21) $(1 \leqslant s \leqslant M(N-1))$. Thus, if we have not yet stopped with $W_{\alpha,\tau}$ and if $z_{\alpha,\tau+1}$ is of type r, then, conditionally on all previous choices, the number of children of type s of $z_{\alpha,\tau+1}$ is stochastically larger than a binomial variable with parameters $(1-\eta)L$ and $\lambda\sigma(r,s)$. The expectation of such a variable is

$$\lambda(1-\eta)L\sigma(r,s) = \lambda(1-\eta) \sum_{K+(s-1)L \leqslant y < K+sL} H_{K,NK}(z_{\alpha,\tau+1}, y). \tag{2.23}$$

We see that the number of vertices obtained in the various generations is bounded below by the $M(N-1)$-type branching process with mean offspring matrix (2.23), as long as this branching process has not been stopped because of (2.20). Since $h(i,j) \geqslant \sigma(r,s)$ we can couple the vertices which are connected to i in $\mathcal{G}_K$ to this branching process, in such a way that any child of i in the branching process (until it is stopped by (2.20)) is connected to i in $\mathcal{G}_K$. It follows that the probability in the left-hand side of (2.16) with $\gamma NK = \eta L$ or $\gamma = \eta/MN$ is at least

$$P\{\text{above branching process does not become extinct}\}. \tag{2.24}$$

However, the largest eigenvalue of the offspring matrix in (2.23) (with $1 \leqslant r,s \leqslant M(N-1)$) is precisely

$$\lambda(1-\eta)\rho(H_{K,NK}) \geqslant (1-\eta)(1+3\eta) \geqslant 1+\eta.$$

It is now easy to see from standard branching-process results (e.g. [7,

Theorem I.7.1] that (2.24) is bounded away from 0, uniformly for large K and $i \in [K, NK-1]$. Thus (2.16) and the lemma are proven. □

3 Extensions and variants

(a) Asymptotic equivalence Simple monotonicity arguments, combined with the zero-one law (1.13), show that (1.6) continues to hold for a function h satisfying (1.2)–(1.4) if we only have

$$\lim_{i,j\to\infty} \frac{p(i,j)}{h(i,j)} = \lambda \tag{3.1}$$

instead of the equality $p(i,j) = \lambda h(i,j)$. Such arguments also show that $\lambda_c = \infty$ if $(i+j)h(i,j) \to 0$ and $\lambda_c = 0$ if $(i+j)h(i,j) \to \infty$. This to some extent justifies the restrictive hypothesis (1.3).

(b) Directed graphs One may want to consider directed or oriented edges, instead of the undirected edges considered so far. One then includes an edge from i to j with probability $p(i,j)$ and another edge from j to i with probability $p(j,i)$, and all these edges are chosen independently of each other. One now wishes to know the value of

$$P\{i \text{ is connected by a directed path to } j \text{ for all } i \neq j\}.$$

For $p(i,j)$ as in (1.1) we denote this probability by $C_d(h,\lambda)$. Many of the arguments in Section 2 go through for this directed case after minor changes only. Crucial use of the symmetry was made in Lemma 1 and we can no longer identify λ_c as θ^{-1}, but only as

$$\lambda_{c,d} := [\lim_{M\to\infty} \rho(H_{1,M})]^{-1} = [\lim_{K\to\infty}\lim_{M\to\infty} \rho(H_{K,M})]^{-1}. \tag{3.2}$$

The last equality is just the equality of the left-hand sides of (2.1) and (2.3). The proof of this did not rely on (1.2) (but does use (1.7), which we assume here). We note in passing that (2.4) still gives us

$$\lambda_{c,d} \geqslant \theta^{-1},$$

even without (1.2).

The result is that, under (1.3), (1.4) and (1.7),

$$C_d(h,\lambda) = \begin{cases} 0 & \text{if } \lambda < \lambda_{c,d}, & (3.3a) \\ 1 & \text{if } \lambda > \lambda_{c,d}. & (3.3b) \end{cases}$$

The proof of $C_d(h,\lambda) < 1$ for $\lambda < \lambda_{c,d}$ is unchanged from Lemma 2. However, we do not know whether the zero–one law (1.13) holds in the directed case, and the proof of (3.3a) therefore must be completed by hand. We now replace (2.12)–(2.14) by the far easier estimate

$$G^n(i,i+1) \leqslant \sup_{k\neq i,i+1} \frac{h(k,i+1)}{h(k,i)} \sum\nolimits^{**} h(i,i_1)\cdots h(i_{n-1},i)$$
$$\leqslant \sup_{k\neq i,i+1} \frac{h(k,i+1)}{h(k,i)} F^n(i,i),$$

where $\sum^{**}$ is the sum over all distinct $i_1,\ldots,i_{n-1}$ which are also distinct from i and $i+1$ and F^n is defined by (2.7) with $T = H_{1,\infty}$. (This is immediate from (2.11) with $j = i+1$; merely replace $h(i_{n-1},i+1)$ by $h(i_{n-1},i)$ under the sum.) Since

$$A := \sup_i \sup_{k\neq i,i+1} \frac{h(k,i+1)}{h(k,i)} < \infty$$

(by (1.3) and (1.4)), we have for $\lambda < \lambda_{c,d}$

$P\{i$ is connected to $i+1\}$

$$\leqslant E\{\text{number of self-avoiding directed paths from } i \text{ to } i+1\}$$
$$\leqslant \sum_{n=1}^{\infty} G^n(i,i+1)\lambda^n$$
$$\leqslant \sum_{n=1}^{M-1} G^n(i,i+1)\lambda^n + A\left[\frac{\lambda}{\lambda_{c,d}}\right]^M \sum_{n=M}^{\infty} F^n(i,i)[\lambda_{c,d}]^n$$
$$\leqslant \sum_{n=1}^{M-1} G^n(i,i+1)\lambda^n + A\left[\frac{\lambda}{\lambda_{c,d}}\right]^M$$

(by (2.9)). Thus if we show

$$\lim_{i\to\infty} G^n(i,i+1) = 0 \quad \text{for fixed } n, \tag{3.4}$$

then for $\lambda < \lambda_{c,d}$

$$C_d(h,\lambda) \leqslant \liminf_{i\to\infty} P\{i \text{ is connected to } i+1\} = 0.$$

However, (3.4) can be shown by brute force, since $h(i,j) = O(\{i+j\}^{-1})$. This proves (3.3a).

The proof of (3.3b) is essentially the same as that of Lemma 3. The symmetry assumption is used in the deduction of $C(h,\lambda) = 1$ from (2.16), because we used that if u and v are connected to the same cluster of $\mathcal{G}_K$, then u and v are connected. This can easily be adapted to the directed case. We now first check the directed edges from u to $[K, NK-1]$. If there is such an edge in $\mathcal{G}$, then choose any $i \in [K, NK-1]$ such that the edge from u to i is present. We then grow the branching process, starting from i as in Lemma 3. If any of the descendants of i has an edge to v, then u is connected to v (via i). This change is sufficient to take care of directedness at this place. We also used symmetry to obtain (2.2) and hence (2.18). However, when (1.7) is assumed we do not need (2.18). We now have

$$\lambda_{c,d} = [\lim_{N\to\infty} \rho(H_{1,N})]^{-1}$$

and in the proof of (3.3b) we can therefore fix N such that

$$\lambda\rho(H_{1,N}) \geqslant 1+4\eta > 1.$$

We may then replace $\boldsymbol{\sigma}$ in Lemma 3 by the σ of (2.5) and take L/K large, instead of small as in Lemma 3. We further replace $\mathcal{G}_K$ by the restriction of $\mathcal{G}$ to $[K, K+NL-1]$, and $H_{K,NK}$ by $H_{K,K+NL}$, and use that (under (1.7))

$$\rho(H_{K,K+NL}) \geqslant (1-\eta)\rho(H_{1,N})$$

for L/K large (as in the proof of (2.3)). With these replacements one can repeat the proof of Lemma 3 (compare also [15, Section 2]). □

References

[1] M. Aizenman, H. Kesten & C. M. Newman, Uniqueness of the infinite cluster and continuity of connectivity functions for short and long range percolation, *Comm. Math. Phys.*, **111** (1987), 505–31

[2] B. Bollobás, *Random Graphs*, Academic Press, London, 1985

[3] P. Erdős & A. Rényi, On random graphs I, *Publ. Math. Debrecen*, **6** (1959), 290–97

[4] P. Erdős & A. Rényi, On the evolution of random graphs, *Publ. Math. Inst. Hung. Acad. Sci.*, **5**, Ser. A (1960), 17–61

[5] G. Grimmett, Random graphs, Chapter 7 in *Selected Topics in Graph Theory 2*, (eds. L. W. Beineke & R. J. Wilson), Academic Press, London, 1983

[6] G. R. Grimmett, M. Keane & J. M. Marstrand, On the connectedness of a random graph, *Math. Proc. Camb. Phil. Soc.*, **96** (1984), 151–66

[7] T. E. Harris, *The Theory of Branching Processes*, Springer, New York, 1963

[8] S. Karlin, *A First Course in Stochastic Processes*, Academic Press, New York, 1966

[9] *Random Graphs* '85, eds. M. Karonski & Z. Palka, North-Holland, Amsterdam, 1987

[10] *Random Graphs* '83, eds. M. Karonski & A. Rucinski, North-Holland, Amsterdam, 1985

[11] S. Kalikow & B. Weiss, When are random graphs connected?, *Isr. J. Math.*, **62** (1988), 257–68

[12] P. Lancaster & M. Tismenetsky, *The Theory of Matrices*, 2nd edn., Academic Press, New York, 1985

[13] E. Palmer, *Graphical Evolution: An Introduction to the Theory of Random Graphs*, John Wiley, New York, 1985

[14] E. Seneta, *Non-negative Matrices and Markov Chains*, 2nd edn., Springer, Heidelberg, 1981
[15] L. A. Shepp, Connectedness of certain random graphs, *Isr. J. Math.*, **67** (1989), 23–33
[16] D. Vere-Jones, Ergodic properties of nonnegative matrices I, *Pacific J. Math.*, **22** (1967), 361–86
[17] E. T. Whittaker & G. N. Watson, *A Course of Modern Analysis*, 4th edn., Cambridge University Press, Cambridge, 1952

Multiplicative functions on arithmetic progressions: III. The large moduli

P. D. T. A. Elliott*

1 Introduction

In this paper I continue the study of the value distribution on arithmetic progressions of multiplicative functions g which satisfy $|g(n)| \leqslant 1$ for all positive integers n. As in the previous papers, no other assumption will be made upon g. This generality is important for applications.

Theorem 1 *Let* $0 < \delta < \frac{1}{2}$. *Then*

$$\sum_{l^4 < p \leqslant x^\delta} (p-1) \max_{(r,p)=1} \max_{y \leqslant x} \left| \sum_{\substack{n \leqslant y \\ n \equiv r \pmod{p}}} g(n) - \frac{1}{p-1} \sum_{n \leqslant y} g(n) \right|^2 \ll \frac{x^2(\log\log x)^4}{(\log x)^2},$$

where p denotes a prime number and $l = \log x$.

A crude upper bound for the multiple sum(s) is $O(x^2 \log\log x)$, over which the present estimate saves about $(\log x)^2$. This may be compared with Theorem 1 in the second paper of this series [6]. According to that theorem there is an inequality of a similar type, with an upper bound $\ll x^2(\log x)^{-c}$ for a positive absolute constant c, but with the summation over the prime moduli extending all the way down to $p = 2$ with at most one exception. Examples show that this last condition is necessary. It would clearly be advantageous to possess a result of this type with as large a value of c as possible, especially in pursuit of a generalisation involving composite moduli. At present every $c < \frac{1}{16}$ is possible.

* Partially supported on NSF contract DMS-8722913.

As in the second paper of this series, I introduce the hybrid functions $g\chi$, where χ is a Dirichlet character, and employ inequalities of large-sieve type. However, rather than with the method of high moments, I treat the Dirichlet series

$$\sum_{n=1}^{\infty} g(n)\chi(n)n^{-s} \qquad (s = \sigma + i\tau),$$

where $\sigma = \mathrm{Re}\, s > 1$, directly. My present aim is then to throw into relief that part of the function g which most limits the size of the error term in Theorem 1. We shall be reduced to an arithmetic problem not directly concerned with multiplicative functions.

For $z \geqslant 2$ define

$$k_1(n) = \sum_{\substack{pm=n \\ p\leqslant z}} g(m)g(p)\log p, \qquad k_2(n) = \sum_{\substack{pr=n \\ r\leqslant z}} g(r)g(p)\log p.$$

Further define

$$K_j = K_j(t,\chi) = \sum_{n\leqslant t} k_j(n)\chi(n) \qquad (j = 1,2).$$

A multiplicative function g is *exponentially multiplicative* if it satisfies $g(p^k) = g(p)^k/k!$ for all primes p and positive integers k.

Theorem 2 *If* $z = l^{6A+15}$, $\sqrt{3} \leqslant y \leqslant x^{\delta}$, *where* $\delta < \frac{1}{2}$, *and* ϵ *is sufficiently small in terms of* δ, *then for exponentially multiplicative* g

$$\sum_{y<d\leqslant y^{1+\epsilon}} \frac{1}{d} \sum_{\chi (\mathrm{mod}\, d)}^{*} \max_{t\leqslant x} \left| \sum_{n\leqslant t} g(n)\chi(n)\log n - K_1 - K_2 \right|$$
$$\ll y^{-1}xl^{2A+8} + xl^{-A},$$

where * *denotes summation over the primitive characters* (mod d).

From this we anticipate that the sums K_j will most affect the error term in Theorem 1 and its analogues. Since K_1 again involves the function g on the integers, one may treat it by introducing an argument by induction. The sum K_2, however, is essentially $g(p)\chi(p)\log p$ summed over the primes p up to t and, beyond the restriction $|g(p)| \leqslant 1$, the genesis of $g(p)$ within a multiplicative function is irrelevant. Thus it is the study of the sum K_2 which lies before any improvement of the error term in Theorem 1.

In an appendix I motivate the proof of Theorem 1 in functional-analytic terms and speculate upon refinements.

I sketched out the results of this paper during a visit in July 1988 to the mathematics department of the University of Ulm, Germany, as a guest of Hans-Peter Schlickewei. I would particularly like to thank him and his wife Heidi for their generous hospitality.

2 Preliminaries

Lemma 1 *The inequality*

$$\sum_{D\leqslant Q}\ \sideset{}{^*}\sum_{\chi\,(\mathrm{mod}\,D)}\left|\sum_{n\leqslant y} a_n\chi(n)\right|^2 \ll (y+Q^2)\sum_{n\leqslant y}|a_n|^2$$

holds uniformly for real $Q \geqslant 0$, $y \geqslant 0$ *and complex* a_n.

Proof This is a standard large sieve. See, for example, Gallagher [8] or Bombieri [1]. □

Lemma 2 *Let* $\epsilon > 0$. *The inequality*

$$\sum_{D\leqslant Q}\ \sideset{}{^*}\sum_{\chi\,(\mathrm{mod}\,D)}\left|\sum_{q\leqslant y} a_q\chi(q)\right|^2 \ll \left(\frac{y}{\log y}+Q^{2+\epsilon}\right)\sum_{q\leqslant y}|a_q|^2,$$

where q *denotes a prime, holds uniformly for real* $Q \geqslant 0$, $y \geqslant 2$ *and complex* a_q.

Proof The proof of the similar result in Elliott [4], Chapter 6, Lemma (6.3), adapts easily to give the present lemma. □

Let $\|\boldsymbol{a}\|$ denote the norm $\sqrt{\sum_{n=-\infty}^{\infty}|a_n|^2}$ on the Hilbert space of square summable sequences of complex numbers.

Lemma 3

$$\sum_{D\leqslant Q}\ \max_{v-u\leqslant H}\ \sideset{}{^*}\sum_{\chi\,(\mathrm{mod}\,D)}\left|\sum_{u<n\leqslant v} a_n\chi(n)\right|^2 \leqslant (H+6Q^2\log Q)\|\boldsymbol{a}\|^2$$

holds uniformly for $Q \geqslant \mathrm{e}$ *and* $\boldsymbol{a}$.

Proof See Elliott [5], Lemma 3. □

Lemma 3 is a maximal version of Lemma 1. The following maximal version of Lemma 2 will be obtained in the next section.

Lemma 4 *The inequality*

$$\sum_{D\leqslant Q}\ \max_{t\leqslant x}\ \sideset{}{^*}\sum_{\chi\,(\mathrm{mod}\,D)}\left|\sum_{q\leqslant t} a_q\chi(q)\right|^2 \ll \left(\frac{x}{\log x}+Q^{2+\epsilon}\right)\sum_{q\leqslant x}|a_q|^2(\log\log x)^2$$

holds uniformly for real $Q > 0$, $x \geqslant 2$ *and complex* a_q *(*q *prime).*

Remark Doubtless the factor $(\log\log x)^2$ does not belong in the upper bound.

It is convenient to note here that for prime moduli p, the sets of primitive and the non-principal Dirichlet characters $(\bmod p)$ coincide. The identity

$$\sum_{\chi(\bmod p)}^{*}\left|\sum_{n\leqslant y} a_n\chi(n)\right|^2 = (p-1)\sum_{r=1}^{p-1}\left|\sum_{\substack{n\leqslant y\\ n\equiv r(\bmod p)}} a_n - \frac{1}{p-1}\sum_{\substack{n\leqslant y\\ (n,p)=1}} a_n\right|^2$$

follows readily from the well-known orthogonality property of characters.

For any sequence of complex numbers a_n $(n = 1,2,\ldots)$ let $A(x)$ denote the sum of those a_n for which n does not exceed x. Let $F(s)$ be the corresponding Dirichlet series $\sum_{n=1}^{\infty} a_n n^{-s}$ $(s = \sigma + i\tau)$. Proceeding formally, an integration by parts shows that

$$s^{-1}F(s) = \int_1^{\infty} A(y)y^{-(s+1)}\,\mathrm{d}y = \int_0^{\infty} A(\mathrm{e}^u)\mathrm{e}^{-\sigma u}\mathrm{e}^{-i\tau u}\,\mathrm{d}u.$$

Hence $(s\sqrt{2\pi})^{-1}F(s)$ as a function of τ and $A(\mathrm{e}^u)\mathrm{e}^{-\sigma u}$ as a function of u are Fourier transforms. The following result is then an application of Plancherel's theorem.

Lemma 5

$$\int_{-\infty}^{\infty}\left|\frac{F(s)}{s}\right|^2\mathrm{d}\tau = 2\pi\int_0^{\infty}\mathrm{e}^{-2\sigma u}|A(\mathrm{e}^u)|^2\,\mathrm{d}u,$$

provided one of the integrals exists in an L^2 sense.

In our present applications $\sigma \geqslant \frac{1}{2}$ will always hold, and the integrand involving τ will clearly be meaningful.

For positive integers m, complex $s = \sigma + i\tau$ with $\sigma > 0$ and real ρ and x which satisfy $1 \leqslant \rho \leqslant x/2m$, I introduce the kernels $K_j(x,s)$ $(0 \leqslant j \leqslant m)$ obtained iteratively by $K_0(x,s) = x^s/s$ and

$$K_{j+1}(x,s) = \frac{1}{\rho}\int_{x-\rho}^{x} K_j(y,s)\,\mathrm{d}y \qquad (0 \leqslant j \leqslant m-1).$$

In particular

$$K_1(x,s) = \frac{x^{s+1}-(x-\rho)^{s+1}}{\rho s(s+1)}.$$

Lemma 6 *For fixed j and m* $(0 \leqslant j \leqslant m)$,

$$K_m(x,s) \ll \frac{x^\sigma}{|s|}\left(\frac{x}{\rho|s|}\right)^j$$

and

$$\int_{\sigma-i\infty}^{\sigma+i\infty} \sup_{2m\rho\leqslant y\leqslant x} |K_m(y,s)|\, d\tau \ll x^\sigma \log\frac{xe}{\sigma\rho}$$

uniformly for $0 < \sigma \leqslant 2$. *The constants depend only upon m.*

In particular, if $\sigma \geqslant \frac{1}{2}$ *and* ρ *exceeds* $x(\log x)^{-B}$ *for some constant B, then the upper bound for the integral is* $\ll x^\sigma \log\log x$.

Proof See Elliott [5], [6]. □

Given a multiplicative function g, we define a corresponding exponentially multiplicative function g_1 by $g_1(p^k) = g(p)^k/k!$. Let h be the further multiplicative function defined by the Dirichlet convolution $g = g_1 * h$.

Lemma 7 *If* $|g(n)| \leqslant 1$ *for all positive n, then*

$$\sum_{r\leqslant x} r^{-1/2}|h(r)| \ll (\log x)^{3/2}$$

uniformly for $x \geqslant 2$.

Proof Examination of the appropriate Euler products shows that for $\sigma > 0$

$$1 + \sum_{k=1}^{\infty} h(p^k)p^{-k\sigma} = \exp\left(\frac{g(p)}{p^\sigma}\right)\left(1 + \sum_{k=1}^{\infty} g(p^k)p^{-k\sigma}\right).$$

Hence

$$h(p^k) = \sum_{u+v=k} \frac{\{-g(p)\}^u}{u!} g(p^v)$$

with the understanding that 0-powers are to be replaced by 1. In particular, $h(p) = 0$, $h(p^2) = g(p^2) - \frac{1}{2}g(p)^2$ and $|h(p^k)| \leqslant e$ for $k \geqslant 3$. Hence the sum to be estimated does not exceed

$$\prod_{p\leqslant x}\left(1 + \sum_{k=1}^{\infty} |h(p^k)|p^{-k/2}\right) \leqslant \exp\left(\sum_{p\leqslant x}\sum_{k=1}^{\infty} |h(p^k)|p^{-k/2}\right)$$

$$\leqslant \exp\left(\frac{3}{2}\sum_{p\leqslant x}\frac{1}{p} + O(1)\right) \ll (\log x)^{3/2},$$

this last step by a well-known estimate from elementary number theory.

3 Basic result

For reals ρ and x $(1 \leqslant \rho \leqslant \frac{1}{2}x)$ and positive integers n, define

$$\gamma_n = \gamma(x,\rho,n) = \begin{cases} 1 & \text{if } n \leqslant x-\rho, \\ \dfrac{x-n}{\rho} & \text{if } x-\rho < n \leqslant x, \\ 0 & \text{if } n > x. \end{cases}$$

Thus $0 \leqslant \gamma_n \leqslant 1$. Note that, for any positive integer d,

$$\gamma(x,\rho,dn) = \gamma\left(\frac{x}{d},\frac{\rho}{d},n\right).$$

For $\sqrt{3} \leqslant y \leqslant x$, $\epsilon > 0$ and $z \geqslant 2$, define

$$h_1(n) = \sum_{\substack{pm=n \\ p\leqslant z,\, m>y^{2+3\epsilon}}} g(m)g(p)\log p,$$

$$h_2(n) = \sum_{\substack{rp=n \\ r\leqslant z,\, p>y^{2+3\epsilon}}} g(r)g(p)\log p,$$

so that the h_j are restrictions of the k_j in Theorem 2. Further define

$$S_j = S_j(t,\chi) = \sum_{n\leqslant t} h_j(n)\chi(n)\gamma(t,\rho,n) \qquad (j = 1,2).$$

Let

$$\Delta = \frac{x^2}{\rho}\sqrt{\frac{l^2\log y}{y^2} + \frac{l\log y\log z}{z}} + y^{1+4\epsilon}x^{1/2}l.$$

Main lemma *For exponentially multiplicative functions* g,

$$\sum_{y<d<y^{1+\epsilon}} \frac{1}{d} \sum_{\chi(\text{mod } d)}^{*} \max_{2\rho\leqslant t\leqslant x} \left| \sum_{n\leqslant t} g(n)\chi(n)\log n\gamma(t,\rho,n) - S_1 - S_2 \right| \ll \Delta.$$

Assuming the validity of this lemma I shall deduce the theorems.

Proof of Lemma 4 I employ the representation

$$\sum_{q\leqslant t} a_q\chi(q)\gamma_q = \frac{1}{2\pi i}\int_{(\alpha)} \sum_{q\leqslant x} a_q\chi(q)q^{-s}K_1(t,s)\, ds,$$

where $\gamma_q = \gamma(t,\rho,q)$ is defined with $\rho = x(\log x)^{-4}$. The integration is taken along the line $\text{Re}(s) = \alpha = (\log x)^{-1}$, from $-\mathrm{i}\infty$ to $\mathrm{i}\infty$. Then, by Lemma 6, the integrand of $\sup|K_1|$ $(2\rho \leqslant t \leqslant x)$ along this line is $\ll \log\log x$.

It follows from an application of the Cauchy–Schwarz inequality for integrals that, if we restrict the maxima in the lemma to be over the range $2\rho \leqslant t \leqslant x$, then the sum to be estimated is

$$\ll \log\log x \int_{(\alpha)} \sum_{D\leq Q} \sum_{\chi(\operatorname{mod} D)}^{*} \left| \sum_{q\leq x} a_q\chi(q)q^{-s}\right|^2 \sup_{2\rho\leq t\leq x} |K_1| \, d\tau.$$

The desired upper bound then follows from Lemma 2.

We may enlarge the range $2\rho \leq t \leq x$ under the maxima, and remove the weights γ_q, by applying the maximal variant of the large sieve given in Lemma 3. This introduces a further error of

$$\ll (\rho + Q^2 \log Q) \sum_{q\leq x} |a_q|^2,$$

which is well within acceptable bounds. □

Remark If we retain the weights γ_n, but restrain the parameters ρ only by a condition $x(\log x)^{-B} \leq \rho \leq x$ for a fixed $B > 0$, then we may invert the order of the maxima and the summation over primitive characters.

It was pointed out in a previous paper (Elliott [5, Lemma 8]) that with Q^5 in place of $Q^{2+\epsilon}$ an inequality of this type holds even without the weights γ_n and the factors $(\log\log x)^2$.

Is there a result of the form

$$\sum_{D\leq Q} \sum_{\chi(\operatorname{mod} D)}^{*} \max_{t\leq x} \left| \sum_{q\leq t} a_q\chi(q)\right|^2 \ll \left(\frac{x}{\log x} + Q^2(\log Q)^{c_1}\right) \sum_{q\leq x} |a_q|^2$$

for some absolute $c_1 \geq 0$?

Proof of Theorem 1 Assume first that g is exponentially multiplicative. For $2 \leq 2\rho \leq t \leq x$ and for each prime $p \leq x^\delta$, choose a reduced residue class $r \pmod p$ and set

$$\begin{aligned} E(p,t) = & \sum_{\substack{n\leq t \\ n\equiv r \pmod p}} \{g(n)\log n - h_1(n) - h_2(n)\}\gamma(t,\rho,n) \\ & - \frac{1}{p-1} \sum_{\substack{n\leq t \\ (n,p)=1}} \{g(n)\log n - h_1(n) - h_2(n)\}\gamma(t,\rho,n). \end{aligned}$$

Clearly $pE(p,t) \ll xl$. By the orthogonality of Dirichlet characters to a fixed modulus,

$$E(p,t) = \frac{1}{p-1} \sum_{\chi(\operatorname{mod} p)}^{*} \bar{\chi}(r) \sum_{n\leq t} \{g(n)\log n - h_1(n) - h_2(n)\}\chi(n)\gamma(t,\rho,n),$$

so that, after the main lemma,

$$\sum_{y<p\leq y^{1+\epsilon}} (p-1) \max_{2\rho\leq t\leq x} |E(p,t)|^2 \ll xl\Delta \qquad (y \geq 1). \tag{1}$$

If we set $z = l^{2A}$ and restrict y not to exceed x^{δ}, then for ϵ sufficiently small this upper bound will be $\ll x^3\rho^{-1}(y^{-1}l^2\sqrt{\log y}+l^{2-A})$.

An application of the maximal form of the large sieve given in Lemma 3 shows that we may replace the weights $\gamma(t,\rho,n)$ in (1) by 1, at an expense of $\ll (\rho+x^{2\delta})xl^2$. At a similar expense we may also extend the maxima to be over the whole interval $0 < t \leq x$. Setting $\rho = c_1x(y^{-1}\sqrt{\log y}+l^{-A})^{1/2}$ for a suitable positive absolute constant c_1, we obtain

$$\sum_{y<p\leq y^{1+\epsilon}} (p-1) \max_{t\leq x} \left| \sum_{\substack{n\leq t \\ n\equiv r (\operatorname{mod} p)}} g(n)\log n - \frac{1}{p-1} \sum_{\substack{n\leq t \\ (n,p)=1}} g(n)\log n - \lambda_1 - \lambda_2 \right|^2$$
$$\ll x^2 l^2 \sqrt{\frac{\sqrt{\log y}}{y} + \frac{1}{l^A}}, \tag{2}$$

where

$$\lambda_j = \sum_{\substack{n\leq t \\ n\equiv r (\operatorname{mod} p)}} h_j(n) - \frac{1}{p-1} \sum_{\substack{n\leq t \\ (n,p)=1}} h_j(n) \qquad (j = 1,2),$$

uniformly for $y \geq 1$ and $x \geq 2$.

As I remarked earlier, we may treat the sum(s) involving λ_1 with an argument by induction. For the present purpose it suffices to apply the Cauchy–Schwarz inequality: $|\lambda_1|^2$ does not exceed

$$\sum_{\substack{q\leq z \\ (q,p)=1}} \frac{(\log q)^2}{q} \sum_{\substack{q\leq z \\ (q,p)=1}} q \left| \sum_{\substack{y^{2+3\epsilon}<m\leq x/q \\ m\equiv \bar{q}r (\operatorname{mod} p)}} g(m) - \frac{1}{p-1} \sum_{\substack{y^{2+3\epsilon}<m\leq x/q \\ (m,p)=1}}{}^{\bullet}\, g(m) \right|^2,$$

where $q\bar{q} \equiv 1 \pmod p$ and q denotes a prime number. Then, by Theorem 1 of Elliott [6], the λ_1 contribute towards the sums at (2) an amount

$$\ll (\log z)^2 \sum_{q\leq z} q\left(\frac{x}{q}\right)^2 \left(\log \frac{x}{q}\right)^{-c} \ll x^2(\log x)^{-c/2}.$$

The sum λ_2 depends upon y only through the agency of the condition $p > y^{2+3\epsilon}$ which is implicit in the definition of $h_2(n)$. A similar application of the Cauchy–Schwarz inequality, followed this time by an application of Lemma 3, shows that for small enough ϵ, and within an even smaller error, we may delete this condition, so rendering the modified $\hat{\lambda}_2$ independent of y.

Let $1 \leqslant y_0 \leqslant x$. Summing the expressions at (2) over the values $y = \exp(1+\epsilon)^\nu$ consistent with $y_0 \leqslant y \leqslant x^\delta$, and setting $A = c+6$, gives

$$\sum_{y_0<p\leqslant x^\delta} (p-1) \max_{t\leqslant x} \left| \sum_{\substack{n\leqslant t \\ n\equiv r \,(\mathrm{mod}\, p)}} g(n)\log n - \frac{1}{p-1} \sum_{\substack{n\leqslant t \\ (n,p)=1}} g(n)\log n - \hat{\lambda}_2 \right|^2$$

$$\ll x^2 l^2 y_0^{-1/2} (\log y_0)^{1/4} + x^2 (\log x)^{-c/3}.$$

To obtain Theorem 1 I specialise y_0 to $(\log x)^4$, but it is clear that a larger value of y_0 decreases the upper bound. The worst contribution comes from $\hat{\lambda}_2$ and, after an application of the Cauchy–Schwarz inequality, does not exceed a bounded multiple of

$$\sum_{y_0<p\leqslant x^\delta} (p-1) \sum_{\substack{w\leqslant z \\ (w,p)=1}} \frac{1}{w}$$

$$\times \sum_{\substack{w\leqslant z \\ (w,p)=1}} w \max_{t\leqslant x/w} \left| \sum_{\substack{q\leqslant t \\ q\equiv r\bar{w} \,(\mathrm{mod}\, p)}} g(q)\log q - \frac{1}{p-1} \sum_{\substack{q\leqslant t \\ (q,p)=1}} g(q)\log q \right|^2,$$

where once again q denotes a prime. Here the only feature of the multiplicative function g which remains is the size restriction $|g(q)| \leqslant 1$. Applications of Lemma 4 show that this expression is $\ll x^2(\log\log x)^4$. We have thus arrived at

$$\sum_{l^4<p\leqslant x^\delta} (p-1) \max_{(r,p)=1} \max_{t\leqslant x} \left| \sum_{\substack{n\leqslant t \\ n\equiv r \,(\mathrm{mod}\, p)}} g(n)\log n - \frac{1}{p-1} \sum_{\substack{n\leqslant t \\ (n,p)=1}} g(n)\log n \right|^2$$

$$\ll x^2(\log\log x)^4.$$

An application of Lemma 3 enables us to replace the weights $\log n$ by $\log x$, at the expense of an error $\ll x \sum_{n\leqslant x} (\log x/n)^2 \ll x^2$, and a division by $(\log x)^2$ completes the proof in this case.

For the general case I adopt the conventions of Lemma 7. Then the innermost sum(s) in the statement of Theorem 1 have the alternative representation

$$\sum_{\substack{d\leqslant y \\ (d,p)=1}} h(d) \left(\sum_{\substack{m\leqslant y/d \\ m\equiv r\bar{d} \,(\mathrm{mod}\, p)}} g_1(m) - \frac{1}{p-1} \sum_{\substack{m\leqslant y/d \\ (m,p)=1}} g_1(m) \right).$$

Let

$$N(t,p) = \max_{(r,p)=1} \left| \sum_{\substack{m \le t \\ m \equiv r (\bmod p)}} g_1(m) - \frac{1}{p-1} \sum_{\substack{m \le t \\ (m,p)=1}} g_1(m) \right| .$$

Further set $\theta = (1-2\delta)/(1+2\delta)$ and $\psi = \frac{1}{4}(1+2\delta)$, so that $x^\delta \le (x/d)^\psi$ uniformly for $1 \le d \le x^\theta$. Note that, since $\delta < \frac{1}{2}$, $\psi < \frac{1}{2}$. The contribution towards the multiple sums in Theorem 1 which ultimately arises from the terms with $d \le x^\theta$ is estimated, by means of an application of the Cauchy–Schwarz inequality, not to exceed

$$\sum_{d \le x^\theta} \frac{|h(d)|}{d} \sum_{d \le x^\theta} d|h(d)| \sum_{l^4 < p \le (x/d)^\psi} (p-1) \max_{t \le x/d} N(t,p)^2.$$

Since g_1 is exponentially multiplicative, the inner sum is

$$\ll d^{-2} \frac{x^2 (\log l)^4}{d^2 l^2},$$

and the two sums over the $d^{-1}|h(d)|$ contribute only bounded extra factors, by Lemma 7.

We proceed in a similar manner when $x^\theta < d \le x$, save that the inner sum, which now runs over the moduli $l^4 < p \le x^\delta$, is estimated by Lemma 3. The corresponding contribution towards the bound in Theorem 1 is

$$\ll \sum_{x^\theta < d \le x} \frac{|h(d)|}{d^{1/2}} \sum_{x^\theta < d \le x} d^{1/2} |h(d)| \left(\frac{x}{d} + x^{2\delta} l \right) \frac{x}{d},$$

which by Lemma 7 is $\ll x^{2-\theta} l^3 + x^{1+2\delta} l^{5/2}$.

The proof of Theorem 1 is complete. □

Proof of Theorem 2 I apply the main lemma with the choice $t = x$, the maxima being deleted. To remove the weights $\gamma(x, \rho, n)$ we need an estimate for the sum

$$\sum_{y < d \le y^{1+\epsilon}} \frac{1}{d} \sum_{\chi (\bmod d)}^{*} \left| \sum_{n \le x} w(n) \chi(n) \{\gamma(x, \rho, n) - 1\} \right|,$$

where $w(n) = g(n) \log n - h_1(n) - h_2(n)$. An application of the Cauchy–Schwarz inequality shows that this sum does not exceed

$$\sqrt{\sum_{d \le y^{1+\epsilon}} \frac{1}{d} \sum_{y < d \le y^{1+\epsilon}} \sum_{\chi (\bmod d)}^{*} \left| \sum_{n \le x} w(n) \chi(n) \{\gamma(x, \rho, n) - 1\} \right|^2},$$

which, by Lemma 3, is $\ll \sqrt{(\rho + y^{2(1+\epsilon)}) x (\log x)^3}$.

Similar arguments enable us to remove the conditions $m > y^{2+3\epsilon}$ and $p > y^{2+3\epsilon}$ in the functions h_1 and h_2, respectively, thus converting them into the k_j, at the expense of an amount $\ll y^{2+3\epsilon} z l^{1/2}$. Altogether

$$\sum_{y<d\leqslant y^{1+\epsilon}} \frac{1}{d} \sum_{\chi \,(\mathrm{mod}\, d)}^{*} \left| \sum_{n\leqslant x} g(n)\chi(n)\log n - K_1(x,\chi) - K_2(x,\chi) \right| \tag{3}$$

does not exceed a constant multiple of $\Delta + \sqrt{(\rho + y^{2(1+\epsilon)})xl^3} + y^{2+3\epsilon} z l^{1/2}$, where we may freely choose z (in Δ) and ρ within the constraints $z \geqslant 2$ and $1 \leqslant \rho \leqslant \frac{1}{2}x$. If we set $z = l^{6A+15}$, $\rho = xl^{-2A-16/3}$ and assume that $y \leqslant x^{\delta}$, where $\delta < \frac{1}{2}$, then, for sufficiently small ϵ, this upper bound is $\ll y^{-1} x l^{2A+7} + x l^{-A-1}$.

To complete the proof of Theorem 2 I argue as in the proof of Lemma 4. Without loss of generality the parameter t may be assumed to be half an odd positive integer. Then with $\rho = \frac{1}{4}$, there is a representation

$$\sum_{n\leqslant t} g(n)\chi(n)\log n - K_1(t,\chi) - K_2(t,\chi) = \frac{1}{2\pi i} \int_{(\alpha)} L(s)\, K_1(t,s)\, ds,$$

where

$$L(s) = \sum_{n\leqslant x} g(n)\chi(n) n^{-s} \log n - \sum_{j=1}^{2} \sum_{n\leqslant x} k_j(n)\chi(n) n^{-s}$$

and $\alpha = (\log x)^{-1}$. This time Lemma 6 shows that the integral of $\sup|K_1(t,s)|$ $(1 \leqslant t \leqslant x)$ taken over the whole line $\mathrm{Re}\, s = \alpha$, is $O(\log x)$. Since our estimate for the expression (3) remains valid when the function $n \mapsto g(n)$ is everywhere replaced by $n \mapsto g(n)n^{-s}$, Theorem 2 is established. □

4 Proof of the Main Lemma

For each prime $p \leqslant x$, define

$$G_p = 1 + \sum_{k=1}^{\infty} g(p^k)\chi(p^k)p^{-ks} = \exp(g(p)\chi(p)p^{-s})$$
$$(s = \sigma + i\tau,\ \sigma > 1).$$

This definition of G_p differs slightly from that employed in the second paper of this series. Define

$$G_1 = \prod_{p\leqslant y^{2+3\epsilon}} G_p, \qquad G_0 = \prod_{y^{2+3\epsilon} < p \leqslant x} G_p$$

and $G = G_1 G_0$. Then $G' = (G_1'/G_1)G + G(G_0'/G_0)$. The first of these two summands I further decompose as

$$\frac{G_1'}{G_1}\sum_{n\leqslant y^{2+3\epsilon}}\frac{g(n)\chi(n)}{n^s}+\left(\frac{G_1'}{G_1}+\sum_{p\leqslant z}\frac{g(p)\chi(p)\log p}{p^s}\right)\left(G-\sum_{n\leqslant y^{2+3\epsilon}}\frac{g(n)\chi(n)}{n^s}\right)$$
$$-\sum_{p\leqslant z}\frac{g(p)\chi(p)\log p}{p^s}\left(G-\sum_{n\leqslant y^{2+3\epsilon}}\frac{g(n)\chi(n)}{n^s}\right)$$
$$=\sum_{m=1}^{2}J_m-\sum_{n=1}^{\infty}h_1(n)\chi(n)n^{-s}.$$

The second I decompose as

$$\left(G-\sum_{n\leqslant z}\frac{g(n)\chi(n)}{n^s}\right)\frac{G_2'}{G_2}+\frac{G_2'}{G_2}\sum_{n\leqslant z}\frac{g(n)\chi(n)}{n^s}=J_3-\sum_{n=1}^{\infty}h_2(n)\chi(n)n^{-s}.$$

In terms of these Dirichlet series there is a representation

$$\sum_{n\leqslant t}g(n)\chi(n)\log n\gamma(t,\rho,n)-S_1-S_2=-\frac{1}{2\pi i}\sum_{m=1}^{3}\int_{(\alpha_m)}J_mK_1(t,\rho,s)\,ds,$$

where the integrals are taken along the lines $\operatorname{Re}s=\alpha_m$, with $\alpha_1=\frac{1}{2}$ and $\alpha_2=\alpha_3=1+(\log x)^{-1}$. The presence of the kernel K_1 ensures that these integrals converge absolutely. The sum to be estimated therefore does not exceed

$$\sum_{m=1}^{3}\sum_{y<d\leqslant y^{1+\epsilon}}\frac{1}{d}\sum_{\chi(\bmod d)}^{*}\int_{(\alpha_m)}|J_m|\sup_{2\rho\leqslant t\leqslant x}|K_1(t,\rho,s)|\,d\tau.$$

Applications of the Cauchy–Schwarz inequality, first for sums and then for integrals, show that the sum involving J_1 does not exceed $\sqrt{M_1M_2}$, where

$$M_1=\sum_{y<d\leqslant y^{1+\epsilon}}\frac{1}{d}\sum_{\chi(\bmod d)}^{*}\int_{(\alpha_1)}\left|\frac{G_1'}{G_1}\right|^2\sup_{2\rho\leqslant t\leqslant x}|K_1(t,\rho,s)|\,d\tau$$

and M_2 is a similar expression, but with G_1'/G_1 replaced by the sum of the $g(n)\chi(n)n^{-s}$ over the integers $n\leqslant y^{2+3\epsilon}$. An application of Lemma 1 shows that, on $\operatorname{Re}s=\frac{1}{2}$,

$$\sum_{d\leqslant y^{1+\epsilon}}\sum_{\chi(\bmod d)}^{*}\left|\frac{G_1'}{G_1}\right|^2\ll(y^{2+3\epsilon}+y^{2(1+\epsilon)})\sum_{p\leqslant y^{2+3\epsilon}}\frac{(\log p)^2}{p}\ll y^{2+4\epsilon},$$

so that, after Lemma 6, $M_1\ll y^{1+4\epsilon}x^{1/2}l$. A similar bound may be obtained for M_2, and therefore for $\sqrt{M_1M_2}$.

In a like manner the terms involving J_2 do not exceed $\sqrt{M_3M_4}$, where

$$M_4 = \sum_{y<d\leqslant y^{1+\epsilon}} \sum_{\chi(\mathrm{mod}\, d)}^{*} \int_{(\alpha_2)} \left| G - \sum_{n\leqslant y^{2+3\epsilon}} g(n)\chi(n)n^{-s} \right|^2 \sup_{2\rho\leqslant t\leqslant x} |K_1| \, d\tau,$$

and M_3 has the square in the integrand replaced by

$$d^{-2}\left| \frac{G_1'}{G_1} + \sum_{p\leqslant z} \frac{g(p)\chi(p)\log p}{p^s} \right|^2.$$

To estimate M_4 I employ the bound $K_1 \ll \rho^{-1}x|s|^{-2}$ and then apply Plancherel's theorem, as in Lemma 5. This gives for M_4 a bound

$$\ll \frac{x^2}{\rho} \int_0^\infty e^{-2\sigma u} \sum_{y<d\leqslant y^{1+\epsilon}} \sum_{\chi(\mathrm{mod}\, d)}^{*} \left| \sum_{y^{2+3\epsilon}<n\leqslant e^u} g(n)\chi(n) \right|^2 du.$$

The multiple sum in this integrand may be estimated by Lemma 1, so that

$$M_4 \ll \frac{x^2}{\rho} \int_{e^u \geqslant y^{2+3\epsilon}} e^{-2\sigma u}(e^u + y^{2(1+\epsilon)}) \sum_{n\leqslant e^u} |g(n)|^2 \, du \ll \frac{x^2 l}{\rho}.$$

The argument for M_3 runs along similar lines until we reach

$$M_3 \ll \frac{x^2}{\rho} \int_0^\infty e^{-2\sigma u} \sum_{y<d\leqslant y^{1+\epsilon}} \frac{1}{d^2} \sum_{\chi(\mathrm{mod}\, d)}^{*} \left| \sum_{z<p\leqslant e^u} g(p)\chi(p)\log p \right|^2 du.$$

It is here convenient to partition the interval $(y, y^{1+\epsilon}]$ into subintervals of the form $(w, 2w]$, with possibly a partial interval at the upper end. For those terms $p > y^{2+3\epsilon}$ we apply Lemma 2, obtaining

$$\sum_{w<d\leqslant 2w} \frac{1}{d^2} \sum_{\chi(\mathrm{mod}\, d)}^{*} \left| \sum_{\substack{z<p\leqslant e^u \\ p>y^{2+3\epsilon}}} g(p)\chi(p)\log p \right|^2 \ll \frac{1}{w^2}\left(\frac{e^u}{u} + w^{2+\epsilon_0} \right) e^u u$$

for each fixed $\epsilon_0 > 0$. The innermost sum is empty unless $e^u > y^{2+3\epsilon}$. Then, if ϵ_0 is sufficiently small in terms of ϵ, this bound is $\ll w^{-2}e^{2u}$. Summing over the various w in $(y, y^{1+\epsilon}]$ we gain towards the bound for M_3 an amount $\ll x^2 l(\rho y^2)^{-1}$. The sums

$$\sum_{w<d\leqslant 2w} \frac{1}{d^2} \sum_{\chi(\mathrm{mod}\, d)}^{*} \left| \sum_{\substack{z<p\leqslant e^u \\ p\leqslant y^{2+3\epsilon}}} g(p)\chi(p)\log p \right|^2$$

we estimate, by Lemma 1, to be $\ll w^{-2}(e^u + w^2)e^u \min(u, \log y)$, giving rise in the bound for M_3 to an amount

$$\ll \frac{x^2}{\rho}\left(\frac{l\log y}{y^2} + \log y \int_{e^u \geq z} e^{-u} u \, du\right) \ll \frac{x^2}{\rho}\left(\frac{l\log y}{y^2} + \frac{\log y \log z}{z}\right).$$

Hence

$$\sqrt{M_3 M_4} \ll \frac{x^2}{\rho}\sqrt{\frac{l^2 \log y}{y^2} + \frac{l \log y \log z}{z}}.$$

There remains the sum involving J_3. This does not exceed $\sqrt{M_5 M_6}$, where

$$M_6 = \sum_{y<d\leq y^{1+\epsilon}} \sum_{\chi(\operatorname{mod} d)}^{*} \int_{(\alpha_3)} \left|\frac{G_2'}{G_2}\right|^2 \sup_{2\rho\leq t\leq x} |K_1| \, d\tau,$$

and M_5 is the similar expression with $|G_2'/G_2|^2$ replaced by

$$d^{-2}\left|G - \sum_{n\leq z} \frac{g(n)\chi(n)}{n^s}\right|^2.$$

These expressions are estimated along the lines for M_3 and M_4, applying Lemma 2 and Lemma 1 respectively, to give

$$\sqrt{M_5 M_6} \ll \frac{x^2}{\rho}\sqrt{\frac{l^2}{y^2} + \frac{l \log y \log z}{z}}.$$

The proof of the main lemma is complete. □

Appendix: Motivation

Let $Q \geq 2$, and let χ_j denote primitive Dirichlet characters to moduli not exceeding Q. Let $x \geq 1$. We may regard the expression

$$\sum_{j=1}^{m} \left|\sum_{n\leq x} \chi_j(n) a_n\right|^2$$

as $\|Ta\|^2$, where T is an appropriate operator on the space $\mathbb{C}^{[x]}$, of complex n-tuples $\boldsymbol{a}$, into the space $\mathbb{C}^m$, both spaces being equipped with the usual Euclidean norm. Thus

$$\|Ta\|^2 = \sum_{k=1}^{m} \lambda_k |c_k|^2,$$

where $c_j = (T^*T\boldsymbol{a}, \boldsymbol{d}_k)$, and the $\boldsymbol{d}_k$ are the eigenvectors of the self-adjoint operator T^*T: $\mathbb{C}^{[x]} \to \mathbb{C}^{[x]}$, with corresponding eigenvalues $\lambda_1 \geq \lambda_2 \geq \cdots \geq \lambda_m \geq 0$. Disregarding maxima over ranges, we might consider the case when the χ_j run through the primitive characters to all prime moduli not exceeding Q. In this case

$$m = \sum_{p \leq Q} (p-1) = \frac{Q^2}{2\log Q} + O\left(\frac{Q^2}{(\log Q)^2}\right).$$

It is not easy to determine the λ_k directly, but we may consider the operator T^*: $\mathbb{C}^m \to \mathbb{C}^{[x]}$, which is adjoint to T, and with it

$$\|T^*b\|^2 = \sum_{n \leq x} \left| \sum_{j=1}^{m} \chi_j(n)\, b_j \right|^2.$$

This expression proves more tractible. The eigenvalues of TT^* are all of the form $x + O(mQ \log Q)$ and, in any case, $\ll x + Q^2$. Note that the matrix representing TT^* has diagonal entries

$$\sum_{n \leq x} |\chi_j(n)|^2 \qquad (j = 1, \ldots, m),$$

so that a largest eigenvalue of TT^* is at least

$$m^{-1} \operatorname{tr} TT^* = m^{-1} \sum_{j=1}^{m} \sum_{n \leq x} |\chi_j(n)|^2 = x + O(m^{-1} xQ \{\log Q\}^{-1}).$$

The non-zero eigenvalues of T^*T and TT^* coincide and, in our particular example, for $Q \ll \sqrt{x}$ the bound $\lambda_1 \ll x$ is essentially best possible.

The operator T^*T has a null space of dimension at least $[x] - m$, which is large for small values of $Qx^{-1/2}$. Let P denote the projection onto this space. Then these considerations of duality enable us to obtain the estimates

$$\|Ta\|^2 = \{x + O(mQ \log Q)\} \|a - Pa\|^2.$$

In general we settle for the bound $\|Ta\|^2 \ll (x + Q^2)\|a\|^2$. This is otherwise represented by

$$\sum_{p \leq Q} \sum_{r=1}^{p-1} \left| \sum_{\substack{n \leq x \\ n \equiv r \,(\mathrm{mod}\, p)}} a_n - \frac{1}{p-1} \sum_{\substack{n \leq x \\ (n,p)=1}} a_n \right|^2 \ll (x + Q^2) \sum_{n \leq x} |a_n|^2. \tag{4}$$

For background see Elliott [2], Forti & Viola [7] and Bombieri [1].

Consider next the modified expression

$$\sum_{j=1}^{m} \left| \sum_{q \leq x} \chi_j(q)\, a_q \right|^2,$$

where the inner summation runs over the prime powers q rather than the positive integers n. We may of course estimate this expression by specializing our earlier bound for $\|Ta\|^2$. However, better can be done.

In this modified situation T is replaced by A, an operator defined on the space $\mathbb{C}^t$ of the smaller dimension

$$t = \sum_{q \leqslant x} 1 = \frac{x}{\log x} + O\left(\frac{x}{(\log x)^2}\right).$$

Calculation of $\operatorname{tr} AA^*$ shows that a maximal eigenvalue of A^*A is at least

$$m^{-1} \sum_{j=1}^{m} \sum_{q \leqslant x} |\chi_j(q)|^2 = \frac{x}{\log x} + O\left(\frac{x}{(\log x)^2}\right).$$

In accordance with the (somewhat widely held) notion that the primes behave more or less as randomly as the integers, one may hope to obtain an inequality of the type (4), but with the leading term x replaced by $x/\log x$. In the event we have

$$\sum_{p \leqslant Q} (p-1) \sum_{r=1}^{p-1} \left| \sum_{\substack{q \leqslant x \\ q \equiv r \pmod p}} a_q - \frac{1}{p-1} \sum_{\substack{q \leqslant x \\ (q,p)=1}} a_q \right|^2 \ll \left(\frac{x}{\log x} + Q^{2+\epsilon}\right) \sum_{q \leqslant x} |a_q|^2,$$

valid for each fixed $\epsilon > 0$.

Similar trace calculations show that under favourable circumstances one may hope to replace the leading coefficients x or $x/\log x$, as the case may be, by the appropriate length of support of the arithmetic function a_n $(n = 1, 2, \ldots)$. For background see Elliott [3], [4, Chapter 6].

An arithmetic function $f(n)$ is *additive* if it satisfies the relation $f(ab) = f(a) + f(b)$ for mutually prime positive integers a and b. Functional-analytic considerations show that it is reasonable to look for a bound of the form

$$\sum_{p \leqslant Q} (p-1) \max_{y \leqslant x} \max_{(r,p)=1} \left| \sum_{\substack{n \leqslant y \\ n \equiv r \pmod p}} f(n) - \frac{1}{p-1} \sum_{\substack{n \leqslant y \\ (n,p)=1}} f(n) \right|^2 \ll \Delta \sum_{q \leqslant x} \frac{|f(q)|^2}{q}. \quad (5)$$

A direct application of (4) together with the Turán–Kubilius inequality shows that if we omit the maxima over $y \leqslant x$, then we may take

$\Delta = x^2+Q^2x$. This is not good enough for certain interesting applications, where it would be desirable to have an estimate $\Delta = x(\log x)^{-c}$ for some $c > 0$, at least when Q is a sufficiently small power of x.

The additive function f may be expressed as a Dirichlet convolution of two arithmetic functions, one of which is supported on the prime powers. In terms of the corresponding Dirichlet series

$$F(s) = \sum_{n=1}^{\infty} f(n)n^{-s}$$

we have $F(s) = \zeta(s)H(s)$, where $H(s) = \sum f(q)q^{-s}$, the sum running over all prime powers q. Accordingly, one might hope to treat the multiple sum at (5) by carefully peeling away the effect of the arithmetic function which is everywhere 1, corresponding to $\zeta(s)$, and reducing oneself to the application of an operator which, like A, is supported on the prime powers. In this way an estimate $\Delta = x^2/\log x+\cdots$ might be expected. As to its dependence upon x, this would then be largely best possible.

Indeed, I could show in this way [5] that one may take $\Delta = x^2(\log\log x)^2(\log x)^{-1}+Q^{2+\epsilon}x$ for any fixed $\epsilon > 0$. The above idea already motivated the treatment of the similar (but slightly weaker) result obtained in Chapter 7 of my book [4]. For background see Elliott [5], [4, Chapter 7].

A direct application of Lemma 3 (c.f. (4)) to the multiplicative function g yields an upper bound $\ll x^2$ in Theorem 1, again weak. If

$$G(s) = \sum_{n=1}^{\infty} g(n)n^{-s}$$

is the Dirichlet series corresponding to g, then the series $-G'(s)$ which corresponds to $n \mapsto g(n)\log n$ may be factorized $-G' = (-G'/G)G$. Thus $g(n)\log n$ is a convolution of two arithmetic functions, one of which is (again) supported on the prime powers. Here $-G'/G$ is essentially $\sum q^{-s}g(q)\log q$. By reducing to an application of the operator A, one might hope to obtain an estimate of the form

$$\sum_j \left| \sum_{n\leqslant x} \chi_j(n)\, g(n)\log n \right|^2 \ll \left(\frac{x}{\log x}+\cdots\right) \sum_{q\leqslant x} |g(q)\log q|^2 \ll x^2,$$

and so obtain in Theorem 1 the upper bound $O(x^2\{\log x\}^{-2})$. This I almost manage. If the maxima over $y \leqslant x$ are removed, then $(\log\log x)^4$ in the upper bound may be reduced to $(\log\log x)^2$. Perhaps, with a suitable modification, the present proof would allow the remaining factor of $(\log\log x)^2$ to be removed.

Ultimately we are reduced to a bound for

$$\max_{|a_q|\leqslant 1} \sum_{p\leqslant Q} (p-1) \max_{(r,p)=1} \left| \sum_{\substack{q\leqslant x \\ q\equiv r (\mathrm{mod}\, p)}} a_q - \frac{1}{p-1} \sum_{\substack{q\leqslant x \\ (q,p)=1}} a_q \right|^2 .$$

Setting $a_q = \exp(2\pi i d_q \theta)$ for distinct integers d_q and integrating with respect to θ over the interval $0 \leqslant \theta \leqslant 1$, Parseval's relation gives for this maximum over the special $\boldsymbol{a}$ a lower bound $\gg \sqrt{xQ/\log x \log Q}$. We have at present no treatment of the operators T and A which leads to error terms of anywhere near this accuracy. Note that $\sqrt{xQ}$ is $\ll x(\log x)^{-c}$ as soon as $Q \ll x(\log x)^{-2c}$. It seems likely that there is a general asymptotic expansion

$$\|T\boldsymbol{a}\|^2 \sim \sum_{r=0}^{\infty} \psi_r x^2 (\log x)^{-r},$$

where the coefficients $\psi_r = \psi_r(\boldsymbol{a})$ $(r = 0, 1)$ become essentially zero when $\boldsymbol{a}$ represents a multiplicative function g which satisfies $|g(n)| \leqslant 1$.

Note that since Theorem 1 deals only with prime moduli, at the expense of an extra factor of $(\log x)^{\epsilon}$ $(\epsilon > 0)$ in the upper bounds, we can determine a $\beta(\epsilon)$ $(0 < \beta < 1)$ so that a modified result is valid with summation running over prime moduli up to $\sqrt{x}\exp(\{\log x\}^{\beta})$.

References

[1] E. Bombieri, Le grand crible dans la théorie analytique des nombres. *Astérisque* **18** *(1974), Société Math. de France, Paris*

[2] P. D. T. A. Elliott, On inequalities of large sieve type. *Acta Arithmetica*, **18** (1971), 405–22.

[3] P. D. T. A. Elliott, Subsequences of primes in arithmetic progressions with large moduli, Turán Memorial Volume, *Studies in Pure Mathematics*, Birkhäuser, Basel–Boston, MA, (1983), 157–64

[4] P. D. T. A. Elliott, *Arithmetic Functions and Integer Products*, Grund. der math. Wiss., **272**, Springer, New York–Berlin–Tokyo, 1985

[5] P. D. T. A. Elliott, Additive arithmetic functions on arithmetic progressions, *Proc. London Math. Soc* (third series), **54** (1987), 15–37.

[6] P. D. T. A. Elliott, Multiplicative functions on arithmetic progressions, II. *Mathematika*, **35** (1988), 38–50

[7] M. Forti & C. Viola. On large sieve type estimates for the Dirichlet series operator, *Proc. Symposia in Pure Math.*, Amer. Math. Soc., XXIV (1973), 31–49.

[8] P. X. Gallagher. The large sieve, *Mathematika*, **14** (1967), 14–20.

Locally finite groups of permutations of $\mathbb{N}$ acting on l^∞

Matthew Foreman*

Abstract

We show that Martin's axiom implies that there is a locally finite group acting on $\mathbb{N}$ admitting a unique invariant mean on l^∞.

1 Introduction

If G is an amenable group acting on a set X, then there is a G-invariant, positive linear functional on $l^\infty(X)$ (the space of bounded real-valued functions with domain X). Such linear functionals are called G-invariant means and each one is uniquely determined by a finite additive G-invariant probability measure $\mu\colon \mathcal{P}(X) \to [0,1]$.

Rosenblatt asked whether G can completely determine the mean.

Yang [3] showed that under the continuum hypothesis there is a locally finite (hence amenable) group with a unique invariant finitely additive probability measure $\mu\colon \mathcal{P}(\mathbb{N}) \to [0,1]$. This measure is onto $[0,1]$. In this paper we show that Martin's axiom (MA) implies that there is a locally finite group with a unique invariant finitely additive probability measure $\mu\colon \mathcal{P}(\mathbb{N}) \to [0,1]$. Moreover, this measure is an ultrafilter, that is, it only takes values in $\{0,1\}$. Further, any given ultrafilter is the unique invariant measure of some locally finite group. It is open under MA whether there is a locally finite (or amenable) group with a unique invariant mean μ and $\mu\colon \mathcal{P}(\mathbb{N}) \xrightarrow[\text{onto}]{} [0,1]$.

It is known that large classes of amenable groups (e.g. solvable groups [2]) do not admit a unique invariant mean. The author has shown that analytic groups of permutations of $\mathbb{N}$ do not admit a unique invariant

* This work was partially supported by NSF Grant # DMS-8701119.

mean. For this and related matters see [1]. The general problem has a long history which we will not attempt to summarize.

2

Let U be an ultrafilter on $\mathbb{N}$. An almost disjoint sequence of infinite sets $\langle A_\alpha : \alpha < \mathfrak{c}\rangle$ *penetrates* U if and only if $\mathcal{P}(\mathbb{N})\backslash U = \langle Y_\alpha : \alpha < \mathfrak{c}\rangle$ and, for all $\alpha < \gamma < \mathfrak{c}$, $A_\gamma \cap Y_\gamma = \varnothing$ and $A_\gamma \cap Y_\alpha$ is finite. (Note that this depends on the enumeration $\langle Y_\alpha : \alpha < \mathfrak{c}\rangle$.)

The following fact is well known.

Proposition (Martin's axiom) *Let U be an ultrafilter with $\mathcal{P}(\mathbb{N})\backslash U = \langle Y_\alpha : \alpha \in \mathfrak{c}\rangle$. Then there is an almost disjoint sequence of sets $\langle A_\alpha : \alpha \in \mathfrak{c}\rangle$ penetrating U.*

We now state our main theorem.

Theorem *Suppose U is an ultrafilter on $\mathbb{N}$ and $\langle A_\alpha : \alpha \in \mathfrak{c}\rangle$ penetrates U. Then there is a locally finite group G of permutations of $\mathbb{N}$ with U as its unique G-invariant finitely additive probability measure.*

Note This is a theorem of ZFC.

Corollary *If Martin's axiom holds then for every ultrafilter U there is an amenable group G acting on $\mathbb{N}$ with a unique G-invariant mean μ on $l^\infty(\mathbb{N})$ and μ is the mean induced by U.*

Proof of the theorem Let $\langle A_\alpha : \alpha \in \mathfrak{c}\rangle$ penetrate U with respect to the enumeration $\langle Y_\alpha : \alpha \in \mathfrak{c}\rangle = \mathcal{P}(\mathbb{N})\backslash U$. For each α partition A_α into disjoint infinite pieces $\langle A_\alpha^n : n \in \mathbb{N}\rangle$ (i.e. $A_\alpha^n \cap A_\alpha^m = \varnothing$ and $\bigcup_n A_\alpha^n = A_\alpha$). For each pair n, α choose

$$s_\alpha^n : A_\alpha^n \xrightarrow[\text{onto}]{1-1} A_\alpha^{n+1}$$

and define

$$\sigma_\alpha^n(k) = \begin{cases} s_\alpha^n(k) & k \in A_\alpha^n, \\ (s_\alpha^n)^{-1}(k) & k \in A_\alpha^{n+1}, \\ k & \text{otherwise.} \end{cases}$$

Then σ_α^n is a permutaion of $\mathbb{N}$.

For each $\gamma < \mathfrak{c}$ we let

$$t_\gamma : Y_\gamma \xrightarrow[\text{onto}]{1-1} A_\gamma^0$$

and let

$$\tau^\gamma(k) = \begin{cases} \tau^\gamma(k) & k \in Y_\gamma, \\ \tau_\gamma^{-1}(k) & k \in A_\gamma^0, \\ k & \text{otherwise.} \end{cases}$$

Then τ_γ is a permutaion of ℕ. □

Main claim

$$\{\sigma_\alpha^n : n \in \mathbb{N}, \alpha \in \mathfrak{c}\} \cup \{\tau_\gamma : \alpha \in \mathfrak{c}\}$$

generates a locally finite subgroup of the permutation group on ℕ.

Proof It is enough to see for each $N \in \mathbb{N}$ and $\alpha_1, \ldots, \alpha_j, \gamma_1, \ldots, \gamma_m \in \mathfrak{c}$ there is a $B \in \mathbb{N}$ which bounds the cardinality of the orbits of ℕ under the group generated by $\{\sigma_{\alpha_i}^n : i \leqslant j, n \leqslant N\} \cup \{\tau_{\gamma_l} : 1 \leqslant l \leqslant m\}$. In other words, if

$$H = \langle \{\sigma_{\alpha_i}^n : i \leqslant j, n \leqslant N\} \cup \{\tau_{\gamma_l} : 1 \leqslant l \leqslant m\} \rangle$$

and $k \in \mathbb{N}$ then $|Hk| \leqslant B$. This uniform bound implies that H can be faithfully embedded in the product of finitely many copies of the permutaion group S_B.

We assume that $\gamma_1 < \gamma_2 < \cdots < \gamma_m$. We prove by induction on $0 \leqslant l \leqslant m$ that

$$H_l = \langle \{\sigma_{\alpha_i}^n : 1 \leqslant i \leqslant j, 0 \leqslant n \leqslant N\} \cup \{\tau_{\gamma_{l'}} : l' \leqslant l\} \rangle$$

has the property that each orbit is bound by some B'.

Consider $l = 0$. In this case we just have the group H_0 generated by $\{\sigma_{\alpha_i}^n : 1 \leqslant i \leqslant j, 0 \leqslant n \leqslant N\}$. We note that $\sigma_{\alpha_i}^n$ is the identity on the complement of $A_{\alpha_i}^n \cup A_{\alpha_i}^{n+1}$.

Let $K \in \mathbb{N}$ be so large that, for $0 \leqslant i \leqslant i' \leqslant j$,

$$A_{\alpha_i} \cap A_{\alpha_{i'}} \subseteq \{0, 1, \ldots, K-1\}.$$

For $k \in \mathbb{N}$, if $H_0 k \cap K = \emptyset$ then there is at most one i with $k \in A_{\alpha_i}$. Hence $|H_0 k| \leqslant N$. Thus it is enough to show that $H_0 K$ is finite. But

$$H_0 K = \bigcup_{\substack{i \leqslant j \\ n \leqslant N}} \sigma_{\alpha_i}^n K$$

since $\sigma_{\alpha_i}^n \restriction \tilde{A}_{\alpha_i}$ is the identity and $A_{\alpha_i} \cap A_{\alpha_{i'}} \subseteq K$. So $B' = j(N+1)(K+1)$ works for the uniform bound.

Assume by induction that H_l is finite. Let $H_{l+1} = \langle H_l \cup \{\tau_{\gamma_{l+1}}\} \rangle$ with $\gamma = \gamma_{l+1} > \gamma_{l'}$, for $l' \leqslant l$. We show that the orbits of H_{l+1} are uniformly bounded.

Let $X \subseteq \mathbb{N}$ be finite, with

$$\begin{aligned} &\text{(a)}\quad X \supseteq Y_{\gamma_{l'}} \cap A_\gamma && (l' \leqslant l); \\ &\text{(b)}\quad X \supseteq A_\gamma \cap A_{\gamma_{l'}} && (l' \leqslant l); \\ &\text{(c)}\quad X \supseteq A_\gamma \cap A_{\alpha_i} && (i \leqslant j,\ \alpha_i \neq \gamma). \end{aligned}$$

If $H_{l+1}k \cap X = \varnothing$ then

$$H_{l+1}k = H_l k \cup \tau_\gamma H_l k \cup H_l \tau_\gamma H_l k \cup \tau_\gamma H_l \tau_\gamma H_l k \cup H_l \tau_\gamma H_l \tau_\gamma H_l k,$$

whence $|H_{l+1}k| \leqslant 5|H_l|^3$. Hence it is enough to show that $H_{l+1}X$ is finite. Let

$$H_l' = \langle \{\sigma^n_{\alpha_i} : 1 \leqslant i \leqslant j,\ n \leqslant N,\ \alpha_i \neq \gamma\} \cup \{\tau_{\gamma_{l'}} : l' \leqslant l\} \rangle$$

and let $S = \langle \{\sigma^n_\gamma : n \leqslant N\} \rangle$ if $\gamma = \alpha_i$ for some i. So $H_l = \langle H_l' \cup S \rangle$. We note that

(a) for $k \in A_\gamma \backslash X$, $H_l' k = \{k\}$;
(b) for $k \notin A^0_\gamma \cup Y_\gamma$, $\tau_\gamma(k) = k$;
(c) if $\tau_\gamma(k) \neq k$ and $\sigma^n_\gamma(k) \neq k$, then $n = 0$ and $k \in A^0_\gamma$;
(d) for $k \notin \bigcup_{n \leqslant N} A^n_\gamma$, $\sigma^n_\gamma(k) = k$;
(e) for $k \in X$, $S \upharpoonright (H_l' k \backslash X)$ is the identity;
(f) for $k \in A^0_\gamma$, $Sk \cap A^0_\gamma = \{k\}$.

So

$$\langle H_l, \tau_\gamma \rangle X = H_l' X \cup S t_\gamma^{-1}(H_l' X \cap Y_\gamma) \cup SX \cup H_l' t_\gamma (SX \cap A^0_\gamma) \cup S t_\gamma^{-1}(H_l' t_\gamma (SX \cap A^0_\gamma) \cap Y_\gamma)$$

(recall that t_γ was used to define τ_γ).

Hence $|\langle H_l, t_\gamma \rangle X| \leqslant 5|H_l'|^3|S|^3|X|$. Thus the orbits of elements of $\mathbb{N}$ under H_{l+1} are uniformly bounded by $\max\{5|H_l'|^3|S|^3|X|, 5|H_l|^3\}$. This proves the main claim. □

Let G be the group generated by $\{\sigma^n_\alpha : \alpha < \mathfrak{c},\ n \in \mathbb{N}\} \cup \{\tau_\gamma : \gamma \in \mathfrak{c}\}$. We note that each σ^n_α and each τ_γ is the identity on a set of the ultrafilter. Hence U yields a G-invariant finitely additive probability measure by setting

$$\mu(X) = \begin{cases} 1 & \text{if } X \in U, \\ 0 & \text{if } X \notin U. \end{cases}$$

To finish the theorem we need only show that there is no other invariant finitely additive probability measure. For this it is sufficient to show that if $Y \notin U$ ($Y \subseteq \mathbb{N}$) then, for every G-invariant probability measure ν, $\nu(Y) = 0$.

If $Y \notin U$ the $Y = Y_\gamma$ for some γ. Since $\tau_\gamma Y = A_\gamma^0$ and

$$\sigma_\gamma^n \circ \sigma_\gamma^{n-1} \circ \cdots \circ \sigma_\gamma^0 \circ \tau_\gamma Y = A_\gamma^{n+1},$$

we know that if ν is any G-invariant probability measure, for all n, $\nu(Y) = \nu(A_\gamma^n)$. But $\{Y\} \cup \{A_\gamma^n : n \in \mathbb{N}\}$ are pairwise disjoint. Hence $\nu(Y) = 0$. This completes the proof of the theorem. □

References

[1] M. Foreman, Amenable group actions on the integers, *Bulletin Amer. Math. Soc.*, **21** (2) (1989)

[2] S. Krasa, The action of a solvable group on an infinite set never has a unique invariant mean, *TAMS*, **305** (1) (Jan. 1988) 369–76

[3] Z. Yang, Uniqueness of invariant means, *J. Functional Analysis*, (to appear)

Hypergraph games and the chromatic number

Fred Galvin*

In this paper a *hypergraph* is a finite collection $\mathcal{H}$ of finite sets H with $|H| \geqslant 2$. The elements of $\mathcal{H}$ are the *edges* and the elements of $V(\mathcal{H}) = \bigcup \mathcal{H}$ are the *vertices* of $\mathcal{H}$. The *rank* of $\mathcal{H}$ is $\|\mathcal{H}\| = \sup\{|H| : H \in \mathcal{H}\}$. A hypergraph is *uniform* if its edges all have the same cardinality. A uniform hypergraph consisting of r-element sets is an *r-graph*: a 2-graph is a *graph*, a 3-graph is a *triple system*, a 4-graph is a *quadruple system*, and so on. The *complete r-graph on a set V* is $[V]^r = \{H \subseteq V : |H| = r\}$. A set $S \subseteq V(\mathcal{H})$ is *independent* if it does not contain an edge. The hypergraph $\mathcal{H}$ is *n-colourable* if $V(\mathcal{H})$ is the union of n independent sets. The *chromatic number* of $\mathcal{H}$ is $\chi(\mathcal{H}) = \min\{n : \mathcal{H} \text{ is } n\text{-colourable}\}$; $\mathcal{H}$ is *n-chromatic* if $\chi(\mathcal{H}) = n$.

Given a hypergraph $\mathcal{H}$ and positive integers m and n, the *weak biased game* of Beck & Csirmaz [1, p. 305], denoted by $G(m, n, \mathcal{H})$, is defined as follows. Two players, called Maker and Breaker, take turns choosing previously unchosen vertices. Maker goes first and chooses m vertices at a time; Breaker goes second and chooses n vertices at a time; the game ends when all the vertices of $\mathcal{H}$ have been chosen. Thus a play of $G(m, n, \mathcal{H})$ results in a partition of $V(\mathcal{H})$ into two subsets: the set M of vertices chosen by Maker and the set B of vertices chosen by Breaker. Maker wins if M contains an edge; Breaker wins if M is independent. Clearly the game is determined, that is, either Maker or Breaker has a winning strategy. We will say that the hypergraph $\mathcal{H}$ is *n-fragile* if the game $G(1, n, \mathcal{H})$ is a win for Breaker.

Note that if a graph (i.e. a 2-graph) is n-fragile then each of its vertices has degree at most n; hence every n-fragile graph is $(n+1)$-

* The author was partly supported by NSF Grant 8802856.

colourable. On the other hand, the complete graph on $n+1$ vertices is n-fragile but not n-colourable; we record for later use the easy generalization of this example to r-graphs.

Lemma 1 *For $n \geqslant 1$ and $r \geqslant 2$, there is an $(n+1)$-chromatic n-fragile r-graph, e.g. the complete r-graph $[V]^r$, where*

$$(r-1)n < |V| \leqslant (r-1)(n+1).$$

Trivially, the only 0-fragile hypergraph is the empty hypergraph $\mathcal{H} = \varnothing$, which is also the only 1-colourable hypergraph. More significantly, it is well known that every 1-fragile hypergraph is 2-colourable; in other words, if $\chi(\mathcal{H}) > 2$, then $G(1,1,\mathcal{H})$ is a win for Maker [1, Proposition 4, p. 301]. In view of this, Beck & Csirmaz raised the natural question [1, p. 306], whether every n-fragile hypergraph is $(n+1)$-colourable. They stated in a footnote [1, p. 306] that Peter Frankl constructed a counterexample. In fact, Frankl's (unpublished) construction gives, for each $n \geqslant 2$, an $(n+2)$-chromatic n-fragile quadruple system.

In this paper we give some stronger counterexamples to the question of Beck & Csirmaz. Namely, we construct $(n+2)$-chromatic n-fragile triple systems for $n \geqslant 2$ (Theorem 2) and we also construct 2-fragile hypergraphs with arbitrarily high chromatic number (Theorem 3). Since the rank of the hypergraph constructed in Theorem 3 is a rapidly increasing function of the chromatic number, it is still conceivable that, for hypergraphs $\mathcal{H}$ of fixed rank $\|H\| = r$, there is a number $N(r)$ such that $G(1,2,\mathcal{H})$ is a win for Maker whenever $\chi(\mathcal{H}) > N(r)$.

Problem 1 *Do there exist 2-fragile triple systems with arbitrarily high chromatic number? Is every 2-fragile triple system 4-colourable?*

We begin by showing that the distinction between uniform and non-uniform hypergraphs is more or less irrelevant to the question considered here: a non-uniform hypergraph can be replaced by a (somewhat larger) uniform hypergraph with the same rank, fragility and chromatic number.

Theorem 1 *Let $r \geqslant 2$ and $1 \leqslant n \leqslant k$. Suppose there is a hypergraph $\mathcal{H}_0$, of rank $\|\mathcal{H}_0\| \leqslant r$, which is n-fragile but not k-colourable. Then there is an n-fragile r-graph $\mathcal{H}$ with $\chi(\mathcal{H}) = k+1$.*

Proof We use induction on k, for fixed r and n. If $k = n$, then by Lemma 1 we can take $\mathcal{H} = [V]^r$, where $|V| = (r-1)n+1$. Now suppose $k > 1$, and let $\mathcal{H}_0$ be an n-fragile but not k-colourable hypergraph with $\|\mathcal{H}_0\| \leqslant r$. By the inductive hypothesis, there is an n-fragile r-graph $\mathcal{H}'$ with $\chi(\mathcal{H}') = k$. Let $t = \max\{r-|H| : H \in \mathcal{H}_0\}$ and let $V = V_0 \cup V_1 \cup \cdots \cup V_t$, where $V_0, \ldots, V_t$ are disjoint sets, $V_0 = V(\mathcal{H}_0)$ and $|V_1| = \cdots = |V_t| = |V(\mathcal{H}')|$. For each $i \in \{1, \ldots, t\}$, let $\mathcal{H}_i$ be an isomorphic copy of $\mathcal{H}'$ with $V(\mathcal{H}_i) = V_i$. Then

$$\mathcal{H}^* = \{H \in [V]^r : H \cap V_0 \in \mathcal{H}_0\} \cup \mathcal{H}_1 \cup \cdots \cup \mathcal{H}_t$$

is an n-fragile r-graph with $\chi(\mathcal{H}^*) > k$. Since deleting an edge can lower the chromatic number by no more than 1, there is a subhypergraph $\mathcal{H} \subseteq \mathcal{H}^*$ with $\chi(\mathcal{H}) = k+1$. □

Lemma 2 *Let $n \geqslant 3$ and let $|V| = 2n^2+n$. There are graphs $\mathcal{G}_1, \mathcal{G}_2, \ldots, \mathcal{G}_n \subseteq [V]^2$ such that*
(a) $\mathcal{G}_2 = \cdots = \mathcal{G}_n$;
(b) each $\mathcal{G}_i$ has maximum degree $n-1$;
(c) for any partition $V = V_1 \cup \cdots \cup V_n$, one has $[V_i]^2 \cap \mathcal{G}_i \neq \varnothing$ for some i.

Proof Let $V = B \cup X_1 \cup Y_1 \cup \cdots \cup X_n \cup Y_n$, where $B, X_1, Y_1, \ldots, X_n, Y_n$ are disjoint n-element sets. Let $B = \{b_1, \ldots, b_n\}$; choose $x_j \in X_j$ and $y_j \in Y_j$ for $j = 1, \ldots, n$. Abbreviating $\{x, y\}$ to xy, we define

$$\mathcal{G}_1 = \bigcup_{j=1}^{n} (\{xy : x_j \neq x \in X_j,\ y_j \neq y \in Y_j\} \cup \{b_j x_j, b_j y_j\})$$

and, for $2 \leqslant i \leqslant n$, we define

$$\mathcal{G}_i = [B]^2 \cup [X_1]^2 \cup [Y_1]^2 \cup \cdots \cup [X_n]^2 \cup [Y_n]^2.$$

Clearly (a) and (b) hold; we have to verify (c). Suppose

$$V = V_1 \cup \cdots \cup V_n.$$

We may assume that each of the sets $B, X_1, Y_1, \ldots, X_n, Y_n$ contains at least one point of V_1: otherwise it contains at least two points of some V_i $(i \geqslant 2)$ and we have $[V_i]^2 \cap \mathcal{G}_i \neq \varnothing$. Thus $b_j \in V_1$ for some j: choose $x \in X_j \cap V_1$ and $y \in Y_j \cap V_1$. Now, if $x \neq x_j$ and $y \neq y_j$, then $\{x, y\} \in [V_1]^2 \cap \mathcal{G}_1$; on the other hand, if (say) $x = x_j$, then $\{b_j, x_j\} \in [V_1]^2 \cap \mathcal{G}_1$. □

Lemma 3 *If $n \geqslant 3$ and $|W| = 2n^2+2n$, then there is an $(n+1)$-chromatic $(n-1)$-fragile hypergraph $\mathcal{H} \subseteq [W]^2 \cup [W]^3$.*

Proof Let $W = V \cup A$, where $V \cap A = \varnothing$, $|V| = 2n^2+2$, $|A| = n$ and $A = \{a_1, \ldots, a_n\}$. Choose $\mathcal{G}_1, \ldots, \mathcal{G}_n \subseteq [V]^2$ satisfying (b) and (c) of Lemma 2, and define $\mathcal{H} = [A]^2 \cup \{\{a_i, x, y\} : \{x, y\} \in \mathcal{G}_i\ (1 \leqslant i \leqslant n)\}$.

Let a partition $W = W_1 \cup \cdots \cup W_n$ be given. Since $[A]^2 \subseteq \mathcal{H}$, we can assume without loss of generality that $a_i \in W_i$ for $i = 1, \ldots, n$. By (c), we have that $[W_i]^2 \cap \mathcal{G}_i \neq \emptyset$ for some i. Choose $\{x,y\} \in [W_i]^2 \cap \mathcal{G}_i$; then $W_i \supseteq \{a_i, x, y\}$. This shows that $\mathcal{H}$ is not n-colourable. In fact, the chromatic number is exactly $n+1$ since W is the union of the independent sets $V, \{a_1\}, \ldots, \{a_n\}$.

In order to win $G(1, n-1, \mathcal{H})$, Breaker must choose $n-1$ vertices from A on his first turn; if possible he will choose the remaining vertex of A on his second turn, thereby winning. To prevent this, Maker must choose some vertex a_i on one of his first two turns. Then they are in effect playing $G(1, n-1, \mathcal{G}_i)$ and this is a win for Breaker because of (b). □

Theorem 2 *For each $n \geqslant 3$, there is an $(n+1)$-chromatic $(n-1)$-fragile triple system.*

Proof This follows from Lemma 3 and Theorem 1. □

The interested reader may verify that the hypergraph in Lemma 3 can be constructed with $n^4 - \frac{1}{2}(n^3 + 3n^2 - 6n)$ edges, of which $\frac{1}{2}n(n-1)$ are pairs and the rest triples, while the triple system of Theorem 2 can have $2n^2 + 4n - 1$ vertices and $n^4 + \frac{1}{6}(11n^3 - 45n^2 + 46n - 6)$ edges.

For technical reasons, we define a game $G^+(1, 2, \mathcal{H})$ which is the same as $G(1, 2, \mathcal{H})$ except that Maker gets to choose an extra vertex at the start, that is, Maker chooses two vertices at his first turn, but only one at every subsequent turn; Breaker chooses two vertices at every turn.

Lemma 4 *Let $k \geqslant 3$, $V_1, \ldots, V_m$ be disjoint k-element sets, $\mathcal{H} = \{V_1, \ldots, V_m\}$ and $\mathcal{I} = \{\{x_1, \ldots, x_m\} : x_1 \in V_1, \ldots, x_m \in V_m\}$.*

(a) If $m \geqslant 2^{k-2}$, then $G(2, 1, \mathcal{H})$ is a win for Maker.

(b) If $m \geqslant 2^{k-2} + 1$, then $G(1, 2, \mathcal{I})$ is a win for Breaker.

(c) If $m \geqslant 2^{k-2} + 2$, then $G^+(1, 2, \mathcal{I})$ is a win for Breaker.

Proof Assertion (a) is easily proved by induction on k; (b) and (c) follow from (a). □

Lemma 5 *For each $n \geqslant 2$, there is an n-chromatic hypergraph $\mathcal{H}$ such that $G^+(1, 2, \mathcal{H})$ is a win for Breaker.*

Proof We use induction on n. For $n = 2$, let $\mathcal{H}$ be a triple system with one edge. For $n = 3$, let $|V| = 7$ and $\mathcal{H} = [V]^4$: alternatively, let $\mathcal{H}$ be the Steiner triple system on 7 vertices.

Now let $n \geqslant 2$, and suppose there is an n-chromatic hypergraph $\mathcal{H}'$ such that $G^+(1, 2, \mathcal{H}')$ is a win for Breaker. Let $k = |V(\mathcal{H}')|$, $m = 2^{k-2} + 2$ and $V = V_1 \cup \cdots \cup V_m$, where $V_1, \ldots, V_m$ are disjoint k-element sets. For each $i \in \{1,\ldots,m\}$ let $\mathcal{H}_i$ be an isomorphic copy of $\mathcal{H}'$ with $V(\mathcal{H}_i) = V_i$. Let $\mathcal{I} = \{\{x_1,\ldots,x_m\} : x_1 \in V_1,\ldots,x_m \in V_m\}$ and $\mathcal{H} = \mathcal{I} \cup \mathcal{H}_1 \cup \cdots \cup \mathcal{H}_m$.

By Lemma 4(c), the game $G^+(1,2,\mathcal{I})$ is a win for Breaker. Now Breaker's winning strategy in $G^+(1,2,\mathcal{I})$ can be combined with his winning strategies in the games $G^+(1,2,\mathcal{H}_i)$ to produce a winning strategy for Breaker in $G^+(1,2,\mathcal{H})$. As long as Maker chooses vertices from different V_i's, Breaker responds according to his strategy in $G^+(1,2,\mathcal{I})$; but whenever Maker chooses a vertex from some V_i from which he has already chosen, the additional vertex in V_i is irrelevant to the game $G^+(1,2,\mathcal{I})$, and so Breaker responds according to his strategy in $G^+(1,2,\mathcal{H}_i)$. (Our reason for defining the game $G^+(1,2,\mathcal{H})$ as we did, with Maker choosing two vertices at his first turn, was just to make possible the combining of strategies.)

It is easy to see that $\mathcal{H}$ is not n-colourable: in order to avoid monochromatic edges in the $\mathcal{H}_i$'s, each of the n colours would have to occur in every V_i; but then $\mathcal{I}$ would have monochromatic edges. On the other hand, $\mathcal{H}$ is $(n+1)$-colourable. Namely, we colour each V_i with n of the $n+1$ colours in such a way that each colour is omitted from at least one of the V_i's; this is possible since $m \geqslant k+1 \geqslant n+1$. Thus $\chi(\mathcal{H}) = n+1$, and this completes the proof of Lemma 5. □

Theorem 3 *There are 2-fragile uniform hypergraphs with arbitrarily high chromatic number.*

Proof This follows from Lemma 5 and Theorem 1. □

For $r \geqslant 2$, let $N(r)$ be the lest number n (if one exists) such that no r-graph with chromatic number greater than n is 2-fragile; equivalently, if $\mathcal{H}$ is any hypergraph with $\|\mathcal{H}\| \leqslant r$ and $\chi(\mathcal{H}) > n$, then $G(1,2,\mathcal{H})$ is a win for Maker. Trivially, $N(2) = 3$; Theorem 2 shows that $N(3) \geqslant 4$ (if it exists). Problem 1 was the case $r = 3$ for the following question.

Problem 2 *Does $N(r)$ exist for $r \geqslant 3$?*

It may be of interest to examine the lower bounds for $N(r)$ that are implicit in the proof of Lemma 5. In other words, we are interested in the upper bounds for

$$R(n) = \min\{\|\mathcal{H}\| : \chi(\mathcal{H}) \geqslant n \text{ and } \mathcal{H} \text{ is 2-fragile}\}.$$

Let us also define

$$A(n) = \min\{|V(\mathcal{H})| : \chi(\mathcal{H}) \geqslant n \text{ and } \mathcal{H} \text{ is 2-fragile}\}.$$

and

$$B(n) = \min\{|V(\mathcal{H})| : \chi(\mathcal{H}) \geqslant n \text{ and } G^+(1,2,\mathcal{H}) \text{ is a win for Breaker}\}.$$

Theorem 4 *For $n \geqslant 2$ the following inequalities hold:*

(a) $B(n+1) \leqslant (2^{B(n)-2}+2)\,B(n)$;

(b) $A(n+1) \leqslant (2^{B(n)-2}+1)\,B(n)$;

(c) $R(n+1) \leqslant 2^{B(n)-2}+1$;

(d) $B(n) \leqslant nA(n)$.

Proof The inequality (a) is clear form the proof of Lemma 5. Inequalities (b) and (c) are shown by an obvious modification of the construction in Lemma 5, using Lemma 4 (b) instead of Lemma 4 (c). For the proof of (d), let $\mathcal{H}_1, \ldots, \mathcal{H}_n$ be vertex-disjoint copies of an n-chromatic 2-fragile hypergraph with $A(n)$ vertices. Now let $\mathcal{H}$ be the hypergraph consisting of all the sets of the form $H \cup K$, where H belongs to one of the $\mathcal{H}_i$'s and K to another; $\mathcal{H}$ has $nA(n)$ vertices, and it is easy to see that $\mathcal{H}$ is n-chromatic and that $G^+(1,2,\mathcal{H})$ is a win for Breaker. □

Now, Frankl's example shows that $A(4) \leqslant 16$. By Theorem 4, it follows that $B(4) \leqslant 64$ and $R(5) \leqslant 2^{62}+1$; hence $N(2^{62}+1) \geqslant 5$.

References

[1] J. Beck & L. Csirmaz, Variations on a game, *J. Combin. Theory Ser. A*, **33** (1982), 297–315

On arithmetic graphs associated with integral domains

K. Győry*

1 Introduction

Let K be a finitely generated extension field of $\mathbb{Q}$, R a subring of K containing 1, S a finitely generated subgroup of the unit group of R with $-1 \in S$ and N a finite non-empty subset of $R \setminus \{0\}$. For each pair of distinct positive integers i, j, we select an element of N, denoted by δ_{ij}, such that $\delta_{ij} = \delta_{ji}$. If $A = \{\alpha_1, \ldots, \alpha_m\}$ is any finite ordered subset of R, then we denote by $\mathcal{G}(A)$ the simple graph with vertex set A whose edges are the (unordered) pairs $[\alpha_i, \alpha_j]$ for which

$$\alpha_i - \alpha_j \notin \delta_{ij} \cdot S.$$

The ordered subsets $A = \{\alpha_1, \ldots, \alpha_m\}$ and $A' = \{\alpha'_1, \ldots, \alpha'_m\}$ are called *S-equivalent* if $A' = \epsilon A + \beta$ for some $\epsilon \in S$ and $\beta \in R$. It is obvious that in this case the graphs $\mathcal{G}(A)$ and $\mathcal{G}(A')$ are isomorphic.

These graphs $\mathcal{G}(A)$ in the number field case (when R is the ring of integers of an algebraic number field) were introduced in 1972, in my paper [6]. Since then the graphs $\mathcal{G}(A)$ and their complements $\overline{\mathcal{G}(A)}$ have been investigated and applied in Győry ([7], [8], [9], [11]), Evertse *et al.* [5] (in the number field case) and in Győry ([10], [12]) and Leutbecher & Niklasch [14] (in the general case considered in the present paper). It has turned out that many problems of diophantine geometry, algebraic number theory and irreducible polynomials can be reduced to the study of ordered subsets $A = \{\alpha_1, \ldots, \alpha_m\}$ of R for which the graphs $\mathcal{G}(A)$ (or $\overline{\mathcal{G}(A)}$) have some specific connectedness properties. In [7] (see also [5]) it was shown (in the number field case and in an effective form) that, for given $m \geqslant 3$ and for all but finitely many S-equivalence classes of

* Research supported in part by Grant 273 from the Hungarian National Foundation for Scientific Research.

ordered subsets $A = \{\alpha_1, \dots, \alpha_m\}$ of R, the graphs $\mathcal{G}(A)$ possess certain well-utilizable connectedness properties. Further, it was proved that the structure of these graphs becomes simpler if m is sufficiently large. In [10] (see also [12]) these results were extended (in a weaker and ineffective form) to the more general situation considered above. The results of [7] (see also [5]) and [10] were applied to diophantine equations (cf. [8], [10]), algebraic number theory (cf. [9], [10]) and irreducible polynomials (cf. [11], [5]).

The main purpose of the present paper is considerably to generalize and refine (in an ineffective, but, in part, quantitative form) the above-mentioned results of [7], [10] and [5] on the graphs $\mathcal{G}(A)$ (cf. Theorems 1 and 2 in Section 2). These enable one to achieve significant generalizations and refinements of some earlier applications to irreducible polynomials and decomposable form equations. The generalized and improved versions of these applications will be published in separate papers. Some new applications to resultants of polynomials and resultant form equations will be established in Section 5 (cf. Theorems 7 and 8).

In Section 3, we shall give an equivalent formulation (cf. Theorem 5) of Theorem 1 in terms of the graphs $\overline{\mathcal{G}(A)}$. Theorem 5 is a diophantine finiteness assertion for certain ordered subsets $A = \{\alpha_1, \dots, \alpha_m\}$, In some applications (e.g. to certain decomposable form equations) it is more convenient to apply this form of Theorem 1.

Our Theorems 1 and 2 will be deduced from some general finiteness theorems (cf. Theorems 3 and 6) on unit equations which were recently obtained by Evertse & Győry [3], [4]. Moreover, we shall point out in Section 3 that Theorem 1 (on graphs $\mathcal{G}(A)$), Theorem 3 (on unit equations) and Theorem 4 (on systems of unit equations) are in fact equivalent. Further, in Section 4 we shall show that Theorem 2 (on graphs $\mathcal{G}(A)$) is equivalent to Theorem 6 (on unit equations). Our Theorems 1 and 2 considerably extend the applicability of unit equations and embrace many different applications.

2 Main results

We adopt the notation of Section 1. Let $A = \{\alpha_1, \dots, \alpha_m\}$ be an ordered subset of R with $m \geqslant 3$. Denote by $\overline{\mathcal{G}(A)}^0$ the *polygon hypergraph* of $\overline{\mathcal{G}(A)}$, i.e. that hypergraph whose vertices are the edges of $\overline{\mathcal{G}(A)}$ and whose edges are the cycles* $\alpha_{i_1}, \dots, \alpha_{i_k}$ $(k \geqslant 3)$ in $\overline{\mathcal{G}(A)}$ such that

* In other words, $\alpha_{i_1}, \dots, \alpha_{i_k}$ $(k \geqslant 3)$ are distinct elements of A such that $[\alpha_{i_1}, \alpha_{i_2}]$, …, $[\alpha_{i_{k-1}}, \alpha_{i_k}]$, $[\alpha_{i_k}, \alpha_{i_1}]$ are all edges in $\overline{\mathcal{G}(A)}$ (cf. [2]).

$$\sum_{j \in J} (\alpha_{i_j} - \alpha_{i_{j+1}}) \neq 0 \quad \text{for each non-empty subset } J \text{ of } \{1,2,\ldots,k-1\}. \tag{1}$$

If, in particular, we consider only cycles of length 3 (i.e. triangles) of $\overline{\mathscr{G}(A)}$ then the hypergraph so obtained is called the *triangle hypergraph* of $\overline{\mathscr{G}(A)}$ (cf. [1], p. 440) and is denoted by $\overline{\mathscr{G}(A)}^{\triangle}$ (cf. [7], [10]).

The number $C(m-1,S)$ appearing in Theorem 1 depends only on m and S. It will be defined in Theorem 3 below.

Theorem 1 *Let $m \geqslant 3$ be an integer. Then for all but at most $[(m+1)!\,C(m-1,S)]^{\binom{m}{2}}$ S-equivalence classes of ordered subsets $A = \{\alpha_1,\ldots,\alpha_m\}$ of R, one of the following cases holds:*

(a) *$\mathscr{G}(A)$ is connected and at least one of $\overline{\mathscr{G}(A)}$ and $\overline{\mathscr{G}(A)}^0$ is not connected;*

(b) *$\mathscr{G}(A)$ has two connected components, $\mathscr{G}_1$ and $\mathscr{G}_2$, say, such that $\overline{\mathscr{G}_1}$ is not connected and $|\mathscr{G}_2| = 1$.**

Furthermore, if $m = 4$,

(c) *$\mathscr{G}(A)$ has two connected components of order 2 and $\overline{\mathscr{G}(A)}^0$ is not connected.*

If S is infinite, then both the cases (a) and (b) can occur for infinitely many S-equivalence classes (cf. [7]). The same holds for case (c) when $m = 4$, $\delta_{24}/\delta_{13} \in S$ and $\delta_{14}/\delta_{23} \in S$. This is, for instance, the case for all but finitely many quadruplets $A = \{0, \delta_{13}\epsilon + \delta_{23}, \delta_{13}\epsilon, \delta_{23}\}$ with ϵ running through the elements of S.

Our Theorem 1 above generalizes and refines (in an ineffective form) Theorem 1 of [7] and Theorem 1 of [10]. In [7] and [10], in statement (a) only the triangle hypergraphs $\overline{\mathscr{G}(A)}^{\triangle}$ were considered instead of $\overline{\mathscr{G}(A)}^0$. The present version considerably extends the scope of the applications to diophantine equations. For instance, we shall show in Section 3 that Theorem 1 in the present form is equivalent to Theorem 3 on unit equations and Theorem 4 on systems of unit equations. Statement (c) in Theorem 1 above is much more precise than the corresponding statements (c) in [7] and [10]. This refinement in (c) is of crucial importance for certain applications to resultants of polynomials (cf. Section 5) and irreducible polynomials.

The independence of $C(m-1,S)$ of N which is very important for certain applications is new with respect to the earlier versions. It is a consequence of the non-explicit character of Theorem 3 that, apart from

* $|\mathscr{G}|$ denotes the order (number of vertices) of a graph $\mathscr{G}$. Moreover, $|A|$ will denote the cardinality of a finite set A.

certain special cases (cf. the Remark after the statement of Theorem 3), we are not able to make explicit $C(m-1,S)$ and hence the upper bound in Theorem 1.

In the number field case, we proved in [7] the existence of a number $C_1 = C_1(R,S,N)$ such that if $|\mathcal{G}(A)| > C_1$ then $\mathcal{G}(A)$ has at most two connected components and one of them is of order at least $|\mathcal{G}(A)|-1$. In [7], C_1 was given explicitely. Later, this explicit constant was improved in [5]. In [10], a weaker and non-explicit version was established in the general case considered in the present paper. Our Theorem 2 below is an explicit extension of the mentioned results of [7] and [5] to the general case.

To state Theorem 2 we need some further notation. Let $\{z_1,\ldots,z_q\}$ be a transcendence basis of K over $\mathbb{Q}$ and let $K_0 = \mathbb{Q}(z_1,\ldots,z_q)$. Then K is a finite extension of K_0 of degree d, say. The poynomial ring $\mathbb{O} = \mathbb{Z}[z_1,\ldots,z_q]$ is a unique factorization domain. To every prime element π of $\mathbb{O}$ there corresponds an (additive) valuation v_π on K_0 with the property that $v_\pi(\pi) = 1$ and $v_\pi(a/b) = 0$ if a and b are elements of $\mathbb{O}$ not divisible by π. Each of these valuations can be extended in at most d different ways to K. Let M_K denote the set of these valuations on K. For any finite subset Γ of M_K we put

$$G_\Gamma = \{\alpha \in K : v(\alpha) = 0 \text{ for all } v \in M_K \backslash \Gamma\}.$$

Then G_Γ is a subgroup of the multiplicative group K^* of non-zero elements of K, and one can show that it is finitely generated. Since the group S is finitely generated, it can be embedded in such a group G_Γ. Let Γ_S be the smallest subset of M_K for which $S \subseteq G_{\Gamma_S}$ and denote by s the cardinality of Γ_S.

Theorem 2 *Let $A = \{\alpha_1,\ldots,\alpha_m\}$ be a finite ordered subset of R. If*

$$m > 12 \times 7^{3d+2s}|N|^2$$

then $\mathcal{G}(A)$ has at most two connected components and one of them is of order at least $m-1$.

In Section 5, Theorem 2 will be applied to resultants of polynomials and resultant form equations. Further, we note that our Theorems 2 and 1 (as well as Theorem 5) can be applied to those subsets A of R for which $\overline{\mathcal{G}(A)}$ is complete. The structure of these subsets A was investigated e.g. by Leutbecher & Niklasch [12], who gave some important applications of their results to Euclidean number fields.

We remark that for groups S which are not finitely generated our Theorems 1 and 2 do not remain valid in general.

3 Proof of Theorem 1; equivalence of Theorems 1, 3, 4 and 5

Let K, R, S and N have the same meanings as in Sections 1 and 2. Let $n \geqslant 1$ be an integer. Points in the vector space K^{n+1} are denoted by $\mathbf{x} = (x_0, x_1, \ldots, x_n)$. If we identify pairwise linearly dependent non-zero points in K^{n+1}, we obtain the n-dimensional projective space $\mathbb{P}^n(K)$. Points in $\mathbb{P}^n(K)$, so-called projective points, are denoted by $\mathbf{x} = (x_0 : x_1 : \ldots : x_n)$, where the homogeneous coordinates are in K and are determined up to a multiplicative constant in K. We denote the subset of $\mathbb{P}^n(K)$ of projective points with all homogeneous coordinates in S by $\mathbb{P}^n(S)$. Let $\delta_0, \delta_1, \ldots, \delta_n \in K^*$ and consider the *unit equation*

$$\delta_0 x_0 + \delta_1 x_1 + \cdots + \delta_n x_n = 0 \qquad \text{in } \mathbf{x} = (x_0 : x_1 : \ldots : x_n) \in \mathbb{P}^n(S). \quad (2)$$

A solution $\mathbf{x}$ of (2) is called *non-degenerate* if $\delta_0 x_0 + \delta_1 x_1 + \cdots \delta_n x_n$ has no proper vanishing subsum (i.e. if $\sum_{i \in I} \delta_i x_i \neq 0$ for each proper non-empty subset I of $\{0, 1, \ldots, n\}$) and *degenerate* otherwise. If S is infinite and if (2) has a degenerate solution then (2) has infinitely many degenerate solutions.

Theorem 3 (Evertse and Győry [4]) *The number of non-degenerate solutions of* (2) *is at most* $C = C(n, S)$, *where* C *is a number depending only on* n *and* S.

Remark The proof in [4] does not make it possible to compute C explicitely. We may, however, assume that in Theorem 3 the number $C(n, S)$ is a monotonic function of the parameter n. An explicit expression for $C(2, S)$ was given by Evertse & Győry [3] (see also Theorem 6 of the present paper). At the last conference on diophantine approximations in Oberwolfach (14–18 March 1988) H. P. Schlickewei announced that, in the special case when $K = \mathbb{Q}$ and S is generated by s distinct prime numbers, $\{8(s+1)\}^{2^{26n+4}(s+1)^6}$ can be taken for $C(n, S)$. □

Let $k \geqslant 1$ be an integer and let $I_1, \ldots, I_k$ be subsets of $\{0, 1, \ldots, n\}$ with $\bigcup_j I_j = \{0, 1, \ldots, n\}$ and $|I_j| \geqslant 2$ for $j = 1, \ldots, k$. Put $|I_j| = n_j + 1$ for each j. Then, obviously, $n_j \leqslant n$. As a generalization of (2), consider the *system of unit equations*

$$\begin{aligned} \sum_{i\in I_1} \delta_{1i}x_i &= 0 \\ &\vdots \qquad \text{in } \boldsymbol{x} = (x_0:x_1:\ldots:x_n) \in \mathbb{P}^n(S), \\ \sum_{i\in I_k} \delta_{ki}x_i &= 0 \end{aligned} \tag{3}$$

where the coefficients δ_{ji} are non-zero elements of K. We shall say that a solution $\boldsymbol{x}$ of (3) is *non-degenerate* if none of the sums $\sum_{i\in I_j} \delta_{ji}x_i$ $(j = 1,\ldots,k)$ has a proper vanishing subsum. If (3) has a degenerate solution and S is infinite then (3) has infinitely many degenerate solutions. Let $\mathcal{H}_{(3)}$ denote the graph whose vertex set is $\{x_0,x_1,\ldots,x_n\}$ and whose edges are the pairs $[x_i,x_j]$ for which there exists an equation in (3) containing both x_i and x_j. The system of unit equations (3) will be called *connected* if the graph $\mathcal{H}_{(3)}$ is connected. It is easy to verify that if (3) is not connected and has a non-degenerate solution then it has infinitely many non-degenerate solutions (provided that S is infinite). Apart from the form of the bounds, the next theorem is equivalent to Theorem 2 of Evertse & Győry [4] on systems of unit equations.

Theorem 4 *Suppose that* (3) *is connected. Then the number of non-degenerate solutions of* (3) *is at most* $C(n_1,S)\cdots C(n_k,S)$, *where* $C(n,S)$ *denotes the number occurring in Theorem 3.*

We note that one can get in general a better upper bound for the number of solutions by applying Theorem 4 to a minimal connected subsystem of (3) which contains all the variables $x_0,x_1,\ldots,x_n$.

For $k = 1$ Theorem 4 reduces to Theorem 3. Conversely, Theorem 4 easily follows from Theorem 3, that is, Theorem 4 and Theorem 3 are equivalent. Indeed, by Theorem 3, the j-th equation in (2) has at most $C(n_j,S)$ non-degenerate solutions in $\mathbb{P}^{n_j}(S)$ for $j = 1,\ldots,k$. By assumption, (3) is connected and $I_1\cup\cdots\cup I_k = \{0,1,\ldots,n\}$. Hence it is easy to see that there are $j_1,\ldots,j_l$ in $\{1,\ldots,k\}$ with the following three properties:

$$I_{j_1}\cup\cdots\cup I_{j_l} = \{0,1,\ldots,n\};$$

for $p = 1,\ldots,l$ the system of equations

$$\begin{aligned} \sum_{i\in I_{j_1}} \delta_{j_1 i}x_i &= 0 \\ &\vdots \qquad \text{in } \mathbb{P}^{n'_p}(S) \text{ with } n'_p = |I_{j_1}\cup\cdots\cup I_{j_p}|-1 \\ \sum_{i\in I_{j_p}} \delta_{j_p i}x_i &= 0 \end{aligned} \tag{4}$$

is connected; and, for $p = 1,\ldots,l-1$, $I_{j_{p+1}}$ has at least one element not contained in $I_{j_1} \cup \cdots \cup I_{j_p}$. Then (4) and the j_{p+1}-th equation have a common variable for $p = 1,\ldots,l-1$, and one of them can be chosen to be 1. It follows by induction on p that the number of non-degenerate solutions of (4) is at most $C(n_{j_1},S)\cdots C(n_{j_p},S)$ for $p = 1,\ldots,l$. Consequently, (3) has at most $C(n_1,S)\cdots C(n_k,S)$ non-degenerate solutions. □

The next theorem is concerned with the complements of the graphs $\mathscr{G}(A)$.

Theorem 5 *Let $m \geqslant 3$ be an integer. There are at most $[(m+1)!\,C(m-1,S)]^{\binom{m}{2}}$ S-equivalence classes of ordered subsets $A = \{\alpha_1,\ldots,\alpha_m\}$ of R, for which both $\overline{\mathscr{G}(A)}$ and $\overline{\mathscr{G}(A)}^0$ are connected. (Here $C(n,S)$ denotes the bound occurring in Theorem 3.)*

We shall deduce Theorem 5 from Theorem 4. We note that Theorem 5 can be proven in a similar manner with $\overline{\mathscr{G}(A)}^0$ replaced by $\overline{\mathscr{G}(A)}^\triangle$ and with $[(m+1)!C(m-1,S)]^{\binom{m}{2}}$ replaced by $[2^mC(2,S)]^{\binom{m}{2}}$. As mentioned above, $C(2,S)$ was given explicitely in [3].

Proof of Theorem 5 Put $M = \{1,2,\ldots,m\}$. For ordered subsets $A = \{\alpha_1,\ldots,\alpha_m\}$ of R, we associate with $\mathscr{G}(A)$ the graph $\mathscr{G}(M)$ with vertex set M such that $[p,q]$ is an edge of $\mathscr{G}(M)$ if and only if $[\alpha_p,\alpha_q]$ is an edge of $\mathscr{G}(A)$. Then, replacing $\alpha_1,\ldots,\alpha_m$ by $1,\ldots,m$, respectively, to $\overline{\mathscr{G}(A)}^0$ there corresponds a hypergraph in an obvious way which will be denoted by $\overline{\mathscr{G}(M)}^0$. Let now $A = \{\alpha_1,\ldots,\alpha_m\}$ be any ordered subset of R for which both $\overline{\mathscr{G}(A)}$ and $\overline{\mathscr{G}(A)}^0$ are connected. Then the corresponding graphs $\overline{\mathscr{G}(M)}$ and $\overline{\mathscr{G}(M)}^0$ are also connected. It is easy to see that there exists a connected subhypergraph $\mathscr{H}_A$ of $\overline{\mathscr{G}(A)}^0$ with vertex set consisting of all the edges of $\overline{\mathscr{G}(A)}$ such that the number of edges of $\mathscr{H}_A$ is at most $\binom{m}{2}$. The subhypergraph of $\overline{\mathscr{G}(M)}^0$ corresponding to $\mathscr{H}_A$ will be denoted by $\mathscr{H}$. Consider an edge of $\mathscr{H}$ and let $\alpha_{i_1},\ldots,\alpha_{i_k}$ $(k \geqslant 3)$ be the corresponding cycle of $\overline{\mathscr{G}(A)}$ with property (1). Then $k \leqslant m$ and

$$(\alpha_{i_1}-\alpha_{i_2})+\cdots+(\alpha_{i_{k-1}}-\alpha_{i_k})+(\alpha_{i_k}-\alpha_{i_1}) = 0. \qquad (5)$$

Further, for each edge $[\alpha_p,\alpha_q]$ in $\overline{\mathscr{G}(A)}$ we have $\alpha_p-\alpha_q = \delta_{pq}x_{pq}$ with some $x_{pq} \in S$. Now (5) and (1) imply that $(x_{i_1i_2},\ldots,x_{i_{k-1}i_k},x_{i_ki_1})$ is a non-degenerate solution of the unit equation

$$\delta_{i_1i_2}x_{i_1i_2}+\cdots+\delta_{i_{k-1}i_k}x_{i_{k-1}i_k}+\delta_{i_ki_1}x_{i_ki_1} = 0. \qquad (6)$$

We can associate in the same way with each edge of $\mathscr{H}$ an equation of type (6), and we obtain in this way a connected system of unit equations in $x_{pq} \in S$ for each pair p, q for which $[\alpha_p, \alpha_q]$ is an edge of $\overline{\mathcal{G}(A)}$. For every further ordered subset $A' = \{\alpha'_1, \ldots, \alpha'_m\}$ of R for which $\mathcal{G}(A')$ and $\mathcal{H}_{A'}$ correspond to $\mathcal{G}(M)$ and $\mathcal{H}$, respectively, the tuple of numbers x'_{pq} defined by $\alpha'_p - \alpha'_q = \delta_{pq} x'_{pq}$ is also a non-degenerate solution of the system of unit equations corresponding to $\mathcal{H}$. But this system of equations consists of at most $\binom{m}{2}$ equations and the number of variables is at most m in each equation. It follows now from Theorem 4 that, for every A associated with $\mathcal{H}$,

$$\alpha_p - \alpha_q = \epsilon \gamma_{pq} \qquad \text{for each } \alpha_p, \alpha_q \text{ with } [\alpha_p, \alpha_q] \text{ an edge in } \overline{\mathcal{G}(A)}, \quad (7)$$

where $\epsilon \in S$ and the tuple (γ_{pq}) belongs to a set of tuples of cardinality at most $C(m-1, S)^{\binom{m}{2}}$ (which set is independent of A). It follows from the connectedness of $\overline{\mathcal{G}(A)}$ that, for every α_p, α_q with $1 \leqslant p, q \leqslant m$ $(p \neq q)$, there is in $\overline{\mathcal{G}(A)}$ a path of length at most m from α_p to α_q. Hence (7) implies that

$$\alpha_p - \alpha_q = \epsilon \rho_{pq} \qquad \text{for each } p, q \text{ with } 1 \leqslant p, q \leqslant m \ (p \neq q),$$

where the tuple (ρ_{pq}) belongs to a set of tuples of cardinality at most $C(m-1, S)^{\binom{m}{2}}$ (which set is independent of A). Putting now $A^* = \{0, \rho_{21}, \ldots, \rho_{m1}\}$, we have that $A = \epsilon A^* + \alpha_1$, where A^* belongs to a finite subset of R^m of cardinality at most $C(m-1, S)^{\binom{m}{2}}$. Thus, for fixed $\mathcal{H}$, there are at most $C(m-1, S)^{\binom{m}{2}}$ pairwise S-inequivalent ordered subsets $A = \{\alpha_1, \ldots, \alpha_m\}$ of R which correspond to $\mathcal{H}$ and for which $\overline{\mathcal{G}(A)}$ and $\overline{\mathcal{G}(A)}^0$ are connected. Hence it is enough to estimate from above the number of possible choices of $\mathcal{H}$. For $m = 3$, the number of choices is 1. Hence we assume that $m \geqslant 4$.

The number of graphs $\mathcal{G}(M)$ defined on M is at most $2^{\binom{m}{2}}$. This is an upper bound for the number of possible choices of the vertex set of $\mathcal{H}$. Further, for fixed $\mathcal{G}(M)$, the number of possible cycles in $\overline{\mathcal{G}(M)}$ is at most

$$\sum_{k=3}^{m} \binom{m}{k} k! \leqslant (m+1)!.$$

But $\mathcal{H}$ can have at most $\binom{m}{2}$ vertices. Hence, fixing the vertex set of $\mathcal{H}$, the number of possible choices for the edge set of $\mathcal{H}$ is at most

$$\sum_{k=1}^{\binom{m}{2}} \binom{(m+1)!}{k} \leqslant \binom{m}{2} \binom{(m+1)!}{\binom{m}{2}}.$$

Consequently, the number of possible $\mathcal{H}$ is at most

$$2^{\binom{m}{2}}\binom{m}{2}\binom{(m+1)!}{\binom{m}{2}} \leqslant (m+1)!^{\binom{m}{2}}$$

and this completes the proof of the theorem. □

We shall now show that Theorem 5 implies Theorem 3 with another upper bound. Hence, apart from the forms of the bounds, Theorems 3, 4 and 5 are equivalent.

For $n = 1$ the assertion of Theorem 3 is trivial, so it suffices to deal with the case $n \geqslant 2$. Let $m = n+1$, let R' be the subring of K generated by $\delta_0, \delta_1, \ldots, \delta_n$ and S over $\mathbb{Z}$, and let $N = \{1, \delta_0, \delta_1, \ldots, \delta_n\}$, where the δ_i are the coefficients occurring in (2). Then S is a subgroup of the unit group of R'. For distinct integers i and j with $1 \leqslant i, j \leqslant m$ we define the numbers δ_{ij} such that $\delta_{ij} = \delta_{ji}$, $\delta_{i+1,i} = \delta_{i-1}$ for $i = 1, \ldots, m-1$, $\delta_{1m} = \delta_n$ and $\delta_{ij} = 1$ otherwise. For every ordered subset $A = \{\alpha_1, \ldots, \alpha_m\}$ of R', consider the graph $\mathcal{G}(A)$ defined in Section 1. If $\boldsymbol{x} = (x_0 : x_1 : \ldots : x_n) \in \mathbb{P}^n(S)$ is an arbitrary non-degenerate solution of (2) with $x_0 = 1$, the numbers

$$\alpha_1 = 0, \qquad \alpha_2 = \delta_0, \qquad \alpha_3 = \delta_0 + \delta_1 x_1, \qquad \ldots,$$
$$\alpha_m = \delta_0 + \delta_1 x_1 + \cdots + \delta_{n-1} x_{n-1}$$

are pairwise distinct elements of R'. Further, it is easy to see that, for this $A = \{\alpha_1, \ldots, \alpha_m\}$, both $\overline{\mathcal{G}(A)}$ and $\overline{\mathcal{G}(A)}^0$ are connected. For distinct non-degenerate solutions $\boldsymbol{x}$ of (2), the corresponding subsets A so defined are S-equivalent. This implies that if $C_1(m, S)$ is an upper bound for the number of S-equivalence classes of ordered sets $A = \{\alpha_1, \ldots, \alpha_m\}$ in R' for which $\overline{\mathcal{G}(A)}$ and $\overline{\mathcal{G}(A)}^0$ are connected then (2) has at most $C_1(n+1, S)$ non-degenerate solutions. □

Before deducing Theorem 1 from Theorem 5, we note that if $A = \{\alpha_1, \ldots, \alpha_m\}$ is an ordered subset of R for which both $\overline{\mathcal{G}(A)}$ and $\overline{\mathcal{G}(A)}^0$ are connected, then for $\mathcal{G}(A)$ none of the statements (a), (b) and (c) of Theorem 1 can hold. Hence Theorem 1 implies Theorem 5. Consequently Theorem 1 and Theorem 5 are equivalent and thus, apart form the forms of the bounds, Theorems 1, 3, 4 and 5 are also equivalent.

Proof of Theorem 1 To prove Theorem 1 we shall use Theorem 5 as well as some of the arguments from [7] and [10]. Let $A = \{\alpha_1, \ldots, \alpha_m\}$ be an arbitrary but fixed ordered subset of R such that at least one of $\overline{\mathcal{G}(A)}$ and $\overline{\mathcal{G}(A)}^0$ is not connected. In view of Theorem 5 it will be enough to show that, for $\mathcal{G}(A)$, (a), (b) or (c) holds. Denote by l the number of connected components of $\mathcal{G}(A)$. It is easily seen that, for $l \geqslant 3$, both $\overline{\mathcal{G}(A)}$ and $\overline{\mathcal{G}(A)^{\triangle}}$ (and hence also $\overline{\mathcal{G}(A)^0}$) would be connected. Thus we have $l \leqslant 2$.

For $l = 1$, (a) holds. Next consider the case $l = 2$. Then $\overline{\mathcal{G}(A)}$ is connected and hence, by assumption, $\overline{\mathcal{G}(A)^0}$ is not connected. Let $\mathcal{G}_1$ and $\mathcal{G}_2$ be connected components of $\mathcal{G}(A)$ with $|\mathcal{G}_1| \geqslant |\mathcal{G}_2|$. If $|\mathcal{G}_2| = 1$ then $\overline{\mathcal{G}_1}$ cannot be connected because otherwise $\overline{\mathcal{G}(A)^{\triangle}}$ (and hence also $\overline{\mathcal{G}(A)^0}$) would be connected. Thus in this case (b) holds. It remains to consider the case when $|\mathcal{G}_2| \geqslant 2$.

First suppose that $|\mathcal{G}(A)| \geqslant 5$. Let A_1' and A_2' be arbitrary subsets of the vertex sets of $\mathcal{G}_1$ and $\mathcal{G}_2$, respectively, such that $2 \leqslant |A_2'| \leqslant |A_1'|$ and $|A_1'| \geqslant 3$. Then obviously $\overline{\mathcal{G}(A_1' \cup A_2')}$ is connected. Further, putting $t = |A_1' \cup A_2'|$, we have $t \geqslant 5$. We shall show by induction on t that $\overline{\mathcal{G}(A_1' \cup A_2')^0}$ is also connected. We prove first this assertion for $t = 5$. Let $A_1' = \{\alpha_{i_1}, \alpha_{i_2}, \alpha_{i_3}\}$ and $A_2' = \{\alpha_{i_4}, \alpha_{i_5}\}$. The sum

$$(\alpha_{i_4} - \alpha_{i_1}) + (\alpha_{i_1} - \alpha_{i_5}) + (\alpha_{i_5} - \alpha_{i_2}) + (\alpha_{i_2} - \alpha_{i_4}) \tag{8}$$

can have a proper vanishing subsum (with summands $(\alpha_{i_4} - \alpha_{i_1})$, $(\alpha_{i_1} - \alpha_{i_5})$, $(\alpha_{i_5} - \alpha_{i_2})$ and $(\alpha_{i_2} - \alpha_{i_4})$) only if

$$0 = (\alpha_{i_4} - \alpha_{i_1}) + (\alpha_{i_5} - \alpha_{i_2}) = (\alpha_{i_1} - \alpha_{i_5}) + (\alpha_{i_2} - \alpha_{i_4}). \tag{9}$$

Similarly, the only possible proper vanishing subsums of

$$(\alpha_{i_4} - \alpha_{i_1}) + (\alpha_{i_1} - \alpha_{i_5}) + (\alpha_{i_5} - \alpha_{i_3}) + (\alpha_{i_3} - \alpha_{i_4}) \tag{10}$$

(with summands $(\alpha_{i_4} - \alpha_{i_1})$, $(\alpha_{i_1} - \alpha_{i_5})$, $(\alpha_{i_5} - \alpha_{i_3})$ and $(\alpha_{i_3} - \alpha_{i_4})$) are

$$0 = (\alpha_{i_4} - \alpha_{i_1}) + (\alpha_{i_5} - \alpha_{i_3}) = (\alpha_{i_1} - \alpha_{i_5}) + (\alpha_{i_3} - \alpha_{i_4}). \tag{11}$$

But $\alpha_{i_2} \neq \alpha_{i_3}$. Hence (9) and (11) cannot hold simultaneously. Consequently, at least one of the quadruplets $\alpha_{i_4}, \alpha_{i_1}, \alpha_{i_5}, \alpha_{i_2}$ and $\alpha_{i_4}, \alpha_{i_1}, \alpha_{i_5}, \alpha_{i_3}$ is a cycle in $\overline{\mathcal{G}(A_1' \cup A_2')}$ with the 'non-vanishing' property (1). We amay assume that $\alpha_{i_4}, \alpha_{i_1}, \alpha_{i_5}, \alpha_{i_2}$ is a cycle with property (1). One can show in a similar way that at least one of the quadruplets $\alpha_{i_4}, \alpha_{i_3}, \alpha_{i_5}, \alpha_{i_1}$ and $\alpha_{i_4}, \alpha_{i_3}, \alpha_{i_5}, \alpha_{i_2}$ is also a cycle in $\overline{\mathcal{G}(A_1' \cup A_2')}$ with property (1). Hence it follows that $\overline{\mathcal{G}(A_1' \cup A_2')^0}$ is connected. Assume now that $t > 5$ and that the assertion has been proved for each integer t' with $5 \leqslant t' \leqslant t-1$. Let again $\alpha_{i_1}, \alpha_{i_2}, \alpha_{i_3} \in A_1'$ and $\alpha_{i_4}, \alpha_{i_5} \in A_2'$. For $|A_1'| > |A_2'|$, let $A_1'' = A_1' \backslash \{\alpha_{i_1}\}$ and $A_2'' = A_2'$ and, for $|A_1'| = |A_2'|$, let $A_1'' = A_2'$ and $A_2'' = A_1' \backslash \{\alpha_{i_1}\}$. Then, by the inductive hypothesis, both $\overline{\mathcal{G}(A_1'' \cup A_2'')^0}$ and $\overline{\mathcal{G}(\{\alpha_{i_1}, \alpha_{i_2}, \alpha_{i_3}, \alpha_{i_4}, \alpha_{i_5}\})^0}$ are connected. It is easy to see that then $\overline{\mathcal{G}(A_1' \cup A_2')}^0$ is also connected. Finally, it follows that $\overline{\mathcal{G}(A)}^0$ is connected, which contradicts the assumption.

It remains to consider the case when $l = 4$, $|\mathcal{G}_1| = |\mathcal{G}_2| = 2$ and $\overline{\mathcal{G}(A)}^0$ is not connected. However, this is just the case (c). □

4 Proof of Theorem 2; equivalence of Theorems 2 and 6

We shall keep the notation and terminology used in Sections 1 and 3. In the proof of Theorem 2 we shall need the following slight generalization of Theorem 1 of Evertse & Győry [3].

Theorem 6 *Let* $\lambda, \mu \in K^*$. *Then the equation*

$$\lambda x + \mu y = 1 \qquad \text{in } x, y \in N \cdot S \tag{12}$$

has at most $4 \times 7^{3d+2s} |N|^2$ *solutions.*

For $N = \{1\}$, Theorem 6 becomes Theorem 1 of [3].

Proof of Theorem 6 Every pair $(x,y) \in N \cdot S \times N \cdot S$ can be expressed as $x = \delta_1 x'$, $y = \delta_2 y'$, where $\delta_1, \delta_2 \in N$ and $x', y' \in S$. Fix a pair $(\delta_1, \delta_2) \in N^2$. If (x,y) is a solution of (12) of the form $x = \delta_1 x'$, $y = \delta_2 y'$ then

$$(\lambda\delta_1)x' + (\mu\delta_2)y' = 1 \qquad \text{with } x', y' \in S. \tag{13}$$

It follows from Theorem 1 of [3] and from the fact that $S \subseteq G_{\Gamma_s}$ that equation (13) has at most $4 \times 7^{3d+2s}$ solutions in $x', y' \in S$. This implies that (12) has at most $4 \times 7^{3d+2s}$ solutions (x,y) of the form $x = \delta_1 x'$, $y = \delta_2 y'$ with $x', y' \in S$. Since N^2 contains $|N|^2$ pairs, this proves Theorem 6. □

Proof of Theorem 2 To prove Theorem 2 we shall combine Theorem 6 with some arguments of the proof of Theorem 2 in [7]. Let $A = \{\alpha_1, \ldots, \alpha_m\}$ be a finite ordered subset of R and let $\mathcal{G}_1, \ldots, \mathcal{G}_l$ be the connected components of $\mathcal{G}(A)$ such that $|\mathcal{G}_1| \geqslant |\mathcal{G}_2| \geqslant \cdots \geqslant |\mathcal{G}_l|$. Suppose that $l \geqslant 3$ or that $l = 2$ and $|\mathcal{G}_2| \geqslant 2$. We select two elements α_{i_1} and α_{i_2} from A in the following way: for $l \geqslant 3$, let α_{i_1} and α_{i_2} belong to the vertex sets of $\mathcal{G}_l$ and $\mathcal{G}_{l-1}$, respectively, and, for $l = 2$ let α_{i_1} and α_{i_2} be elements of the vertex set of $\mathcal{G}_2$. Further, denote by A' the union of the vertex sets of $\mathcal{G}_1, \ldots, \mathcal{G}_{l-2}$ if $l \geqslant 3$ and the vertex set of $\mathcal{G}_1$ if $l = 2$. Then we have

$$\alpha_{i_2}-\alpha_{i_1} = (\alpha_{i_2}-\alpha_j)+(\alpha_j-\alpha_{i_1}) \tag{14}$$

for each element α_j of A'. Furthermore, $\alpha_{i_2}-\alpha_j, \alpha_j-\alpha_{i_1} \in N\cdot S$ for each $\alpha_j \in A'$. Hence Theorem 6 can be applied to (14) and we obtain that

$$|A'| \leqslant 4\times 7^{3d+2s}|N|^2. \tag{15}$$

On the other hand, we have $|A'| \geqslant \frac{1}{3}m$. Comparing this with (15), we conclude that if

$$m > 12\times 7^{3d+2s}|N|^2$$

then $l = 1$ or $l = 2$ and $|\mathcal{G}_2| = 1$. This completes the proof of Theorem 2. □

We shall now show that, conversely, Theorem 2 implies Theorem 6 with a slightly weaker bound. Therefore Theorem 2 and Theorem 6 are in fact equivalent. First we derive an upper bound for the number of solutions of (13) by using Theorem 2. Fix δ_1 and δ_2 in N and let R' be the subring of K generated by S, $\lambda' := \lambda\delta_1$ and $\mu' := \mu\delta_2$. Suppose that (13) is solvable and let (x_i', y_i') $(i = 1,\ldots,t)$ be distinct solutions of (13). Put $m = t+2$ and

$$\alpha_1 = 0, \qquad \alpha_2 = -1, \qquad \alpha_3 = -1+\lambda' x_1', \qquad \ldots, \qquad \alpha_m = -1+\lambda' x_t'.$$

For all distinct integers i and j with $1 \leqslant i, j \leqslant m$, define numbers δ_{ij} such that $\delta_{ij} = \delta_{ji}$, $\delta_{i_2} = \lambda'$, $\delta_{1i} = \mu'$ for $i = 3,\ldots,m$ and $\delta_{ij} = 1$ otherwise. Consider the graph $\mathcal{G}(A)$ with vertex set $A = \{\alpha_1,\ldots,\alpha_m\}$ and with N replaced by the set consisting of the elements δ_{ij} just defined. By assumption

$$\lambda' x_i' + \mu' y_i' = 1 \qquad \text{for } i = 1,\ldots,t,$$

from which it follows that $\alpha_1 = 0$ and $\alpha_2 = -1$ are isolated vertices of $\mathcal{G}(A)$. But then $\mathcal{G}(A)$ cannot have a connected component of order at least $|\mathcal{G}(A)|-1$. Thus Theorem 2 implies that

$$t+2 \leqslant 12\times 7^{3d+2s}\times 3^2,$$

whence

$$t \leqslant 108\times 7^{3d+2s}-2.$$

It was pointed out in the proof or Theorem 6 that (12) is equivalent to $|N|^2$ equations of the form (13). Therefore (12) has at most $(108\times 7^{3d+2s}-2)|N|^2$ solutions.

5 Applications to resultants of polynomials and resultant form equations

Let K be as above and let now R be an integrally closed integral domain with quotient field K. Suppose that R is finitely generated over $\mathbb{Z}$. The resultant of two polynomials $f, g \in R[x]$ will be denoted by $r(f,g)$. Let $\delta \in R\backslash\{0\}$. In the case $K = \mathbb{Q}$, $R = \mathbb{Z}$ and, for a given f, several results were established which imply the finiteness of the number of polynomials g satisfying the equation

$$r(f,g) = \delta; \tag{16}$$

see e.g. Wirsing [17] and Schmidt [16]. As an application of Theorems 1 and 2 we shall now deduce some finiteness results for (16) in the case when neither f nor g are fixed but all their roots lie in a fixed finite extension, say L, of K.

If f and g are monic polynomials in $R[x]$, $f^*(x) = f(x+a)$ and $g^*(x) = g(x+a)$ for some $a \in R$, then the pairs f, g and f^*, g^* have the same resultant. Such pairs (f,g) and (f^*,g^*) of polynomials in $R[x]$ will be called R-equivalent.

Theorem 7 *There are only finitely many pairwise R-inequivalent pairs of monic polynomials* $f, g \in R[x]$ *with* $\deg(f) \geqslant 2$, $\deg(g) \geqslant 2$ *and* $\deg(f) + \deg(g) \geqslant 5$, *without multiple roots and with splitting field L, such that (16) holds.*

It is easy to show that the finiteness assertion of Theorem 7 does not remain valid in general if f or g is linear or if $\deg(f)+\deg(g) = 4$.

Equation (16) can be expressed as a polynomial diophantine equation. Let m and n be positive integers, and consider the decomposable form

$$r(\boldsymbol{x}) = \prod_{\substack{i=1,\dots,m \\ j=m+1,\dots,m+n}} (x_i - x_j) \qquad \text{in } \boldsymbol{x} = (x_1, \dots, x_m, \dots, x_{m+n}).$$

$r(\boldsymbol{x})$ is the resultant of the polynomials

$$\prod_{i=1}^{m} (Y - x_i) \quad \text{and} \quad \prod_{j=m+1}^{m+n} (Y - x_j).$$

Hence $r(\boldsymbol{x})$ will be called a *resultant form*, and the decomposable form equation

$$r(\boldsymbol{x}) = \delta \qquad \text{in } \boldsymbol{x} \in R^{m+n} \tag{17}$$

a *resultant form equation*. If $\boldsymbol{x}$ is a solution of (17) then so is $\boldsymbol{x}+\boldsymbol{a}$ for each $\boldsymbol{a} = (a, \dots, a) \in R^{m+n}$. Such solutions will be called *equivalent*.

Let $\hat{R} = R[\delta^{-1}]$. Then $\hat{R}^*$, which is the unit group of $\hat{R}$, is finitely generated (cf. [13]). Let M_K and G_Γ be as in Section 2, and let Γ be the smallest subset of M_K for which $\hat{R}^* \subseteq G_\Gamma$. Then Γ is finite. Let t denote the cardinality of Γ. Further, denote by $N(m,n)$ the maximal number of pairwise inequivalent solutions $\boldsymbol{x}$ of (17) with pairwise distinct $x_1,\ldots,x_m,\ldots,x_{m+n}$. For $L = K$, the next theorem is a more precise version of Theorem 7.

Theorem 8 *Let m and n be integers with $m \geqslant 2$, $n \geqslant 2$ and $m+n \geqslant 5$. If $N(m,n) > 0$ then*

$$m+n \leqslant 12\times 7^{3d+2t} \tag{18}$$

and

$$N(m,n) \leqslant mn\{(m+n+1)!\,C(m+n,\hat{R}^*)\}^{\binom{m+n}{2}},$$

where $C(\cdot,\cdot)$ denotes the number occurring in Theorem 3.

It will be clear form the proof that in Theorem 8 it is not necessary to assume that R is integrally closed. Similarly to Theorem 7, the assumptions $\min\{m,n\} \geqslant 2$ and $m+n \geqslant 5$ in Theorem 8 are also necessary in general.

First we shall deduce Theorem 8 from Theorems 1 and 2. Then we shall prove Theorem 7 using Theorem 8. We note that from Theorem 8 one can easily deduce Theorem 7 with an upper bound for the number of solutions of (16).

Proof of Theorem 8 Every equivalence class of solutions of (17) contains a single solution $\boldsymbol{x}$ for which $x_1 = 0$. Suppose that $N(m,n) > 0$ and let $\boldsymbol{x} = (x_1 = 0, x_2, \ldots, x_{m+n})$ be an arbitrary solution of (17) in R^{m+n} with distinct $x_1,\ldots,x_{m+n}$. Since $\delta \in \hat{R}^*$, it follows from (17) that $x_i - x_j \in \hat{R}^*$ for each i and j with $1 \leqslant i \leqslant m$ and $m+1 \leqslant j \leqslant m+n$. Let $\delta_{uv} = 1$ for all distinct integers u and v with $1 \leqslant u, v \leqslant m+n$ and, for $A = \{x_1,\ldots,x_{m+n}\}$ consider the graph $\mathcal{G}(A)$ with the choice $S = \hat{R}^*$ and $N = \{1\}$. In view of $m \geqslant 2$ and $n \geqslant 2$, this graph cannot have any connected component of order at least $m+n-1$. By applying Theorem 2 with R and S replaced by $\hat{R}$ and $\hat{R}^*$, (18) follows. Further, if $m+n \geqslant 5$, Theorem 1 implies that

$$x_i = \epsilon\alpha_i^* + \gamma \qquad (i = 1,\ldots,m+n) \tag{19}$$

for some $\epsilon \in \hat{R}^*$, $\gamma \in \hat{R}$ and $(\alpha_1^*,\ldots,\alpha_{m+n}^*) \in \hat{R}^{m+n}$, where the number of possible tuples $(\alpha_1^*,\ldots,\alpha_{m+n}^*)$ is at most

$$C' := \{(m+n+1)!\,C(m+n,\hat{R}^*)\}^{\binom{m+n}{2}}.$$

(19) gives

$$x_i - x_j = \epsilon(\alpha_i^* - \alpha_j^*)$$

for all distinct i and j with $1 \leq i, j \leq m+n$. We see from (17) and (20) that, for fixed $\alpha_1^*, \ldots, \alpha_{m+n}^*$, ϵ^{mn} is also fixed, that is, ϵ can assume at most mn distinct values. Further, for fixed $\alpha_1^*, \ldots, \alpha_{m+n}^*$ and ϵ, the γ is uniquely determined by (19), since $x_1 = 0$. Therefore the number of solutions $(x_1 = 0, x_2, \ldots, x_{m+n})$ of (17) with distinct $x_1, \ldots, x_{m+n}$ is at most mnC', which completes the proof of Theorem 8. □

Proof of Theorem 7 Suppose that (16) has a solution $f, g \in R[x]$ with the properties specified in Theorem 7. Let $(x_1, \ldots, x_m)$ and $(x_{m+1}, \ldots, x_{m+n})$ be the root sets of f and g in L, respectively. Then $x_i \in \hat{R}_L$ for $i = 1, \ldots, m+n$, where $\hat{R}_L$ denotes the integral closure of $\hat{R}$ in L. Thus $\boldsymbol{x} = (x_1, \ldots, x_{m+n})$ is a solution of (17) in $\hat{R}_L^{m+n}$. $\hat{R}_L$ is an integrally closed integral domain with quotient field L. Further, both $\hat{R}_L$ and $\hat{R}_L^*$ (the unit group of $\hat{R}_L$) are finitely generated (cf. [15], [13]). Let f', g' be another solution of (16) with $\deg(f') = m$, $\deg(g') = n$ and with the required properties such that the pairs f, g and f', g' are not R-equivalent. Let $(x_1', \ldots, x_m')$ and $(x_{m+1}', \ldots, x_{m+n}')$ be arbitrary but fixed permutations of the root sets of f' and g', respectively. Then $(x_1', \ldots, x_{m+n}')$ is also a solution of (17) in $\hat{R}_L^{m+n}$. We show now that this solution cannot be equivalent to $(x_1, \ldots, x_{m+n})$. Indeed, assuming the contrary, we have $x_i' - x_i = a$ for $i = 1, \ldots, m+n$ with some $a \in \hat{R}_L$. This implies that

$$\sum_{i=1}^{m+n} x_i' - \sum_{i=1}^{m+n} x_i = na.$$

But $\sum_{i=1}^{m+n} x_i'$ and $\sum_{i=1}^{m+n} x_i$ are coefficients of $f \cdot g$ and $f' \cdot g'$, respectively, whence $a \in K$. Further, the x_i' and x_i are integral over R, whence a is also integral over R. Since by assumption R is integrally closed in K, we get that $a \in R$. This is, however, impossible because the pairs f', g' and f, g were supposed to be R-inequivalent. We can now apply Theorem 8 to the root sets of the solutions f, g of (16). By Theorem 8, equation (17) has only finitely many pairwise inequivalent solutions $\boldsymbol{x} \in \hat{R}_L^{m+n}$ with distinct components and thus the assertion follows. □

I thank J. H. Evertse for his remarks on an earlier draft of this paper.

References

[1] C. Berge, *Graphs and Hypergraphs*, North-Holland, Amsterdam–Holland–New York, 1973

[2] B. Bollobás, *Extremal Graph Theory*, Academic Press, London–New York–San Francisco, 1978

[3] J. H. Evertse & K. Győry, On unit equations and decomposable form equations, *J. Reine Angew. Math.*, **358** (1985), 6–19

[4] J. H. Evertse & K. Győry, On the numbers of solutions of weighted unit equations, *Compositio Math.*, **66** (1988), 329–54

[5] J. H. Evertse, K. Győry, C. L. Stewart & R. Tijdeman, *S*-unit equations and their applications, *New Advances in Transcendence Theory* (ed. A. Baker), Cambridge, 1988, 110–74

[6] K. Győry, Sur l'irréductibilité d'une classe de polynômes II, *Publ. Math. Debrecen*, **19** (1972), 293–326

[7] K. Győry, On certain graphs composed of algebraic integers of a number field and their applications I, *Publ. Math. Debrecen*, **27** (1980), 229–42

[8] K. Győry, On the representation of integers by decomposable forms in several variables, *Publ. Math. Debrecen*, **28** (1981), 89–98

[9] K. Győry, On discriminants and indeces of integers of an algebraic number field, *J. Reine Angew. Math.*, **324** (1981), 114–26

[10] K. Győry, On certain graphs associated with an integral domain and their applications to deophantine problems, *Publ. Math. Debrecen*, **29** (1982), 79–94

[11] K. Győry, On the irreducibility of a class of polynomials III, *J. Number Theory*, **15** (1982), 164–81

[12] K. Győry, Graphs associated with an integral domain and their applications, *Coll. Math. Soc. J. Bolyai*, **37**, Finite and Infinite Sets, Eger (Hungary), 1981, 349–58

[13] S. Lang, *Fundamentals of Diophantine Geometry*, Springer, Berlin–Heidelberg–New York, 1983

[14] A. Leutbecher & G. Niklasch, On cliques of exceptional units and Lenstra's construction of Euclidean fields, *Number Theory, Ulm 1987*, Springer Lecture Notes in Mathematics, **1380** (1989), 150–78

[15] M. Nagata, A general theory of algebraic geometry over Dedekind domains I, *Amer. J. Math.*, **78** (1956), 78–116

[16] W. M. Schmidt, Inequalities for resultants and for decomposable forms, *Diophantine Approximation and its Applications*, Academic Press, New York, 1973, 235–53

[17] E. Wirsing, On approximations of algebraic numbers by algebraic numbers of bounded degree, *Proc. Symp. in Pure Math.* XX (1971) (1969 Number Theory Institute), 213–47

On the number of certain subgraphs of graphs without large cliques and independent subsets

A. Hajnal, Z. Nagy and L. Soukup*

Abstract

A graph $g = \langle V, E\rangle$ without cliques or independent subsets of size $|V|$ is called non-trivial. We say that $G = \langle V, E\rangle$ is almost smooth iff it is isomorhphic to $C[W]$ whenever $W \subset V$ with $|V \setminus W| < |V|$. Given a graph $G = \langle V, E\rangle$, denote by $I(G)$ the set of all isomorphism classes of induced subgraphs of cardinality $|V|$. It is shown that

1. $|I(G)| \geqslant 2^{\omega}$ for each non-trivial graph $G = \langle \omega_1, E\rangle$;
2. under $\diamondsuit^+$ there is a non-trivial graph $G = \langle \omega_1, E\rangle$ with $|I(G)| = \omega_1$;
3. the existence of a nontrivial, almost smooth graph on ω_1 is consistent with different set-thoretic assumptions;
4. under $\diamondsuit^+$ there exists a family $\mathcal{F} \subset [\omega_1]^{\omega}$ which is non-trivial in a certain sense and which is isomorphic to $\mathcal{F} \restriction A$ whenever $A \subset \omega_1$ is an uncountable set.

1 Introduction

In 1981 R. Jamison posed the following problem: 'If an n-uniform hypergraph $G = \langle V, E\rangle$ is isomorphic to each of its induced subgraphs of cardinality $|V|$, must G be either empty or complete?' H. A. Kierstead & P. J. Nyikos [2] gave an affirmative answer. They raised several new problems. In this paper we answer some of their questions.

We will use the standard set-theoretical notation throughout, cf. [1]. Given sets A and B we write $[A, B] = \{\{a, b\} : a \in A \wedge b \in B\}$ and

* The preparation of this paper was supported by the Hungarian Foundation for Scientific Research grant no. 1805.

$A \triangle B = (A\backslash B) \cup (B\backslash A)$. Consider a hypergraph G Write $G = \langle V(G), E(G)\rangle$. If $H \subset V(G)$ we define $G[H]$ to be $\langle H, E(G) \cap [H]^{<\omega}\rangle$. If $\mathcal{T} = \langle T, <_{\mathcal{T}}\rangle$ is a tree and $Y \subset T$ then write $\mathcal{T} \restriction Y$ for $\langle Y, <_{\mathcal{T}} \cap (Y \times Y)\rangle$. Given $x, y \in T$, we write $x \|_{\mathcal{T}} y$ to mean $x <_{\mathcal{T}} y$ or $y <_{\mathcal{T}} x$.

We recall the definition of $\diamondsuit^+$: 'there is a sequence $\langle S_\alpha : \alpha < \omega_1\rangle$ of countable sets such that, for each $X \subset \omega_1$, we have a closed unbounded $C \subset \omega_1$ satisfying $X \cup \nu \in S_\nu$ and $C \cap \nu \in S_\nu$ for each $\nu \in C$'.

We write 'club' for 'closed unbounded'. Given a set X we define $\mathrm{TC}(X)$ to be the transitive closure of X. If κ is a cardinal, set

$$H_\kappa = \{X : |\mathrm{TC}(X)| < \kappa\} \quad \text{and} \quad \mathcal{H}_\kappa = \langle H_\kappa, \in\rangle.$$

Given a set $Y \subset On$ and an ordinal α write

$$[Y]^\alpha = \{X \subset Y : |X| = |\alpha|\} \quad \text{and} \quad (Y)^\alpha = \{X \subset Y : \mathrm{typ}(X) = \alpha\}.$$

Given a finite set I we denote by $H(I)$ the set of all functions mapping a finite subset of I into 2.

A finite hypergraph $G = \langle V, E\rangle$ is called *non-trivial* iff G contains no clique or independent subset of size $|V|$. We say that a graph $G = \langle V, E\rangle$ is *strongly non-trivial* provided each family $\mathcal{F}$ of pairwise disjoint, finite subsets of V with $|\mathcal{F}| = |V|$ contains four distinct elements a, b, c and d with $[a, b] \subset E$ and $[c, d] \cap E = \varnothing$. Given a hypergraph $G = \langle V, E\rangle$, let $I(G)$ be the set of all isomorphism classes of induced subgraphs of G with size $|V|$. Now we can formulate the theorem of K. A. Kierstead & P. J. Nyikos as follows: 'If G is a non-trivial infinite hypergraph then $|I(G)| > 1$.'

In Section 2 we show that $|I(G)| \geqslant 2^\omega$ for each non-trivial graph G on ω_1. This gives a partial answer to Problem 1 of [2]. In this section we also present a negative solution for Problem 2: under $\diamondsuit^+$ there exists a non-trivial graph on ω_1 with $|I(G)| = \omega_1$.

Consider the graph $G = \langle V, E\rangle$. We say that G is *almost smooth* if it is isomorphic to $G[W]$ whenever $W \subset V$ with $|V\backslash W| < |V|$. The graph G is called *quasi smooth* iff it is isomorphic either to $G[W]$ or to $G[V\backslash W]$ whenever $W \subset V$. In Section 3, using the continuum hypothesis, we construct a non-trivial, almost smooth graph on ω_1. In Section 4 we show that the existence of a strongly non-trivial, quasi-smooth graph on ω_1 is consistent with ZFC. It is also proved that a non-trivial, almost smooth graph on ω_1 can exist under Martin's axiom. The result of Sections 3 and 4 give partial answers for Problem 3 of [2].

An ω-uniform hypergraph is a pair $\langle A, \mathcal{A}\rangle$ with $\mathcal{A} \subset [A]^\omega$. In Section 5 we show that the theorem of H. A. Kierstead & P. J. Nyikos about

n-uniform hypergraphs cannot be generalized to ω-uniform hypergraphs. It is proved that under $\diamondsuit^+$ there is an ω-uniform hypergraph $\langle \omega_1, \mathcal{A}\rangle$ which is isomorphic to $\langle A, \mathcal{A}\cap [A]^\omega\rangle$ whenever $A \in [\omega_1]^{\omega_1}$ and which is non-trivial in the following sense: for each $A \in [\omega_1]^{\omega_1}$ we have both $(A)^\omega \cap \mathcal{A} \neq \emptyset$ and $(A)^\omega \setminus \mathcal{A} \neq \emptyset$.

2

In this section we investigate the problem: 'Given a non-trivial graph G on ω_1, what can be said about the cardinality of $I(G)$?'

The theorem which follows answers Problem 1 of [2].

Theorem 2.1 *If G is a non-trivial graph on ω_1 then $|I(G)| \geqslant 2^\omega$.*

Proof Let $E = E(G)$. Given $x \in \omega_1$, we take

$$G(x) = \{y \in \omega_1 : \{x,y\} \in E\}.$$

Write

$$\mathcal{B} = \{B \in [\omega_1]^{\omega_1} : \forall\, b \in B\ |G(b)\cap B| = |B\setminus G(b)| = \omega_1\}.$$

Then, for each $A \in [\omega_1]^{\omega_1}$, there is a set $B \in \mathcal{B}$ with $B \subset A$. Indeed, by non-triviality of G, the set

$$C = \{a \in A : |G(a)\cap A| < \omega_1 \text{ or } |A\setminus G(a)| < \omega_1\}$$

is at most countable. Thus $B = A\setminus C$ works. This observation implies that, for each $f \in {}^\omega 2$ and $A \in [\omega_1]^{\omega_1}$, there exists a sequence

$$X_f = \{x_n : n < \omega\} \subset A$$

such that, for each $n < m < \omega$, we have $\{x_n, x_m\} \in E$ iff $f(n) = 1$. This fact was first observed by P. Erdös and H. Hajnal. Since $f \neq g$ implies that $G[X_f]$ and $G[X_g]$ are not isomorphic, we obtain that the following claim holds.

Claim *Every uncountable induced subgraph of G contains 2^ω-many pairwise non-isomorphic countable induced subgraphs.*

Now fix an $x \in \omega_1$ with $|G(x)\cap\omega_1| = |\omega_1\setminus G(x)| = \omega_1$. Next choose $A, B \in \mathcal{B}$ with $A \subset G(x)$ and $B \subset \omega_1\setminus\{x\}\setminus G(x)$. We distinguish two cases.

Case 1 $\exists\, C \in [B]^{\omega_1}\ \forall\, c \in C\ |A\setminus G(c)| = \omega_1$.

Then, whenever $C' \in [C]^\omega$, one can recognize $\{x\}$, C' and A in $G[D]$, where $D = A \cup \{x\} \cup C'$. Indeed, we know that

$$\{x\} = \{y \in D : |D \backslash G(y)| \leqslant \omega\}$$

and so $A = \{a \in D : \{a,x\} \in E\}$ and $C' = D \backslash \{x\} \backslash A$. Thus, given $C', C'' \in [C]^\omega$, if the graphs $G[C' \cup \{x\} \cup A]$ and $G[C'' \cup \{x\} \cup A]$ are isomorphic then $G[C']$ and $G[C'']$ must also be isomorphic. But, by the claim, there are 2^ω-many pairwise non-isomorphic induced countable subgraphs of $G[C]$. This implies $|I(G)| \geqslant 2^\omega$ in case 1.

Case 2 $\forall\, b \in B\ |A \backslash G(b)| < \omega_1$.

Then, whenever $B' \in [B]^\omega$, one can recognize B' and A in $G[D]$, where $D = A \cup B'$. Indeed, we know that

$$B' = \{b \in D : |D \backslash G(y)| < \omega_1\}$$

and so $A = D \backslash B'$. Thus given $B', B'' \in [B]^\omega$, if the graphs $G[B' \cup A]$ and $G[B'' \cup A]$ are isomorphic then $G[B']$ and $G[B'']$ must also be isomorphic. But, by the claim, there are 2^ω-many pairwise non-isomorphic induced countable subgraphs of $G[B]$. This implies $|I(G)| \geqslant 2^\omega$ in case 2 as well.

This completes the proof of Theorem 2.1. □

Note that the same proof yields the following more general result. Assume that $\kappa \geqslant \omega$, $2^{<\kappa} = \kappa$ and $\mathcal{G}$ is a non-trivial graph on κ^+. Then $I(\mathcal{G}) \geqslant 2^\kappa$. We do not investigate the cases $\kappa > \omega_1$ for the rest of the problems.

Now we show that the statement of Theorem 2.1 is the strongest one which can be proved in ZFC.

Theorem 2.2 *Assume* $\diamondsuit^+$. *Then there exists a non-trivial graph on* ω_1 *with* $|I(G)| = \omega_1$.

The statement of this theorem easily follows from the following theorem.

Theorem 2.3 *Assume* $\diamondsuit^+$. *Then there is a Suslin tree* $\mathcal{T} = \langle T, <_{\mathcal{T}} \rangle$ *such that every uncountable* $Y \subset T$ *contains a countable* $X \subset Y$ *which is an initial segment of* Y *in* $\mathcal{T}$ *such that* $\mathcal{T}$ *and* $\mathcal{T} \restriction (Y \backslash X)$ *are isomorphic.*

Proof of Theorem 2.2 from Theorem 2.3 We begin the proof with fixing a Suslin tree $\mathcal{T} = \langle \omega_1, <_{\mathcal{T}} \rangle$ satisfying the requirements of Theorem 2.3. We define G to be the comparability graph of $\mathcal{T}$, that is, $G = \langle \omega_1, E \rangle$, where $E = \{\{x,y\} \in [\omega_1]^2 : x \parallel_{\mathcal{T}} y\}$. Let T_0 be the 0-th level of $\mathcal{T}$.

Now, for each $Y \in [\omega_1]^{\omega_1}$ we can choose a countable set $X_Y \subset Y$ and a function f_Y such that the X_Y is the initial segment of Y and f_Y is an isomorphism between $\mathcal{T}$ and $\mathcal{T} \restriction (Y \backslash X_Y)$.

To show that $|I(G)| = \omega_1$ it is enough to observe that if $Y, Z \in [\omega_1]^{\omega_1}$ with $X_Z = X_Y$ and $f_Y \restriction T_0 = f_Z \restriction T_0$ then $\mathcal{T} \restriction Y$ and $\mathcal{T} \restriction Z$ are isomorphic. But this is clear, because in this case the function $(\mathrm{id} \restriction X_Y) \cup (f_Z \circ f_Y^{-1})$ is an isomorphism between $\mathcal{T} \restriction Y$ and $\mathcal{T} \restriction Z$, where id denotes the identity function. □

The proof of Theorem 2.3 is based on lemmas given below. Consider a tree $\mathcal{T} = \langle T, <_{\mathcal{T}} \rangle$. Given $t \in T$ and $X \subset T$, take

$$b_{\mathcal{T}}(t) = \{t' \in T : t' <_{\mathcal{T}} t\}, \qquad h_{\mathcal{T}}(t) = \mathrm{typ}(b_{\mathcal{T}}(t), <_{\mathcal{T}}),$$

$$\hat{b}_{\mathcal{T}}(t) = b_{\mathcal{T}}(t) \cup \{t\}, \qquad \mathcal{T}[X] = \{y \in T : \exists\, x \in X\, (x \parallel_{\mathcal{T}} y)\}.$$

We usually omit the subscript $\mathcal{T}$ when it does not cause confusion.

We set $T_\alpha = \{t \in T : h_{\mathcal{T}}(t) = \alpha\}$, $T_{<\alpha} = \{t \in T : h_{\mathcal{T}}(t) < \alpha\}$ and $T_{\leqslant\alpha} = \{t \in T : h_{\mathcal{T}}(t) \leqslant \alpha\}$. The height of $\mathcal{T}$ is denoted by $h(\mathcal{T})$.

Definition 2.4 A tree $\mathcal{T} = \langle T, <_{\mathcal{T}} \rangle$ is called *ω-branching* iff (a) and (b) below are satisfied:

(a) $\forall\, x \in T\ |\{y \in T : b_{\mathcal{T}}(x) = b_{\mathcal{T}}(y)\}| = \omega$;
(b) $\forall\, x \in T\ \forall\, \alpha < h(\mathcal{T})\ \exists\, y \in T_\alpha\ (x \parallel_{\mathcal{T}} y)$.

Lemma 2.5 *If $\mathcal{T}_1$ and $\mathcal{T}_2$ are countable, ω-branching trees with $h(\mathcal{T}_1) = h(\mathcal{T}_2)$ then they are isomorphic.*

Proof Straightforward by a zigzag argument. □

Definition 2.6 Given a tree $\mathcal{T} = \langle T, <_{\mathcal{T}} \rangle$, we say that a partial function f from T to T is *almost increasing* iff, for each $\alpha < h(\mathcal{T})$, we have a $\beta < h(\mathcal{T})$ such that both $h(f(x)) > \alpha$ and $h(f^{-1}(x)) > \alpha$ whenever $x \in T$ with $h(x) > \beta$.

We write 'a.i.' for 'almost increasing'.

Definition 2.7 Given a tree $\mathcal{T} = \langle T, <_{\mathcal{T}} \rangle$, a set $Y \subset T$ is called *large* iff $\mathcal{T} \restriction Y$ is an ω-branching tree and Y is cofinal in $\mathcal{T} \restriction \mathcal{T}[Y]$.

Definition 2.8 Given an ω-branching tree $\mathcal{T} = \langle T, <_{\mathcal{T}} \rangle$, let $\mathscr{E}(\mathcal{T})$ be the set of all a.i. functions which are order-preserving bijections between large subsets of $\mathcal{T}$.

Definition 2.9 Given a tree $\mathcal{T} = \langle T, <_{\mathcal{T}} \rangle$, we define $\max(\mathcal{T})$ to be the set of all maximal antichains of $\mathcal{T}$. We define $\mathrm{Cof}(\mathcal{T})$ to be the set of all cofinal subsets of $\mathcal{T}$.

Lemma 2.10 *Assume that $\mathcal{T} = \langle T, <_{\mathcal{T}} \rangle$ is a countable, ω-branching tree with limit height and that $\mathscr{A} \subset \max(\mathcal{T})$ and $\mathscr{F} \subset \mathscr{E}(\mathcal{T})$ are countable* *families. Then there exists a countable ω-branching tree $\mathcal{T}^*$ with $h(\mathcal{T}^*) = h(\mathcal{T})+1$ such that $\mathcal{T}$ is an initial segment of $\mathcal{T}^*$, $\mathcal{A} \subset \max(\mathcal{T}^*)$ and $\mathcal{F} \subset \{f \restriction T : f \in \mathcal{E}(\mathcal{T}^*)\}$.*

Proof We may assume that all levels of $\mathcal{T}$ are in $\mathcal{A}$ and that $f^{-1} \in \mathcal{F}$ for each $f \in \mathcal{F}$. Write

$$\mathcal{A} = \{A_m : m < \omega\}, \qquad \mathcal{F} = \{f_m : m < \omega\}, \qquad T = \{t_m : m < \omega\}.$$

To simplify the notation of the following construction we extend $\mathcal{T}$, adding a new point z to it in such a way that z will be considered as the new root of $\mathcal{T}$, that is, the point z is put below all the points of $\mathcal{T}$.

Let ω^* be the set of all finite non-empty sequences of natural numbers. Fix a bijection $h: \omega \to \omega^* \times \omega \times \omega$. If $f \in \mathcal{F}$ and $t \in T$, take

$$f[t] = \bigcup \{b(f(y)) : y \in \hat{b}(x) \cap \operatorname{dom}(f)\}.$$

If $m \in \omega$, let

$$D_{2m} = \operatorname{dom}(f_m) \cup (T \setminus \mathcal{T}[\operatorname{dom}(f_m)]),$$

$$D_{2m+1} = \{t \in T : \hat{b}(t) \cap A_m \neq \emptyset\}.$$

Write $\mathcal{D} = \{D_k : k < \omega\}$. Clearly $\mathcal{D} \subset \operatorname{Cof}(\mathcal{T})$.

If $x, y \in T$ and $i \in \omega$, we write '$x >^i y$, for '$x \notin \mathcal{T}[\operatorname{dom}(f_i)]$ or $y \notin \mathcal{T}[\operatorname{ran}(f_i)]$ or $y \in f_i[x]$'. We construct now the family

$$\{t_s^n : s \in \omega^*, n \in \omega\} \subset T,$$

by induction on n, satisfying the following conditions:

(a) $t^0_{\langle m \rangle} = t_m$;
(b) $t_s^{n-1} \leqslant_{\mathcal{T}} t_s^n$ for each $s \in \omega^*$ and $n \in \omega$;
(c) $t_s^n >^j t^n_{s^\frown\langle j \rangle}$ for each $s \in \omega^*$, $j \in \omega$ and $n \in \omega$;
(d) if $h(n-1) = \langle s, m, i \rangle$ then $(b(t_s^n) \setminus b(t_s^{n-1}) \cap D_m \neq \emptyset$.

If $n = 0$ then simply take

$$t_s^0 = \begin{cases} t_m & \text{if } s = \langle m \rangle \text{ for some } m \in \omega, \\ z & \text{otherwise.} \end{cases}$$

Assume that we have constructed $\{t_s^k : s \in \omega^*, k < n\}$. To follow the construction we need the following claim.

Claim *If $x_0 >^{j_1} x_1 >^{j_2} \cdots >^{j_k} x_k$ and D is cofinal in $\mathcal{T}$ above x_k then there is a sequence $y_0, y_1, \ldots, y_k$ such that $y_i \geqslant_{\mathcal{T}} x_i$ for $i \leqslant k$, $y_0 >^{j_1} y_1 >^{j_2} \cdots >^{j_k} y_k$ and $(b(y_k) \setminus b(x_k)) \cap D \neq \emptyset$.*

Proof of the claim This is by induction on k. If $k = 0$ then the claim is trivial.

Assume that we know it for $k-1$. We can assume that $x_k \in f_{j_k}[x_{k-1}]$. Indeed, if $x_x \notin f_{j_k}[x_{k-1}]$ then choose y_k with $x_k <_{\mathcal{T}} y_k$ and $(b(y_k)\backslash b(x_k)) \cap D \neq \emptyset$. Taking $y_i = x_i$ for $i < k$, the sequence $y_0, y_1, \ldots, y_k$ satisfies the requirements of the claim because $x_{k-1} >^{j_k} y_k$.

Thus we have $x_k \in f_{j_k}[x_{k-1}]$. Take

$$B = \{v \in \operatorname{dom}(f_{j_k}) : x_{k-1} <_{\mathcal{T}} v \wedge \big(b(f_{j_k}(v))\backslash b(x_k)\big) \cap D \neq \emptyset\}.$$

We show that B is cofinal in $\mathcal{T}$ above x_{k-1}. Indeed, let $y \in T$ with $x_{k-1} <_{\mathcal{T}} y$. Then $y \in \mathcal{T}[\operatorname{dom}(f_{j_k})]$ because $\hat{b}(x_{k-1}) \cap \operatorname{dom}(f_{j_k}) \neq \emptyset$. Pick $u \in \operatorname{dom}(f_{j_k})$ with $y <_{\mathcal{T}} u$. Now, $x_k <_{\mathcal{T}} f_{j_k}(u)$. Since $\operatorname{ran}(f_{j_k})$ is cofinal in $\mathcal{T}[\operatorname{ran}(f_{j_k})]$, we can choose an $e \in \operatorname{ran}(f_{j_k})$ with $f_{j_k}(u) <_{\mathcal{T}} e$ and $\big(b(e)\backslash b(f_{j_k}(u))\big) \cap D \neq \emptyset$. Then $f_{j_k}^{-1}(e) \in B$ with $y <_{\mathcal{T}} f_{j_k}^{-1}(e)$, which shows that B is cofinal in $\mathcal{T}$ above x_{k-1}.

Now apply the inductive hypothesis for the sequence $x_0 >^{j_1} x_1 >^{j_2} \cdots >^{j_{k-1}} x_{k-1}$ and for B. We obtain the sequence $y_0, y_1, \ldots, y_{k-1}$. Choose a $v \in B \cap (b(y_{k-1})\backslash b(x_{k-1}))$. Let $y_k = f_{j_k}(v)$. Then

$$y_k \in f_{j_k}[y_{k-1}] \qquad \text{and} \qquad x_k <_{\mathcal{T}} y_k.$$

Thus the sequence $y_0, y_1, \ldots, y_{k-1}, y_k$ satisfies the requirements of the claim. The claim is proved. □

We now return to the proof of Lemma 2.10. We assume that $\{t_s^k : s \in \omega^*,\, k < n\}$ are constructed. Write $h(n-1) = \langle s, m, i\rangle$ and $s = \langle j_0, \ldots, j_k\rangle$. Now apply the claim for the sequence

$$t^{n-1}_{\langle j_0\rangle} >^{j_1} t^{n-1}_{\langle j_0, j_1\rangle} >^{j_2} \cdots >^{j_k} t^{n-1}_{\langle j_0,\ldots,j_k\rangle}$$

and for D_m. We get the sequence $y_0, \ldots, y_k$. Now we are ready to define $\{t_u^n : u \in \omega^*\}$. Take

$$t_u^n = \begin{cases} y_l & \text{if } u = \langle j_0,\ldots,j_l\rangle \text{ for some } l \leqslant k, \\ t_u^{n-1} & \text{otherwise.} \end{cases}$$

It is clear form the construction that $\{t_u^m : u \in \omega^*,\, m < n+1\}$ satisfies 2.10 (a)–(d). The construction is complete.

Now for each $s \in \omega^*$ take $b_s = \bigcup\{b(t_s^n) : n \in \omega\}$. Then b_s meets every level of $\mathcal{T}$ because the levels are elements of $\mathcal{D}$. Thus every b_s is a cofinal branch of $\mathcal{T}$. Let $\mathcal{B} = \{b_s : s \in \omega^*\}$.

Now we extend $\mathcal{T}$ to $\mathcal{T}^*$, putting ω-many new points above each branch of $b \in \mathcal{B}$. More precisely, we take $\mathcal{T}^* = \langle \mathcal{T}^*, <_{\mathcal{T}^*}\rangle$, where

$$\mathcal{T}^* = T \cup (\mathcal{B}\times\omega); \qquad <_{\mathcal{T}^*} = <_{\mathcal{T}} \cup \{\langle x, \langle b, i\rangle\rangle : b \in \mathcal{B},\, i \in \omega \text{ and } x \in b\}.$$

Since every $b \in \mathcal{B}$ meets each $A \in \mathcal{A}$, we have $\mathcal{A} \subset \max(\mathcal{T}^*)$. To complete the proof of the lemma we must show that

$$\mathcal{F} \subset \{f \restriction T : f \in \mathcal{E}(\mathcal{T}^*)\}.$$

Fix $f_j \in \mathcal{F}$. Then $f_j^{-1} \in \mathcal{F}$. Write $f_k = f_j^{-1}$.

Take

$$D = \{b \in \mathcal{B} : b \cap \mathrm{dom}(f_j) \text{ is cofinal in } b\}$$

and

$$R = \{b \in \mathcal{B} : b \cap \mathrm{ran}(f_j) \text{ is cofinal in } b\}.$$

Then, whenever $b \in D$ and $b = b_s$ we have that $b_{s^\frown\langle j\rangle} \in R$ and $f_j[b_s] = b_{s^\frown\langle j\rangle}$ and, vice versa, whenever $b \in R$ and $b = b_u$ we have that $b_{u^\frown\langle k\rangle} \in D$ and $f_k[b_u] = b_{u^\frown\langle k\rangle}$, that is, $f_j[b_{u^\frown\langle k\rangle}] = b_u$. This implies that there is a function f^* satisfying (a)–(d) below:

(a) $\mathrm{dom}(f^*) = \mathrm{dom}(f_j) \cup D \times \omega$;

(b) $\mathrm{ran}(f^*) = \mathrm{ran}(f_j) \cup R \times \omega$;

(c) $f^* \supset f$;

(d) if $b \in D$ then $f^*(\langle b, i\rangle) = \langle f_j[b], i\rangle$.

Then clearly $f = f^* \restriction T$. To show that $f^* \in \mathcal{E}(\mathcal{T}^*)$ it is enough to show that both $\mathrm{dom}(f^*)$ and $\mathrm{ran}(f^*)$ are large subsets of $\mathcal{T}^*$. We need check the condition 2.4 (b) only. But that is clear, provided that we can show that for each $t \in \mathrm{dom}(f)$ there is a $b \in D$ with $t \in b$. Let $t = t_m$. By the construction the set $b_{\langle m\rangle} \cap D_{2j}$ is cofinal in $b_{\langle m\rangle}$. Since $t \in b_{\langle m\rangle} \cap \mathrm{dom}(f_j)$ we have that $b_{\langle m\rangle} \cap (T \setminus \mathcal{T}[\mathrm{dom}(f_j)]) = \emptyset$. Thus $b_{\langle m\rangle} \cap \mathrm{dom}(f_j) = b_{\langle m\rangle} \cap D_{2j}$, whence $b_{\langle m\rangle} \cap \mathrm{dom}(f_j)$ is cofinal in $b_{\langle m\rangle}$. Thus $b_{\langle m\rangle} \in D$, which completes the proof of Lemma 2.10. □

Proof of Theorem 2.3 First we construct the tree $\mathcal{T}$. Fix a large enough cardinal κ, a $\prec$ well ordering of H_κ and a $\diamondsuit^+$-sequence

$$S = \langle S_\alpha : \alpha < \omega_1\rangle.$$

Next choose a sequence $\langle N_\alpha : \alpha < \omega_1\rangle$ of countable, elementary submodels of $\langle H_\kappa, \in, \prec\rangle$ with $S \in N_0$ and $\langle N_\alpha : \alpha < \beta\rangle \in N_\beta$ for $\beta < \omega_1$.

Now we build a sequence $\langle \mathcal{T}^\alpha : \alpha < \omega_1\rangle$ of countable, ω-branching trees, by induction on α, such that $h(\mathcal{T}^\alpha) = \alpha$, $\langle \mathcal{T}^\beta : \beta \leqslant \alpha\rangle \in N_\alpha$ for $\alpha < \omega_1$ and $\mathcal{T}^\alpha$ is an end-extension of $\mathcal{T}^\beta$ for $\beta < \alpha < \omega_1$.

Set $\mathcal{T}^0 = \langle \emptyset, \emptyset\rangle$. If α is a limit take $\mathcal{T}^\alpha = \bigcup\{\mathcal{T}^\beta : \beta < \alpha\}$. Since, as we shall see, the construction of $\langle \mathcal{T}^\beta : \beta < \alpha\rangle$ was done in such a uniform way that it can be repeated in N_α, we get $\langle \mathcal{T}^\beta : \beta < \alpha\rangle \in N_\alpha$ and so $\langle \mathcal{T}^\beta : \beta \leqslant \alpha\rangle \in N_\alpha$ too.

If $\alpha = \beta+1$ and β is a successor ordinal, then let $\mathcal{T}^\alpha$ be the $<$-minimal end-extension of T^β which is an ω-branching tree with height α.

If $\alpha = \beta+1$ and β is a limit ordinal, then apply Lemma 2.10 for $\mathcal{T} = \mathcal{T}^\beta$, $\mathcal{F} = \mathcal{E}(\mathcal{T}^\beta) \cap N_\beta$ and $\mathcal{A} = \max(\mathcal{T}^\beta) \cap N_\beta$. Define $\mathcal{T}^\alpha$ to be the $<$-minimal element of N_α satisfying the requirement of Lemma 2.10.

The construction is done. Take $\mathcal{T} = \bigcup\{\mathcal{T}^\alpha : \alpha < \omega_1\}$. Since $S_\alpha \subset N_\alpha$ for $\alpha < \omega_1$, we have that $\mathcal{T}$ is an ω-branching Suslin tree.

Since $\mathcal{T}$ is Suslin, it is easy to see that each $X \in [T]^{\omega_1}$ contains a countable initial segment Z_X such that $X \setminus Z_X$ is cofinal in $\mathcal{T}[X \setminus Z_X]$. Thus to complete the proof it is enough to prove the following lemma.

Lemma 2.11 *Assume that $Y \in [T]^{\omega_1}$ and Y is cofinal in $\mathcal{T}[Y]$. Then $\mathcal{T}$ and $\mathcal{T} \restriction Y$ are isomorphic.*

Proof First remark that in this case $\mathcal{T} \restriction Y$ must be ω-branching. Let $\mathcal{Y} = \mathcal{T} \restriction Y$. Write $\mathcal{Y} = \langle Y, <_y \rangle$. Let

$$B = \{\beta \in \omega_1 : Y_{<\beta} = Y \cap T_{<\beta} \text{ and } Y_{<\beta} \text{ is cofinal in } \mathcal{T}[Y] \cap \mathcal{T}_{<\beta}\}.$$

Clearly B is a club subset of ω_1. By $\diamondsuit^+$ we have a club $C \subset B$ such that, for each $\gamma \in C$, we have $\{\mathcal{T}_{<\gamma}, \mathcal{Y}_{<\gamma}, C \cap \gamma\} \subset N_\gamma$. Write $C = \{\gamma_\nu : \nu < \omega_1\}$.

Now we construct a sequence of functions, $\langle \pi_\nu : \nu < \omega_1 \rangle$, by induction on ν, satisfying (a)–(c) below:

(a) π_ν is an isomorphism between $\mathcal{T}_{\leqslant \gamma_\nu}$ and $\mathcal{Y}_{\leqslant \gamma_\nu}$;
(b) $\pi_\nu \subset \pi_\mu$ for $\nu < \mu < \omega_1$;
(c) $\langle \pi_\nu : \nu < \mu \rangle \in N_{\gamma_\mu}$.

Assume that $\langle \pi_\nu : \nu < \mu \rangle$ is already constructed and we know (a)–(c) for $\mu' < \mu$.

Case 1 $\mu = \xi+1$.

Then both $\mathcal{T}_{\leqslant \gamma_\mu} \setminus \mathcal{T}_{\leqslant \gamma_\xi}$ and $\mathcal{Y}_{\leqslant \gamma_\mu} \setminus \mathcal{Y}_{\leqslant \gamma_\xi}$ are countable, ω-branching trees with height $(\gamma_\mu+1)-(\gamma_\xi+1)$. Thus, by Lemma 2.5, we can define π_μ to be the $<$-minimal element of $N_{\gamma_\mu+1}$ which extends π_ξ and which is an isomorphism between $\mathcal{T}_{\leqslant \gamma_\mu}$ and $\mathcal{Y}_{\leqslant \gamma_\mu}$.

Case 2 *μ is a limit.*

As one can see, the construction of $\langle \pi_\nu : \nu < \mu \rangle$ can be carried out in N_{γ_μ}. Thus $\langle \pi_\nu : \nu < \mu \rangle \in N_{\gamma_\mu}$. Take

$$\pi = \bigcup\{\pi_\nu : \nu < \mu\}.$$

Then $\pi \in N_{\gamma_\mu}$ and $\pi: \mathcal{T}_{<\gamma_\nu} \to \mathcal{Y}_{<\gamma_\nu}$ is an isomorphism. Thus $\pi \in \mathcal{E}(\mathcal{T}_{<\gamma_\nu})$. Hence, by definition of $\mathcal{T}$, there is a $\rho \in \mathcal{E}(\mathcal{T}_{\leqslant\gamma_\mu})$ with $\rho \supset \pi$. But ran(ρ) is the largest subset of $\mathcal{T}_{\leqslant\gamma_\mu}$ and $\mathcal{Y}_{<\gamma_\mu}$ is cofinal in $\mathcal{T}[Y] \cap \mathcal{T}_{<\gamma_\mu}$. Thus

$$\operatorname{ran}(\rho) \cap \mathcal{T}_{\gamma_\mu} = \mathcal{T}[Y_{<\gamma_\nu}] \cap T_{\gamma_\mu} = \mathcal{T}[Y] \cap T_{\gamma_\mu}.$$

This implies that, for each $u \in T_{\gamma_\mu}$ we have

$$|\{z \in Y_{\gamma_\mu} : \rho'' b_{\mathcal{T}}(u) = b_{\mathcal{Y}}(z)\}| = \omega$$

and, vice versa, for each $z \in Y_{\gamma_\mu}$

$$|\{u \in T_{\gamma_\mu} : \rho'' b_{\mathcal{T}}(u) = b_{\mathcal{Y}}(z)\}| = \omega.$$

Thus there is a bijection $\pi^*: \mathcal{T}_{\gamma_\mu} \to \mathcal{Y}_{\gamma_\mu}$ with $\rho'' b_{\mathcal{T}}(u) = b_{\mathcal{Y}}(\pi^*(u))$ for each $u \in T_{\gamma_\mu}$. But $\pi \cup \pi^*$ is an isomorphism between $T_{\leqslant\gamma_\mu}$ and $\mathcal{Y}_{\leqslant\gamma_\mu}$. Since we used only parameters from $N_{\gamma_\mu+1}$ to construct π^*, we can define π_μ to be the $\prec$-minimal element of $N_{\gamma_\mu+1}$ which extends π and is an isomorphism between $T_{\leqslant\gamma_\mu}$ and $\mathcal{Y}_{\leqslant\gamma_\mu}$.

This completes the construction. Then $\pi = \bigcup\{\pi_\mu : \mu < \omega_1\}$ is the required isomorphism between $\mathcal{T}$ and $\mathcal{T} \restriction Y$. Lemma 2.11 is proved. □

This completes the proof of Theorem 2.3. □

Before concluding this section, we make some further remarks. First we mention that we do not konw if it is consistent with CH that $I(\mathcal{G}) \geqslant 2^{\omega_1}$ holds for every non-trivial graph on ω_1. (Of course $2^\omega = 2^{\omega_1}$ implies this, by Theorem 2.1.) However, there are strengthenings of non-triviality which imply this, e.g. the following theorem.

Theorem 2.12 *Assume $\mathcal{G} = \langle\omega_1, E\rangle$ is such that neither $\mathcal{G}$ nor its complement contains K_{ω,ω_1}, that is, for each $A \in [\omega_1]^\omega$ and $B \in [\omega_1]^{\omega_1}$, $[A, B] \not\subset E$ and $[A, B] \cap E \neq \emptyset$. Then $I(\mathcal{G}) \geqslant 2^{\omega_1}$.*

The theorem follows easily from the following lemma.

Lemma 2.13 *Assume $2^\omega < 2^{\omega_1}$, $\mathcal{G} = \langle\omega_1, E\rangle$ is a graph, $A \in [\omega_1]^\omega$ and $|\{\mathcal{G}(x) \cap A : x \in \omega_1\}| = \omega_1$. Then $I(\mathcal{G}) \geqslant 2^{\omega_1}$.*

Proof Let $B \in [\omega_1 \setminus A]^{\omega_1}$ be such that $\mathcal{G}(x) \cap A \neq \mathcal{G}(y) \cap A$ for $x \neq y \in B$. For $F \in [B]^{\omega_1}$, let $\mathcal{G}_F = \mathcal{G}[A \cup F]$. If $I(\mathcal{G}) < 2^{\omega_1}$ then there are a $\mathcal{G}'$ of size ω_1 and an $\mathcal{F} \subset [B]^{\omega_1}$ of size $(2^\omega)^+$ such that $\mathcal{G}' \simeq \mathcal{G}_F$ for $F \in \mathcal{F}$. Then there are $F_0 \neq F_1 \in \mathcal{F}$ and an isomorphism π between $\mathcal{G}_{F_0}$ and $\mathcal{G}_{F_1}$ such that $\pi \restriction A = \mathrm{id}$. Hence $\pi'' F_0 = F_0 \neq F_1$. □

Lemma 2.13 also explains why we were looking for comparability graphs of Suslin trees to establish Theorem 2.2

One final remark. Theorem 2.3 yields the consistency of existence of a non-trivial graph $\mathcal{G} = \langle \omega_1, E \rangle$ such that for each pair $A, B \in [\omega_1]^{\omega_1}$ there are countable sets $X \subset A$ and $Y \subset B$ such that $\mathcal{G}[A \setminus X]$ and $\mathcal{G}[B \setminus Y]$ are isomorphic. If $2^{\omega} < 2^{\omega_1}$ then we know that X and Y cannot be chosen finite in this statement.

3 Construction under CH

Given two graphs $\mathcal{H}$ and $\mathcal{G}$ we write $\mathcal{H} \subset \mathcal{G}$ iff $\mathcal{H} = \mathcal{G}[V(\mathcal{H})]$. Let $\mathrm{Aut}(\mathcal{G})$ be the set of all automorphisms of the graph $\mathcal{G}$. We say that a function i is a *partial automorphism* iff $\mathrm{dom}(i) \cup \mathrm{ran}(f) \subset V(\mathcal{G})$ and i is an automorphism between $\mathcal{G}[\mathrm{dom}(i)]$ and $\mathcal{G}[\mathrm{ran}(i)]$. The set of all partial automorphisms of a graph $\mathcal{G}$ will be denoted by $\mathrm{Aut}_{\mathrm{p}}(\mathcal{G})$.

The main result of this section can be formulated as follows.

Theorem 3.1 *Assume CH. Then there exists a non-trivial, almost smooth graph $\mathcal{G}$ with $|V(\mathcal{G})| = \omega_1$.*

The statement of this theorem follow easily from the following lemma.

Lemma 3.2 (ZFC) *Let $\mathcal{G}$ be a graph with $|V(\mathcal{G})| = \omega$. Let*

$$\{i_n : 0 < n < \omega\} \subset \mathrm{Aut}_{\mathrm{p}}(\mathcal{G}), \qquad H \subset V(\mathcal{G}), \qquad \{X_n : n < \omega\} \subset [V(\mathcal{G})]^{\omega}.$$

Then there exist a graph $\mathcal{G}' \supset \mathcal{G}$ with $|V(\mathcal{G}')| = \omega$ and a sequence $\{i'_n : n < \omega\} \subset \mathrm{Aut}_{\mathrm{p}}(\mathcal{G}')$ satisfying (a)–(c) *below:*

(a) *$\{x\} \cup X_n$ is neither a clique nor an independent subset in $\mathcal{G}'$ for each $x \in V(\mathcal{G}') \setminus V(\mathcal{G})$ and $n < \omega$;*

(b) *$\mathrm{dom}(i'_0) = V(\mathcal{G}')$ and $\mathrm{ran}(i'_0) = V(\mathcal{G}') \setminus H$;*

(c) *$i'_n \supset i_n$ and $i'_n \setminus i_n \in \mathrm{Aut}(\mathcal{G}'[V(\mathcal{G}') \setminus V(\mathcal{G})])$ for each $0 < n < \omega$.*

Proof of Lemma 3.2 First we define the set of vertices of $\mathcal{G}'$. Denoting by $\mathbb{Z}$ the set of integers, take $V(\mathcal{G})$ to be the set of all sequences

$$\langle g, \langle \langle n_j, m_j \rangle : j < k \rangle \rangle \in V(\mathcal{G}) \times {}^{<\omega}(\omega \times (\mathbb{Z} \setminus \{0\}))$$

satisfying the conditions that $n_0 = 0$, $n_j \neq n_{j+1}$ for $j < k$ and that, if $k > 0$ and $g \in H$, then $m_0 > 0$. Every $g \in V(\mathcal{G})$ will be identified with $\langle g, \emptyset \rangle \in V(\mathcal{G}')$.

Now we define the action of i'_n on $V(\mathcal{G}')$ as follows: If $n > 0$ then $i'_n(\langle g, \emptyset \rangle)$ is defined iff $i_n(g)$ is defined and, in this case, take $i'_n(\langle g, \emptyset \rangle) = \langle i_n(g) \emptyset \rangle$.

Otherwise, if $x = \langle g, \langle\langle n_j, m_j\rangle : j < k\rangle\rangle \in V(\mathcal{G}')$ and $n = 0$ or $k > 0$, then let

$$i'_n(x) = \begin{cases} \langle g, \langle\langle n_j, m_j\rangle : j < k\rangle\rangle^\wedge\langle n, 1\rangle & \text{if } n \neq n_{k-1}, \\ \langle g, \langle\langle n_j, m_j\rangle : j < k-1\rangle\rangle^\wedge\langle n_{k-1}, m_{k-1}+1\rangle & \text{if } n = n_{k-1} \text{ and } m_{m-1} \neq 1, \\ \langle g, \langle\langle n_j, m_j\rangle : j < k-1\rangle\rangle & \text{if } n = n_{k-1} \text{ and } m_{m-1} = 1. \end{cases}$$

It is easy to see that the functions i'_n are 1–1 and that the requirements for the domain, range and restriction are satisfied.

Now we have to define the edges of $\mathcal{G}'$ so that the mappings i'_n become partial automorphisms. Let

$$I = \{i'_n, (i'_n)^{-1} : n \in \omega\} \quad \text{and} \quad I^+ = I \setminus \{i'_0, (i'_0)^{-1}\}.$$

Let J and J^+ be the transitive closures of I and I^+, respectively.

Next we define an equivalence relation on $[V(\mathcal{G}')]^2$: let $\{g_0, g_1\} \approx \{h_0, h_1\}$ iff there is a $j \in J$ with $j''\{g_0, g_1\} = \{h_0, h_1\}$. The functions i'_n (and so members of J) are in Aut($\mathcal{G}'$) iff for every equivalence class of $\approx$ the elements of the class are all edges or all non-edges of $\mathcal{G}'$.

We distinguish two kinds of equivalence classes. For the first kind the equivalence class contains an element of $[V(\mathcal{G})]^2$. In this case for all the elements of the equivalence class the second coordinates of the pair are equal. If two elements of $[V(\mathcal{G})]^2$ are in one of these equivalence classes then there is a $j \in J^+$ mapping one of the pair into the other; therefore they are both edges or non-edges of $\mathcal{G}$. This means that the following definition is correct: let the elements of such an equivalence class be edges of $\mathcal{G}'$ iff only the pairs from $[V(\mathcal{G})]^2$ are edges of $\mathcal{G}$.

For the equivalence classes of the second kind we are at liberty to define the set of edges. We order the pairs

$$\{\langle g, n\rangle : g \in V(\mathcal{G}')\setminus V(\mathcal{G}),\ n \in \omega\}$$

in type ω and do the definition by induction. For the pair $\langle g, n\rangle$ we selct h_1 and h_2 in such a way that $h_1, h_2 \in X_n$ and $\{g, h_1\}$ and $\{g, h_2\}$ are in different equivalence classes neither of which is fixed previously as edges or non-edges of $\mathcal{G}'$. Fix the equivalence class of $\{g, h_0\}$ as edges of $\mathcal{G}'$ and the class of $\{g, h_1\}$ as non-edges. The selection of h_0 and h_1 can be done as we can prove for any $g \in V(\mathcal{G}')\setminus V(\mathcal{G})$ and any equivalence class E of $\approx$ that the set $\{h \in V(\mathcal{G}) : \{g, h\} \in E\}$ has at most one element. Assume that $\{g, h_0\} \approx \{g, h_1\}$ and $j \in J$ with

$j(g) = g$ and $j(h_1) = h_2$, where $h_1, h_2 \in V(\mathcal{G})$. Then $j \in J^+$ because j takes an element of $V(\mathcal{G})$ into another element of $V(\mathcal{G})$, and this property characterizes the elements of J^+. The images of g by $j \in J^+$ have different second coordinates, so $j(g) \neq g$. Thus this case is impossible. On the other hand assume that $\{g, h_0\} \approx \{g, h_1\}$ and $j \in J$ shows it in such a way that $j(g) = h_2$ and $j(h_1) = g$. Then $(j \circ j)(h_1) \in V(\mathcal{G})$ and so $j \circ j \in J^+$. But then $j \in J^+$. Since $h_1 \in V(\mathcal{G})$ we have $g = j(h_1) \in V(\mathcal{G})$. This contradiction shows that we can finish the construction. The lemma is proved. □

Proof of Theorem 3.1 Fix an enumeration $\{H_\nu : \nu \in \mathrm{Lim}(\omega_1)\}$ of $[\omega_1]^\omega$ with $H_\nu \subset \nu$ and an enumeration $\{A_\nu : \nu \in \mathrm{Lim}(\omega_1)\}$ of $[\omega_1]^{\leq\omega}$ with $A_\nu \subset \nu$, where $\mathrm{Lim}(\omega_1) = \{\nu \in \omega_1 : \nu \text{ is a limit ordinal}\}$.

We define graphs G_ν on ν and partial automorphisms $\{i_{\nu,\xi} : \xi < \nu\}$ on G_ν by induction on ν, where $\nu \in \mathrm{Lim}(\omega_1)$, in such a way that $G_\eta = G_\nu[\eta]$, $i_{\eta,\xi} = i_{\nu,\xi}[\eta]$ and $i_{\nu,\xi}$ is an isomorphism between G_ν and $G_\nu[\nu \backslash A_\xi]$ whenever $\xi < \eta < \nu < \omega_1$.

If ν is a limit of limits simply take

$$G_\nu = \bigcup \{G_\eta : \eta \in \nu \cap \mathrm{Lim}(\omega_1)\} \quad \text{and} \quad i_{\nu,\xi} = \bigcup \{i_{\eta,\xi} : \xi < \eta < \nu\}.$$

If $\nu = \eta + \omega$ we apply Lemma 3.2 where $G = G_\eta$, $\{i_n : 0 < n < \omega\}$ enumerates $\{i_{\eta,\xi} : \xi < \eta\}$, $\{X_n : n \in \omega\}$ enumerates $\{H_\xi : \xi < \eta\}$ and $H = A_\eta$. We obtain the graph $\mathcal{G}'$ and the sequence $\{i'_n : n < \omega\}$. We identify $V(\mathcal{G}') \backslash V(\mathcal{G})$ with $\nu \backslash \eta$ and define G_η to be $\mathcal{G}'$, $i_{\nu,\xi}$ to be i'_n provided $i_{\eta,\xi} = i_n$, and $i_{\nu,\xi}$ to be i'_0. Then the inductive hypothesis will be satisfied.

Take

$$G = \bigcup \{G_\nu : \nu \in \mathrm{Lim}(\omega_1)\} \quad \text{and} \quad i_\xi = \bigcup \{i_{\nu,\xi} : \xi < \nu \in \mathrm{Lim}(\omega_1)\}.$$

Then G is a graph on ω_1 and i_ξ is an isomorphism between G and $G[\omega_1 \backslash A_\xi]$; thus G is almost smooth. If a subgraph of G contains H_ν then it can be complete or empty only if this subgraph is a subset of ν. Therefore G is non-trivial, which completes the proof of Theorem 3.1. □

Note that a similar but technically more complicated argument yields the existence of a strongly non-trivial almost smooth graph under CH, as well. We prefered to give a simpler proof instead of a slight improvement. This we could not do in the next section since strong non-triviality is indispensable to carry out the proof needed there.

4 Generic construction

Theorem 4.1 *There is a c.c.c. poset P with cardinality 2^ω such that*

$V^P \models$ *'there is a strongly non-trivial, almost smooth graph $H = \langle \omega_1, E\rangle$'.*

Theorem 4.2 *There is a c.c.c. poset P with cardinality 2^{ω_1} such that*

$V^P \models$ *'there is a strongly non-trivial, quasi-smooth graph $H = \langle \omega_1, E\rangle$'.*

Since CH implies the existence of a non-trivial, almost smooth graph on ω_1, it is natural to raise the question: what about Martin's axiom? Since under Martin's axiom (MA) there are no strongly non-trivial graphs on ω_1, one can hope that MA also excludes the existence of non-trivial, almost smooth ones. But, as it turns out, this is not true.

Theorem 4.3 *If ZF is consistent then so is ZFC + 'Martin's axiom' + 'there exists a non-trivial almost smooth graph on ω_1'.*

The proofs of these theorems are based on several lemmas to be given below. If Q is a poset, $p, q \in Q$, then we write $p \parallel_Q q$ to mean that p and q are compatible in Q.

Definition 4.4 Let $G = \langle V, e\rangle$ be a graph with $|V| = \omega_1$.

(a) A set $A \subset V$ is called *dense in G* iff, for each $B \in [V]^{<\omega}$ and $f \in {}^B2$, we have $a \in A$ such that $\{a, b\} \in E$ iff $f(b) = 1$ for each $b \in B$.

(b) A partition $\{A_\nu : \nu < \omega_1\}$ of V is called a *good partition* iff every A_ν is countable and dense in G.

Lemma 4.5 *There is a c.c.c. poset Q with $|Q| = \omega_1$ such that $V^Q \models$ 'there exists a graph $G = \langle \omega_1, E\rangle$ having a good partition and satisfying the property (*) below':*

(*) *for each $n \in \omega$, for each 1–1 function $F: \omega_1 \times n \to \omega_1$ and for each function $f \in {}^{n\times n}2$, we have $\nu < \mu < \omega_1$ such that $f(i, j)$ iff $\{F(\nu, i), F(\mu, j)\} \in E$ for $i, j \in n$.*

Proof Let $Q = \langle H([\omega_1]^2), \supseteq\rangle$. Then Q clearly satisfies c.c.c. and has cardinality ω_1. Let $\mathcal{H}$ be Q-generic over V. Take $h = \bigcup \mathcal{H}$ and $E = g^{-1}\{1\}$. It is well known that $G = \langle \omega_1, E\rangle$ satisfies (*). To see that G has a good partition it is enough to observe that, whenever A is an infinite subset of ω_1 with $A \in V$, then, for each $B \in [\omega_1]^{<\omega}$ and $g \in {}^B2$, the set

$$D_{A,g} = \{q \in Q : (\exists\, a \in \mathrm{dom}(q) \cap A)(\forall\, b \in B)\ q(a, b) = g(b)\}$$

is dense in Q. It completes the proof of Lemma 4.5. □

Lemma 4.6 *Consider a graph $G = \langle V, E \rangle$ having a good partition. Then, for each $X \subset V$, either $G[X]$ or $G[V \setminus X]$ has a good partition.*

Proof of Lemma 4.6 First fix a good partition $\bar{A} = \{A_\alpha : \alpha < \omega_1\}$ of G.

Claim *If $A \subset V$ is dense in G then either $A \cap X$ is dense in $G[X]$ or $A \setminus X$ is dense in $G[V \setminus X]$.*

Assume, on the contrary, that the claim fails. Then we have finite functions $f \in H(X)$ and $g \in H(V \setminus X)$, showing that $A \cap X$ is not dense in $G[X]$ and $A \setminus X$ is not dense in $G[V \setminus X]$, respectively. But $f \cup g \in H(V)$ and so there is an $a \in A$ such that $(f \cup g)(b) = 1$ iff $\{a, b\} \in E$ for each $b \in \mathrm{dom}(f) \cup \mathrm{dom}(g)$. Then either $a \in X$ or $a \in V \setminus X$. We may assume that $a \in X$. But in this case f does not show that $A \cap X$ is not dense in $G[X]$. This contradiction proves the claim. □

Now we return to the proof of the lemma. Take

$$I = \{\alpha < \omega_1 : A_\alpha \cap X \text{ is dense in } G[X]\};$$
$$J = \{\alpha < \omega_1 : A_\alpha \setminus X \text{ is dense in } G[V \setminus X]\}.$$

By the claim we have that $I \cup J = \omega_1$. Thus we may assume that I is uncountable. Write $I = \{\alpha_\nu : \nu < \omega_1\}$. Now consider a partition $\bar{B} = \{B_\mu : \mu < \omega_1\}$ of X into countable pieces such that, for each $\mu < \omega_1$ with $A_{\alpha_\nu} \cap X \subset B_\mu$. Then every B_μ is dense and thus the partition $\bar{B}$ is good. Thus Lemma 4.6 is proved. □

Definition 4.7 Let $G = \langle V, E \rangle$ and $H = \langle W, F \rangle$ be finite graphs. Let $\bar{A} = \{A_\nu : \nu < \omega_1\}$ and $\bar{B} = \{B_\nu : \nu < \omega_1\}$ be partitions of V and W, respectively. Define the poset $Q(G, \bar{A}, H, \bar{B})$ as follows. The underlying set of the poset consists of all finite 1–1 functions f mapping a finite subset of V to W satisfying (a) and (b) below:

(a) $\{x, y\} \in E \Leftrightarrow \{f(x), f(y)\} \in E$;
(b) $f''A_\nu \subset B_\nu$ for each $\nu < \omega_1$.

We take $f \leqslant g$ iff $f \supseteq g$.

Lemma 4.8 *Assume that the graph $G = \langle W, E \rangle$ is strongly non-trivial with $|W| = \omega_1$ and $X \in [W]^{\omega_1}$. If $\bar{A} = \{A_\alpha : \alpha < \omega_1\}$ is a good partition of G and $\bar{B} = \{B_\beta : \beta < \omega_1\}$ is a good partition of $G[X]$ then the poset $Q(G, \bar{A}, G[X], \bar{B})$ satisfies the c.c.c. and*

$$V^{Q(G, \bar{A}, G[X], \bar{B})} \models \text{'}G \text{ is strongly non-trivial'}.$$

Proof For simplicity we write $Q = Q(G, \bar{A}, G[X], \bar{B})$. Using a standard argument, it is clear that the following claim implies the statement of Lemma 4.8.

Claim *If* $\{f_\alpha : \alpha < \omega_1\} \subset Q$ *and* $\{s_\alpha : \alpha < \omega_1\} \subset [W]^{<\omega}$ *is a family of pairwise disjoint finite subsets of* W, *then there are four distinct ordinals,* α, β, γ *and* δ *satisfying* (a) *and* (b) *below:*

(a) f_α *and* f_β *are compatible in* Q *and* $[s_\alpha, s_\beta] \subset E$;

(b) f_γ *and* f_δ *are compatible in* Q *and* $[s_\gamma, s_\delta] \cap E = \emptyset$.

It is enough to find suitable α and β. To start with we take $c(f) = \mathrm{dom}(f) \cup \mathrm{ran}(f)$ for $f \in Q$. We can assume that $\{c(f_\alpha : \alpha \in \omega_1\}$ forms a Δ-system with kernel c. Choose a countable $D \subset W$ such that $c \subset D$ and $(A_\nu \cup B_\nu) \cap D \neq \emptyset$ implies that $(A_\nu \cup B_\nu) \subset D$ for $\nu < \omega_1$. If $\nu < \omega_1$ take $c_\nu = c(f) \backslash c$. Since D is countable, we can asume that $c_\nu \cap D = \emptyset$ for $\nu < \omega_1$. But in this case we have $f_\nu'' c \subset c$ and $f_\nu'' c_\nu \subset c_\nu$. Thus, we can thin out the sequence so that $f_\nu \restriction c = f_\mu \restriction c$ for $\nu < \mu < \omega_1$. Since both the family $\{s_\alpha : \alpha < \omega_1\}$ and the family $\{c_\alpha : \alpha < \omega_1\}$ consist of pairwise disjoint finite sets, we can find an uncountable $I \subset \omega_1$ with $(s_\alpha \cup c_\alpha) \cap (s_\beta \cup c_\beta) = \emptyset$ for each $\alpha \neq \beta \in I$. But we know that G is strongly non-trivial and thus there are $\alpha \neq \beta \in I$ with $[(s_\alpha \cup c_\alpha), (s_\beta \cup c_\beta)] \subset E$. We must show that f_α and f_β are compatible. Take $f = f_\alpha \cup f_\beta$. Clearly f is a 1–1 function because $f_\alpha \restriction c = f_\beta \restriction c$ and $c_\alpha \cap c_\beta = \emptyset$. Condition 4.7 (b) obviously holds. We need check only 4.7 (a). Choose $x, y \in \mathrm{dom}(f)$. We can assume that $x \in \mathrm{dom}(f_\alpha) \backslash \mathrm{dom}(f_\beta)$ and $y \in \mathrm{dom}(f_\beta) \backslash \mathrm{dom}(f_\alpha)$ or 4.7 (a) is trivial. But in this case $x, f(x) \in c_\alpha$ and $y, f(y) \in c_\beta$ and thus both $\{x, y\} \in E$ and $\{f(x), f(y)\} \in E$. This means that f is a common extension of f_α and f_β. Thus α and β are suitable ordinals, which was to be proved. □

Lemma 4.9 *Assume that* $G = \langle W, E \rangle$ *and* $H = \langle Z, F \rangle$ *are graphs with good partitions* $\bar{A}$ *and* $\bar{B}$, *respectively,. Let* $Q = Q(G, \bar{A}, H, \bar{B})$. *Then*

(a) *if* $w \in W$ *and* $z \in Z$ *then both the set* $C_w = \{f \in Q : w \in \mathrm{dom}(f)\}$ *and the set* $D_z = \{f \in Q : z \in \mathrm{ran}(f)\}$ *are dense in* Q;

(b) $V^Q \models$ *'G and H are isomorphic'.*

The proof of this lemma is straightforward and so we omit it. □

Lemma 4.10 *Assume that* $H = \langle W, E \rangle$ *is a graph with* $|W| = \omega_1$ *and* α *is a limit ordinal. If* $\langle P_\gamma : \gamma \leqslant \alpha \rangle$ *is an* α*-stage iteration of c.c.c. posets with finite support and, for each* $\gamma < \alpha$, *we have* $V^{P_\gamma} \models$ *'H is strongly non-trivial' then* $V^{P_\alpha} \models$ *'H is strongly non-trivial', too.*

Proof Assume that $p \in P_\alpha$ and

$p \Vdash$ '$\{\bar{s}_\nu : \nu < \omega_1\}$ is a family of pairwise disjoint, finite subsets of W'.

For each $\nu < \omega_1$ choose $p_\nu \leqslant p$ and $t_\nu \in [W]^{<\omega}$ with $p_\nu \Vdash \bar{s}_\nu = t_\nu$. Now we can find an $I \in [\omega_1]^{\omega_1}$, $t \in [W]^{<\omega}$ and $d \in [\alpha]^{<\omega}$ such that $\{t_\nu : \nu \in I\}$ forms a Δ-sequence with kernel t and $\{\text{supp}(p_\nu) : \nu \in I\}$ forms a Δ-system with kernel d. Since P_α satisfies c.c.c., we have $t = \emptyset$. Fix $\beta < \alpha$ with $d \subset \beta$ and $\text{supp}(p) \subset \beta$.

But P_β satisfies c.c.c. and so we can choose a condition $q \in P_\beta$ ($q \leqslant p$) such that

$$q \Vdash \text{`the set } \{\nu \in \omega_1 : p_\nu \upharpoonright \beta \in \mathcal{G}^\beta\} \text{ is uncountable'},$$

where $\mathcal{G}^\beta$ is the canonical name for the P_β-generic filter. Since $V^{P_\beta} \models$ 'H is strongly non-trivial', we have an $r \in P_\beta$ ($r \leqslant q$) and we have distinct ordinals ν, μ, ξ and η such that

$$r \Vdash \{p_\nu \upharpoonright \beta, p_\mu \upharpoonright \beta, p_\xi \upharpoonright \beta, p_\eta \upharpoonright \beta\} \subset \mathcal{G}^\beta,$$

$$[t_\nu, t_\mu] \subset E, \qquad [t_\xi, t_\eta] \cap E = \emptyset.$$

But in this case r, p_ν, p_μ, p_ξ and p_η have a common extension u not only in P_β but also in P_α. Then

$$u \Vdash \text{`}[s_\nu, s_\mu] \subset E \text{ and } [s_\xi, s_\eta] \cap E = \emptyset\text{'},$$

which completes the proof of Lemma 4.10. □

Proof of Theorem 4.1 We define a 2^ω-stage iteration of c.c.c. posets $\langle P_\nu : \nu \leqslant 2^\omega \rangle$, with finite support, in such a way that each P_ν has cardinality $\leqslant 2^\omega$.

First take P_0 to be the poset satisfying the requirements of Lemma 4.5. Then, in V^{P_0}, we fix a strongly non-trivial graph $H = \langle \omega_1, E \rangle$ having a good partition $\bar{A}$. We want to preserve the strong non-triviality of H. It is not distroyed in the limit steps because we take direct limits and so we can apply Lemma 4.10. In the $(\nu+1)$-th step we pick a countable $X_\nu \subset \omega_1$ with $X_\nu \in V^{P_\nu}$. By Lemma 4.6, the graph $H[\omega_1 \backslash X_\nu]$ has a good partition $\bar{B}$. Then take

$$P_{\nu+1} = P_\nu * Q(H, \bar{A}, H[\omega_1 \backslash X_\nu], \bar{B}).$$

By Lemmas 4.8 and 4.9 (b), the poset $P_{\nu+1}$ satisfies c.c.c and

$$V^{P_{\nu+1}} \models \text{`}H \text{ is strongly non-trivial and is isomorphic to } H[\omega_1 \backslash X_\nu]\text{'}.$$

Clearly $|P_{\nu+1}| \leqslant 2^\omega$. Using a bookkeeping function we can achieve that

$$\{X_\nu : \nu < 2^\omega\} = [\omega_1]^{\leqslant \omega} \cap V^{P_{2^\omega}}.$$

But this implies that P_{2^ω} satisfies the requirements of Theorem 4.1. □

Proof of Theorem 4.2 We define a 2^{ω_1}-stage iteration of c.c.c. posets $\langle P_\nu : \nu \leqslant 2^{\omega_1}\rangle$, with finite support, in such a way that each P_ν has cardinality $\leqslant 2^{\omega_1}$.

First take P_0 to be the poset satisfying the requirements of Lemma 4.5. Then, in V^{P_0}, we fix a strongly non-trivial graph $H = \langle \omega_1, E\rangle$ having a good partition $\bar{A}$. We want to preserve the strong non-triviality of H. It is not distroyed in the limit steps because we take direct limits and so we can apply Lemma 4.10. In the $(\nu+1)$-th step we pick a countable $X_\nu \subset \omega_1$ with $X_\nu \in V^{P_\nu}$. Take

$$Y_\nu = \begin{cases} X_\nu & \text{if } H[X_\nu] \text{ has a good partition,} \\ \omega_1 \setminus X_\nu & \text{otherwise.} \end{cases}$$

By Lemma 4.6, the graph $H[Y_\nu]$ has a good partition $\bar{B}$. Then take

$$P_{\nu+1} = P_\nu * Q(H, \bar{A}, H[Y_\nu], \bar{B}).$$

By Lemmas 4.8 and 4.9 (b), the poset $P_{\nu+1}$ satisfies c.c.c and

$$V^{P_{\nu+1}} \models \text{`}H \text{ is strongly non-trivial and is isomorphic to } H[Y_\nu]\text{'}.$$

Clearly $|P_{\nu+1}| \leqslant 2^{\omega_1}$. Using a bookkeeping function we can achieve that

$$\{X_\nu : \nu < 2^{\omega_1}\} = \mathcal{P}(\omega_1) \cap V^{P_{2^{\omega_1}}}.$$

But this implies that $P_{2^{\omega_1}}$ satisfies the requirements of Theorem 4.2. □

To present the proof of Theorem 4.3 we must recall some definitions and a theorem of [3].

Definition 4.11 A pair $\langle e, T\rangle$ is called *correct* iff it satisfies (a)–(d) below:

(a) e is a function and $T \subset [\mathrm{dom}(e)]^2$;
(b) $e(x) \neq e(y)$ for each $\{x, y\} \in T$;
(c) $\forall\, A \in [\omega_1]^{\omega_1}\ \exists\, B \in [A]^{\omega_1}$ either $[B]^2 \subset T$ or $[B]^2 \cap T = \emptyset$;
(d) $\forall\, A, B \in [\omega_1]^{\omega_1}$ if $[A]^2 \cup [B]^2 \subset T$, $A = \langle a_\alpha : \alpha < \omega_1\rangle$ and $B = \langle b_\beta : \beta < \omega_1\rangle$ then $\exists\, I \in [\omega_1]^{\omega_1}\ \forall\, \alpha \neq \beta \in I\ \{a_\alpha, b_\beta\} \in T$.

Definition 4.12 If e is a function and $m \in \omega$ write

$$S_m(e) = \{s \in [\mathrm{dom}(e)]^m : |s| = |e''s|\}.$$

Definition 4.13 Let $\langle e, T\rangle$ be a correct pair and $R \subset [\mathrm{dom}(e)]^2$.

(a) Let $m \in \omega$. We say that R is *m-complicated for* $\langle e, T\rangle$ iff for each sequence $\langle s_\alpha : \alpha < \omega_1\rangle \subset S_m(e)$ we have $\alpha < \beta < \omega_1$ with $[s_\alpha, s_\beta] \subset R$ or $[s_\alpha, s_\beta] \not\subset T$.

(b) R is called *strongly complicated for* $\langle e,T\rangle$ iff it is m-complicated for $\langle e,t\rangle$ for each $m \in \omega$.

We denote by $C(\omega_1)$ the filter of closed unbounded subsets of ω_1.

Theorem 4.14 *Assume that* $C(\omega_1)$ *is* 2^{ω_1}*-complete. Let* $\langle e,T\rangle$ *be a correct pair and* R *be strongly complicated for* $\langle e,T\rangle$*. Then for each natural number* m *there is a c.c.c. poset* P_m *with cardinality* 2^{ω_1} *such that*

$V^{P_m} \models$ *'The* m*-complicatedness of* R *for* $\langle e,T\rangle$ *is c.c.c. indestructible'.*

See [3, Theorem 4.3]. This theorem clearly implies the following corollary.

Corollary 4.15 *If* $2^{\omega_1} = \omega_2$, $\langle e,T\rangle$ *is a correct pair and* R *is strongly complicated for* $\langle e,T\rangle$ *then there is a c.c.c. poset* P *with cardinality* 2^{ω_1} *such that*

$V^P \models$ *'R is 1-complicated for* $\langle e,T\rangle$ *and Martin's axiom holds'.*

Proof of Theorem 4.3 Let $2^{\omega_1} = \omega_2$ in the ground model. First take P_0 to be the poset satisfying the conditions of Lemma 4.5. Then, in V^{P_0}, we fix a graph $H = \langle\omega_1, E\rangle$ having a good partition $\bar{A} = \{A_\alpha : \alpha < \omega_1\}$ and satisfying (*).

From now on we work in V^{P_0}. Here $2^{\omega_1} = \omega_2$, whence Corollary 4.15 can be applied. Take $Q^* = Q(H,\bar{A},H,\bar{A})$. Define the function $h\colon \omega_1 \to \omega_1$ by $\nu \in A_{h(\nu)}$. If $f \in Q^*$ take $u(f) = \mathrm{dom}(f) \cup \mathrm{ran}(f)$. Set

$$D = \omega_1 \times Q^* \times 2.$$

If $x \in D$, write $x = \langle \nu_x, f_x, j_x\rangle$. Define the function $e\colon D \to \omega_1$ by $e(d) = h(\nu_d)$. Take

$$T = \{\{x,y\} \in D^2 : e(x) \neq e(y) \text{ and } u(f_x) \cap u(f_y) = \emptyset\}$$

and

$$R = \{\{x,y\} \in T : f(x) \cup f(y) \in Q^* \text{ and } (j_x = j_y = 1 \Leftrightarrow \{\nu_x, \nu_y\} \in E)\}.$$

Standard arguments show that the pair $\langle e,T\rangle$ is correct. □

Lemma 4.16 *R is strongly complicated for* $\langle e,T\rangle$.

Proof Fix a natural number m and a sequence $\langle s_\alpha : \alpha < \omega_1\rangle \subset S_m(e)$. We may assume that $[s_\alpha, s_\beta] \subset T$ for each $\alpha < \beta < \omega_1$. For $\alpha < \omega_1$ write $s_\alpha = \{s_{\alpha,i} : i < m\}$ $(s_{\alpha,i} = \langle \nu_i^\alpha, f_i^\alpha, j_i^\alpha\rangle)$, $\xi_i^\alpha = h(\nu_i^\alpha)$ and $c_\alpha = \{\xi_i^\alpha : i < m\} \cup \bigcup\{u(f_i^\alpha) : i < m\}$. By thinning out the sequence we can assume that $\max(c_\alpha) < \min(c_\beta)$ for each $\alpha < \beta < \omega_1$. Recall that

$s_\alpha \in S_m(e)$ implies that, for each $\rho < \omega_1$, there is at most one $i < m$ with $\nu_i^\alpha \in A_\rho$. Thus, taking into account that H satisfies (*), we can find ordinals $\alpha < \beta < \omega_1$ satisfying (•) below:

(•) for each $\rho < \eta < \omega_1$, $[c_\alpha \cap A_\rho, c_\beta \cap A_\eta] \subset E$ if there are $i, k < m$ with $\nu_i^\alpha \in A_\rho$, $\nu_k^\beta \in A_\eta$ and $j_i^\alpha = j_k^\beta = 1$; otherwise
$$[c_\alpha \cap A_\rho, c_\beta \cap A_\eta] \cap E = \emptyset.$$

We prove that $[r_\alpha, r_\beta] \subset R$. To show $\{s_{\alpha,i}, s_{\beta,k}\} \in R$ we need only check that $f_i^\alpha \cup f_k^\beta \in Q^*$, because the other requirements are clearly satisfied. Write $f = f_i^\alpha \cup f_k^\beta$. Sine $c_\alpha \cap c_\beta = \emptyset$ and $f_i^\alpha, f_k^\beta \in Q^*$, the conditions 4.7 (a) and 4.7 (c) hold. To check 4.7 (b) choose $x, y \in \text{dom}(f)$. We can assume that $x \in \text{dom}(f_i^\alpha)$ and $y \in \text{dom}(f_k^\beta)$, or 4.7 (b) is trivial. But in this case we have $x, f(x) \in A_{h(x)}$ and $y, f(y) \in A_{h(y)}$, and so $\{x, y\} \in E$ iff $\{f(x), f(y)\} \in E$ by (•). This completes the proof of Lemma 4.16. □

We now follow the proof of Theorem 4.3. Applying Corollary 4.15 for $\langle e, T\rangle$ and for R in V^{P_0}, we obtain a c.c.c. poset P with cardinality 2^{ω_1} such that
$$V^{P_0 * P} \models \text{'}R \text{ is 1-complicated for } \langle e, T\rangle \text{ and Martin's axiom holds'}.$$
To show that the model $V^{P_0 * P}$ satisfies the requirements of Theorem 4.3, we prove that H is almost smooth and non-trivial in it. Thus the following two lemmas complete the proof.

Lemma 4.17 *If R is 1-complicated in a model $W \supset V^{P_0}$ then $W \models$ 'H is non-trivial'.*

Lemma 4.18 *If R is 1-complicated in a model $W \supset V^{P_0}$ then $W \models$ 'For each countable $X \subset \omega_1$ the graph $H[\omega_1 \backslash X]$ has a good partition $\bar{B}$ such that $Q(H, \bar{A}, H[\omega_1 \backslash X], \bar{B})$ satisfies c.c.c.'.*

Proof of Lemma 4.17 Assume on the contrary that – for example – $Y \in [\omega_1]^{\omega_1}$ with $[Y]^2 \subset E$. We can assume that $|Y \cap A_\nu| \leqslant 1$ for each $\nu < \omega_1$. Write $Y = \{y_\alpha : \alpha < \omega_1\}$. Take $s_\alpha = \langle y_\alpha, \emptyset, 0\rangle$ for $\alpha < \omega_1$. Then $s_\alpha \in S_1(e)$ and $[s_\alpha, s_\beta] \subset T$ for $\alpha < \beta < \omega_1$. Thus there are $\alpha < \beta < \omega_1$ with $[s_\alpha, s_\beta] \subset R$. But, by definition of R, this means that $\{y_\alpha, y_\beta\} \notin E$. Contradiction; the lemma is proved. □

Proof of Lemma 4.18 Fix a $\mu < \omega_1$ with $X \subset \bigcup\{A_\nu : \nu < \mu\}$. Choose an $\eta < \omega_1$ with $\mu + \eta = \eta$. Write $Y = \omega_1 \backslash X$. Define the partition $\bar{B} = \{B_\xi : \xi < \omega_1\}$ of Y as follows:

$$B_\xi = \begin{cases} \bigcup\{A_\nu : \nu < \mu\}\backslash X & \text{if } \xi = 0, \\ A_{\mu+\xi} & \text{if } 1 \leq \xi < \eta, \\ A_\xi & \text{otherwise.} \end{cases}$$

Since each B_ξ contains some A_χ, the partition $\bar{B}$ is a good partition of $H[Y]$. In order to show that $Q = Q(H, \bar{A}, H[Y], \bar{B})$ satisfies c.c.c., let us fix an arbitrary $\{f_\alpha : \alpha < \omega_1\} \subset Q$. If $\alpha < \omega_1$ take

$$D_\alpha = \{\xi : A_\xi \cap \operatorname{dom}(f_\alpha) \neq \emptyset\}.$$

We may assume that $\{D_\alpha : \alpha < \omega_1\}$ forms a Δ-system with kernel D. Choose a countable ordinal $\rho > \eta$ with $D \subset \rho$. We can assume that $(D_\alpha \backslash D) \cap \rho = \emptyset$. Let

$$B = \bigcup\{A_\xi \cup B_\xi : \xi < \rho\}.$$

Then we have $f''_\alpha B \subset B$ for each $\alpha < \omega_1$. So we can thin out our sequence so that $f_\alpha \restriction B$ is independent of α. This means that $f_\alpha \parallel_Q f_\beta$ iff $f_\alpha \restriction (\omega_1 \backslash B) \parallel_Q f_\beta \restriction (\omega_1 \backslash B)$ for $\alpha < \beta < \omega_1$. Thus $D = \emptyset$ can be assumed. By the definition of $\bar{B}$ and by the choice of ρ we get $f_\alpha \in Q^*$ for $\alpha < \omega_1$. Write $s_\alpha = \{\langle \nu_\alpha, f_\alpha, 1\rangle\}$ for $\alpha < \omega_1$, where $\nu_\alpha = \min B_\alpha$. Then $s_\alpha \in S_1(e)$ and $[s_\alpha, s_\beta] \subset T$ for $\alpha < \beta < \omega_1$. By the 1-complicatedness of R we have $\alpha < \beta < \omega_1$ with $[s_\alpha, s_\beta] \subset R$. Then $f_\alpha \cup f_\beta \in Q^*$. Thus $f_\alpha \cup f_\beta \in Q$, that is, f_α and f_β are compatible in Q, which was to be proved. □

The proof of Theorem 4.3 is complete. □

5 ω-uniform hypergraphs

An ω-uniform hypergraph is a pair $\langle A, \mathscr{A}\rangle$ with $\mathscr{A} \subset [A]^\omega$. We say that $\langle A, \mathscr{A}\rangle$ is *non-trivial* iff both $(Y)^\omega \cap \mathscr{A} \neq \emptyset$ and $(Y)^\omega \backslash \mathscr{A} \neq \emptyset$ for each $Y \subset A$ with $|A| = |Y|$. Two ω-uniform hypergraphs, $\langle A, \mathscr{A}\rangle$ and $\langle B, \mathscr{B}\rangle$, are said to be *isomorphic* iff there is a bijection $f: A \to B$ with $\mathscr{B} = \{f''X : X \in \mathscr{A}\}$. The main result of this section can be formulated as follows.

Theorem 5.1 *Assume* $\diamondsuit^+$. *Then there is a non-trivial* ω*-uniform graph* $\langle \omega_1, \mathscr{B}\rangle$ *with* $\mathscr{B} \subset (\omega_1)^{<\omega \cdot 2}$ *such that it is isomorphic to* $\langle Y, \mathscr{B} \cap [Y]^\omega\rangle$ *whenever* $Y \in [\omega_1]^{\omega_1}$.

This theorem easily follows from the next one, whose formulation demands a few definitions.

Given an ω-uniform hypergraph $\langle A, \mathcal{A}\rangle$, a function f and a set Y, we write

$$f_*(\mathcal{A}) = \{f''X : X \in \mathcal{A}\};$$
$$\mathcal{A} \restriction Y = \mathcal{A} \cap [Y]^\omega;$$
$$\mathcal{A}f \restriction^* Y = \{X \cap Y : X \in \mathcal{A} \text{ and } X \subset^* Y\};$$
$$\langle A, \mathcal{A}\rangle^* = \langle A, \{X \Delta Y : X \in \mathcal{A}, Y \in [A]^{<\omega}\}\rangle.$$

Definition 5.2 Given two ω-uniform hypergraphs, $\langle C, \mathcal{C}\rangle$ and $\langle D, \mathcal{D}\rangle$, and a function f, we write

(a) $\langle C, \mathcal{C}\rangle \cong_f \langle D, \mathcal{D}\rangle$ iff $f: C \xrightarrow[\text{onto}]{1-1} D$ with $\mathcal{D} = f_*(\mathcal{C})$;

(b) $\langle C, \mathcal{C}\rangle \cong \langle D, \mathcal{D}\rangle$ iff there is an onto function g with $\langle C, \mathcal{C}\rangle \cong_g \langle D, \mathcal{D}\rangle$.

(c) $\langle C, \mathcal{C}\rangle \cong^*_f \langle D, \mathcal{D}\rangle$ iff $f: C \xrightarrow[\text{onto}]{1-1} D$ with $\langle D, \mathcal{D}\rangle^* = \langle D, f_*(\mathcal{C})\rangle^*$;

(d) $\langle C, \mathcal{C}\rangle \cong^* \langle D, \mathcal{D}\rangle$ iff there is an onto function g with $\langle C, \mathcal{C}\rangle \cong^*_g \langle D, \mathcal{D}\rangle$.

Theorem 5.3 *Assume* $\diamondsuit^+$. *Then there is an almost disjoint family* $\mathcal{A} \subset (\omega_1)^\omega$ *satisfying* (a)–(c) *below:*

(a) $\mathcal{A} \cap (Y)^\omega \neq \emptyset$ *for each* $Y \in [\omega_1]^{\omega_1}$;

(b) $|\mathcal{A} \cap (\alpha)^\omega| \leq \omega$ *for each* $\alpha < \omega_1$;

(c) $\langle \omega_1, \mathcal{A}\rangle \cong^* \langle Y, \mathcal{A}f \restriction^* Y\rangle$ *for each* $Y \in [\omega_1]^{\omega_1}$.

Proof of Theorem 5.1 using Theorem 5.3 Choose $\mathcal{A} \subset (\omega_1)^\omega$ satisfying the requirements of Theorem 5.3. Define $\langle \omega_1, \mathcal{B}\rangle$ to be $\langle \omega_1, \mathcal{A}\rangle^*$. Clearly $\mathcal{B} \subset (\omega_1)^{<\omega \cdot 2}$. Consider a $Y \in [\omega_1]^{\omega_1}$. Then $(Y)^\omega \cap \mathcal{B} \neq \emptyset$ because $\mathcal{A} \subset \mathcal{B}$. Pick an $\alpha < \omega_1$ with $|Y \cap \alpha| = \omega$. Then the set $\{X \in \mathcal{B} : X \subset Y \cap \alpha\}$ is countable because $\mathcal{A}$ satisfies 5.3 (b). On the other hand, the set $(Y \cap \alpha)^\omega$ is uncountable and thus $(Y)^\omega \not\subset \mathcal{B}$. This means that $\langle \omega_1, \mathcal{B}\rangle$ is non-trivial. To prove $\langle \omega_1, \mathcal{B}\rangle \cong \langle Y, \mathcal{B} \restriction Y\rangle$, fix a function f with $\langle \omega_1, \mathcal{A}\rangle \cong^*_f \langle Y, \mathcal{A}f \restriction^* Y\rangle$. Then

$$\langle \omega_1, \mathcal{A}\rangle^* \cong_f \langle Y, \mathcal{A}f \restriction^* Y\rangle^*.$$

But $\langle Y, \mathcal{B} \restriction Y\rangle = \langle Y, \mathcal{A}f \restriction^* Y\rangle^*$. Thus $\langle \omega_1, \mathcal{B}\rangle \cong_f \langle Y, \mathcal{B} \restriction Y\rangle$, which completes the proof. □

The proof of Theorem 5.3 is based on several lemmas to be given below.

Lemma 5.4 *If* $\mathcal{A}, \mathcal{B} \subset [\omega]^\omega$ *are disjoint families with* $|\mathcal{A}| = |\mathcal{B}| = \omega$, *then* $\langle \omega, \mathcal{A}\rangle \cong^* \langle \omega, \mathcal{B}\rangle$.

Proof Write $\mathscr{A} = \{A_n : n < \omega\}$ and $\mathscr{B} = \{B_n : n < \omega\}$. By induction on n define

$$C_n = \left(A_n \setminus \bigcup\{C_m : m < n\}\right) \cup \left\{\min\left(\omega \setminus \bigcup\{C_m : m < n\}\right)\right\}$$

and

$$D_n = \left(B_n \setminus \bigcup\{D_m : m < n\}\right) \cup \left\{\min\left(\omega \setminus \bigcup\{D_m : m < n\}\right)\right\}$$

for $n < \omega$.

Then both $\mathscr{C} = \{C_n : n < \omega\}$ and $\mathscr{D} = \{D_n : n < \omega\}$ are partitions of ω into infinitely many infinite pieces and thus there is a function f with $\langle \omega, \mathscr{C} \rangle \cong_f \langle \omega, \mathscr{D} \rangle$. But $\langle \omega, \mathscr{A} \rangle \cong^*_{\text{id}} \langle \omega, \mathscr{C} \rangle$ and $\langle \omega, \mathscr{B} \rangle \cong^*_{\text{id}} \langle \omega, \mathscr{D} \rangle$, where id is the identity function. Thus $\langle \omega, \mathscr{A} \rangle \cong_f \langle \omega, \mathscr{B} \rangle$, which was to be proved. $\square$

Given an ordinal α and a function f with $\text{dom}(f) \cup \text{ran}(f) \subset \text{On}$, we say that f is *almost increasing in* α iff $\sup(f) = \alpha$ and $\sup(f''X) = \alpha$ for each $X \subset \text{dom}(f)$. We define $\mathscr{A}(\alpha)$ to be the set of all functions which are almost increasing in α. Take $\text{Cof}(\alpha) = \{A \subset \alpha : \sup A = \alpha\}$.

Lemma 5.5 *Let $\alpha < \omega_1$ be a limit ordinal. If $\mathscr{F} \subset \mathscr{A}(\alpha)$ and $\mathscr{B} \subset \text{Cof}(\alpha)$ are countable sets and the elements of $\mathscr{B}$ are pairwise disjoint then there is a countable, almost disjoint family $\mathscr{C} \subset \text{Cof}(\alpha) \cap (\alpha)^\omega$ satisfying* (a) *and* (b) *below:*

(a) *for each $B \in \mathscr{B}$ we have a $C \in \mathscr{C}$ with $C \subset B$;*
(b) *$\langle \text{dom}(f), \mathscr{C}f \restriction^* \text{dom}(f) \rangle \cong^*_f \langle \text{ran}(f), \mathscr{C}f \restriction^* \text{ran}(f) \rangle$ for each $f \in \mathscr{F}$.*

Proof Write $\omega^* = \{s_n : n < \omega\}$, $\mathscr{F} = \{f_n : n < \omega\}$ and $\mathscr{B} = \{B_n : n < \omega\}$. If $s \in \omega^*$ set $\|s\| = \min\{i : s_i = s\}$. If $s = \langle j_0, \ldots, j_{n-1} \rangle$ and $t = \langle i_0, \ldots, i_{m-1} \rangle$ are in ω^* then we define the functions $h_{s,t}$, provided $i_0 = j_0$, taking

$$h_{s,t} = f_{i_{m-1}} \circ f_{i_{m-2}} \circ \cdots \circ f_{i_k} \circ f_{j_k}^{-1} \circ \cdots \circ f_{j_{n-2}}^{-1} \circ f_{j_{n-1}}^{-1},$$

where $k = \min\{l : i_l \neq j_l\}$. For example, $h_{s,s}$ is the identity on α.

By induction on n, one can easily construct families $\langle D^n_t : t \in \omega^* \rangle$ satisfying (a)–(d) below:

(a) $D^0_t = \begin{cases} B_k & \text{if } t = \langle k \rangle, \\ \alpha & \text{otherwise;} \end{cases}$
(b) $D^n_t \subset D^{n-1}_t$ *and* $D^n_t \in \text{Cof}(\alpha)$;
(c) $f''_k D^n_s \subset D^n_{s^\frown\langle k \rangle}$;
(d) *If $\|s\|, \|t\| < n$ and $h_{s,t}$ is defined then either $\text{dom}(h_{s,t}) \cap D^n_s = \emptyset$ or $D^n_t = h''_{s,t} D^n_s$ and, in case $D^n_t = h''_{s,t} D^n_s$, either $h_{s,t}(x) = x$ for each $x \in D^n_s$ or $h_{s,t}(x) \neq x$ for each $x \in D^n_s$.*

Next fix an increasing sequence of ordinals $\{\alpha_n : n < \omega\}$ with $\sup\{\alpha_n : n < \omega\} = \alpha$ and a bijection $h\colon \omega \to \omega^*$.

We construct systems of finite subsets $\{C_s^n : s \in \omega^*\}$ of α, with $C_s^n = \emptyset$ for $\|s\| \geqslant n$, by induction on n. Take $C_s^0 = \emptyset$ for $s \in \omega^*$. Assume that $\{C_s^{n-1} : s \in \omega^*\}$ is done. Take

$$C^{n-1} = \bigcup \{C_s^{n-1} : s \in \omega^*\}.$$

Then C^{n-1} is finite. Let $h(n-1) = \langle s, k\rangle$. If $\|s\| \geqslant n$ then we take $C_t^n = C_t^{n-1}$ for $t \in \omega^*$.

Suppose that $\|s\| < n$. Write

$$D^n = \{h_{u,v}(x) : u, v \in \omega^*, \|u\| < n, \|v\| < n \text{ and } x \in C^{n-1}\}.$$

Clearly the set D^n is finite. We say that $t \in \omega^*$ is *good* iff $\|t\| < n$, $h_{s,t}$ is defined and $h''_{s,t} D_s^n = D_t^n$. Whenever t is good, the function $h_{s,t}$ is almost increasing in α. Since there are only finitely good t, we can pick a $d_n \in D_s^n \setminus D^n$ satisfying $h_{s,t}(d_n) > \alpha_n$ for each good t. Take

$$C_t^n = \begin{cases} C_t^{n-1} \cup \{h_{s,t}(d_n)\} & \text{if } t \text{ is good,} \\ C_t^{n-1} & \text{otherwise.} \end{cases}$$

The construction is done. Now take $C_s = \bigcup \{C_s^n : n \in \omega\}$ for each $s \in \omega^*$. Clearly $C_s \in \mathrm{Cof}(\alpha)$.

Claim *If $s, t \in \omega^*$ then either $C_s =^* C_t$ or $C_s \cap C_t = \emptyset$.*

Proof of the claim If $n \in \omega$ write $h(n) = \langle u_n, k_n\rangle$. Assume that $c \in C_s \cap C_t$. Since $C^{n-1} \cap (C^n \setminus C^{n-1}) = \emptyset$, we obtain that there is an $n \in \omega$ with $h_{u_n,s}(d_n) = c = h_{u_n,t}(d_n)$. Then

$$h_{s,t}(c) = h_{u_n,t} \circ h_{s,u_n}(c) = h_{u_n,t}(d_n) = c.$$

So, by (d), we have $h''_{s,t} D_s^n = D_t^n$ and $h_{s,t} = x$ for each $x \in D_s^n$. Thus $C_s \Delta C_t \subset C^{n-1}$, that is, $C_s =^* C_t$, which was to be proved. □

Set $\mathscr{C}' = \{C_s : s \in \omega^*\}$. Choose a maximal disjoint family $\mathscr{C} \subset \mathscr{C}'$ with $\mathscr{C} \supset \{C_{\langle k\rangle} : k < \omega\}$. By the claim we have $\langle \alpha, \mathscr{C}\rangle^* = \langle \alpha, \mathscr{C}'\rangle^*$. By the construction, we have that $f''_k C_s =^* C_{s^\frown\langle k\rangle}$ and $C_{\langle k\rangle} \subset B_k$. This means that $\mathscr{C}$ satisfies the requirements of Lemma 5.5. □

Lemma 5.6 *Let $\alpha < \omega_1$ be a limit ordinal. If $\mathscr{F} \subset \mathscr{A}(\alpha)$ and $\mathscr{B} \subset \mathrm{Cof}(\alpha)$ are countable sets then there is a countable, almost disjoint family $\mathscr{C} \subset \mathrm{Cof}(\alpha) \cap (\alpha)^\omega$ satisfying* (a) *and* (b) *below:*

(a) *for each $B \in \mathscr{B}$ the set $\{C \in \mathscr{C} : C \subset B\}$ is infinite;*

(b) $\langle \mathrm{dom}(f), \mathscr{C}f\restriction^* \mathrm{dom}(f)\rangle \cong^*_f \langle \mathrm{ran}(f), \mathscr{C}f\restriction^* \mathrm{ran}(f)\rangle$ *for each $f \in \mathscr{F}$.*

The statement of this lemma follows easily from Lemma 5.5. □

Now we turn to the proof of Theorem 5.3.

Proof of Theorem 5.3 Fix a large enough regular cardinal κ, a $<$ well-ordering of H_κ and a $\diamondsuit^+$ sequence $\mathcal{S} = \langle S_\alpha : \alpha < \omega_1 \rangle$. Next choose a sequence $\langle N_\alpha : \alpha < \omega_1 \rangle$ of countable, elementary submodels of $\langle H_\kappa, \in, < \rangle$ with $\mathcal{S} \in N_0$ and $\langle N_\alpha : \alpha < \beta \rangle \in N_\beta$ for $\beta < \omega_1$.

By induction on α we build a sequence $\langle \mathcal{A}_\alpha : \alpha < \omega_1 \rangle$, where $\mathcal{A}_\alpha \subset \mathrm{Cof}(\omega\alpha) \cap (\omega\alpha)^\omega$ with $|\mathcal{A}_\alpha| = \omega$.

Assume that $\langle \mathcal{A}_\gamma : \gamma < \alpha \rangle$ is done. We distinguish two cases.

Case 1 $\alpha = \beta + 1$.

Define $\mathcal{A}_\alpha$ to be the $<$-minimal almost disjoint

$$\mathcal{A}' \in N_\alpha \cap [\mathrm{Cof}(\omega\alpha \backslash \omega\beta]^\omega.$$

Case 2 *α is a limit.*

Take $\mathcal{F}_\alpha = \mathcal{A}(\omega\alpha) \cap N_\alpha$ and $\mathcal{B}_\alpha = \mathrm{Cof}(\omega\alpha) \cap N_\alpha$. Apply Lemma 5.6, choosing $\omega\alpha$ as α, $\mathcal{F}_\alpha$ as $\mathcal{F}$ and $\mathcal{B}_\alpha$ as $\mathcal{B}$. We define $\mathcal{A}_\alpha$ to be the $<$-minimal family $\mathcal{C}$ satisfying the requirements of Lemma 5.6.

The construction is done. Now take $\mathcal{A} = \bigcup \{A_\alpha : \alpha < \omega_1\}$. By the construction of $\mathcal{A}$, condition 5.3 (b) is clear. Consider a $Y \in [\omega_1]^{\omega_1}$. Since the set $E = \{\nu < \omega_1 : \nu = \omega\nu \wedge \sup(Y \cap \omega\nu) = \omega\nu\}$ is club in ω_1, there is a limit ordinal $\nu \in E$ with $Y \cap \omega\nu \in S_\nu$. Then we have $A \in \mathcal{A}_\nu$ with $A \subset Y \cap \omega\nu$ because $S_\nu \subset N_\nu$. Thus $\emptyset \neq \mathcal{A}_\nu \cap (Y)^\omega \subset \mathcal{A} \cap (Y)^\omega$, that is, condition 5.3 (c) is also satisfied.

Lastly we show that $\langle \omega_1, \mathcal{A} \rangle \cong^* \langle Y, \mathcal{A}f \restriction^* Y \rangle$. Write

$$B = \{\alpha < \omega_1 : \omega\alpha = \alpha,\ Y \cap \alpha \in \mathrm{Cof}(\alpha) \text{ and } \mathcal{A}_\alpha \cap (Y \cap \alpha)^\omega \text{ is infinite}\}.$$

Then B is club. Since $\diamondsuit^+$ holds, we have a club $C \subset B$ satisfying

$$\{G \cap \gamma, \mathcal{A}f \restriction^* \gamma, \mathcal{A}f \restriction^* (Y \cap \gamma), Y \cap \gamma\} \subset N_\gamma$$

for $\gamma \in C$. Write $C = \{\gamma_\nu : \nu < \omega_1\}$. If $\xi < \omega_1$, take

$$A_{<\xi} = \bigcup \{\mathcal{A}_\eta : \eta < \xi\}.$$

Now we construct a sequence of functions $\langle \pi_\nu : \nu < \omega_1 \rangle$, by induction on ν, satisfying (a)–(c) below:

(a) $\langle \gamma_\mu, \mathcal{A} \uparrow \gamma_\mu \rangle \cong^*_{\pi_\mu} \langle \gamma_\mu \cap Y, \mathcal{A}f \restriction^* (\gamma_\mu \cap Y) \rangle$ for each $\mu < \omega_1$;

(b) $\pi_\nu \subset \pi_\mu$ for $\nu < \mu < \omega_1$;

(c) $\langle \pi_\nu : \nu < \mu \rangle \in N_{\gamma_\mu}$.

Assume that $\langle \pi_\nu : \nu < \mu \rangle$ is already constructed and that we know (a)–(c) for $\mu' < \mu$. We distinguish two cases.

Case 1 $\mu = \nu + 1$.

Then both $\mathscr{A}f\restriction^*(\gamma_\mu \backslash \gamma_\nu)$ and $\mathscr{A}f\restriction^*((\gamma_\mu \backslash \gamma_\nu) \cap Y)$ are infinite countable families of infinite subsets of α and thus, by Lemma 5.4, there is a π' with

$$(\circ) \quad \langle \gamma_\mu \backslash \gamma_\nu, \mathscr{A}f\restriction^*(\gamma_\mu \backslash \gamma_\nu) \rangle \cong^*_{\pi'} \langle ((\gamma_\mu \backslash \gamma_\nu) \cap Y), \mathscr{A}f\restriction^*((\gamma_\mu \backslash \gamma_\nu) \cap Y) \rangle.$$

Then define π to be the $<$-minimal $\pi' \in N_{\gamma_\mu}$ satisfying $(\circ)$ above. Take $\pi_\mu = \pi_\nu \cup \pi$. Then (a)–(c) hold for μ.

Case 2 *μ is a limit.*

Then we simply take $\pi_\mu = \bigcup \{\pi_\nu : \nu < \mu\}$. As we can see, the construction of $\langle \pi_\nu : \nu < \mu \rangle$ uses only parameters $C \cap \gamma_\mu$, $\mathscr{A}f\restriction^*\gamma_\mu$, $\mathscr{A}f\restriction^*(Y \cap \gamma_\mu)$, $Y \cap \gamma_\mu$ and $\langle N_\nu : \nu < \mu \rangle$. But, by the choice of C, all these parameters are in N_{γ_μ} and so we have $\langle \pi_\nu : \nu < \mu \rangle \in N_{\nu_\mu}$. Hence $\pi_\mu \in N_{\gamma_\mu}$. This means that (c) is satisfied. Condition (b) is trivial. Then $(\circ\circ)$ below is also satisfied:

$$(\circ\circ) \qquad \langle \gamma_\mu, \mathscr{A}_{<\gamma_\mu} \rangle \cong^*_{\pi_\mu} \langle \gamma_\mu \cap Y, \mathscr{A}_{<\gamma_\mu} f\restriction^*(\gamma_\mu \cap Y) \rangle.$$

The function π_μ is an almost increasing bijection between γ_μ and $Y \cap \gamma_\mu$ because $\pi''_\mu \gamma_\nu = Y \cap \gamma_\nu$ for $\nu < \mu$. Then $\pi_\mu \in \mathscr{A}(\gamma_\mu) \cap N_{\gamma_\nu}$. Thus, by the construction of $\mathscr{A}_\mu$, we have

$$(\circ\circ\circ) \qquad \langle \gamma_\mu, \mathscr{A}_{\gamma_\mu} \rangle \cong^*_{\pi_\mu} \langle \gamma_\mu \cap Y, \mathscr{A}_{\gamma_\mu} f\restriction^*(\gamma_\mu \cap Y) \rangle.$$

But $(\circ\circ)$ and $(\circ\circ\circ)$ together are nothing else than condition (a).

Thus the construction of $\langle \pi_\mu : \mu < \omega_1 \rangle$ is finished. Taking $\pi = \bigcup \{\pi_\mu \,;\, \mu < \omega_1\}$, we have

$$\langle \omega_1, \mathscr{A} \rangle \cong^*_\pi \langle Y, \mathscr{A}f\restriction^* Y \rangle.$$

This means that $\mathscr{A}$ satisfies condition 5.3 (c), too, which completes the proof of Theorem 5.3. □

References

[1] T. Jech, *Set Theory*, Academic Press, New York, 1978

[2] H. A. Kierstead & P. J. Nyikos, *Hypergraphs with Finitely Many Isomorphism Subtypes,* preprint

[3] L. Soukup, Martin Axiómájával Konzisztens Tulajdonságok, Ph.D. dissertation, Budapest, 1988

Sets of multiples and Behrend sequences

R. R. Hall

1 Introduction

I am concerned here with sequences $A = \{a_i\}$ of integers, where $2 \leqslant a_1 < a_2 < \cdots$ and such that $B(A)$, the set of multiples of A, has logarithmic density 1. An immediate consequence of a theorem of Davenport & Erdős [2] is that the asymptotic density $dB(A)$ is also 1; equivalently, almost all integers have at least one divisor belonging to A. I shall refer to these sequences as *Behrend sequences*.

Behrend [1] proved the fundamental inequality (see Halberstam & Roth [7], Chapter 5)

$$T(a_1, a_2, \ldots, a_l) \geqslant T(a_1, a_2, \ldots, a_k)\, T(a_{k+1}, a_{k+2}, \ldots, a_l), \tag{0}$$

where (for any integers, not necessarily distinct) $T(a_1, a_2, \ldots, a_l)$ denotes the density of the integers n not divisible by any of $a_1, a_2, \ldots, a_l$. A consequence of this for Behrend sequences is that it is necessary that

$$\sum_{i=1}^{\infty} \frac{1}{a_i} = \infty. \tag{1}$$

We remark that (1) follows equally well from a weaker inequality due to Heilbronn [13] and Rohrbach [14] obtained ten years earlier which states that

$$T(a_1, a_2, \ldots, a_l) \geqslant \prod_{i=1}^{l} \left(1 - \frac{1}{a_i}\right). \tag{2}$$

If the a_i are pairwise relatively prime there is equality in (2) and, in this case, (1) is also sufficient for A to be a Behrend sequence. To this extent (1) is final.

In general (1) is rather weak. A random Behrend sequence (this is undefined, but we may imagine the elements chosen according to some approximate growth condition without any multiplicative constraint) will be significantly thicker than (1) requires. To see this heuristically, recall that, according to Hardy & Ramanujan [12], for almost all n the divisor function $\tau(n)$ satisfies

$$(\log n)^{\log 2-\epsilon} < \tau(n) < (\log n)^{\log 2+\epsilon}. \tag{3}$$

Now a Behrend sequence should normally occupy a proportion greater than $1/\tau(n)$ of the logarithmic interval $(0, \log n]$; that is, we expect

$$\sum_{a_i \leqslant n} \frac{1}{a_i} > (\log n)^{1-\log 2-\epsilon} \tag{4}$$

or

$$\sum_{i=1}^{\infty} a_i^{-1}(\log a_i)^{\log 2-1+\epsilon} = \infty. \tag{5}$$

Of course there are many Behrend sequences for which (5) is false (e.g. any of those consisting solely of primes) but, nevertheless, my intention is to give a necessary condition for these sequences, stronger than (1) and motivated by (5).

Theorem 1 *Let A be a Behrend sequence. Then, for all y $(0 < y \leqslant 1)$ and $\beta > y-1-\log y$, we have*

$$\sum_{i=1}^{\infty} a_i^{-1}(\log a_i)^{\beta} y^{\Omega(a_i)} = \infty. \tag{6}$$

This is best possible in the following sense: given a particular y $(0 < y \leqslant 1)$ and $\beta < y-1-\log y$, there exists a Behrend sequence $A = A(y, \beta)$ such that

$$\sum_{i=1}^{\infty} a_i^{-1}(\log a_i)^{\beta} y^{\Omega(a_i)} < \infty. \tag{7}$$

Here Ω denotes the number of prime factors counted multiply. When $y = 1$ we may allow $\beta = 0$ in (6) because of (1): it is an open question whether $\beta = y-1-\log y$ is permissible for $y < 1$.

Let us see how this applies to sequences

$$A = \bigcup \{\mathbb{Z}^+ \cap (u_j, v_j]\}, \tag{8}$$

made up of fairly long blocks, for which

$$v_j > u_j + u_j^c \tag{9}$$

for some fixed $c > 0$. By a theorem of Shiu [15], for $y \leqslant 1$,

$$\sum_{u_j<a\leqslant v_j} y^{\Omega(a)} \ll_c (v_j-u_j)(\log u_j)^{y-1}$$
$$\ll_c \sum_{u_j<a\leqslant v_j} (\log a)^{y-1}.$$

For these special sequences, (6) becomes

$$\sum_{i=1}^{\infty} a_i^{-1}(\log a_i)^{\gamma} = \infty \qquad (\gamma > 2y-2-\log y). \tag{10}$$

On setting $y = \frac{1}{2}$ we obtain (5). *In this sense a block sequence is multiplicatively random.*

In 1988 Erdős made a conjecture, referred to in Hall & Tenenbaum [11] as the $\boldsymbol{B}(\lambda)$-conjecture, which states that, for λ sufficiently small but greater than 1, the sequence

$$\boldsymbol{A}_0(\lambda) := \bigcup_{j=1}^{\infty} \{\mathbb{Z}^+ \cap (\exp j^{\lambda}, 2\exp j^{\lambda}]\} \tag{11}$$

is (what I now call) a Behrend sequence. For $\lambda < \lambda_0 = 1.31457\ldots$ this has been established by Hall & Tenenbaum [10]. A heuristic argument suggests that $\boldsymbol{A}_0(\lambda)$ should be Behrend for $\lambda < \lambda_1 = 1/(1-\log 2)$ and not for $\lambda > \lambda_1$, the case of equality being left in doubt. Since $\boldsymbol{A}(\lambda)$ is a block sequence, it is now immediate from (5) that $\boldsymbol{A}_0(\lambda)$ is not Behrend for $\lambda > \lambda_1$; moreover, if we could take $\beta = y-1-\log y$ in (6), the equality case would also be settled negatively.

The harder part of Theorem 1 is the construction of the special Behrend sequence which demonstrates that the condition $\beta > y-1-\log y$ in (6) is the correct one (apart from the question of equality). This depends on probabilistic group theory, a branch of mathematics founded by Erdős, Rényi and Turán. In Section 2 we prove the first part of the theorem, which also depends on a probabilistic idea of Erdős.

2 Proof of Theorem 1 (first part)

For an arbitrary $\boldsymbol{A} \subset \mathbb{Z}^+$, let

$$\tau(n,\boldsymbol{A}) := \operatorname{card}\{a : a\,|\,n, a\in\boldsymbol{A}\};$$
$$\chi(n,\boldsymbol{A}) := \min\{1, \tau(n,\boldsymbol{A})\}.$$

If $\boldsymbol{A}$ is a Behrend sequence, we have $\chi(n,\boldsymbol{A}) = 1$ for almost all n and so

$$\lim_{\sigma\to 1+} (\sigma-1) \sum_{n=1}^{\infty} n^{-\sigma}\chi(n,A) = 1.$$

This applies equally well to any tail $A_l := \{a_i, i = l, l+1, l+2, \ldots\}$ of A.

We may assume that $y < 1$; for $y = 1$ our theorem is simply the assertion that (1) holds. Let $\beta > y-1-\log y$ be given and

$$\sum_{i=1}^{\infty} a_i^{-1}(\log a_i)^{\beta} y^{\Omega(a_i)} < \infty. \tag{12}$$

We are going to deduce that A is not a Behrend sequence by showing that, for suitable l, we have

$$\limsup_{\sigma\to 1+} (\sigma-1) \sum_{n=1}^{\infty} n^{-\sigma}\chi(n,A) \leqslant \tfrac{2}{3}. \tag{13}$$

We need a weak form of the law of the iterated logarithm for prime factors of Erdős [3]. For a statement of this and background see Hall & Tenenbaum [11], Chapter 1. We just use Exercise 10 from that chapter.

Lemma 1 *Let $B(\kappa, t_0)$ denote the sequence of integers n for which*

$$\sup_{t\geqslant t_0}\left|\frac{\Omega(n,t)}{\log\log t} - 1\right| > \kappa \qquad (0<\kappa<1),$$

where

$$\Omega(n,t) := \sum(\alpha : p^{\alpha}\|n,\, p \leqslant t).$$

Then the upper asymptotic density of $B(\kappa, t_0)$ satisfies

$$\bar{d}B(\kappa,t_0) \ll_{\kappa} (\log t_0)^{-K}, \tag{14}$$

where $K := (1+\kappa)\log(1+\kappa)-\kappa$, uniformly for $t_0 \geqslant 3$.

In fact, this result was stated for $\omega(n,t)$ (counting distinct prime factors) but, provided $\kappa < 1$ as we may assume, the proof is unaltered.

We begin by fixing κ to satisfy $\beta = y-1-(1+\kappa)\log y$, so that κ and, hence, $K > 0$. By (14) we may now fix t_0 so that $\bar{d}B(\kappa,t_0) \leqslant \frac{1}{3}$. Next we fix l so that $a_l \geqslant t_0$. We have a choice of majorants for $\chi(n, A_l)$. For $n \in B(\kappa, t_0)$ we have $\chi(n, A_l) \leqslant 1$ and

$$\limsup_{\sigma\to 1+} (\sigma-1) \sum_{n\in B(\kappa,t_0)} n^{-\sigma}\chi(n,A) \leqslant \tfrac{1}{3}. \tag{15}$$

On the other hand, for $n \notin B(\kappa, t_0)$ we have

$$\chi(n,A_l) \leqslant \tau(n,A_l) \leqslant \sum_{\substack{a_i|n\\ i\geqslant l}} y^{\Omega(n,a_i)-(1+\kappa)\log\log a_i}.$$

Therefore

$$
\begin{aligned}
&\sum_{n\notin B} n^{-\sigma}\chi(n,A_l)\\
&\quad\leqslant \sum_{i\geqslant l} (\log a_i)^{-(1+\kappa)\log y} \sum_{n\equiv 0\,(\mathrm{mod}\, a_i)} n^{-\sigma}y^{\Omega(n,a_i)}\\
&\quad\leqslant \sum_{i\geqslant l} a_i^{-\sigma}(\log a_i)^{-(1+\kappa)\log y}y^{\Omega(a_i)}\sum_{m=1}^{\infty} m^{-\sigma}y^{\Omega(m,a_i)}\\
&\quad\leqslant \zeta(\sigma)\sum_{i\geqslant l} a_i^{-\sigma}(\log a_i)^{-(1+\kappa)\log y}y^{\Omega(a_i)}\prod_{p\leqslant a_i}\frac{1-p^{-\sigma}}{1-yp^{-\sigma}}\\
&\quad\ll (\sigma-1)^{-1}\sum_{i\geqslant l} a_i^{-\sigma}(\log a_i)^{-(1+\kappa)\log y}y^{\Omega(a_i)}\exp\left(\sum_{p\leqslant a_i}(y-1)p^{-\sigma}\right),
\end{aligned}
$$

where the constant implied by $\ll$ is absolute. Now $y-1<0$ and therefore

$$
\begin{aligned}
\sum_{p\leqslant a_i}(y-1)p^{-\sigma} &\leqslant \sum_{p\leqslant a_i}(y-1)p^{-1}+(\sigma-1)\sum_{p\leqslant a_i}p^{-1}\log p\\
&\leqslant (y-1)\log\log a_i+(\sigma-1)\log a_i+\text{constant},
\end{aligned}
$$

where the constant is again absolute. Hence

$$(\sigma-1)\sum_{n\notin B} n^{-\sigma}\chi(n,A_l)\ll\sum_{i\geqslant l} a_i^{-1}(\log a_i)^{\beta}y^{\Omega(a_i)}.$$

From (12), the sum on the right is the tail of a convergent series. If l is large enough this does not exceed $\frac{1}{3}$: together with (15) this gives (13). □

3 Some special Behrend sequences

Theorem 2 *For each t $(0<t<1)$ the sequence $A_1(t)$ is a Behrend sequence, where*

$$
\begin{aligned}
A_1(t) := \{d\in\mathbb{Z}^+ : {}&\Omega(d)\leqslant t\log\log d,\\
&0<\|\log d\|<(\log d)^{-\mu(t)}\exp(\psi(d)\sqrt{\log\log d})\},
\end{aligned}
\tag{16}
$$

in the case $0<t<\frac{1}{2}$, (as usual $\|u\| := \min_{m\in\mathbb{Z}}|u-m|$);

$$
\begin{aligned}
A_1(t) := \{d\in\mathbb{Z}^+ : {}&\Omega(d)\geqslant t\log\log d,\\
&0<\|\log d\|<(\log d)^{-\mu(t)}\exp(\psi(d)\sqrt{\log\log d})\}
\end{aligned}
\tag{17}
$$

in the case $\frac{1}{2}<t<1$, and

$$A_1(\tfrac{1}{2}) := \{d \in \mathbb{Z}^+ : 0 < \|\log d\| < (\log d)^{-\log 2} \exp(\psi(d)\sqrt{\log\log d})\}, \tag{18}$$

provided in each case that $\psi(d) \to \infty$ as $d \to \infty$, and where

$$\mu(t) := -t\log t - (1-t)\log(1-t). \tag{19}$$

Notice that $\mu(t)$ increases from 0 to its maximum value, $\log 2$, as t increases form 0 to $\frac{1}{2}$ and then decreases again, so that the (mod 1) condition is strongest in (18) above and becomes progressively weaker as t approaches either 0 or 1. On the other hand, the compensating condition on $\Omega(d)$ which appears in (16) and (17) reflects the principle that *almost all divisors have approximately* $\frac{1}{2}\log\log d$ *prime factors.* We know from the Hardy–Ramanujan theorem that almost all integers have approximately $\log\log n$ prime factors and, for $d|n$, $\Omega(d)$ has essentially (ignoring multiple factors) a binomial distribution with mean $\frac{1}{2}\Omega(n)$. Of course, the iterated logarithm increases so slowly that $\log\log d$ and $\log\log n$ are effectively indistinguishable. Thus the ability of a divisor to divide is not simply measured by $1/d$: the principle set out above as a deliberate paradox is the basis of the proof (Hall [8]) that divisor density and logarithmic density are independent.

When $t = \frac{1}{2}$ the theorem that (18) is a Behrend sequence is contained in Erdős & Hall [4], Theorem 2. It depends on a theorem in probabilistic group theory of Erdős & Rényi [6]. In the case $t \neq \frac{1}{2}$, the number-theoretic part of the proof goes much as in [4], but a new theorem in probabilistic group theory is needed. In one respect I have made this more precise than what is required here, and this result, which may have further applications, is presented separately in Section 4.

We conclude this section by applying Theorem 2 to the second part of Theorem 1. Let $y < 1$ and $\beta < y - 1 - \log y$. We put $t = 1 - y$ and claim that

$$\sum_{d \in A_1(t)} d^{-1}(\log d)^{\beta} y^{\Omega(d)} < \infty. \tag{20}$$

We consider only the case $y > \frac{1}{2}$, the others being treated similarly. Evidently we just estimate, for each $k \in \mathbb{Z}^+$, the sum

$$\sum_{\substack{|d-e^k| < e^k k^{-\mu(t)} \exp(\psi\sqrt{\log k}) \\ \Omega(d) \leqslant t \log\log d}} d^{-1}(\log d)^{\beta} y^{\Omega(d)}$$

$$\ll e^{-k} k^{\beta - t\log z} \sum_{|d-e^k| < e^k k^{-\mu(t)} \exp(\psi\sqrt{\log k})} (yz)^{\Omega(d)}$$

for any $z \leqslant 1$. By Shiu's theorem, this is

$$\ll k^{\beta - t\log z - \mu(t) + yz - 1}\exp(\psi\sqrt{\log k})$$

and we choose $z = t/y = (1-y)/y < 1$. The exponent of k is then

$$\beta - t\log\frac{t}{y} - \mu(t) + t - 1 = \beta + \log y - y < -1$$

and we may suppose that $\psi(d) = o(\sqrt{\log\log d})$: it must, of course, tend to infinity. This establishes (20). □

Conversely, Theorem 1 shows that the exponent $\mu(t)$ appearing in (16) and (17) is best possible.

4 Probabilistic group theory

Theorem (Erdős & Rényi 1965) *Let G be an Abelian group of order N and let l elements $g_1, g_2, \ldots, g_l$ be chosen from G, with repetitions allowed. Let $\epsilon > 0$ be given and*

$$1 \geqslant \frac{\log N}{\log 2} + \frac{\log\log N}{\log 2} + \frac{2\log(1/\epsilon)}{\log 2} + \frac{\log(32/\log 2)}{\log 2}. \qquad (21)$$

Then, for all but at most ϵN^l of these choices, every element $g \in G$ may be represented in the form

$$g = \delta_1 g_1 + \delta_2 g_2 + \cdots + \delta_l g_l \qquad (\text{each } \delta_i = 0 \text{ or } 1). \qquad (22)$$

Erdős and Rényi stated that it is most natural to prove results of this sort via probability methods. Thus we choose the l elements randomly and independently from G: any particular element has probability $1/N$ of being chosen. If $R(g)$ denotes the number of representations (22), the conclusion becomes

$$\operatorname{Prob}(\min_g R(g) > 0) \geqslant 1 - \epsilon.$$

The deduction from this that $A_1(\frac{1}{2})$ is a Behrend sequence may be found in Erdős & Hall [4]. We showed that, for any real numbers b and c, the asymptotic density of the integers n having a divisor d such that

$$0 < \|\log d - b\| < 2^{-\log\log n - c\sqrt{\log\log n}}$$

is

$$\frac{1}{\sqrt{2\pi}}\int_c^\infty \exp(-\tfrac{1}{2}u^2)\,du.$$

Here c was allowed to converge to $\pm\infty$; we need $c = -\psi(n)/\log 2$ for (18).

The idea of the proof is as follows. We consider the integers $n \leqslant x$, and let G be the cyclic group with N the integer part of

$$(\log x)^{\log 2} \exp\{-\tfrac{1}{2}\psi(x)\sqrt{\log\log x}\}. \tag{23}$$

Associated with each prime factor p_i of n is a group element

$$g_i \equiv [N \log p_i] \pmod{N}. \tag{24}$$

For technical reasons only prime factors in a certain interval $I(x)$ are admitted. The sums $\delta_1 g_1 + \delta_2 g_2 + \cdots + \delta_l g_l$ do not exactly correspond to divisors but, nevertheless, when (22) represents the zero in G, n must have a divisor d such that $\|\log d\| < (l+1)/N$. We have to show that for $x+o(x)$ integers $n \leqslant x$, the choice of the group elements arising from (24) is in effect random.

In order to show that A_1 is a Behrend sequence for $t \neq \frac{1}{2}$ the proof needs modification. First, N becomes smaller, the $\log 2$ in (23) being replaced by $\mu(t)$. Second, the sums (22) have to be restricted to contain

$$\begin{cases} \leqslant tl & (t < \frac{1}{2}) \\ \geqslant tl & (t > \frac{1}{2}) \end{cases}$$

summands, where $l = [\log\log x - \frac{1}{3}\psi(x)\sqrt{\log\log x}]$ will be a lower bound for the number of prime factors in $I(x)$ of any of the $x+o(x)$ integers $n \leqslant x$ under consideration. It is no more difficult in practice to obtain a theorem in probabilistic group theory of this sort in which the *exact* number of summands allowed in (22) is specified rather than an inequality constraint.

The argument outlined here shows that as $x \to \infty$, $x+o(x)$ of the integers $n \leqslant x$ have a divisor d as specified in (16)–(18) except that the bounds for $\Omega(d)$ and $\|\log d\|$ involve $\log x$ rather than $\log d$. In the case $t \geqslant \frac{1}{2}$ these constraints are actually stronger than (17) and (18), but when $t < \frac{1}{2}$ the constraint $\Omega(d) \leqslant t \log\log x$ $(d|n,\ n \leqslant x)$, is weaker than (16) requires, so that $A_1(t)$ is a narrower sequence than that considered in the foregoing (sketch) proof. To cope with this, let $P^+(n)$ denote the greatest prime factor of n. We may assume that $P^+(n) > x^{\xi}$, where $\xi = \xi(x) \to 0$ very slowly. We may also assume that $P^+(n)|d$, so that $\log\log d > \log\log x + \log \xi(x)$, since, if $g^* \equiv [N \log P^+(n)]$ is the group element associated with the prime factor $P^+(n)$, we need only use (22) to represent $-g^*$ instead of zero and then multiply the divisor d' so constructed by $P^+(n)$. Provided $\xi(x) \to 0$ sufficiently slowly so that we know that $P^+(n)$ is large, it will lie outside $I(x)$, whence $P^+(n) \nmid d'$. For the details of this proof and definition of $I(x)$ the reader should see [4]. We conclude by stating our variant of the Erdős–Rényi theorem.

Theorem 3 *Let $\xi > 0$ and $\eta \in (0, \frac{1}{2}]$ be given, and let*

$$\mu(\eta) := -\eta \log \eta - (1-\eta)\log(1-\eta);$$
$$\kappa(\eta) := \tfrac{1}{2} - \frac{\log(\eta(1-\eta))}{\log 2};$$
$$\xi(\epsilon, \eta) := 2\log\frac{1}{\epsilon} + \tfrac{1}{2}\log\frac{1}{\mu(\eta)} + 9.$$

let G be an Abelian group of order N and let l elements $g_1, g_2, \ldots, g_l$ be chosen randomly and independently from G. The probability that any particular element should be chosen is $1/N$. Let $N \geqslant N_0(\epsilon, \eta)$ and

$$l \geqslant \frac{1}{\mu(\eta)}\{\log N + \kappa(\eta)\log\log N + \xi(\epsilon, \eta)\}, \tag{25}$$
$$\min(r, 1-r) \geqslant \eta l \quad (\eta < \tfrac{1}{2}), \qquad r = [\tfrac{1}{2}l] \quad (\eta = \tfrac{1}{2}). \tag{26}$$

Then, with probability $\geqslant 1-\epsilon$, every $g \in G$ can be represented in the form

$$g = g_{i_1} + g_{1_2} + \cdots + g_{i_r} \qquad (1 \leqslant i_1 < i_2 < \cdots < i_r \leqslant l).$$

Of course the coefficient $1/\mu(\eta)$ of $\log N$ in (25) is best possible, as may be seen by minimizing the binomial coefficient $\binom{l}{r}$ subject to (25) and (26) using Stirling's formula. Plainly this binomial coefficient must not be less than N.

References

[1] F. A. Behrend, Generalizations of an inequality of Heilbronn and Rohrbach, *Bull. Amer. Math. Soc.*, **54** (1948), 681–4

[2] H. Davenport & P. Erdős, On sequences of positive integers, *Acta Arith.*, **2** (1937), 147–51

[3] P. Erdős, On some applications of probability to analysis and number theory, *J. London Math. Soc.*, **39** (1964), 692–6

[4] P. Erdős & R. R. Hall, Some contribution problems concerning the divisors of integers, *Acta Arith.*, **26** (1974), 175–88

[5] P. Erdős & R. R. Hall, Probabilistic methods in group theory II, *Houston J. of Math.*, **2** (1976), 173–80

[6] P. Erdős & A. Rényi, Probabilistic methods in group theory, *J. D'Analyse Math.*, **14** (1965), 127–38

[7] H. Halberstam & K. F. Roth, *Sequences*, Vol. 1, Oxford University Press, Oxford, 1966

[8] R. R. Hall, A new definition of the density of an integer sequence, *J. Australian Math. Soc., Ser. A*, **26** (1978), 487–500

[9] R. R. Hall, Extensions of a theorem of Erdős–Rényi in probabilistic group theory, *Houston J. of Math.*, **3** (1977), 225–34

[10] R. R. Hall & J. Tenenbaum, Les ensembles de multiples de la densité divisorielle, *J. of Number Theory*, **22** (1986), 308–33

[11] R. R. Hall & J. Tenenbaum, *Divisors*, Cambridge Tract No. 90, Cambridge University Press, Cambridge, 1988

[12] G. H. Hardy & S. Ramanujan, The normal number of prime factors of a number *n*, *Quart. J. Math.*, **48** (1917), 76–92

[13] H. Heilbronn, On an inequality in the elementary theory of numbers, *Proc. Camb. Phil. Soc.*, **33** (1937), 207–9

[14] H. Rohrbach, Beweis einer Zahlentheoretischen Ungleichung, *J. Reine Angew. Math.*, **177** (1937), 193–6

[15] P. Shiu, A Brun–Titchmarsh theorem for multiplicative functions, *J. Reine Angew. Math.*, **313** (1980), 161–70

A functional equation arising from the mortality tables

W. K. Hayman and A. R. Thatcher

To Paul Erdős on his 75th birthday with affection and in the hope that he will find the results reassuring

1 Introduction

For a given population of humans, we denote by l_x the number of members of the population who survive to reach age x. We also write

$$d_x = l_x - l_{x+1}, \qquad q_x = \frac{d_x}{l_x}, \qquad p_x = 1 - q_x = \frac{l_{x+1}}{l_x}.$$

Thus q_x is the proportion of those reaching age x who die before reaching age $x+1$ and p_x is the proportion of those reaching age x who survive to reach age $x+1$.

In a study of mortality tables, Gompertz [2] discovered a remarkable relationship: the first differences of $\log l_x$ are very close to a geometric progression. Nowadays his 'law of mortality' is usually expressed in the continuous form

$$-\frac{\mathrm{d}}{\mathrm{d}x}(\log l_x) = Ac^x, \tag{1.1}$$

where A and c are constants. Mortality rates have changed greatly since 1825 but Gompertz' law still holds, approximately, from middle age onwards.

More complicated 'laws' have since been proposed, in attempts to fit the data in mortality tables more accurately and for a wider range of ages. In particular, Heligman & Pollard [3] have proposed a law covering all ages from birth onwards. Their full expression has three terms and eight parameters. However, above age 50 the first two terms can be neglected and we are left with just the term

$$\frac{q_x}{p_x} = GH^x, \tag{1.2}$$

where G and H are constants.

Table 1. *Heligman and Pollard's Law compared with English Life Table No. 14*

	Heligman & Pollard			ELT No. 14		
	l_x	q_x	e_x	l_x	q_x	e_x
Males 1981						
Age 50	1.000	0.007	24.3	1.000	0.006	24.3
55	0.958	0.011	20.3	0.961	0.011	20.1
60	0.895	0.018	16.5	0.897	0.018	16.4
65	0.801	0.029	13.2	0.801	0.029	13.0
70	0.671	0.046	10.2	0.667	0.047	10.1
75	0.505	0.072	7.7	0.497	0.074	7.7
80	0.321	0.111	5.7	0.312	0.113	5.8
85	0.158	0.169	4.1	0.153	0.166	4.3
90	0.052	0.247	2.9	0.053	0.227	3.3
Females 1981						
Age 50	1.000	0.004	29.3	1.000	0.004	29.4
55	0.978	0.006	24.8	0.977	0.006	25.0
60	0.943	0.010	20.7	0.940	0.010	20.9
65	0.887	0.016	16.8	0.886	0.015	17.0
70	0.802	0.027	13.3	0.807	0.024	13.4
75	0.679	0.044	10.2	0.692	0.041	10.2
80	0.516	0.071	7.6	0.532	0.070	7.5
85	0.329	0.112	5.6	0.336	0.119	5.4
90	0.159	0.174	4.0	0.153	0.185	3.9

l_x is the proportion of a population who survive from age 50 to age x, q_x is the probability at age x of dying before reaching age $x+1$ and e_x is the expectation of life at age x. The Heligman & Pollard values are calculated from equation (1.3), taking $G = 5.6467\times10^{-5}$ and $H = 1.1011$ for males, and $G = 2.1563\times10^{-5}$ and $H = 1.1075$ for females. The last three columns are from *English Life Tables*, No. 14 (HMSO, London, 1987).

We note that q_x/p_x is the odds on dying between age x and age $x+1$ and, according to (1.2), these odds increase exponentially with age. The law is numerically very similar to that of Gompertz; indeed, at ages 50–90 the results of (1.1) and (1.2), if we take $c = H$ and use suitably chosen values of A and G, are hard to distinguish. At still higher ages, (1.2) gives a slightly lower mortality than (1.1) and, to fit the mortality tables at these ages, we need to put c in (1.1) slightly lower than H in (1.2). There are, however, considerable difficulties in analysing the scanty data which are available at the very highest ages (cf. Thatcher [4]).

The Heligman–Pollard constant H is close to 1.1 for both sexes and also for past as well as present populations. The table shows some numerical values given by (1.2), when G and H are chosen to fit the data in the latest English Life Tables.

The equation (1.2) reduces to

$$p_x = \frac{l_{x+1}}{l_x} = (1+GH^x)^{-1} \tag{1.3}$$

and it is this equation which we propose to study.

2 A uniqueness theorem

It is convenient to write

$$G = H^{-\alpha}, \qquad z = x-\alpha, \qquad F(z) = l_{\alpha+z}. \tag{2.1}$$

The quantity α has some interest and represents the age x at which $q_x/p_x = 1$, so that a member having reached age x has an even chance of surviving for another year. It is reassuring that α appears to be about 105 for females and over 101 even for males.

With the normalization (2.1) the Heligman–Pollard law (1.2) or (1.3) can be written in the form

$$F(z+1) = (1+H^z)^{-1}F(z). \tag{2.2}$$

Suppose that $F_1(z)$ and $F(z)$ are two solutions of (2.2). We write

$$P(z) = \frac{F(z)}{F_1(z)} \tag{2.3}$$

and deduce that

$$P(z+1) = P(z). \tag{2.4}$$

Thus $P(z)$ is periodic with period 1. Conversely, if $F_1(z)$ is a solution of (2.2) and $P(z)$ is periodic with period 1 then $F(z) = F_1(z)P(z)$ also satisfies (2.2). Thus the general solution of (2.2) is obtained from any particular solution by multiplication with a periodic factor $P(z)$. This leads us to ask whether there is a solution which in some sense is 'natural'. The following result supplies an answer to this question.

Theorem 1 *If $H > 0$ then the function*

$$F_1(z) = \prod_{n=1}^{\infty} (1+H^{z-n})^{-1} \tag{2.5}$$

is the unique solution of (2.2) which is monotonic decreasing for all real z and satisfies

$$F_1(-\infty) = \lim_{z\to-\infty} F_1(z) = 1. \tag{2.6}$$

Further, $F_1(z)$ is positive for real z and meromorphic in the complex plane $\mathbb{C}$ with simple poles at

$$z = z_{k,n} = n + \frac{(2k+1)\pi i}{\log H}, \tag{2.7}$$

where n is a natural number and k is an integer. Also $F_1(z) \neq 0$ in $\mathbb{C}$.

Proof Suppose that X_0 is a fixed real number and that $z = X + iY \in \mathbb{C}$, where $X \leqslant X_0$. Then

$$|H^{z-n}| = H^{X-n} \leqslant \frac{H^{X_0}}{H^n}.$$

If $N = [X_0] + 2$, we deduce that

$$|H^{z-n}| \leqslant H^{N-n-1} \leqslant \frac{1}{H} \qquad (n \geqslant N).$$

Thus the product

$$\Pi_N(z) = \prod_{n=N}^{\infty} (1 + H^{z-n})$$

is uniformly and absolutely convergent for $X \leqslant X_0$ to a function $\Pi_N(z)$ which is entire and such that $\Pi_N(z) \neq 0$, since none of the factors $1 + H^{z-n}$ vanishes for $n \geqslant N$. Thus

$$F_1(z)^{-1} = \Pi_N(z) \prod_{n=1}^{N-1} (1 + H^{z-n})$$

is entire with simple zeros at the points $z_{k,n}$ given by (2.7), so that $F_1(z)$ is meromorphic and non-zero in $\mathbb{C}$, with simple poles at $z_{k,n}$. In particular, the product in (2.5) is uniformly convergent for $X \leqslant X_0$, $Y = 0$ and so continuous on the real axis together with the point $X = -\infty$, where all factors reduce to 1. Thus the product is 1 for $z = -\infty$ and this yields (2.6). Also H^{z-n} increases with z for real z and fixed n and so does $1 + H^{z-n}$. Thus $F_1(z)^{-1}$ also increases with z and $F(z)$ decreases strictly with increasing z. Clearly $F_1(z) > 0$ when z is real. Thus $F_1(z)$ has all the properties asserted in Theorem 1.

It remains to prove the uniqueness. Suppose that $F(z)$ is a second solution of (2.2) for real z which is monotonic near $z = -\infty$ and satisfies (2.6). We define $P(z)$ by (2.3), so that $P(z)$ is periodic. Since $F(z)$ and $F_1(z)$ satisfy (2.6), we deduce that $P(z)$ does the same. Thus, given ϵ, we have

$$1-\epsilon < P(z) < 1+\epsilon \tag{2.8}$$

for $z < X_0(\epsilon)$. Using the periodicity (2.4), we deduce that (2.8) holds for all real z. Since ϵ can be chosen as small as we please, we deduce that $P(z) \equiv 1$. This proves uniqueness. □

We remark that the only solutions of (2.2) which are monotonic in some interval $X < X_0$ of the real axis are functions of the form

$$F(z) = CF_1(z),$$

where C is a real constant. For we deduce from the monotonicity that

$$C = \lim_{z\to-\infty} F(z) \tag{2.9}$$

exists and $-\infty \leqslant C \leqslant +\infty$. However, C cannot be infinite. For if $P(z)$ is defined by (2.3) and n is an integer such that $n \leqslant X_0$, we deduce, for $n-1 \leqslant z \leqslant n$, from the monotonicity that

$$|F(z)| \leqslant \max\{|F(n)|, |F(n-1)|\},$$

so that

$$\begin{aligned}|P(z)| &\leqslant |F_1(z)|^{-1}\max\{|F(n)|, |F(n-1)|\} \\ &\leqslant |F_1(n)|^{-1}\max\{|F(n)|, |F(n-1)|\} = K,\end{aligned}$$

say. Since $P(z)$ is periodic, we deduce that $|P(z)| \leqslant K$ for all real z. Thus $|C| \leqslant K$. Now we deduce form (2.9) and (2.6) that

$$P(z) = \frac{F(z)}{F_1(z)} \to C \qquad \text{as } z \to -\infty.$$

Since $P(z)$ is periodic, we deduce, just as above, that $P(z) \equiv C$. Hence $F(z) = CF_1(z)$, where C is a real constant. Evidently the functions $CF_1(z)$ are monotonic for all real z.

We remark that the study of the functions of the form (2.5) has a very long history going back to Euler [1]. They are sometimes called partition functions and are constituents of theta functions (Whittaker & Watson [5]). We return to this subject in Section 4.

3 Behaviour of solutions near $+\infty$

Theorem 1 gives us little information about the behaviour of solutions near $z = +\infty$. Since $F_1(X)$ is positive decreasing we deduce that

$$F_1(+\infty) = \lim_{X\to+\infty} F_1(X)$$

exists and, since $F_1(X) \leqslant (1+H^{X-1})^{-1}$, we have that $F_1(+\infty) = 0$. Hence, if $F(z)$ is any other locally bounded solution and $P(z)$ is defined by (2.3), $P(z)$ is also locally and so globally bounded, so that

$$F(z) \to 0 \qquad \text{as } z \to +\infty.$$

However, we can prove much more than this.

Theorem 2 *The function*

$$F_2(z) = H^{\frac{1}{2}z(1-z)} F_1(1-z)^{-1} = H^{\frac{1}{2}z(1-z)} \prod_{n=0}^{\infty} (1+H^{-z-n}) \qquad (3.1)$$

is the unique solution of (2.2) such that

$$\log F_2(z) = \tfrac{1}{2}z(1-z)\log H + o(1) \qquad \text{as } z \to +\infty. \qquad (3.2)$$

Proof We note that $F(z) = F_2(z)$, defined by (3.1), does indeed satisfy (2.2). In fact,

$$\frac{F_2(z+1)}{F_2(z)} = H^{\frac{1}{2}(z+1)(-z)-\frac{1}{2}z(1-z)} \frac{F_1(-z+1)}{F_1(-z)}$$
$$= H^{-z}(1+H^{-z})^{-1} = (1+H^{z})^{-1},$$

so that (2.2) holds. Again, by (2.6),

$$F_2(z) H^{-\frac{1}{2}z(1-z)} = F_1(1-z)^{-1} \to 1 \qquad \text{as } z \to +\infty.$$

Thus (3.2) also holds. Next, if $F(z)$ is any other solution of (2.2) which satisfies (3.2), we may write $F(z) = P(z)F_2(z)$, where $P(z)$ is periodic with period 1 and positive at ∞. Also, by (3.2),

$$\log P(z) = \log F(z) - \log F_2(z)$$
$$= \tfrac{1}{2}z(1-z)\log H - \tfrac{1}{2}z(1-z)\log H + o(1)$$
$$= o(1) \qquad \text{as } z \to +\infty.$$

Thus

$$P(z) \to 1$$

and, using the periodicity of $P(z)$ just as in the proof of Theorem 1, we deduce that $P(z) \equiv 1$. This proves Theorem 2. □

We note that if $F(z)$ is a continuous and positive solution of (3.2) then $P(z) = F(z)/F_2(z)$ is also continuous and positive, so that for $0 \leqslant z \leqslant 1$ we have

$$C_1 \leqslant P(z) \leqslant C_2,$$

where C_1 and C_2 are positive constants. Since $P(z)$ is periodic, these inequalities remain valid for all real z. In particular, we deduce that

$$\begin{aligned}\log F(z) &= \log F_2(z) + O(1)\\ &= \tfrac{1}{2}z(1-z)\log H + O(1) \qquad \text{as } z \to +\infty.\end{aligned} \tag{3.3}$$

By considering $F(z)/F_1(z)$, we deduce similarly from (2.6) that

$$\log F(z) = O(1) \qquad \text{as } z \to -\infty. \tag{3.4}$$

We note that a periodic function $P(z)$ can only tend to a limit C as $z \to +\infty$, or as $z \to -\infty$, if $P(z) \equiv C$ (we recall the argument after (2.8)). It follows that if

$$F(z)\,H^{\frac{1}{2}z(1-z)} \to C \qquad \text{as } z \to +\infty$$

then $F(z) = CF_2(z)$, while if

$$F(z) \to C \qquad \text{as } z \to -\infty$$

then $F(z) = CF_1(z)$. For other solutions $F(z)$, the term $O(1)$ on the right-hand side of (3.3) must oscillate and so must $F(z)$ as $z \to -\infty$.

In the above analysis we have only assumed that $F(z)$ is positive and continuous. If $F(z) \in C^p$, that is, if $F(z)$ has continuous derivatives of orders up to p, we can formally differentiate the relations (3.3) and (3.4), respectively, p times. For, in this case, if $P(z) = F(z)/F_1(z)$, or $P(z) = F(z)/F_2(z)$, the derivative

$$\psi(z) = \left(\frac{\mathrm{d}}{\mathrm{d}z}\right)^p \log P(z) \tag{3.5}$$

is continuous on the real axis and periodic and, consequently, bounded. The function $F_1(z)$ is regular on the real axis by Theorem 1 and so has derivatives of all orders. Using (3.1) we see that $F_2(z)$ has the same property. Using these facts we prove the following theorem.

Theorem 3 *Suppose that $F(z)$ satisfies (2.2), that $F(z) \in C^p$ on the real axis and that $F(z) > 0$. Then*

$$\phi(z) = \left(\frac{\mathrm{d}}{\mathrm{d}z}\right)^p \log F(z) \tag{3.6}$$

satisfies

$$\phi(z) = O(1) \qquad \text{as } z \to -\infty \tag{3.7}$$

and

$$\phi(z) = \left(\frac{\mathrm{d}}{\mathrm{d}z}\right)^p \{\tfrac{1}{2}z(1-z)\log H\} + O(1) \qquad \text{as } z \to +\infty. \tag{3.8}$$

If $p \geqslant 1$ and

$$\phi(z) \to l \qquad \text{as } z \to -\infty \tag{3.9}$$

then $F(z) = KF_1(z)$, where K is a constant, and so $l = 0$.

Similarly, if

$$\phi(z) - \left(\frac{\mathrm{d}}{\mathrm{d}z}\right)^p \{\tfrac{1}{2}z(1-z)\log H\} \to l \qquad \text{as } z \to +\infty \tag{3.10}$$

then $F(z) = KF_2(z)$ and $l = 0$.

Proof We need the following simple lemma.

Lemma 1 *If $F_1(z)$ is the function of Theorem 1 and $p \geqslant 0$, then*

$$\left(\frac{\mathrm{d}}{\mathrm{d}z}\right)^p \log F_1(z) = O(H^X), \tag{3.11}$$

uniformly as $X \to -\infty$, where $z = X + \mathrm{i}Y$.

Proof of Lemma 1 We write $z = X + \mathrm{i}Y$ and assume that $X \leqslant 0$. We write

$$\phi(z) = \log F_1(z) = -\sum_{n=1}^{\infty} \log(1 + H^{z-n}),$$

where the principal value of the logarithm is taken on the right-hand side. This is possible since

$$|H^{z-n}| = H^{X-n} \leqslant H^{-n} \leqslant \frac{1}{H},$$

by hypothesis. Writing $\zeta = H^{z-n}$, we have

$$|\log(1+\zeta)| \leqslant |\zeta| + |\zeta|^2 + \cdots + |\zeta|^n + \cdots = \frac{|\zeta|}{1-|\zeta|} \leqslant \frac{H}{H-1}|\zeta|.$$

Thus

$$|\phi(z)| \leqslant \frac{H}{H-1}\sum_{n=1}^{\infty} H^{-n+X} = \frac{H^{X-1}}{(H-1)(1-H^{-1})} = \frac{H^X}{(H-1)^2}.$$

This proves Lemma 1 when $p = 0$. If p is a positive integer, we deduce from Cauchy's inequality that, if $z_0 = X_0 + \mathrm{i}Y_0$ and $X_0 \leqslant -1$,

$$|\phi^p(z_0)| \leqslant p! \max_{|z-z_0|=1} |\phi(z) \leqslant \frac{p!\,H^{X_0+1}}{(H-1)^2}.$$

Thus Lemma 1 is proved in all cases. □

We can now prove (3.7). We write $F(z) = P(z)F_1(z)$, so that

$$\phi(z) = \left(\frac{\mathrm{d}}{\mathrm{d}z}\right)^p \log F_1(z) + \psi(z),$$

where $\psi(z)$ is given by (3.5) and so $\psi(z)$ is bounded as $z \to -\infty$. Now (3.7) follows from (3.11).

To prove (3.8) we writer $F(z) = F_2(z)P(z)$ and use (3.1). Thus

$$\phi(z) = \left(\frac{\mathrm{d}}{\mathrm{d}z}\right)^p \log F_2(z) + \psi(z),$$

where $\psi(z)$ is given by (3.5) and so is bounded as $z \to +\infty$. Also

$$\left(\frac{\mathrm{d}}{\mathrm{d}z}\right)^p \log F_2(z) = -\left(\frac{\mathrm{d}}{\mathrm{d}z}\right)^p \log F_1(1-z) + \left(\frac{\mathrm{d}}{\mathrm{d}z}\right)\{\tfrac{1}{2}z(1-z)\log H\},$$

by (3.1). It follows from Lemma 1 that the first term on the right-hand side tends to zero as $z \to +\infty$, so that $1-z \to -\infty$. This yields (3.8).

To conclude the proof of Theorem 3, we suppose that (3.9) holds as $z \to -\infty$. We write $F(z) = P(z)F_1(z)$ and, using Lemma 1, deduce that

$$\psi(z) \to l \qquad \text{as } z \to -\infty, \tag{3.12}$$

where $\psi(z)$ is given by (3.5). The function $\psi(z)$ is periodic. We now suppose that $p \geqslant 1$. Then

$$\psi_1(z) = \left(\frac{\mathrm{d}}{\mathrm{d}z}\right)^{p-1} \log P(z)$$

is also periodic. Thus

$$\int_z^{z+1} \psi(\zeta)\,\mathrm{d}\zeta = \psi_1(z+1) - \psi_1(z) = 0. \tag{3.13}$$

By (3.12) the left-hand side tends to l as $z \to -\infty$, so that $l = 0$. Also, by (3.13), $\psi_1(n) = C$, a constant, for integral values of n. If X is real we choose n so that $n \leqslant X < n+1$. Thus

$$\psi_1(X) - \psi_1(n) = \int_n^X \psi(\zeta)\,\mathrm{d}\zeta \to 0,$$

by (3.12) with $l = 0$. Letting X and n tend to $-\infty$, we deduce that

$$\psi_1(X) \to C \qquad \text{as } X \to -\infty. \tag{3.14}$$

If $p \geqslant 2$, we now use (3.14) instead of (3.12) and eventually conclude that $\log P(z) \to l$ and so, that $\log P(z) \equiv l$. The argument near $z = +\infty$ is similar. This concludes the proof of Theorem 3. □

4 A comparison of $F_1(z)$ and $F_2(z)$

It is not easy to decide which of the two functions $F_1(z)$ and $F_2(z)$ constitutes the natural solution, since $F_1(z)$ is characterized by tending to the limit 1 as $z \to -\infty$, while (3.2) and (3.10) provide similar characterizations for $F_2(z)$. (However, the monotinicity gives $F_1(z)$ the edge.) The ratio $F_2(z)/F_1(z)$ is not constant, unfortunately. In fact,

$$\frac{F_2(z)}{F_1(z)} = H^{\frac{1}{2}z(1-z)} \prod_{n=0}^{\infty} (1+H^{-z-n}) \prod_{n=1}^{\infty} (1+H^{z-n}) \tag{4.1}$$

by (2.5) and (3.1). The right-hand side of (4.1) is an entire function with simple zeros only at the points $z_{k,n}$ of (2.7) for integral values of k and n. Thus $F = F_2/F_1$ is not constant and so, for any constant C, the equation $F = C$ has only isolated solutions.

On the other hand, the ratio is constant for all practical (and, in particular, actuarial) purposes.

Theorem 4 *If $1 < H < \infty$ we have, for real z,*

$$\frac{F_2(z)}{F_1(z)} = C\{1+\epsilon(z)\}, \tag{4.2}$$

where

$$C = H^{\frac{1}{8}}\left(\frac{1}{2\pi}\log H\right)^{-\frac{1}{2}} \prod_{n=1}^{\infty} (1-H^{-n})^{-1} \tag{4.3}$$

and

$$\epsilon(z) < \frac{2q'}{1-q'^3}, \quad \text{where } q' = \exp\left(-\frac{2\pi^2}{\log H}\right). \tag{4.4}$$

In particular, if $H = 1.1$ then $|\epsilon(z)| < 10^{-89}$.

We need some facts about theta functions which are put together in the following lemma.

Lemma 2 *Suppose that $|q| < 1$ and $Z \neq 0$, and define*

$$\theta(Z, q) = 1 + \sum_{n=1}^{\infty} q^{n^2}(Z^n + Z^{-n}). \tag{4.5}$$

Then

$$\theta(Z, q) = G \prod_{n=1}^{\infty} (1+q^{2n-1}Z)(1+q^{2n-1}/Z), \tag{4.6}$$

where

$$G = \prod_{n=1}^{\infty} (1-q^{2n}). \tag{4.7}$$

Further, if

$$q' = \exp\left(-\frac{\pi^2}{\log(1/q)}\right), \qquad Z' = \exp\left(\pi i \frac{\log Z}{\log q}\right), \tag{4.8}$$

we have

$$\theta(Z,q) = \left(\frac{1}{\pi}\log\frac{1}{q}\right)^{-\frac{1}{2}} \exp\left(-\frac{(\log Z)^2}{4\log q}\right)\theta(Z',q'). \tag{4.9}$$

Proof The results of Lemma 2 are all contained, with slight change of notation, in Whittaker & Watson [5]. In fact, these authors [5, p. 464] define

$$\vartheta_3(z,q) = 1 + 2\sum_{n=1}^{\infty} q^{n^2}\cos 2nz$$

and prove [5, p. 469] that

$$\vartheta_3(z,q) = G\prod_{n=1}^{\infty} (1+q^{2n-1}e^{2iz})(1+q^{2n-1}e^{-2iz}),$$

where [5, p. 473] G is given by (4.7). Thus, in our notation,

$$\vartheta_3(z,q) = \theta(e^{2iz},q) \tag{4.10}$$

and then (4.5), (4.6) and (4.7) hold.

Next [5, p. 475], if

$$q = e^{\pi i\tau}, \qquad q' = e^{-\pi i/\tau}$$

then

$$\vartheta_3(z,q) = (-i\tau)^{-\frac{1}{2}}\exp\left(\frac{z^2}{\pi i\tau}\right)\vartheta_3\left(\frac{z}{\tau},q'\right). \tag{4.11}$$

Writing $Z = e^{2iz}$ and $Z' = e^{2iz/\tau}$ in (4.10), we deduce (4.9) from (4.11). □

We can now prove Theorem 4. We write

$$q = H^{-\frac{1}{2}}, \qquad Z = H^{\frac{1}{2}-z} = q^{2z-1}. \tag{4.12}$$

Then

$$\begin{aligned}\frac{F_2(z)}{F_1(z)} &= q^{z(z-1)} \prod_{n=1}^{\infty} (1+H^{z-n})(1+H^{1-z-n}) \\ &= q^{z(z-1)} \prod_{n=1}^{\infty} (1+H^{z-\frac{1}{2}}H^{-n+\frac{1}{2}})(1+H^{\frac{1}{2}-z}H^{-n+\frac{1}{2}}) \\ &= q^{z(z-1)} \prod_{n=1}^{\infty} (1+Zq^{2n-1})(1+Z^{-1}q^{2n-1}) \\ &= \frac{q^{z(z-1)}}{G}\theta(Z,q). \end{aligned} \tag{4.13}$$

Since $Z > 0$ for real z, (4.8) and (4.12) yield

$$q' = \exp\left(-\frac{2\pi^2}{\log H}\right), \qquad |Z'| = 1. \tag{4.14}$$

Also (4.9) and (4.12) yield

$$\theta(Z,q) = \left(\frac{1}{2\pi}\log H\right)^{-\frac{1}{2}} \exp\{-\tfrac{1}{4}(2z-1)^2 \log q\}\, \theta(Z',q').$$

Substituting this in (4.13) we obtain

$$\frac{F_2(z)}{F_1(z)} = H^{\frac{1}{8}}\left(\frac{1}{2\pi}\log H\right)^{-\frac{1}{2}} G^{-1}\theta(Z',q') = C\{1+\epsilon(z)\},$$

where C is given by (4.3) and

$$\epsilon(z) = \theta(Z',q')-1 = \sum_{n=1}^{\infty} (q')^{n^2}\{(Z')^n+(Z')^{-n}\}.$$

Using (4.14), we deduce that

$$|\epsilon(z)| \leqslant 2\sum_{n=1}^{\infty} (q')^{3n-2} = \frac{2q'}{1-(q')^3}.$$

This yields (4.4) and completes the proof of Theorem 4. □

5 The expectation e_x

Suppose that l_x is defined as in the introduction. The expectation e_x is the number of further years that a person may expect to live, having reached the age x. The probability of dying between ages $x+t_1$ and $x+t_2$ is

$$\frac{1}{l_x}(l_{x+t_1}-l_{x+t_2}).$$

Thus

$$e_x = \frac{1}{l_x}\int_{t=0}^{\infty} -t\,\mathrm{d}l_{x+t} = \frac{1}{l_x}\int_0^{\infty} l_{x+t}\,\mathrm{d}t. \tag{5.1}$$

If $l_x = F_2(x-\alpha)$, we can obtain a complete asymptotic expansion for e_x. In this section we write

$$g(t) \sim \sum_{n=0}^{\infty} a_n t^{-n} \qquad \text{as } t \to \infty \tag{5.2}$$

if, for each fixed N, we have

$$g(t) = \sum_{n=0}^{N} a_n t^{-n} + r_N(t), \tag{5.3}$$

where

$$r_N(t) = O(t^{-(N+1)}) \qquad \text{as } t \to \infty. \tag{5.4}$$

With this terminology we have the following theorem.

Theorem 5 *Suppose that $F(z)$ is a solution of (2.2) and that α and x are obtained by (2.1). If $F(z) = F_2(z)$ we have*

$$e_x \sim \sum_{n=0}^{\infty} \frac{(-\frac{1}{2})^n (2n)!}{n!\,(\log H)^{n+1}} (x-\alpha-\tfrac{1}{2})^{-2n-1} \qquad \text{as } x \to \infty. \tag{5.5}$$

If $F(z)$ is continuous and positive for real z, we still have

$$e_x = \frac{1+o(1)}{x\log H} = \frac{1+o(1)}{|\log p_x|} \qquad \text{as } x \to \infty, \tag{5.6}$$

while, if $F(z)$ is continuously differentiable and positive on the real axis and, in particular, if $F(z) = F_1(z)$, we have

$$e_x = \frac{1}{|\log p_x| + O(1)} \qquad \text{as } x \to \infty. \tag{5.7}$$

Proof We need the following lemma.

Lemma 3 *Suppose that*

$$I(c) = \int_0^{\infty} \exp\left(-s - \frac{s^2}{c}\right) \mathrm{d}s, \tag{5.8}$$

where $c > 0$. Then

$$I(c) \sim \sum_{n=0}^{\infty} \frac{(-1)^n (2n)!}{n!} c^{-n} \qquad \text{as } c \to \infty. \tag{5.9}$$

Proof of Lemma 3 We have from Taylor's theorem that, if $t > 0$,

$$e^{-t} = \sum_{n=0}^{N} \frac{(-1)^n t^n}{n!} + r_N(t),$$

where, for some τ such that $0 < \tau < t$, we have

$$|r_n(t)| = \frac{e^{-\tau} t^{n+1}}{(n+1)!} < \frac{t^{n+1}}{(n+1)!}.$$

Thus

$$e^{-s^2/c} = \sum_{n=0}^{N} \frac{(-1)^n s^{2n}}{n!\,c^n} + R_N(s), \tag{5.10}$$

where

$$|R_N(s)| < \frac{s^{2N+2}}{(N+1)!\,c^{N+1}}.$$

We multiply both sides of (5.10) by e^{-s} and integrate from 0 to ∞. This yields

$$\begin{aligned}\left| I(c) - \sum_{n=0}^{N} \frac{(-1)^n}{n!\,c^n} \int_0^\infty s^{2n} e^{-s}\, ds \right| &= \left| \int_0^\infty R_N(s) e^{-s}\, ds \right| \\ &< \frac{1}{(N+1)!\,c^{N+1}} \int_0^\infty s^{2n+2} e^{-s}\, ds \\ &= \frac{(2N+2)!}{(N+1)!\,c^{N+1}}.\end{aligned}$$

This gives (5.9) in the stronger form that the error in the series is always numerically less than the next term. □

We proceed to prove (5.5). We have

$$e_x = \frac{1}{l_x} \int_0^\infty l_{x+t}\, dt = \frac{1}{F(z)} \int_0^\infty F(z+t)\, dt,$$

where $z = x - \alpha$. Suppose first that $F(z) = F_2(z)$. Then (3.1) and (3.11) with $p = 0$ yield, as $z \to +\infty$,

$$\begin{aligned}\log F(z) &= \tfrac{1}{2} z(1-z) \log H - \log F_1(1-z) \\ &= \tfrac{1}{2} z(1-z) \log H + O(H^{-z})\end{aligned}$$

uniformly as $z \to +\infty$. We replace z by $z+t$, where $t \geqslant 0$, and subtract. This yields, as $z \to \infty$ uniformly for positive t,

$$\log \frac{l_{x+t}}{l_x} = \log \frac{F(z+t)}{F(z)}$$
$$= \tfrac{1}{2}(\log H)\{(z+t)(1-z-t)-z(1-z)\}+O(H^{-z}). \quad (5.11)$$

Thus, by substituting $s = t(z-\frac{1}{2})\log H$, we obtain

$$\int_0^\infty \frac{l_{x+t}}{l_x}\, dt = \{1+O(H^{-z})\} \int_0^\infty \exp\{(\tfrac{1}{2}\log H)[t(1-2z)-t^2]\}\, dt$$
$$= \frac{1+O(H^{-z})}{(\log H)(z-\frac{1}{2})} I(c),$$

where

$$c = 2(z-\tfrac{1}{2})^2 \log H,$$

and $I(c)$ is given by (5.8). Using Lemma 3, we obtain (5.5).

Suppose next that $F(z)$ is a continuous positive solution of (2.2). We write $F(z) = P(z)F_2(z)$, where $P(z)$ is continuous, positive and periodic. Thus

$$\frac{l_{x+t}}{l_x} = \exp\left(\log \frac{P(z+t)}{P(z)} + \log \frac{F_2(z+t)}{F_2(z)}\right).$$

We suppose given a positive ϵ and choose δ so small that

$$\left|\log \frac{P(z+t)}{P(z)}\right| < \epsilon \qquad (0 < t < \delta,\ z \text{ real}). \quad (5.12)$$

The choice is possible, since $\log P(z)$ is continuous and periodic, and so uniformly continuous, on the real axis. Also

$$\left|\log \frac{P(z+t)}{P(z)}\right| < K \quad (5.13)$$

for all real z and t, where K is a positive constant. We write

$$\int_0^\infty \frac{l_{x+t}}{l_x}\, dt = \int_0^\delta + \int_\delta^\infty = I_1 + I_2, \text{ say.}$$

Here we have for all large z, by (5.11) and (5.12),

$$I_1 = e^\theta \int_0^\delta \exp\{(\tfrac{1}{2}\log H)[t(1-2z)-t^2]\}\, dt,$$

where $|\theta| < 2\epsilon$. Substituting $s = t(z-\frac{1}{2})\log H$, as above, we obtain

$$I_1 = \frac{e^\theta}{(z-\frac{1}{2})\log H} \int_0^{\delta(z-\frac{1}{2})\log H} \exp\left(-s-\frac{s^2}{c}\right) ds,$$

where $c \to \infty$ as $z \to \infty$. Thus, for large z, we have

$$e^{-3\epsilon} < I_1(z-\tfrac{1}{2})\log H < e^{3\epsilon}.$$

Similarly we have for large z, using (5.13),

$$|I_2| < \frac{e^K}{(z-\frac{1}{2})\log H}\int_{\delta(z-\frac{1}{2})\log H}^{\infty} \exp\left(-s-\frac{s^2}{c}\right) ds,$$

$$= O\left(\frac{H^{-\delta z}}{z}\right).$$

Thus, as $z \to \infty$, we have

$$I(z)(z-\tfrac{1}{2})\log H = I_1(z-\tfrac{1}{2})\log H + O(H^{-\delta z}) \to 1.$$

This proves (5.6). If $F(z)$ is continuously differentiable, so that $\log P(z)$ has a bounded derivative, we choose $\delta = 1$ in the above argument. Then we have (5.13) as before, but (5.12) can be sharpened to

$$\left|\log \frac{P(z+t)}{P(z)}\right| < K't \qquad (0 < t < 1).$$

This yields

$$\{1+O(H^{-z})\}\int_0^1 \exp\{-t[(z-\tfrac{1}{2})\log H + K'] - \tfrac{1}{2}t^2\log H\}\, dt$$

$$< I_1$$

$$< \{1+O(H^{-z})\}\int_0^1 \exp\{-t[(z-\tfrac{1}{2})\log H - K'] - \tfrac{1}{2}t^2\log H\}\, dt.$$

Thus, setting $s = t[(z-\frac{1}{2})\log H \pm K']$ in turn, we deduce that

$$I_1 = \frac{1}{z\log H + O(1)},$$

while again

$$I_2 = O\left(\frac{H^{-z}}{z}\right).$$

This yields (5.7), and the proof of Theorem 5 is complete. □

References

[1] L. Euler, *Introductio in Analisin Infinitorum, I*, §304, Lausanne, 1748

[2] B. Gompertz, On the nature of the function of the law of human mortality etc, *Phil. Trans. R. Soc.*, **115** (1825), 513–85

[3] L. Heligman & J. H. Pollard, The age pattern of mortality, *J. Inst Act.*, **107** (1980), 49–80
[4] A. R. Thatcher, Mortality at the highest ages, *J. Inst Act.*, **114** (1987), 327–38
[5] E. T. Whittaker & G. N. Watson, *A Course of Modern Analysis*, Cambridge University Press, Cambridge, 1920

The differences between consecutive primes, IV

D. R. Heath-Brown

1 Introduction

This paper is concerned with the frequency of large gaps $p_{n+1}-p_n$ between consecutive primes. It is conjectured that $p_{n+1}-p_n = O(\log^2 p)$. More significantly, it appears that there is a prime between any two squares, so that

$$p_{n+1}-p_n < 4\sqrt{p_n} \qquad (n = 1,2,3,\ldots). \tag{1.1}$$

However, such estimates are quite out of reach at present. If one assumes the Riemann hypothesis one can show that

$$p_{n+1}-p_n \ll \sqrt{p_n}\log p_n, \tag{1.2}$$

but it seems very hard to remove the extraneous factor $\log p_n$. One can prove, however, that the exceptions to (1.1) must be very rare. Indeed, Selberg [6] showed, assuming the Riemann hypothesis, that

$$\sum_{p_n \leqslant x} (p_{n+1}-p_n)^2 \ll x(\log x)^3, \tag{1.3}$$

so that (1.1) must hold for all but $O(\log^3 x)$ primes $p_n \leqslant x$. With such results in mind, Erdős asked whether one could obtain unconditional bounds of the form

$$\sum_{\substack{p_n \leqslant x \\ p_{n+1}-p_n \geqslant \sqrt{x}}} (p_{n+1}-p_n) \ll x^{1-\delta},$$

with a positive constant δ. This was achieved by Wolke [8], with $\delta = \frac{1}{30}$, and various authors improved his bound. The best result of this type to date is

$$\sum_{\substack{p_n \leqslant x \\ p_{n+1}-p_n \geqslant \sqrt{x}}} (p_{n+1}-p_n) \ll x^{\frac{3}{4}+\epsilon},$$

for any constant $\epsilon > 0$, in the third paper [2] of this series. It seems possible that further improvements could be made by incorporating the sieve ideas of Iwaniec & Jutila [4].

The purpose of the present paper is to develop further the method used in the second and third articles [1], [2] of this series. Since the techniques are now rather more complex we shall make the simplifying assumption of the Lindelöf hypothesis: that is to say, we shall assume that the Riemann zeta function satisfies

$$\zeta(\sigma+it) \ll t^{\epsilon} \qquad (\sigma \geqslant \tfrac{1}{2},\, t \geqslant 2), \tag{1.4}$$

for any fixed $\epsilon > 0$. The estimate (1.4) would be a simple consequence of the Riemann hypothesis. It appears that sieve methods are no longer likely to help once the Lindelöf hypothesis is assumed. When one adopts the assumption (1.4) one can prove an estimate

$$p_{n+1}-p_n \ll p_n^{\frac{1}{2}+\epsilon} \tag{1.5}$$

for any fixed $\epsilon > 0$ (see Ingham [3]). This is only slightly weaker than (1.2), which required the Riemann hypothesis. However, one no longer obtains a bound of the same strength as (1.3). In [1] it was shown that

$$\sum_{p_n \leqslant x} (p_{n+1}-p_n)^2 \ll x(\log x)^{\frac{7}{6}+\epsilon}, \tag{1.6}$$

and

$$\sum_{\substack{p_n \leqslant x \\ p_{n+1}-p_n \geqslant \sqrt{x}}} (p_{n+1}-p_n) \ll x^{\frac{5}{8}+\epsilon},$$

(for any fixed $\epsilon > 0$), on the Lindelöf hypothesis, and it is the second of these estimates, the one that relates most closely to Erdős' original question, which will concern us here.

To state our results we put

$$S(x,H) = \sum_{\substack{p_n \leqslant x \\ p_{n+1}-p_n \geqslant H}} (p_{n+1}-p_n).$$

We then have the following bounds.

Theorem *On the Lindelöf hypothesis one has*

$$S(x,H) \ll x^{\frac{1}{2}l+2^{-l}+\epsilon}H^{1-l-2^{-l}} \tag{1.7}$$

for any fixed $\epsilon > 0$ *and* $l \in \mathbb{N}$. *In particular,*

$$S(x,\sqrt{x}) = \sum_{\substack{p_n \leqslant x \\ p_{n+1}-p_n \geqslant \sqrt{x}}} (p_{n+1}-p_n) \ll x^{\frac{1}{2}+\epsilon}.$$

Thus (1.1) would hold for all but $O(x^\epsilon)$ primes $p_n \leqslant x$. For $l = 1,2$, one obtains

$$S(x,H) \ll x^\epsilon \min(xH^{-\frac{1}{2}}, x^{\frac{5}{4}}H^{-\frac{5}{4}}),$$

which were, in effect, the bounds derived by Heath-Brown [1]. The higher values $l \geqslant 3$ become relevant only when $H \geqslant x^{\frac{3}{7}}$. Unfortunately one gets no improvement on (1.6), for which the worst range is $H \approx x^{\frac{1}{3}}$.

The proof follows the method of [1] in its initial stages, but the key additional idea is to use higher moments. One therefore encounters a new counting function $N^{(k)}(\sigma, T)$. This is defined to be the number of ordered sets $(\rho_1, \ldots, \rho_{2k})$ of zeros of $\zeta(s)$ (counted according to multiplicity) for which

$$\sigma \leqslant \beta_j \leqslant 1, \qquad |\gamma_j| \leqslant T,$$

and

$$|\gamma_1 + \cdots + \gamma_k - \gamma_{k+1} - \cdots - \gamma_{2k}| \leqslant 1,$$

where $\rho_j = \beta_j + i\gamma_j$. Thus $N^{(1)}(\sigma, T)$ is essentially the familiar function $N(\sigma, T)$, since one has

$$N(\sigma, T) \ll N^{(1)}(\sigma, T) \ll N(\sigma, T) \log T,$$

and $N^{(2)}(\sigma, T)$ is the function $N^*(\sigma, T)$ introduced in [1]. We shall prove the following result.

Lemma 1 *On the Lindelöf hypothesis one has*

$$N^{(k)}(\sigma, T) \ll T^{4k(1-\sigma)-1+(2\sigma-1)^k+\eta} \qquad (\tfrac{1}{2} \leqslant \sigma \leqslant \tfrac{3}{4}+\eta) \tag{1.8}$$

and

$$N^{(k)}(\sigma, T) \ll T^{\eta} \qquad (\tfrac{3}{4}+\eta \leqslant \sigma \leqslant 1). \tag{1.9}$$

for any fixed $\eta > 0$ *and* $k \in \mathbb{N}$.

When $k = 1$ these reduce to the familiar bounds

$$N(\sigma, T) \ll T^{2-2\sigma+\eta} \qquad (\tfrac{1}{2} \leqslant \sigma \leqslant \tfrac{3}{4}+\eta) \tag{1.10}$$

and

$$N(\sigma, T) \ll T^{\eta} \qquad (\tfrac{3}{4}+\eta \leqslant \sigma \leqslant 1). \tag{1.11}$$

see Titchmarsh [7, Theorem 9.18], with $c = \eta$ and $c' = 0$, and Montgomery [5, Theorem 12.3], with $\alpha = \frac{1}{2}$, respectively. The estimates (1.10) and (1.11) form the basis for the proof of Lemma 1. Indeed, one trivially has

$$N^{(k)}(\sigma,T) \ll N(\sigma,T)^{2k-1}\log T, \qquad (1.12)$$

since $\rho_1,\ldots,\rho_{2k-1}$ determine ρ_{2k} up to $O(\log T)$ choices. Thus (1.9) follows from (1.11) on redefining η. For $k=2$, the bound (1.8) reduces essentially to that of [1, (35)].

We may observe that

$$N^{(k)}(\sigma,T) \gg T^{-1}N(\sigma,T)^{2k},$$

as one can show without too much difficulty. Thus, apart from the term $(2\sigma-1)^k$ in the exponent, which is very small for large k, the bound (1.8) is as good as one could hope for without improving the fundamental estimate (1.10).

2 Preliminaries

In this section we borrow and generalize various results from [1]. All the results will be independent of the Lindelöf hypothesis. We begin by extending [1, Lemma 1] to higher moments, to produce the following lemma.

Lemma 2 *Let $N \geqslant \log x$. Then*

$$\psi\left(y+\frac{y}{\tau}\right)-\psi(y)-\frac{y}{\tau} = E(y,\tau) + \sum_{0\leqslant m,n<N} S(m,n;y,\tau),$$

where

$$E(y,\tau)\ll N, \qquad S(m,n;y,\tau) \ll x^{m/N}\min(\tau^{-1},3^{-n})N\left(\frac{m}{N},3^n\right),$$

and

$$\int_x^{2x} |S(m,n;y,\tau)|^{2k}\,dy \ll x^{1+2km/N}N\min(\tau^{-1},3^{-n})^{2k}N^{(k)}\left(\frac{m}{N},3^n\right),$$

uniformly in m, n, x, y and τ, for $1\leqslant\tau\leqslant x$ and $2\leqslant x\leqslant y\leqslant 2x$.

Next we follow the argument of [1, Section 3] to show that

$$\sum_{\substack{x\leqslant p_n\leqslant 2x\\ 4x/\tau\leqslant p_{n+1}-p_n\leqslant 8x/\tau}} (p_{n+1}-p_n) \ll \frac{x}{\tau} + \sum_{m,n}{}^{(1)} \min_{1\leqslant k\leqslant K}\left\{\left(\frac{\tau\log^3 x}{x}\right)^{2k} x^{1+2km/N} \times N\min(\tau^{-1},3^{-n})^{2k}N^{(k)}\left(\frac{m}{N},3^n\right)\right\}, \qquad (2.1)$$

where $N = 1+[\log x]$ and K is any fixed integer. Here the ranges of summation in $\sum^{(1)}$ are $0 \leqslant m \leqslant N(1-10^{-7})$ and $0 \leqslant n < N$. We shall write $\Delta = 10^{-7}$ for convenience of notation. The proof of (2.1) proceeds exactly on the lines given in [1, Section 3], save that we may now invoke Lemma 2 to take advantage of the higher moment estimates.

We shall now assume that there are bounds

$$N^{(k)}(\sigma, T) \ll T^{A(k,\sigma)}, \qquad (1 \leqslant k \leqslant K, 0 \leqslant \sigma \leqslant 1),$$

in which $A(k,\sigma) \leqslant 2k$. Since we may use the same value of k for each n, we may sum over n to estimate (2.1) as

$$\ll \frac{x}{\tau} + (\log x)^{6K+3} \max_{0 \leqslant \sigma \leqslant 1-\Delta} \min_{1 \leqslant k \leqslant K} x^{1+2k(\sigma-1)} \tau^{A(k,\sigma)}. \tag{2.2}$$

For $0 \leqslant \sigma \leqslant \frac{1}{2}$ one may take $A(k,\sigma) = 2k-1+\eta$, for any $\eta > 0$. Thus,

$$\min_{1 \leqslant k \leqslant K} x^{1+2k(\sigma-1)} \tau^{A(k,\sigma)}$$

will be largest at $\sigma = \frac{1}{2}$. Moreover, we must have $A(k, \frac{1}{2}) \geqslant 2k-1$. The range $0 \leqslant \sigma \leqslant 1-\Delta$ in (2.2) may therefore be reduced to $\frac{1}{2} \leqslant \sigma \leqslant 1-\Delta$, with the loss of a factor x^{η}. Finally, we must relate our original sum $S(x,H)$ to the expression on the left-hand side of (2.1), by the familiar dyadic division technique. We then conclude as follows.

Lemma 3 *For any fixed* $\eta > 0$ *and* $K \in \mathbb{N}$, *we have*

$$S(x,H) \ll x^{\eta} \left\{ H_0 + \max_{\frac{1}{2} \leqslant \sigma \leqslant 1-\Delta} \min_{1 \leqslant k \leqslant K} x^{1+2k(\sigma-1)} \left(\frac{x}{H} \right)^{A(k,\sigma)} \right\}$$

uniformly for $1 \leqslant H \leqslant x$. *Here*

$$H_0 = \max\{p_{n+1} - p_n : p_n \leqslant x\}. \tag{2.3}$$

3 Proof of Lemma 1

As already observed in Section 1, the bound (1.9) is immediate and the case $k = 1$ of (1.8) is already known. This will allow us to use an induction argument in proving (1.8) for general k. As in the previous section, we shall make extensive use of results from [1]. We shall assume that $\sigma \geqslant \frac{1}{2} + \sqrt{\eta}$. Of course, if $\frac{1}{2} \leqslant \sigma \leqslant \frac{1}{2} + \sqrt{\eta}$ then the trivial bound (1.12) yields

$$N^{(k)}(\sigma, T) \ll T^{2k-1} (\log T)^{2k},$$

and (1.8) follows, on redefining η.

We choose parameters $X = T^{3\eta}$ and $Y = T^{4\eta/(2\sigma-1)}$, and then extend the argument of [1, Section 4] up to [1, (23)] in the obvious way. We take

$$a_n = \sum_{\substack{d|n \\ d\leqslant X}} \mu(d),$$

so that $|a_n| \leqslant d(n)$, and we define $\mathcal{S}^{(m)}$ to be the set of zeros $\rho = \beta + i\gamma$ of $\zeta(s)$ for which $\beta \geqslant \sigma$, $|\gamma| \leqslant T$ and

$$\left|\sum_{N<n\leqslant 2N} a_n n^{-\rho}\right| \geqslant \frac{C}{\log T}, \tag{3.1}$$

where C is an appropriate positive constant, and

$$\tfrac{1}{2}X \leqslant N = 2^m \leqslant Y(\log T)^2.$$

We then find that there is some m for which

$$N^{(k)}(\sigma, T) \ll (\log T)^{4k}\{E_k + (\#\mathcal{S}_1^{(0)})^{2k}\}, \tag{3.2}$$

where

$$E_k = \#\left\{(t_1, \ldots, t_{2k}) : \sum_{j=1}^{k} t_j = \sum_{j=k+1}^{2k} t_j\right\}, \tag{3.3}$$

and t_j runs over the set

$$\mathcal{T} = \{[\gamma] : \beta + i\gamma \in \mathcal{S}^{(m)}\}.$$

The set $\mathcal{S}_1^{(0)}$, defined in [1, Section 4], satisfies $\#\mathcal{S}_1^{(0)} \ll \log^3 T$ with our choice of X and Y (*vide* [1, (27)]).

As in [1, Section 4], we raise the sum on the left of (3.1) to the power r, where r is a positive integer depending on N such that $(2N)^r \leqslant T^2$. Since $N \gg X \gg T^{3\eta}$, r will be bounded in terms of η. We write $N^r = M$ and $(2N)^r = P$; then

$$\left|\sum_{M<n\leqslant P} b_n n^{-\rho}\right| \geqslant \left(\frac{C}{\log T}\right)^r,$$

where $|b_n| \leqslant d_{2r}(n)$. We now put

$$S(y, t) = \sum_{M<n\leqslant y} b_n n^{-\sigma-it}.$$

Then, as in [1, (24)], we find that

$$1 \ll (\log T)^r\left(|S(P, t)| + \int_M^P \frac{|S(y, t)|}{y}\, dy\right) \tag{3.4}$$

for each $t \in \mathcal{T}$. To estimate E_k we define

$$n_k(t) = \#\left\{(t_1,\ldots,t_k) : t_j \in \mathcal{T},\ \sum_{j=1}^{k} t_j = t\right\}, \tag{3.5}$$

so that

$$\begin{aligned} E_k &= \sum_{t\in\mathcal{T}} \sum_{t=u-v} n_k(u)\,n_{k-1}(v) \\ &\ll \sum_{u,\,v\in\mathbb{Z}} n_k(u)\,n_{k-1}(v)\left\{(\log T)^r\left(S(P,u-v) + \int_M^P \frac{|S(y,u-v)|}{y}\,\mathrm{d}y\right)\right\}^2 \\ &\ll (\log T)^{2r+1} \sum_{u,\,v\in\mathbb{Z}} n_k(u)\,n_{k-1}(v)\left\{S(P,u-v) + \int_M^P \frac{|S(y,u-v)|^2}{y}\,\mathrm{d}y\right\}, \end{aligned} \tag{3.6}$$

by (3.4). However,

$$\begin{aligned} &\sum_{u,v} n_k(u)\,n_{k-1}(v)\,|S(y,u-v)|^2 \\ &\quad= \sum_{u,v} n_k(u)\,n_{k-1}(v)\left|\sum_{M<n\leqslant y} b_n n^{-\sigma-\mathrm{i}u+\mathrm{i}v}\right|^2 \\ &\quad= \sum_{M<m,n\leqslant y} b_m \overline{b_n}(mn)^{-\sigma}\left\{\sum_u n_k(u)\left(\frac{n}{m}\right)^{\mathrm{i}u}\right\}\left\{\sum_v n_{k-1}(v)\left(\frac{m}{n}\right)^{\mathrm{i}v}\right\} \\ &\quad\ll G \sum_{M<m,n\leqslant P} (mn)^{-\frac{1}{2}}\left|\sum_u n_k(u)\left(\frac{n}{m}\right)^{\mathrm{i}u}\right|\left|\sum_v n_{k-1}(v)\left(\frac{m}{n}\right)^{\mathrm{i}v}\right|, \end{aligned}$$

where

$$G = \max_{M<n\leqslant P} d_{2r}(n)^2 n^{1-2\sigma} \ll P^{1-2\sigma+\eta}.$$

On applying Cauchy's inequality we find that

$$\sum_{u,v} n_k(u)\,n_{k-1}(v)\,|S(y,u-v)|^2 \ll P^{1-2\sigma+\eta}\sqrt{\Sigma_k\Sigma_{k-1}}, \tag{3.7}$$

where

$$\begin{aligned} \Sigma_k &= \sum_{M<m,n\leqslant P} (mn)^{-\frac{1}{2}}\left|\sum_u n_k(u)\left(\frac{n}{m}\right)^{\mathrm{i}u}\right|^2 \\ &= \sum_{u,v} n_k(u)\,n_k(v) \sum_{M<m,n\leqslant P} (mn)^{-\frac{1}{2}}\left(\frac{n}{m}\right)^{\mathrm{i}u-\mathrm{i}v} \\ &= \sum_{u,v} n_k(u)\,n_k(v)\left|\sum_{M<n\leqslant P} n^{-\frac{1}{2}+\mathrm{i}(u-v)}\right|^2 \end{aligned}$$

and similarly for Σ_{k-1}. The inner sum here can be estimated as

$$\sum_{M<n\le P} n^{-\frac{1}{2}-it} \ll T^{\eta}\left(1+\frac{\sqrt{P}}{1+|t|}\right)$$

(see [1, (28)]). Here one uses the Lindelöf hypothesis in an essential way. Now

$$\begin{aligned}\Sigma_k &\ll T^{2\eta}\sum_{u,v} n_k(u)n_k(v)\left(1+\frac{P}{1+|u-v|^2}\right)\\ &\ll T^{2\eta}\left(\sum_u n_k(u)\right)^2 + T^{2\eta}P\sum_w \frac{1}{1+w^2}\sum_u n_k(u)n_k(u+w). \qquad (3.8)\end{aligned}$$

However,

$$\sum_u n_k(u) = (\#\mathcal{T})^k \ll N(\sigma,T)^k \ll T^{2k(1-\sigma)+\eta} \qquad (3.9)$$

by (1.10). Moreover,

$$\begin{aligned}\sum_{u\in\mathbb{Z}} n_k(u)n_k(u+w) &\le \tfrac{1}{2}\sum_{u\in\mathbb{Z}}[n_k(u)^2+n_k(u+w)^2]\\ &= \sum_{u\in\mathbb{Z}} n_k(u)^2\\ &= E_k \qquad (3.10)\end{aligned}$$

by (3.3) and (3.5). Now, on combining the estimates (3.6), (3.7), (3.8), (3.9) and (3.10), we deduce that

$$\begin{aligned}E_k &\ll T^{3\eta}P^{1-2\sigma}\sqrt{\Sigma_k\Sigma_{k-1}}\\ &\ll T^{3\eta}P^{1-2\sigma}\sqrt{T^{2\eta}T^{4k(1-\sigma)+2\eta}+T^{2\eta}PE_k}\\ &\qquad\times\sqrt{T^{2\eta}T^{4(k-1)(1-\sigma)+2\eta}+T^{2\eta}PE_{k-1}}\\ &\ll T^{7\eta}P^{1-2\sigma}\{T^{(4k-2)(1-\sigma)}+T^{2k(1-\sigma)}\sqrt{PE_{k-1}}\\ &\qquad+T^{2(k-1)(1-\sigma)}\sqrt{PE_k}+P\sqrt{E_kE_{k-1}}\}.\end{aligned}$$

Thus either

$$E_k \ll T^{7\eta}P^{1-2\sigma}\{T^{(4k-2)(1-\sigma)}+T^{2k(1-\sigma)}\sqrt{PE_{k-1}}\} \qquad (3.11)$$

or

$$E_k \ll T^{7\eta}P^{1-2\sigma}\{T^{2(k-1)(1-\sigma)}\sqrt{P}+P\sqrt{E_{k-1}}\}\sqrt{E_k}. \qquad (3.12)$$

In the latter case we have

$$E_k \ll T^{14\eta}P^{2-4\sigma}\{T^{4(k-1)(1-\sigma)}P+P^2E_{k-1}\},$$

and so we may conclude that

$$E_k \ll T^{14\eta}\{T^{(4k-2)(1-\sigma)}P^{1-2\sigma}+T^{2k(1-\sigma)}P^{\frac{3}{2}-2\sigma}E_{k-1}^{\frac{1}{2}} + T^{(4k-4)(1-\sigma)}P^{3-4\sigma}+P^{4-4\sigma}E_{k-1}\}, \quad (3.13)$$

whichever of (3.11) or (3.12) holds.

We now choose an integer r such that $P = (2N)^r$ lies in the range

$$T^{1-(2\sigma-1)^{k-1}} < P \leqslant T^{1-(2\sigma-1)^{k-1}}(2N).$$

We have

$$\begin{aligned} N \leqslant Y(\log T)^2 &= T^{4\eta/(2\sigma-1)}(\log T)^2 \\ &\leqslant T^{2\sqrt{\eta}}(\log T)^2 \ll T^{3\sqrt{\eta}} \end{aligned}$$

for $\sigma \geqslant \frac{1}{2}+\sqrt{\eta}$, so that

$$T^{1-(2\sigma-1)^{k-1}} < P \leqslant T^{1-(2\sigma-1)^{k-1}+3\sqrt{\eta}} \leqslant T^2. \quad (3.14)$$

We now prove, by induction on k, that

$$E_k \ll T^{4k(1-\sigma)-1+(2\sigma-1)^k+A\sqrt{\eta}} \quad (3.15)$$

for some $A = A_k$ and any small positive η. The bound (3.15) follows at once from (1.10) for $k = 1$, since $E_1 \ll N(\sigma, T)$. We therefore suppose that (3.15) holds with k replaced by $k-1$ and use (3.13), together with the bounds (3.14), to deduce that

$$E_k \ll T^{14\eta+A\sqrt{\eta}+6\sqrt{\eta}}\{T^{4k(1-\sigma)-1+(2\sigma-1)^k}+T^{4k(1-\sigma)-1+(2\sigma-1)^{k-1}(4\sigma-3)}\},$$

from which (3.15) follows. Finally Lemma 1 is a consequence of (3.2) and (3.15), on choosing a new value for η. □

4 Completion of the proof

It remains to combine Lemmas 1 and 3. In doing this we shall choose η sufficiently small in terms of ϵ and l, and K sufficiently large, in terms of η, ϵ and l. We know, in view of the bound (1.5), that $S(x, H)$ vanishes for $H \gg x^{\frac{1}{2}+\eta}$, and so we shall assume that $H \ll x^{\frac{1}{2}+\eta}$ for the proof of the theorem. Moreover, we may take $H_0 \ll x^{\frac{1}{2}+\eta}$ in (2.3). Then

$$\begin{aligned} x^\eta H_0 \ll x^{\frac{1}{2}+2\eta} &\ll x^{\frac{1}{2}+2\eta}\left(\frac{x^{\frac{1}{2}+\eta}}{H}\right)^{l-1}\left(\frac{x}{H}\right)^{2^{-l}} \\ &= x^{\frac{1}{2}l+2^{-l}+(l+1)\eta}H^{1-l-2^{-l}}, \end{aligned} \quad (4.1)$$

which is good enough for (1.7) on choosing $\eta \leqslant \epsilon/(1+l)$. Finally, if $\frac{3}{4}+\eta \leqslant \sigma \leqslant 1-\Delta$, we may use the bound (1.9), so that the corresponding contribution to $S(x,H)$ is

$$\ll x^{2\eta} \max_{\sigma} \min_{k} x^{1+2k(\sigma-1)} = x^{2\eta} \max_{\sigma} x^{1+2K(\sigma-1)} = x^{2\eta+1-2K\Delta}.$$

As in (4.1) this will be good enough, providing that K is chosen to be at least $(4\Delta)^{-1}$.

We have now to examine

$$\max_{\frac{1}{2}\leqslant\sigma\leqslant\frac{3}{4}+\eta}\ \min_{1\leqslant k\leqslant K} x^{1+2k(\sigma-1)}\left(\frac{x}{H}\right)^{4k(1-\sigma)-1+(2\sigma-1)^k}.$$

For brevity we put

$$T_k(\sigma) = T_k = x^{1+2k(\sigma-1)}\left(\frac{x}{H}\right)^{4k(1-\sigma)-1+(2\sigma-1)^k}.$$

If $H \geqslant x^{\frac{1}{2}-\frac{1}{8}\eta/K}$ we merely choose $k = K$, so that

$$\begin{aligned} T_K^2 &= x^{1+(2\sigma-1)^K}\left(\frac{x}{H^2}\right)^{4K(1-\sigma)-1+(2\sigma-1)^K} \\ &\leqslant x^{1+(2\sigma-1)^K}\left\{\left(\frac{x}{H^2}\right)^{4K}+1\right\} \\ &\ll x^{1+(2\sigma-1)^K+\eta}. \end{aligned}$$

However, since $\frac{1}{2} \leqslant \sigma \leqslant \frac{3}{4}+\eta \leqslant \frac{7}{8}$, the exponent $(2\sigma-1)^K$ tends to zero as K tends to infinity, uniformly in σ. It follows that $T_K \ll x^{\frac{1}{2}+\frac{3}{4}\eta}$ if K is sufficiently large. As in (4.1) this is satisfactory.

We may now assume that

$$H \leqslant x^{\frac{1}{2}-\frac{1}{8}\eta/K}.$$

We observe that $T_k < T_{k+1}$ if and only if

$$\left(\frac{x}{H}\right)^{(2\sigma-1)^k} < \frac{x}{H^2}. \tag{4.2}$$

We therefore wish to use the smallest value of k for which (4.2) holds. Since

$$\left(\frac{x}{H}\right)^{(2\sigma-1)^K} < x^{(2\sigma-1)^K} < x^{(3/4)^K} \leqslant x^{\eta/4K} \leqslant \frac{x}{H^2}$$

for sufficiently large K, we always have a suitable integer $k \leqslant K$. This k is necessarily positive, as $H \geqslant 1$. For our choice of k we have

$$\left(\frac{x}{H}\right)^{(2\sigma-1)^k} \leqslant \frac{x}{H^2} \leqslant \left(\frac{x}{H}\right)^{(2\sigma-1)^{k-1}},$$

which defines an interval $I_k = [\sigma_{k-1}, \sigma_k]$ for σ. (We take $\sigma_0 = \frac{1}{2}$.) On each I_k the function $T_k(\sigma)$ is increasing, since

$$\frac{\mathrm{d}}{\mathrm{d}\sigma}T_k(\sigma) = 2kT_k(\sigma)\log\frac{\left(\frac{x}{H}\right)^{(2\sigma-1)^{k-1}}}{\frac{x}{H^2}} \geqslant 0.$$

Thus

$$\max_{\frac{1}{2}\leqslant\sigma\leqslant\frac{3}{4}+\eta}\ \min_{1\leqslant k\leqslant K} T_k$$

is achieved at $\sigma = \frac{3}{4}+\eta$. Moreover, by our choice of k,

$$T_k(\tfrac{3}{4}+\eta) \leqslant T_l(\tfrac{3}{4}+\eta),$$

and, finally,

$$T_l(\tfrac{3}{4}+\eta) \leqslant x^{\frac{1}{2}l}H^{1-l}\left(\frac{x}{H}\right)^{(\frac{1}{2}+2\eta)^l} \leqslant x^{\frac{1}{2}l}H^{1-l}\left(\frac{x}{H}\right)^{(\frac{1}{2})^l} x^{\frac{1}{2}\epsilon}$$

if η is chosen small enough. Thus

$$\max_{\frac{1}{2}\leqslant\sigma\leqslant\frac{3}{4}+\eta}\ \min_{1\leqslant k\leqslant K} T_k(\sigma) \leqslant x^{\frac{1}{2}l+2^{-l}+2\eta+\frac{1}{2}\epsilon}H^{1-l-2^{-l}}$$

which is satisfactory. This completes the proof of the Theorem. □

References

[1] D. R. Heath-Brown, The differences between consecutive primes, II, *J. London Math. Soc.* (2), **19** (1979), 207–20

[2] D. R. Heath-Brown, The differences between consecutive primes, III, *J. London Math. Soc.* (2), **20** (1979), 177–8

[3] A. E. Ingham, On the difference between consecutive primes, *Quart. J. Math. Oxford Ser.*, **8** (1937), 255–66

[4] H. Iwaniec & M. Jutila, Primes in short intervals, *Arkiv för Mat.*, **17** (1979), 167–76

[5] H. L. Montgomery, *Topics in Multiplicative Number Theory*, Springer, Berlin, 1971

[6] A. Selberg, On the normal density of primes in small intervals, and the difference between consecutive primes, *Arch. Math. Naturvid.*, **47** (1943) (6), 87–105

[7] E. C. Titchmarsh, *The Theory of the Riemann Zeta-function*, Clarendon Press, Oxford, 1951

[8] D. Wolke, Grosse Differenzen zwischen aufeinandergefolgenden Primzahlen, *Math. Ann.*, **218** (1975) (6), 269–71

On the cofinality of countable products of cardinal numbers

Thomas Jech*

Abstract

Let $\lambda_1 < \lambda_2 < \cdots < \lambda_n < \cdots$ be an increasing sequence of regular cardinals. The product $\prod_{n=1}^{\infty} \lambda_n$ is partially ordered under the coordinate-wise ordering. The *cofinality* of $\Pi_n \lambda_n$ is the least size of a family F of functions in $\Pi_n \lambda_n$ such that for each $g \in \Pi_n \lambda_n$ there is some $f \in F$ with $g < f$.

We compute the cofinalities of such countable products in the models of Prikry and Magidor. For example, we show that in a Magidor model where $2^{\aleph_\omega} = \aleph_{\omega+2}$, the cofinality of $\prod_{n=1}^{\infty} \aleph_{3n}$ is $\aleph_{\omega+1}$, while the cofinality of $\prod_{n=1}^{\infty} \aleph_n$ is $\aleph_{\omega+2}$.

1 Introduction

In this paper we consider the following problem: let

$$\lambda_1 < \lambda_2 < \cdots < \lambda_n < \cdots$$

be an increasing sequence of (infinite) regular cardinals, with limit κ, and assume that κ is a strong limit; i.e. $2^\alpha < \kappa$ for all $\alpha < \kappa$. The product $\prod_{n=1}^{\infty} \lambda_n$ is partially ordered by coordinate-wise ordering: $f < g$ if $f(n) < g(n)$ for all n.

A set $F \subset \Pi_n \lambda_n$ is *cofinal* in $\Pi_n \lambda_n$ if for every $f \in \Pi_n \lambda_n$ there exists some $g \in F$ such that $f < g$. The *cofinality* of $\Pi_n \lambda_n$,

$$\operatorname{cof} \Pi_n \lambda_n$$

is the least cardinality of a cofinal family.

* Partially supported by an NSF grant.

An easy diagonalization (Lemma 1.1 below) shows that the cofinality of $\Pi_n \lambda_n$ is at least κ^+ and so we have

$$\kappa^+ \leqslant \operatorname{cof} \Pi_n \lambda_n \leqslant 2^\kappa.$$

Thus the problem of cofinality of countable products is related to the singular-cardinals problem. In this paper we calculate the cofinalities in two kinds of models for the singular-cardinal problem: in models of Prikry [3] and Magidor [2].

In Section 2 we investigate Prikry's model and prove the following theorem.

Theorem 2.1 *Let κ be a measurable cardinal with normal measure U and let $\langle \kappa_n \rangle_{n=0}^\infty$ be a Prikry sequence for U. For every regular cardinal λ such that $\kappa^+ \leqslant \lambda \leqslant 2^\kappa$, there exists an increasing sequence $\langle \lambda_n \rangle_{n=0}^\infty \in V[G]$ of regular cardinals, with limit κ, such that*

$$\operatorname{cof} \Pi_n \lambda_n = \lambda.$$

We also calculate the cofinality of $\Pi_n \kappa_n$ in $V[G]$ and show that it is equal to the cardinal $\operatorname{cf} j(\kappa)$, where j is the elementary embedding associated with the ultrapower by U.

In Section 3 we consider the case of $\kappa = \aleph_\omega$ and investigate a model of Madigor's in which $2^{\aleph_\omega} = \aleph_{\omega+k}$, where $2 \leqslant k < \omega$. We calculate the cofinalities of the infinite products of the $\aleph_n$ and prove the following result (see Section 3 for the relevant definitions).

Theorem 3.1 *Let κ be a supercompact cardinal such that $2^\kappa = \kappa^{+k}$. Let N be Magidor's extension of V and let $\langle \kappa \rangle_{n=0}^\infty$ be the corresponding Prikry sequence. For each $m = 1, \ldots, k$ we have (in N)*

$$\operatorname{cof} \Pi_n \kappa_n^{+m} = \aleph_{\omega+m}.$$

The proof gives a little more information about the cofinalities of infinite products in Magidor's model than what is stated in Theorem 3.1. Firstly, the Prikry sequence in Magidor's model satisfies

$$\kappa_{n+1} = \kappa_n^{+k+1}.$$

Secondly, if A and B are infinite subsets of ω and $A \subseteq B$, then easily (see Lemma 1.2 below)

$$\operatorname{cof} \prod_{n \in A} \aleph_n \leqslant \operatorname{cof} \prod_{n \in B} \aleph_n.$$

Let us say that $\operatorname{cof} \prod_{n \in A} \aleph_n = \lambda$ *hereditarily*, if $\operatorname{cof} \prod_{n \in X} \aleph_n = \lambda$ for every infinite $X \subseteq A$. The proof of Theorem 3.1 shows that the cofinalities $\mathrm{cof}\,\Pi_n \kappa_n^{+m} = \aleph_{\omega+m}$ are hereditary.

Finally, we compute the cofinality of $\Pi_n \kappa_n$ to be (hereditarily) equal to the cardinal $\mathrm{cf}\, i(\kappa)$ (in V), where i is the elementary embedding associated with the normal measure on κ induced by the supercompact embedding (the projected measure). This enables us to compute in N the cofinality of $\mathrm{cof} \prod_{n\in A} \aleph_n$ for every infinite set A.

Note that when $k = 2$ (and $\kappa_0 = \aleph_2$ in N) we have the result stated in the abstract.

Before we proceed further, let me prove Lemmas 1.1 and 1.2 mentioned above.

Lemma 1.1 *If $F \subset \Pi_n \lambda_n$ and $|F| \leqslant \kappa$ then there exists some $g \in \Pi_n \lambda_n$ such that there is no $f \in F$ with $g < f$.*

Proof Let $\{F_n\}_{n=0}^{\infty}$ be such that

$$\bigcup_{n=0}^{\infty} F_n = F, \qquad F_0 \subset F_0 \subset \cdots \subset F_n \subset \cdots$$

and $|F_n| < \lambda_n$ for every n. For each n, let $g(n)$ be an ordinal less than λ_n such that $g(n) > f(n)$ for all $f \in F_n$. The function g will do, as for each $f \in F$ there is n_0 such that $g(n) > f(n)$ for all $n \geqslant n_0$. □

Lemma 1.2 *If X is an infinite subset of ω then*

$$\mathrm{cof} \prod_{n\in X} \lambda_n \leqslant \mathrm{cof} \prod_{n\in\omega} \lambda_n .$$

Proof Let $F \subset \Pi_n \lambda_n$ be a cofinal family. Let $G = \{f \restriction X : f \in F\}$. Then G is a cofinal family in $\prod_{n\in X} \lambda_n$. □

The computation of the cofinalities in Prikry's and Magidor's models will use cofinal families of particular form; we shall introduce a slightly different partial ordering for ordinal functions on ω.

Definition 1.3 Let f and g be ordinal functions on ω: we say $f < g$ if $f(n) < g(n)$ for all but finitely many $n \in \omega$ (this relation is often referred to as *eventual domination*).

Lemma 1.4 *Let λ be a regular cardinal and assume that $\{f_\alpha : \alpha < \lambda\}$ is a family of functions in $\Pi_n \lambda_n$ such that*

(a) *if $\alpha < \beta < \lambda$ then $f_\alpha < f_\beta$;*

(b) *for every $f \in \Pi_n \lambda_n$ there is some $\alpha < \lambda$ such that $f < f_\alpha$.*

Then $\mathrm{cof}\,\Pi_n \lambda_n = \lambda$.

Proof First we show that the cofinality is at least λ. If $|F| < \lambda$, pick for each $f \in F$ some $\alpha = \alpha_f < \lambda$ with $f < f_\alpha$ and let $g = f_\beta$, where $\beta < \lambda$ is greater than all the α_f. Then $f < g$ for every $f \in F$, and so F is not cofinal in $\Pi_n \lambda_n$.

Now we show that there is a cofinal family of size λ. Let F be the family of all finite modifications of the f_α's, that is, f belongs to F just in case for some $\alpha < \lambda$, $f(n) = f_\alpha(n)$ for all but finitely many n. Then F is cofinal in $\Pi_n \lambda_n$ and $|F| = \lambda$. □

Definition 1.5 A family $\{f_\alpha : \alpha < \lambda\}$ that satisfies (a) and (b) in Lemma 1.4 is a *scale* of length λ in $\Pi_n \lambda_n$.

As all computations of cofinalities in Sections 2 and 3 use scales, we shall henceforth use the symbol $<$ solely for eventual domination defined in 1.3.

Regardless of the size of 2^κ, there is always a sequence $\langle \lambda_n \rangle_{n=0}^\infty$ for which $\operatorname{cof} \Pi_n \lambda_n = \kappa^+$, as proved by Shelah.

Theorem (Shelah) *Let κ be a strong-limit singular cardinal of cofinality ω. There exists an increasing sequence $\langle \lambda_n : n < \omega \rangle$ of regular cardinals with limit κ such that* $\operatorname{cof} \Pi_n \lambda_n = \kappa^+$.

This theorem, as well as its proof, are implicit in Shelah's paper [4], and the theorem is explicitly stated in the paper [5] of Veličković. We give Shelah's proof in the appendix. It is this theorem that inspired the present investigations.

2 Prikry's model

Let κ be a measurable cardinal and let U be a normal measure on κ. We consider the following notion of forcing due to Karel Prikry [3].

A forcing condition is a pair (s, A) such that s is a finite increasing sequence of ordinals less than κ and $A \in U$. A condition (t, B) is stronger than (s, A) if $s \subseteq t$, $B \subseteq A$ and $t - s \in A^{<\omega}$.

If G is a generic set of conditions then

$$\langle \kappa_n \rangle_{n=0}^\infty = \bigcup \{s : (s, A) \in G \text{ for some } A\}$$

is an increasing sequence with limit κ. We shall call it a *Prikry sequence* for U. It is a known fact on Prikry forcing that $V[\langle \kappa_n \rangle_n] = V[G]$.

Prikry forcing changes the cofinality of κ to ω while preserving cardinals. In fact, no bounded subsets of κ are added and all regular cardinals other than κ remain regular.

Theorem 2.1 *Let κ be a measurable cardinal with normal measure U and let $\langle\kappa_n\rangle_{n=0}^{\infty}$ be a Prikry sequence for U. For every regular cardinal λ such that $\kappa^+ \leqslant \lambda \leqslant 2^\kappa$ there exists an increasing sequence $\langle\lambda_n\rangle_{n=0}^{\infty} \in V[G]$ of regular cardinals, with limit κ, such that*

$$\operatorname{cof} \Pi_n \lambda_n = \lambda.$$

In addition to proving Theorem 2.1 we shall also calculate the cofinality of $\Pi_n \kappa_n$.

Theorem 2.2 *The cofinality of $\Pi_n \kappa_n$ in $V[G]$ is the cardinal $\operatorname{cf} j(\kappa)$, where j is the elementary embedding associated with the ultrapower by U.*

Theorems 2.2 and 2.3 follow from the technical Lemma 2.6 stated below. First we state two simple properties of Prikry forcing. For any function f on κ let $[f]$ be the element of the ultrapower represented by f.

Lemma 2.4 *If f and g are, in V, ordinal functions on κ, then*

$$[f] = [g] \quad \text{iff} \quad \exists\, k \;\; \forall\, n \geqslant k \;\; f(\kappa_n) = g(\kappa_n);$$
$$[f] < [g] \quad \text{iff} \quad \exists\, k \;\; \forall\, n \geqslant k \;\; f(\kappa_n) < g(\kappa_n).$$

Proof For every $A \subseteq \kappa$ (in V),

$$A \in U \quad \text{iff} \quad \exists\, k \;\; \forall\, n \geqslant k \;\; \kappa_n \in A. \qquad \square$$

Lemma 2.5 *If $f\colon \kappa \to \kappa$ and $f \in V$, then*

$$\exists\, k \;\; \forall\, n \geqslant k \;\; f(\kappa_n) < \kappa_{n+1}.$$

Proof Let (s, A) be a condition. For each α let

$$A_\alpha = \{\xi : \xi > f(\alpha)\}$$

and let $B = A \cap \Delta_\alpha A_\alpha$. Then $(s, B) \leqslant (s, A)$ and, for all $\xi, \eta \in B$, if $\xi < \eta$ then $f(\xi) < \eta$. Thus (s, B) forces that if $n \geqslant |s|$ then $f(\kappa_n) < \kappa_{n+1}$. $\square$

Lemma 2.6 *If $f \in V[G]$ is a function on ω such that $f(n) < \kappa_{n+1}$ for all n, then there exists a collection $\{Y_\alpha : \alpha < \kappa\}$ in V such that $|Y_\alpha| \leqslant \alpha$ for all α and, for every n, $f(n) \in Y_{\kappa_n}$.*

Before proving Lemma 2.6, we shall deduce from it Theorems 2.2 and 2.3.

Proof of Theorem 2.2 Let λ be a regular cardinal such that $\kappa^+ \leqslant \lambda \leqslant 2^\kappa$. Let F be a function (in V) such that $[F] = \lambda$. Since $\lambda \leqslant 2^\kappa$, there is such a function with values less than κ, and we assume that F is increasing and that, for each α, $F(\alpha)$ is a regular cardinal greater than α (because $\lambda \geqslant \kappa^+$). For each n, let $\lambda_n = F(\kappa_n)$.

We shall show that $\operatorname{cof} \Pi_n \lambda_n = \lambda$ by finding a scale of length λ in $\Pi_n \lambda_n$. For each $\xi < \lambda$, let g_ξ be a function (in V) such that $[g_\xi] = \xi$ and let $f_\xi \in V[G]$ be a function on ω defined by $f_\xi(n) = g_\xi(\kappa_n)$.

By Lemma 2.4 we have $f_\xi < f_\eta$ whenever $\xi < \eta$. To show that $\langle f_\xi : f < \lambda \rangle$ is a scale, let $f \in \Pi_n \lambda_n$ be arbitrary. Let $\langle Y_\alpha : \alpha < \kappa \rangle$ be a family given by Lemma 2.6. For each α, let $g(\alpha) = \sup(Y_\alpha \cap F(\alpha))$; since $F(\alpha) > |Y_\alpha|$ is regular, we have $g(\alpha) < F(\alpha)$. Thus $[g] = [g_\xi]$ for some $\xi < \lambda$ and, consequently, $f(n) \leqslant g(\kappa_n) = g_\xi(\kappa_n) = f_\xi(n)$ for all but finitely many n. $\square$

Proof of Theorem 2.3 It is enough to find a scale of length $j(\kappa)$ in $\Pi_n \kappa_{n+1}$. For each $\xi < j(\kappa)$, let g_ξ be such that $[g_\xi] = \xi$ and let $f_\xi(n) = g_\xi(\kappa_n)$. By Lemma 2.5, each f_ξ (or rather its finite modification) belongs to $\Pi_n \kappa_{n+1}$. To show that $\langle f_\xi : \xi < j(\kappa) \rangle$ is a scale, we use Lemma 2.6. If $f \in \Pi_n \kappa_{n+1}$, let $\langle Y_\alpha : \alpha < \kappa \rangle$ be a family given by Lemma 2.6 and let $g(\alpha) = \sup Y_\alpha$. There is a $\xi < j(\kappa)$ such that $[g] = \xi$ and so $f(n) \leqslant g(\kappa_n) = g_\xi(\kappa_n) = f_\xi(n)$ for all but finitely many n. $\square$

The rest of this section is devoted to the proof of Lemma 2.6.

Proof of Lemma 2.6 Let $\dot{f}$ be a name and let p be a condition such that

$$p \Vdash \forall n \, \dot{f}(n) < \kappa_{n+1}.$$

We shall find a stronger condition q and a family $\{Y_\alpha : \alpha < \kappa\}$ such that $|Y_\alpha| \leqslant \alpha$ for each α and such that

$$q \Vdash \forall n \, \dot{f}(n) \in Y_{\kappa_n}.$$

Let $p = (s_0, A_0)$.

First, fix $\alpha < \kappa$ such that $\alpha > \max(s_0)$ and consider the set

$$T_\alpha = \{t \in \kappa^{<\omega} : t \supseteq s \text{ and } \max(t) = \alpha\}.$$

Now fix a $t \in T_\alpha$ and let n be such that $t(n) = \alpha$. Define a function $F_t \colon A_0^{<\omega} \to \kappa$ as follows: for $s \in A_0^{<\omega}$ let

$$F_t(s) = \eta$$

if $\alpha < \min(s)$ and if for some $A \subseteq A_0$

$$(ts, A) \Vdash \dot{f}(n) = \eta$$

(clearly such an η is unique if it exists); otherwise let $F_t(s) = 0$. Note that (ts, A) forces that $\kappa_{n+1} = \min(s)$ and so $F_t(s) < \min(s)$, i.e. F_t is a regressive function on $A_0^{<\omega}$.

By normality of the measure there exists a set $A_t \in U$, with $A_t \subseteq A_0$, and a countable set H_t such that $F_t(s) \in H_t$ for all $s \in A_t^{<\omega}$.

Claim 1 $(t, A_t) \Vdash \dot{f}(n) \in H_t$.

Proof If not, then there exists some $\eta \notin H_t$ and a stronger condition (ts, B) that forces $\dot{f}(n) = \eta$. But then $F_t(s) = \eta$ by definition and, since $s \in A_t^{<\omega}$, we get a contradiction. □

Now we let

$$A_\alpha = \bigcap \{A_t : t \in T_\alpha\}, \qquad Y_\alpha = \bigcup \{H_t : t \in T_\alpha\}, \qquad B_\alpha = \alpha \cup \{\alpha\} \cup A_\alpha.$$

The sets A_α and B_α are in U and $|Y_\alpha| \leqslant \alpha$.

Claim 2 $(s_0, B_\alpha) \Vdash \forall\, n$ *(if* κ_n *then* $f(n) \in Y_\alpha$*)*.

Proof If not, then there exists some n, a stronger condition (ts, A) and some $\eta \notin Y_\alpha$ such that $\max(t) = t(n) = \alpha$ (and so $(ts, A) \Vdash \kappa_n = \alpha$), and $(ts, A) \Vdash \dot{f}(n) = \eta$. It follows that $t \in T_\alpha$ and $F_t(s) = \eta$; since $s \in B_\alpha^{<\omega}$ and $\alpha < \min(s)$, we have $s \in A_\alpha^{<\omega}$, and so $s \in A_t^{<\omega}$ and $\eta \in H_t \subseteq Y_\alpha$, a contradiction. □

Finally, we let $A = \bigcap \{B_\alpha : \alpha < \kappa\}$. Note that for each $\xi < \kappa$

$$\begin{aligned} \xi \in A \quad & \text{iff} \quad \forall\, \alpha \;\; \xi \in B_\alpha \\ & \text{iff} \quad \forall\, \alpha < \xi \;\; \xi \in A_\alpha \end{aligned}$$

and so $A = \Delta_\alpha A_\alpha \in U$. Thus (s_0, A) is a condition and is stronger than each (s_0, B_α). Therefore

$$(s_0, A) \Vdash \forall\, n\, \dot{f}(n) \in Y_{\kappa_n}. \qquad \square$$

3 Magidor's model

In [2], Menachem Magidor proved that it is consistent that $2^{\aleph_\omega} > \aleph_{\omega+1}$, while $\aleph_\omega$ is a strong limit. His construction uses a supercompact cardinal, which is changed by forcing into $\aleph_\omega$. In this section we investigate this model and calculate cofinalities of infinite products of the $\aleph_n$.

Specifically, let κ be a supercompact cardinal, let $k \geqslant 2$ be a natural number and assume that $2^\kappa = \kappa^{+k}$. Let $j\colon V \to M$ be an elementary embedding attesting that κ is κ^{+k}-supercompact and let U be the corresponding normal measure on $P_\kappa(\kappa^{+k-1})$. We consider Magidor's extension of the ground model V that uses U for its forcing conditions (we give the definition below).

We prove the following theorem.

Theorem 3.1 *Let κ be a supercompact cardinal such that $2^\kappa = \kappa^{+k}$. Let N be Magidor's extension of V and let $\langle \kappa_n \rangle_{n=0}^\infty$ be the corresponding Prikry sequence. For each $m = 1,\dots,k$ we have (in N)*

$$\operatorname{cof} \Pi_n \kappa_n^{+m} = \aleph_{\omega+m}.$$

Moreover, the cofinality of each product is hereditary.

In addition to proving Theorem 3.1 we also calculate the cofinality of $\Pi_n \kappa_n$.

Theorem 3.2 *The cofinality of $\Pi_n \kappa_n$ in $V[G]$ is (hereditarily) the cardinal $\operatorname{cf} i(\kappa)$, where i is the elementary embedding associated with the ultrapower of V by the projected measure D on κ.*

We shall first describe Magidor's model. Let $j\colon V \to M$ be an elementary embedding with the critical point κ, such that M is closed under κ^{+k}-sequences. Let U be the normal measure on $P_\kappa(\kappa^{+k-1})$ defined from j: let $\lambda = \kappa^{+k-1}$ and, for each $X \in P_\kappa(\lambda)$, let

$$X \in U \quad \text{iff} \quad j''\lambda \in j(X).$$

For each $P \in P_\kappa(\lambda)$, let $\kappa_P = P \cap \kappa$ and λ_P be the order type of P. There exists a set $A_0 \in U$ such that, for all $P, Q \in A_0$,

(3.3) (a) κ_P is an accessible cardinal and $2^{\kappa_P} = \kappa_P^{+k}$;
(b) $\lambda_P = \kappa_P^{+k-1}$;
(c) if $P \subset Q$ then $\lambda_P < \kappa_Q$.

Madigor's forcing notion consists of conditions

$$p = \langle P_1, f_1, \dots, P_n, f_n, A, G \rangle,$$

where
(a) $P_1 \subset \cdots \subset P_n$ are elements of A_0;
(b) for $i = 1,\dots,n-1$, $f_i \in \operatorname{Col}(\lambda_{P_i}^+, \kappa_{P_{i+1}})$, the Lévy collapse forcing for making $\kappa_{P_{i+1}}$ the successor of $\lambda_{P_i}^+$, and $f_n \in \operatorname{Col}(\lambda_{P_n}^+, \kappa)$;
(c) $A \in U$, $A \subseteq A_0$ and $\forall\, Q \in A\ Q \supset P_n$;
(d) G is a function on A such that $G(P) \in \operatorname{Col}(\lambda_P^+, \kappa)$ for all $P \in A$.

A forcing condition $\bar{p}$ is stronger than p if

$$\bar{p} = \langle P_1, \bar{f}_1, \dots, P_n, \bar{f}_n, \bar{P}_{n+1}, \bar{f}_{n+1}, \dots, \bar{P}_m, \bar{f}_m, \bar{A}, \bar{G} \rangle,$$

where

$$\bar{f}_1 \supseteq f_1, \quad \ldots, \quad \bar{f}_n \supseteq f_n, \qquad \bar{P}_{n+1}, \ldots, \bar{P}_m \in A,$$
$$\bar{f}_{n+1} \supseteq G(\bar{P}_{n+1}), \quad \ldots, \quad \bar{f}_m \supseteq G(\bar{P}_m), \qquad \bar{A} \subseteq A,$$

and

$$\bar{G}(P) \supseteq G(P) \qquad \text{for all } P \in \bar{A}.$$

In the generic extension, let, for each n,

$$\kappa_n = \kappa_{P_n}$$

and let F_n be the generic for the Lévy collapse $\mathrm{Col}(\lambda_n^+, \kappa_{n+1})$.

Magidor's model N is the model

$$V[\langle \kappa_n \rangle_{n=1}^{\infty}, \langle F_n \rangle_{n=1}^{\infty}].$$

In Magidor's model, $\kappa_n, \ldots, \kappa_n^{+k} = \lambda_n^+$ are preserved and κ_{n+1} is the successor of λ_n^+ for each n. Cardinals greater or equal to κ are preserved and $2^\kappa = \kappa^{+k}$. The cardinal κ is a strong limit and $\kappa = \kappa_1^{+\omega}$; upon collapsing below κ_1 (making $\kappa_1 = \aleph_k$, for example), κ becomes $\aleph_\omega$.

Let D be the projection of U to κ: for $X \subseteq \kappa$ let

$$X \in D \quad \text{iff} \quad \{P \in P_\kappa(\lambda) : \kappa_P \in X\} \in U.$$

D is a normal measure on κ: let $i\colon V \to \mathrm{Ult}_D$ be the associated elementary embedding.

Note that $\langle \kappa_n \rangle_{n=1}^{\infty}$ is a Prikry sequence for D: a set $X \subseteq \kappa$ (in V) is in D if and only if all but finitely many κ are in X.

The main technical lemma that we use in the proof of Theorems 3.1 and 3.2 is the following analogue of Lemma 2.6.

Lemma 3.4 *If $f \in N$ is a function on ω such that $f(n) < \kappa_{n+1}$ for all n, there exists a collection $\{Y_P : P \in A_0\}$ in V such that $|Y_P| \leqslant \kappa_P$ for all P and, for every n,*

$$f(n) \in Y_{P_n}.$$

Proof Let $\dot{f}$ be a name for f; since $f \in V[\langle \kappa_n \rangle_n, \langle F_n \rangle_n]$, we may assume that $\pi(\dot{f}) = \dot{f}$ for every automorphism π of the forcing notion. Let p be a condition such that

$$p \Vdash \forall n\, \dot{f}(n) < \kappa_{n+1}.$$

We shall find a stronger condition q and a family $\{Y_P : P \in A_0\}$ such that $|Y_P| \leqslant \kappa_P$ for each P, and that

$$q \Vdash \forall n\, \dot{f}(n) < Y_{P_n}.$$

To simplify matters somewhat, we assume that the finite part $P_1, f_1, \ldots, P_n, f_n$ of the condition p is empty and that $p = \langle A_0, G_0\rangle$; the condition q will have the form $q = \langle A, G\rangle$, where $A \subseteq A_0$ and $G(P) \supseteq G_0(P)$ for all $P \in A$.

We shall construct, for every $P \in A_0$,

(a) a set $A_P \in U$ such that $A_P \subseteq A_0 \cap \hat{P}$, where $\hat{P} = \{Q : P \subseteq Q\}$;

(b) a function G_P on $A_P \cup \{P\}$ such that, for any P, Q and R (for which the values are defined), $G_P(R)$ and $G_Q(R)$ are compatible;

(c) a set $Y_P \subset \kappa$ of size κ_P.

Let us well order A_0 so that if $P \subset Q$ then $P < Q$. We do the construction by transfinite induction on $<$. Let $P \in A_0$ and assume we have A_Q, G_Q and Y_Q for all $Q < P$, in particular, for all $Q \subset P$ in A_0. Let

$$A = A_0 \cap \bigcap \{A_Q : Q < P\} \cap \hat{P}$$

and, for every $R \in A \cup \{P\}$,

$$G(R) = \bigcup \{G_Q(R) : Q \in A_0,\ Q < P \text{ and } Q \subset R\}.$$

By the induction hypothesis, each $G(R)$ is the union of mutually compatible conditions in $\mathrm{Col}(\lambda_R^+, \kappa)$. For each $Q \subset R$, $\lambda_Q = |Q| < \kappa_R$, so that $|G(P)| \leqslant \lambda_R$, and so $G(R) \in \mathrm{Col}(\lambda_R^+, \kappa)$. Also, if $Q < P$ and if $G_Q(R)$ is defined, then $G_Q(R) \subseteq G(R)$ (thus (b) will hold for P if G_P is constructed such that $G_P(R) \supseteq G(R)$).

We shall construct $A_P \subseteq A$, $G_P(R) \supseteq G(R)$ for all $R \in A_P$, and $Y_P \subset \kappa$. Let

$$B = \{R \in A : \forall\, Q \in A \text{ (if } Q \subset R \text{ then } \sup G(Q) < \kappa_R)\}.$$

By normality, B is in U.

Now let $\alpha = \kappa_P$ and let $\langle u_\xi : \xi < \alpha\rangle$ be an enumeration of all finite sequences $\langle \alpha_1, f_1, \ldots, \alpha_{n-1}, f_{n-1}\rangle$ such that $\alpha_i = \kappa_{P_i}$ $(i = 1, \ldots, n-1)$, where $t = \langle P_1, f_1, \ldots, P_{n-1}, f_{n-1}\rangle$ is the finite part of some forcing condition with $P_{n-1} \subset P_n$ and $\sup(f_{n-1}) < \alpha$. We shall construct A_P, G_P and Y_P by induction on $\xi < \alpha$. We start with $B_0 = B$, $G_0 = G$ and $g_0 = G(P)$; at stage ξ we construct $B_{\xi+1} \subseteq B_\xi$, $G_{\xi+1}(R) \supseteq G_\xi(R)$ and $g_{\xi+1} \supseteq g_\xi$, and add ω new elements to Y_P, and take limits at limit stages. At the end we let

$$A_P = \bigcap_{\xi<\alpha} B_\xi, \qquad G_P(R) = \bigcup_{\xi<\alpha} G_\xi(R), \qquad G_P(P) = \bigcup_{\xi<\alpha} g_\xi.$$

We now pause to state (a version of) Magidor's main lemma (see Theorem 2.6 in [2]).

Lemma 3.5 *Let $\langle t, P, f, B, g\rangle$ be a condition, where*

$$t = \langle P_1, f_1, \ldots, P_{n-1}, f_{n-1}\rangle.$$

There exist $g \supseteq f$, $C \subseteq B$ and $H \supseteq G$ such that, if some condition $\langle t', P, g', \vec{Q}, \vec{h}, C', H'\rangle$ stronger than $\langle t, P, g, C, H\rangle$ decides $\dot{f}(n)$, then already $\langle t', P, g, \vec{Q}, H(\vec{Q}), C, H\rangle$ decides $\dot{f}(n)$.

Back to our construction. At stage ξ, we are given B_ξ, G_ξ and g_ξ and want to construct $B_{\xi+1}$, $G_{\xi+1}$ and $g_{\xi+1}$ and add ω elements to Y_P. Consider the finite sequence $u_\xi = \langle \alpha_1, f_1, \ldots, \alpha_{n-1}, f_{n-1}\rangle$ and let us choose $P_1 < \cdots < P_{n-1}$ such that $\alpha_1 = \kappa_{P_1}, \ldots, \alpha_{n-1} = \kappa_{P_{n-1}}$.

Let $t_\xi = \langle P_1, f_1, \ldots, P_{n-1}, f_{n-1}\rangle\rangle$. Look at the the condition $\langle t_\xi, P, g_\xi, B_\xi, G_\xi\rangle$. By Magidor's lemma there exist g, C, and H such that if some stronger $\langle t_\xi, P, g', \vec{Q}, \vec{h}, C', H'\rangle$ decides $\dot{f}(n)$, then $\langle t_\xi, P, g, \vec{Q}, H(\vec{Q}), C, H\rangle$ decides $\dot{f}(n)$.

For each $\vec{Q} \in C^{<\omega}$, let $F(\vec{Q})$ be the η (if it exists) such that $\langle t_\xi, P, g, \vec{Q}, H(\vec{Q}), C, H\rangle \Vdash \dot{f}(n) = \eta$. F is a regressive function on $C^{<\omega}$, as it is forced that $\dot{f}(n) < \kappa_{n+1} = \kappa_Q$. Since U has the partition property, there exists a homogeneous set $B_{\xi+1} \in U$, with $B_{\xi+1} \subseteq B_\xi$, and a countable set $\{\eta_m : m \in \omega\}$ such that, for every $\vec{Q} \in [B_{\xi+1}]^m$, $F(\vec{Q}) = \eta_m$.

We add the η_m $(m = 1, 2, \ldots,)$ to Y_P and let $G_\xi(Q) = H(Q)$ for $Q \in B_{\xi+1}$, and $g_{\xi+1} = g$.

At the end of the construction we let

$$B_\alpha = \bigcap_{\xi<\alpha} B_\xi, \qquad g_\alpha = \bigcup_{\xi<\alpha} g_\xi, \qquad G_\alpha(Q) = \bigcap_{\xi<\alpha} G_\xi(Q) \quad (\text{for } Q \in B_\alpha).$$

Finally, we let

$$A_P = \{Q \in B_\alpha : Q \supset P \text{ and } \kappa_Q > \sup g_\alpha\}$$

$$G_P(P) = g_\alpha, \qquad G_P(Q) = G_\alpha(Q) \quad (\text{for } Q \in A_P).$$

This completes the construction of A_P, G_P and Y_P for all $P \in A_0$. We let

$$A = \{P \in A_0 : P \in A_Q \text{ for all } Q \subset P\}$$

and, for each $P \in A$,

$$G(P) = G_P(P).$$

Let q be the condition (A, G); q is stronger than p.

It remains to show that $q \Vdash \forall\, n\, \dot{f}(n) \in Y_{P_n}$. This follows from the next lemma.

Lemma 3.6 *If* $\langle P_1, f_1, \ldots, P_n, f_n, B, H\rangle$ *is a condition stronger than* $\langle A, G\rangle$ *then it forces* $\dot{f}(n) \in Y_{P_n}$.

Proof If the assertion fails then some even stronger condition $r = \langle P_1, g_1, P_n, g_n, \vec{Q}, \vec{g}, D, K\rangle$ forces $\dot{f}(n) = \eta$ and $\eta \notin Y_{P_n}$. Let $u = \langle \kappa_{P_1}, g_1, \ldots, \kappa_{P_{n-1}}, g_{n-1}\rangle$, let $P = P_n$ and let $\xi < \alpha = \kappa_P$ be such that $u = u_\xi$. Look back at the stage ξ of the construction of A_P, G_P and Y_P, above. We have

$$g_n \supseteq G(P) = G_P(P) \supseteq g_{\xi+1} \supseteq g, \qquad \vec{Q} \cup D \subseteq A \cap \hat{P} \subseteq A_P \subseteq B_{\xi+1},$$

and, for all $Q \in \vec{Q} \cup D$, $Q \supset P$ and

$$K(Q) \supseteq G(Q) = G_Q(Q) \supseteq G_P(Q) \supseteq H(Q).$$

We have

$$t_\xi = \langle P_1', g_1, \ldots, P_{n-1}', g_{n-1}\rangle,$$

where $\kappa_{P_i'} = \kappa_{P_i}$ for $i = 1, \ldots, n-1$. Let π be a permutation of $P - \kappa$ such that $\pi(P_1) = P_1', \ldots, \pi(P_{n-1}) = P_{n-1}'$. Since $\pi(\dot{f}) = \dot{f}$,

$$\pi(r) = \langle t_\xi, P, g_n, \vec{Q}, \vec{g}, D, K\rangle \Vdash \dot{f}(n) = \eta.$$

By Magidor's lemma $\langle t_\xi, P, g, \vec{Q}, H(\vec{Q}), C, H\rangle$ decides $\dot{f}(n)$. By the definition of F, we have $F(\vec{Q}) = \eta$ and, since $\vec{Q} \in B_{\xi+1}^{<\omega}$, it follows that $\eta \in Y_P$, a contradiction. □

We are now ready to prove Theorems 3.1 and 3.2. For the proof of Theorem 3.1 we need the following lemma.

Lemma 3.7 (a) *For all* $m \leqslant k$, κ^{+m} *is represented in* $\mathrm{Ult}_U(V)$ *by the function* κ_P^{+m}.

(b) *For all* $m \leqslant k$, κ^{+m} *is represented in* $\mathrm{Ult}_D(V)$ *by the function* α^{+m}.

(c) *If* $f\colon \kappa \to \kappa$ *and* $g\colon P_\kappa(\lambda) \to \kappa$ *are such that* $[f]_D = [g]_U$, *then, for almost all* P (mod U), $g(P) \leqslant f(\kappa_P)$.

Proof (a) $\mathrm{Ult}_D(V)$ contains all subsets of λ and so it has the same cardinals as V up to $\lambda^+ = \kappa^{+k}$.

(b) By the definition of D, for all $X \subset \kappa$,

$$X \in D \quad \text{iff} \quad \kappa \in j(X),$$

where $j\colon V \to M$ is a κ^{+k}-supercompact embedding. Since $2^\kappa = \kappa^{+k}$, we have $2^\alpha = \alpha^{+k}$ for almost all α (mod D). Hence, in Ult_D, $2^\kappa = \kappa^{+k}$. On the other hand, all subsets of κ are in Ult_D, so that $(2^\kappa)^{\mathrm{Ult}} \geqslant 2^\kappa$ and we have $(\kappa^{+k})^{\mathrm{Ult}} \geqslant 2^\kappa = \kappa^{+k}$. Hence $(\kappa^{+m})^{\mathrm{Ult}} = \kappa^{+m}$ for all $m \leqslant k$.

(c) Let $i = j_D$ and $j = j_U$ be the elementary embeddings associated with the ultrapowers by D and U, respectively. Since D is the projection of U to κ, there is some k: $\mathrm{Ult}_D \to \mathrm{Ult}_U$ such that $j = k \circ i$ and $k(\kappa) = \kappa$. It follows that $(if)(\kappa) \leq (jf)(\kappa)$.

Now $\kappa_{j''\lambda} = j(\kappa) \cap j''\lambda = \kappa$, and (c) follows from

$$(jg)(j''\lambda) = [g]_U = [f]_U = (if)(\kappa) \leq (jf)(\kappa) = (jf)(\kappa_{j''\lambda}). \quad \square$$

Proof of Theorem 3.1 Let m be a natural number ($1 \leq m \leq k$). By Lemma 3.7 (b), the function α^{+m} represents κ^{+m} in the ultrapower $\mathrm{Ult}_D(V)$. For each $\xi < \kappa^{+m}$ let g_ξ be a function on κ such that $[g_\xi]_D = \xi$ and such that $g_\xi(\alpha) < \alpha^{+m}$ for all α. Let f_ξ ($\xi < \kappa^{+m}$) be the function on ω defined by $f_\xi(n) = g_\xi(\kappa_n)$. We show that $\langle f_\xi : \xi < \kappa^{+m} \rangle$ is (in N) a scale in $\prod_n \kappa_n^{+m}$.

Since $\langle \kappa_n \rangle_{n=1}^\infty$ is a Prikry sequence for D, we have $f_\xi < f_\eta$ whenever $\xi < \eta$. It remains to show that, if $f \in N$ is a function on ω such that $f(n) < \kappa_n^{+m}$ for all n, then $f < f_\xi$ for some ξ.

Let $\dot{f}$ be a name for f and let p be a condition that forces $\forall n\, \dot{f}(n) < \kappa_n^{+m}$. As in Lemma 3.4, we assume that $p = \langle A_0, G_0 \rangle$; by Lemma 3.4 there exists a stronger condition $\langle A, G \rangle$ and a collection $\{Y_P : P \in A\}$ such that $|Y_P| \leq \kappa_P$ for each P and

$$\langle A, G \rangle \Vdash \forall n\, \dot{f}(n) \in Y_{P_n}.$$

For every $P \in A$ let

$$g(P) = \sup(Y_P \cap \kappa_P^{+m}).$$

We have $\langle A, G \rangle \Vdash \forall n\, \dot{f}(n) \leq g(P_n)$. Because $|Y_P| \leq \kappa_P$, we have $g(P) < \kappa_P^{+m}$ for all $P \in A$ and, by Lemma 3.7 (a), g represents in Ult_D an ordinal $\nu < \kappa^{+m}$.

Consider the function g_ν on κ. By Lemma 3.7 (c) there exists a set $B \subseteq A$ in U such that $g(P) \leq g_\nu(\kappa_P)$ for every $P \in B$. It follows that

$$\langle B, G \restriction B \rangle \Vdash \forall n\, \dot{f}(n) \leq f_\nu(n). \quad \square$$

Finally we prove Theorem 3.2.

Proof of Theorem 3.2 For each $\xi < i(\kappa)$ let g_ξ: $\kappa \to \kappa$ be the function representing ξ in Ult_D and let $f_\xi(n) = g_\xi(\kappa_n)$. Since $\langle \kappa_n \rangle_{n=1}^\infty$ is a Prikry sequence, we have as in Lemma 2.5 that each f: $\kappa \to \kappa$ satisfies $f(\kappa_n) < \kappa_{n+1}$ for all but finitely many n. Thus the f_ξ are (essentially) functions in $\Pi_n \kappa_{n+1}$, and we claim that in N they form a scale (of length $i(\kappa)$). To verify that, it suffices to prove the following lemma.

Lemma 3.8 *If $f \in N$ is a function on ω such that $f(n) < \kappa_{n+1}$ for all n, then there exists a function $h: \kappa \to \kappa$ in V such that, for every n, $f(n) < h(\kappa_n)$.*

Proof As in Lemma 3.4, let $\dot{f}$ be a name for f and let $\langle A_0, G_0\rangle$ be a condition such that $\langle A_0, G_0\rangle \Vdash \forall n\, \dot{f}(n) < \kappa_{n+1}$. By Lemma 3.4 there exists a stronger condition $\langle A, G\rangle$ and a collection $\{Y_P : P \in A\}$ such that $|Y_P| \leq \kappa_P$, for all $P \in A$, and $\langle A, G\rangle \Vdash \forall n\, \dot{f}(n) \in Y_{P_n}$.

Let $\tilde{A} = \{\kappa_P : P \in A\} \in D$. For each $\alpha \in \tilde{A}$ let

$$Y_\alpha = \bigcup \{Y_P : \kappa_P = \alpha\}.$$

Lemma 3.9 *For each $\alpha \in \tilde{A}$, $|Y_\alpha| < \kappa$.*

Once we have proved Lemma 3.9, the proof of Lemma 3.8 will be complete, as we can let $h(\alpha) = \sup Y_\alpha$ for $\alpha \in \tilde{A}$ and then we have $\langle A, G\rangle \Vdash \forall n\, \dot{f}(n) \leq h(\kappa_n)$.

Proof Assume that, for some $\alpha \in \tilde{A}$, $|Y_\alpha| = \kappa$. There exist κ P's in A, P_i $(i < \kappa)$, such that, for every i, $\kappa_{P_i} = \alpha$ and there is some $\eta_i \in Y_{P_i}$ such that $\eta_i \notin Y_{P_j}$ for all $j < i$. For each $i < \kappa$ let $f_i = G(P_i)$; we have $F_i \in \mathrm{Col}(\alpha^{+k}, \kappa)$. Since the Lévy collapse has the κ-chain condition, there are j and i $(j < i)$ such that f_j and f_i are compatible. Let $g = f_i \cup f_j$. We will show that $\eta_i \in Y_{P_j}$, a contradiction.

The reason why we put η_i in Y_{P_i} was that, for some n, some t and some m,

$$\langle t, P_i, g, \vec{Q}, G(\vec{Q}), A, G\rangle \Vdash \dot{f}(n) = \eta_i$$

for all $\vec{Q} \in [A \cap \hat{P}_i]^m$. There exists a permutation π of λ such that π is the identity on $\kappa \cup (\lambda - P_i - P_j)$ and $\pi(P_i) = P_j$. Since $f \in N$, we may assume that $\pi(\dot{f}) = \dot{f}$. Let $B = A \cap (P_i \cup P_j)^\wedge$. Then, if $\vec{Q} \in [B]^m$, we have

$$\langle \pi(t), P_j, g, \vec{Q}, G(\vec{Q}), B, G \restriction B\rangle \Vdash \dot{f}(n) = \eta_i,$$

which guarantees that η_i was put into Y_{P_j} at the stage corresponding to $\pi(t)$. □

Appendix

In [5], Veličković states the following theorem fo Shelah from [4]. The proof of the theorem is implicit in [4] and we reproduce it here for the benefit of the reader.

Theorem (Shelah) *Let κ be a strong-limit singular cardinal of cofinality ω. There exists an increasing sequence $\langle\lambda_n : n < \omega\rangle$ of regular cardinals with limit κ such that the cofinality of $\Pi_n \lambda_n$ is κ^+.*

Proof For ordinal functions f and g on ω we define

$$f < g \quad \text{iff} \quad f(n) < g(n) \text{ for all but finitely many } n \in \omega.$$

We shall find the λ_n and a κ^+-sequence $\langle h_\xi : \xi < \kappa^+\rangle$ of functions in $\Pi_n \lambda_n$ such that $h_\xi < h_\eta$ whenever $\xi < \eta$ and such that for every $f \in \Pi_n \lambda_n$ there exists some ξ with $f < h_\xi$. This will prove the theorem because on one hand, the family of all finite modifications of the function h_ξ is cofinal in $\Pi_n \lambda_n$, and on the other hand, no smaller family can be cofinal because some h_ξ is an upper bound in $<$.

First we choose any increasing sequence κ_n $(n < \omega)$ of regular cardinals with limit κ. By induction we construct a transfinite sequence $\langle f_\xi : \xi < \kappa^+\rangle$ of functions in $\Pi_n \kappa_n$ such that $f_\xi < f_\eta$ when $\xi < \eta$ (at each stage of the induction, the number of functions constructed so far is κ, and $\Sigma_n \kappa_n = \kappa$).

Let $F = \langle f_\xi : \xi < \kappa^+\rangle$. A function g *dominates* F if $f_\xi < g$ for every ξ. We define another relation among ordinal functions:

$$f \lessdot g \quad \text{iff} \quad f(n) \leqslant g(n) \text{ for all but finitely many } n \text{ and}$$
$$f(n) < g(n) \text{ for infinitely many } n.$$

Lemma *There exists a function $g \colon \omega \to \kappa$ that dominates F and is $\lessdot$ minimal among all functions that dominate F.*

Proof of the Lemma By induction, we construct a maximal transfinite $\lessdot$ decreasing sequence $\langle g_\nu\rangle$ of functions dominating F, starting with $g_0 = \langle\kappa_n : n < \omega\rangle$. It suffices to show that the length of the sequence $\langle g_\nu\rangle$ is not a limit ordinal: for then the last function is $\lessdot$ minimal.

So let θ be a limit ordinal and let $\langle g_\nu : \nu < \theta\rangle$ be a $\lessdot$ decreasing sequence of functions dominating F. We shall find a dominating function g such that $g \lessdot g_\nu$ for all $\nu < \theta$.

First we claim that $|\theta| \leqslant 2^{\aleph_0}$. This is proved using the Erdős–Rado theorem [1]:

$$(2^{\aleph_0})^+ \to (\aleph_0)^2_\omega.$$

For assume that $|\theta| \geqslant (2^{\aleph_0})^+$ and consider the partition $G \colon [\theta]^2 \to \omega$ defined as follows: if $\alpha < \beta$ then

$$G(\alpha, \beta) = \text{the least } n \text{ such that } g_\alpha(n) > g_\beta(n).$$

By Erdős–Rado, there is an infinite set $\alpha_0 < \alpha_1 < \alpha_2 < \cdots$ and some n such that $g_{\alpha_0}(n) > g_{\alpha_1}(n) > g_{\alpha_2}(n) > \cdots$, a contradiction.

Now let

$$A = \bigcup \{\text{range}(g_\nu) : \nu < \theta\}$$

and let S be the set of all functions from ω into A. Since $|\theta| \leqslant 2^{\aleph_0}$, we have $|S| \leqslant 2^{\aleph_0}$. For every function $g \in S$, if g does not dominate F, let ξ_g be such that $f_{\xi_g} < g$ fails. Since $|S| \leqslant 2^{\aleph_0}$, there is some $\eta < \kappa^+$ greater than all the ξ_g.

We now define a function $g: \omega \to A$ as follows:

$$g(n) = \text{the least } \gamma \in A \text{ such that } \gamma > f_\eta(n).$$

The function g dominates F: if not, then $f_{\xi_g} < g$ fails, but $f_{\xi_g} < f_\eta$ and $f_\eta < g$. We complete the proof of the lemma by showing that $g \ll g_\nu$ for all $\nu < \theta$. If $\nu < \theta$ then $g_\nu(n) > f_\eta(n)$ for all but finitely many n and, since $g_\nu(n) \in A$, we have $g_\nu(n) \geqslant g(n)$ for all but finitely many n. □

Let g be the function from the lemma. We claim that for every function $f < g$ there is some $\xi < \kappa^+$ such that $f < f_\xi$. If not, let f be a conterexample and, for each ξ, let A_ξ be the infinite set of all n such that $f(n) > f_\xi(n)$. Since $2^{\aleph_0}$ is small, there exists an infinite set A such that for κ^+ many f_ξ, $f(n) > f_\xi(n)$ on A. Because these ξ are unbounded in κ^+, it follows that $f \restriction A > f_\xi \restriction A$ for *every* ξ. Now, if we let

$$g' = f \restriction A \cup g \restriction (\omega - A),$$

then g' dominates F and $g' \ll g$, a contradiction.

Now, if g is increasing with limit κ and if every $g(n)$ is a regular cardinal, we let $\lambda_n = g(n)$ and are done. In general, all but finitely many $g(n)$ are limit ordinals; so, without loss of generality, all $g(n)$ are limit ordinals. For each n let Y_n be a closed unbounded subset of $g(n)$ such that the order type of Y_n is a regular cardinal γ_n. Note that $\sup \gamma_n = \kappa$; otherwise, the set $\Pi_n Y_n$ is small and, hence, dominated by some f_ξ, a contradiction. So let $\langle k_n : n < \omega \rangle$ be an increasing sequence of numbers such that $\langle \gamma_{k_n} : n < \omega \rangle$ is increasing with limit κ and let $\lambda_n = \gamma_{k_n}$.

We shall now define a family H of functions in $\Pi_n Y_{k_n}$: for each $f \in F$, let h_f be defined by

$$h_f(n) = \text{the least } \alpha \in Y_{k_n} \text{ such that } \alpha \geqslant f(k_n),$$

and let $H = \{h_f : f \in F\}$. Clearly, for every $f \in \Pi_n Y_{k_n}$, there is some $h \in H$ such that $f < h$. Also, the size of H is κ^+ since every smaller set of functions is dominated by some f_ξ. In fact, we can find in H a transfinite sequence $\langle h_\xi : \xi < \kappa^+ \rangle$ with the property that $h_\xi < h_\eta$ when $\xi < \eta$ and that, for every $f \in \Pi_n Y_{k_n}$, there is a ξ with $f < h_\xi$. By copying $\Pi_n Y_n$ onto $\Pi_n \lambda_n$, we get a sequence $\langle h_\xi : \xi < \kappa^+ \rangle$ with the required properties. □

References

[1] P. Erdős & R. Rado, Combinatorial theorems on classifications of subsets of a given set, *Proc. London Math. Soc.*, **2** (1952), 417–39

[2] M. Magidor, On the singular cardinals problem I, *Israel J. Math.*, **28** (1977), 1–31

[3] K. Prikry, Changing measurable into accessible cardinals, *Diss. Math.*, **68** (1970), 5–52

[4] S. Shelah, Jonsson algebras in successor cardinals, *Israel J. Math.*, **30** (1978), 57–84

[5] B. Veličković, Forcing axioms and stationary sets, to appear

On σ-centred posets

I. Juhász* and K. Kunen†

Abstract

We give several examples of σ-centred partial ordes that are not σ-filtered, that is, cannot be written as unions of countably many filters.

Let us start by recalling that a subset S of a partial order $\langle P, \leqslant \rangle$ is centred if any finitely many elements $p_1, \ldots, p_n$ of S have a common lower bound $q \in P$. Moreover, $\langle P, \leqslant \rangle$ is said to be σ-centred if $P = \bigcup_{n \in \omega} S_n$, where each S_n is centred. We say that $F \subset P$ is a filter if for any $p_1, \ldots, p_n \in F$ there is even a $q \in F$ with $q \leqslant p_i$ for all i. It is a natural question then whether a σ-centred poset will necessarily be σ-filtered, that is, the union of countably many filters. Let us note that the answer is clearly 'yes' if our poset is 'well met', that is, if any two elements have a greatest lower bound; moreover, if MA holds then, as is well known, any c.c.c. partial order of cardinality less than $2^{\aleph_0}$ is σ-filtered.

Our aim in this note is to show, however, that the answer to the above question is none the less negative. To do this, we first introduce some notation.

For the poset P we let cent(P) [filt(P)] denote the smallest cardinal κ such that P can be written as the union of κ many centred subsets [filters]. Clearly we have cent$(P) \leqslant$ filt(P) for all P and our question generalizes to asking whether $<$ could occur for some P. We further note that for separated partial orders we also have $|P| \leqslant 2^{\text{cent}(P)}$, whence filt$(P) \leqslant 2^{\text{cent}(P)}$ as well.

* Research partially supported by OTKA grant no. 1805.
† Research partially supported by NSF grant no. DMS-8501521.

The examples that we shall produce make use of certain ideals which include both the ideals of measure-zero sets and nowhere-dense sets on the reals; hence they may have some independent interest, too.

Definition 1 An ideal $\mathcal{I}$ on a set X is said to be *ω-determined* if $\mathcal{I} \supset [X]^{<\omega}$ and there exists a sequence $\langle \mathcal{A}_n : n \in \omega \rangle$ of families of subsets of X such that

(a) $|\mathcal{A}_0| \leqslant \omega$, $\bigcup \mathcal{A}_0 = X$ and, for each $n \in \omega$, we have $\mathcal{A}_{n+1} \subset \mathcal{A}_n$;
(b) if $F \in [X]^{<\omega}$, $A \in \mathcal{A}_0$ and $F \subset A$ then, for each $n \in \omega$, there is some $A' \in \mathcal{A}_n$ with $F \subset A' \subset A$;
(c) if $Y \subset X$ and, for every $n \in \omega$, there is some $A \in \mathcal{A}_n$ with $Y \subset A$ then $Y \in \mathcal{I}$.

If $\mathcal{I}$ is an ω-determined ideal with $\langle \mathcal{A}_n : n \in \omega \rangle$ being the sequence witnessing this then we can define a σ-centred partial order $Q(\mathcal{I})$ as follows.

Definition 2 Put

$$Q(\mathcal{I}) = \{\langle F, A \rangle \in [X]^{<\omega} \times \mathcal{A}_0 : F \subset A \text{ and } A \in \mathcal{A}_{|F|}\};$$

moreover, $\langle F, A \rangle \leqslant \langle F', A' \rangle$ if and only if $F \supset F'$ and $A \subset A'$.

Now, our examples will be of the form $Q(\mathcal{I})$. To get more precise results, we introduce the following piece of notation: if $\mathcal{I}$ is an ideal in X then we denote by $\mathrm{cov}(\mathcal{I})$ the smallest cardinal κ such that there is some $\mathcal{J} \subset \mathcal{I}$ with $|\mathcal{J}| = \kappa$ satisfying $\bigcup \mathcal{J} = X$.

We may now formulate our main result.

Theorem 3 *If $\mathcal{I}$ is an ω-determined ideal on X then* $\mathrm{cent}(Q(\mathcal{I})) \leqslant \omega$ *(i.e. $Q(\mathcal{I})$ is σ-centred) but*

$$\mathrm{filt}(Q(\mathcal{I})) \geqslant \mathrm{cov}(\mathcal{I});$$

hence $Q(\mathcal{I})$ is not σ-filtered if $\mathrm{cov}(\mathcal{I}) > \omega$.

Proof For each $A \in \mathcal{A}_0$ let us put

$$Q_A = Q(\mathcal{I}) \cap ([X]^{<\omega} \times \{A\});$$

clearly $Q(\mathcal{I}) = \bigcup \{Q_A : A \in \mathcal{A}_0\}$. Hence $\mathrm{cent}(Q(\mathcal{I})) \leqslant \omega$ is proven if we show that Q_A is centred for each $A \in \mathcal{A}_0$. Thus let $\langle F_1, A \rangle, \ldots, \langle F_n, A \rangle \in Q_A$ and put $F = \bigcup_{i=1}^n F_i$. Then, by (b), there is some $A' \in \mathcal{A}_{|F|}$ with $F \subset A' \subset A$, whence we have $\langle F, A' \rangle \in Q(\mathcal{I})$ and $\langle F, A' \rangle \leqslant \langle F_i, A \rangle$ for each $i = 1, \ldots, n$, that is, Q_A is indeed centred.

To prove the second statement, let $\mathcal{F}$ be any filter in $Q(\mathcal{I})$ and put

$$Y(\mathcal{F}) = \bigcup\{F : \exists\, A \in \mathcal{A}_0\ (\langle F, A\rangle \in \mathcal{F})\}.$$

We claim that $Y(\mathcal{F}) \in \mathcal{I}$.

Indeed, if $Y(\mathcal{F})$ is finite this is immediate; so let us assume that $|Y(\mathcal{F})| \geqslant \omega$. Fix $n \in \omega$ and pick distinct elements $x_1, \dots, x_n \in Y(\mathcal{F})$. For each $i \leqslant n$ let $F_i \subset [X]^{<\omega}$ and $A_i \in \mathcal{A}_0$ be such that $x_i \in F_i$ and $\langle F_i, A_i\rangle \in \mathcal{F}$. Since $\mathcal{F}$ is a filter there is some $\langle F, A\rangle \in \mathcal{F}$ such that

$$\bigcup_{i=1}^{n} F_i \subset F \subset A \subset \bigcap_{i=1}^{n} A_i.$$

Since $\langle F, A\rangle \in Q(\mathcal{I})$, we have $A \in \mathcal{A}_{|F|} \subset \mathcal{A}_n$. But we also have $Y(\mathcal{F}) \subset A$! In fact, for any $y \in Y(\mathcal{F})$ we may pick $\langle G, B\rangle \in \mathcal{F}$ with $y \in G$ and, since $\mathcal{F}$ is a filter, some $\langle H, C\rangle \in \mathcal{F}$ with

$$y \in F \cup G \subset H \subset C \subset A \cap B.$$

Thus we have established that for each $n \in \omega$ there is some $A \in \mathcal{A}_n$ with $Y(\mathcal{F}) \subset A$; hence $Y(\mathcal{F}) \in \mathcal{I}$ by (c).

Finally, it is obvious from (a) and (b) that if $Q(\mathcal{I}) = \bigcup\{\mathcal{F}_\alpha : \alpha \in \kappa\}$, where each $\mathcal{F}_\alpha$ is a filter then

$$X = \bigcup\{Y(\mathcal{F}_\alpha) : \alpha \in \kappa\}.$$

Hence, indeed, $\operatorname{cov}(\mathcal{I}) \leqslant \operatorname{filt}(Q(\mathcal{I}))$. □

Examples 4 As mentioned above, the ideal $\mathcal{N}$ of null sets on $\mathbb{R}$ is ω-determined. Indeed, it is easy to see that if $\mathcal{A}_0$ is the set of all finite unions of intervals with rational endpoints and $\mathcal{A}_n$ (for $n > 0$) is the set of all $A \in \mathcal{A}_0$ with $\lambda(A) < 1/n$, where λ denotes the Lebesgue measure, then $\langle \mathcal{A}_n : n \in \omega\rangle$ satisfies (a)–(c). Consequently, $Q(\mathcal{N})$ is a σ-centered poset with $\operatorname{filt}(Q(\mathcal{N})) \geqslant \operatorname{cov}(\mathcal{N}) > \omega$.

Next let X be a second-countable Hausdorff space with no isolated points and let $\mathcal{B}$ be a countable base for X closed under finite unions. We may write

$$\mathcal{B} = \{B_n : n \in \omega\}.$$

Let $\mathcal{I}$ be the ideal of nowhere-dense subsets of X. To see that $\mathcal{I}$ is ω-determined, let us define $\mathcal{A}_n$ as the set of those $B \in \mathcal{B}$ for which $\operatorname{Int}(B_i \backslash B) \neq \varnothing$ for all $i < n$. Then (a) and (c) of Definition 1 are obviously true. To show (b), consider $F \in [X]^{<\omega}$ and $B \in \mathcal{B} = \mathcal{A}_0$ with $F \subset B$, and fix $n \in \omega$. For each $i < n$ pick $x_i \in B_i \backslash F$ such that $x_i \neq x_j$ if $i \neq j$. Then, using the Hausdorff property of X, choose pairwise disjoint neighbourhoods about the points of $\{x_i\,;\, i < n\} \cup F$, say $\{U_i : i < n\}$ with $x_i \in U_i$ and $\{V_y : y \in F\} \subset \mathcal{B}$ with $y \in V_y \subset B$ for $y \in F$. Then

$$V = \bigcup \{V_y : y \in F\} \in \mathcal{A}_n$$

and $F \subset V \subset B$.

Thus we may conclude that the σ-centered partial order $Q(\mathcal{I})$ satisfies

$$\operatorname{filt}(Q(\mathcal{I})) \geqslant N(X),$$

where $N(X)$ $(= \operatorname{cov}(\mathcal{I}))$ is the Novak number of the space X, that is uncountable if and only if X has the Baire property.

The above examples provide us with a large variety of σ-centred partial orders that are not σ-filtered. However, they leave the following natural question open.

Problem 5 *Is there a ZFC example of a partial order P such that* $\operatorname{cent}(P) \leqslant \omega$ *but* $\operatorname{filt}(P) = 2^\omega$*?*

A related problem is whether the ideal $[\mathfrak{c}]^{<\mathfrak{c}}$ (or even $[\mathfrak{c}]^\omega$) is ω-determined. The answer to this question is affirmative if CH holds (i.e. $\mathfrak{c} = \omega_1$) as is shown by the nowhere-dense ideal on a Luzin set X.

Let us now consider some higher-cardinal versions of our problem. One such version is of course the following question, for a fixed cardinal $\kappa > \omega$.

Problem 6 *Does there exist a ZFC example of a partial orde $\mathcal{P}$ with*

$$\operatorname{cent}(\mathcal{P}) = \kappa < \operatorname{filt}(\mathcal{P})?$$

We do not know the answer to this question even for $\kappa = \omega_1$.

A natural variation of our original problem for higher cardinals is where, for given κ, we restrict our attention to κ-closed partial orders; moreover, centred collections and filters are replaced by κ-centred collections and κ-closed filters, respectively. Clearly, this problem is only of interest if κ is a regular cardinal. Moreover, if $\mathcal{P}$ is κ-closed and atomless then the assumption that $\mathcal{P}$ is the union of κ many κ-centred collections implies that $\mathcal{P}$ satisfies the κ^+-CC, which, in turn, implies that $2^{<\kappa} = \kappa$. Hence the latter is a natural assumption in the following result.

Theorem 7 *Let κ be a regular cardinal and assume that $2^{<\kappa} = \kappa$. Then there is a κ-closed (and separated and hence atomless) partial order $\mathcal{P}$ which is the union of κ many κ-centred collections, but is not the union of κ many κ-closed filters.*

Proof The proof of this uses a higher-cardinal version of both Theorem 3 and the example obtained from an appropriate nowhere-dense ideal. First, a higher-cardinal version of κ-determined ideals is given.

Definition 8 A κ-complete ideal $\mathcal{I}$ on a set is called *κ-determined* if $\mathcal{I} \supset [X]^{<\kappa}$ and there is a sequence $\langle \mathcal{A}_\alpha : \alpha \in \kappa \rangle$ such that

(a) $\mathcal{A}_0 \subset \mathcal{P}(X)$, $|\mathcal{A}_0| = \kappa$ and $\bigcup \mathcal{A}_0 = X$; moreover, $\mathcal{A}_\alpha \supset \mathcal{A}_\beta$ if $\alpha < \beta < \kappa$;

(b) if $F \in [X]^{<\kappa}$, $A \in \mathcal{A}_0$ and $F \subset A$ then, for each $\alpha \in \kappa$, there is some $A' \in \mathcal{A}_\alpha$ with $F \subset A' \subset A$;

(c) if $Y \subset X$ and, for every $\alpha \in \kappa$, there is some $A \in \mathcal{A}_\alpha$ with $Y \subset A$ then $Y \in \mathcal{I}$.

Now assume that $\mathcal{I}$ is a κ-determined ideal on X and let $\prec$ be a fixed well-ordering of X. We may then define a κ-closed partial order as follows:

$$Q_\kappa(\mathcal{I}) = \{\langle F, A\rangle \in [X]^{<\kappa} \times \mathcal{A}_0 : F \subset A \text{ and } A \in \mathcal{A}_{\mathrm{tp}(F)}\},$$

where, of course, $\mathrm{tp}(F)$ denotes the order type of F under the well-ordering $\prec$. Moreover, we set $\langle F, A\rangle \leqslant \langle F', A'\rangle$ if and only if $F \supset F'$ and $A \subset A'$. (Clearly, $Q_\omega(\mathcal{I}) = Q(\mathcal{I})$.)

It is straightforward to check that $Q_\kappa(\mathcal{I})$ is κ-closed and the union of κ many κ-centred collections.

Finally, it is also easy to check that, if we consider the κ-complete (or $G_{<\kappa}$) topology on 2^κ, then the ideal $\mathcal{I}$ of nowhere-dense subsets of this space is κ-determined and this space has the κ-Baire property, that is, $\mathrm{cov}(\mathcal{I}) > \kappa$. (This is where the assumption $2^{<\kappa} = \kappa$ is used essentially.) We leave it to the reader to verify that $Q_\kappa(\mathcal{I})$ is not the union of κ many κ-closed filters; the verification closely parallels the case for $\kappa = \omega$. $\square$

A Galvin–Hajnal conjecture on uncountably chromatic graphs

P. Komjáth

Abstract

If the existence of a measurable cardinal is consistent then so is the existence of an uncountably chromatic graph which cannot be decomposed into uncountably many uncountably chromatic graphs.

0 Introduction

The following question was originally asked by F. Galvin. Is it true that if X is a graph on vertex set V, with $\mathrm{Chr}(X) > \omega$, then there is a decomposition $V = V_1 \cup V_2$ into disjoint sets such that $\mathrm{Chr}(X|V_i) > \omega$, where $i = 1,2$. The answer is affirmative, as was shown by A. Hajnal in [2]. Hajnal then modified the problem to the question of whether there exists such a decomposition into uncountably many parts. We show that the answer to this latter question is consistently not, assuming the consistency of a measurable cardinal.

In what follows

$$[A]^{\mu} = \{X \subseteq A : |X| = \mu\}, \qquad [A]^{<\mu} = \{X \subseteq A : |X| < \mu\}.$$

A *graph* is a pair $X = (V,E) = (V(X), E(X))$, with $E \subseteq [V]^2$. The *chromatic number*, $\mathrm{Chr}(X)$, is the smallest cardinal μ onto which a *good colouring*, i.e. an $f\colon V \to \mu$ with $f(x) \neq f(y)$ for $\{x,y\} \in E$, exists. For $W \subseteq V$, $X|W = (W, E \cap [W]^2)$.

Returning to the Galvin–Hajnal question, we notice that the answer is positive if $\mathrm{Chr}(X) = \omega_1$. In fact, if X is a counter-example, $\mathrm{Chr}(X) = \kappa$, $f\colon V \to \kappa$ is a good colouring and

$$I = \{A \subseteq \kappa : \mathrm{Chr}(X|f^{-1}(A)) \leqslant \omega\},$$

then I is an ω_1-complete, ω_1-saturated ideal; so κ is at least as large as the first (weakly) Mahlo cardinal. From Hajnal's theorem we know that, if $A \notin I$, then there exist $A_1, A_2 \subseteq A$ with $A_1 \cap A_2 = \emptyset$ and $A_1, A_2 \notin I$. This implies, by an old theorem of Tarski's, that $\kappa \leqslant 2^\omega$ [5].

Another relevant problem is the following question of Galvin's. Is it true that, if $X = (V, E)$ is a graph with $\mathrm{Chr}(X) > \omega$, then there exists an $E' \subseteq E$ with $\mathrm{Chr}(Y) = \omega_1$, where $Y = (V, E')$? Clearly, a 'yes' answer implies that the answer to the Galvin–Hajnal problem is affirmative, too. However, it was shown in [3] that a negative answer is consistent to the Galvin problem. S. Shelah recently found several results concerning Galvin's problem, under $V = L$.

1 The result

Theorem *If the existence of a measurable cardinal is consistent then so is the existence of an uncountably chromatic graph which cannot be edge or vertex decomposed into uncountably many uncountably chromatic graphs.*

Proof Assume that κ is a measurable cardinal and that I is a κ-complete normal ideal on κ. Let P be the notion of forcing which adds a graph X on κ with finite conditions, that is, $p = (s, g) \in P$ iff $s \in [\kappa]^{<\omega}$, $g \subseteq [s]^2$ and (s', g') extends (s, g) whenever $s' \supseteq s$ and $g = g' \cap [s]^2$.

For every $A \in I$, we let $Q(A)$ be the notion of forcing, in V^P, which adds a good colouring with ω colours of X on A, by finite conditions. This means that a function f is in $Q(A)$ if $\mathrm{Dom}(f) \in [A]^{<\omega}$, $\{x, y\} \in E$ implies $f(x) \neq f(y)$, and $f' \leqslant f$ whenever f' extends f. We now define, in V^P, Q as the finite support product of $Q(A)$ when A runs through the members of I, i.e. $A \in V$ always holds. Our final model is V^{P*Q}.

Definition $(p, q) \in P*Q$ is *determined* if there exist $A_1, \ldots, A_{n-1}$, $g_1, \ldots, g_{n-1}$, s and g such that $p = (s, g)$, $p \Vdash \mathrm{supp}(q) = \{A_1, \ldots, A_{n-1}\}$, $p \Vdash q(A_i) = g_i$ and $\mathrm{Dom}(g_i) \subseteq s$.

Lemma 1 *The determined conditions form a dense subset of $P*Q$.*

Proof Every (p, q) has an extension of the form (p', q) which is determined. □

Lemma 2 *$P*Q$ is c.c.c.*

Proof Assume that $(p_\alpha, q_\alpha) \in P*Q$, where $\alpha < \omega_1$. We assume that all (p_α, q_α) are determined, $p_\alpha = (s_\alpha, g_\alpha)$, $s_\alpha \cap s_\beta = t$ and $g_\alpha \cap [t]^2 =$

$g_\beta \cap [t]^2$ for $\alpha < \beta < \omega_1$. We also assume that $p \Vdash \text{supp}(q_\alpha) = S \cup T_\alpha$ with $T_\alpha \cap T_\beta = \emptyset$, $S = \{A_1, \ldots, A_{n-1}\}$ and that the $q_\alpha(A_i)$ are compatible as functions. If we now take

$$p' = (s_\alpha \cup s_\beta, g_\alpha \cup g_\beta),$$

$$q'(A_i) = q_\alpha(A_i) \cup q_\beta(A_i), \qquad q'(A) = \begin{cases} q_\alpha(A) & (A \in T_\alpha), \\ q_\beta(A) & (A \in T_\beta), \end{cases}$$

then (p', q') is as condition extending (p_α, q_α) and (p_β, q_β). □

Lemma 3 *In* V^{P*Q}, $\text{Chr}(X) = \kappa$.

Proof Assume that $(p, q) \Vdash f \colon \kappa \to \gamma$ is a good colouring, where $\gamma < \kappa$. For a measure-one set of α, there is a $(p_\alpha, q_\alpha) \leqslant (p, q)$ forcing $f(\alpha) = \tau$, where $\tau < \gamma$. By shrinking, one may assume that $p_\alpha = (s_\alpha, g_\alpha)$, $s_\alpha \cap s_\beta = t$, $\alpha \in s_\alpha - t$, $g_\alpha \cap [t]^2 = g_\beta \cap [t]^2$, $p \Vdash \text{supp}(q_\alpha) = S \cup T_\alpha$, with $T_\alpha \cap T_\beta = \emptyset$, $S = \{A_1, \ldots, A_{n-1}\}$, that the $g_\alpha(A_i)$ are compatible and that $\alpha \notin A_1 \cup \cdots \cup A_{n-1}$ (as this latter set is in I). If we now take (p', q'), where

$$p' = (s', g'), \qquad s' = s_\alpha \cup s_\beta,$$

$$g' = g_\alpha \cup g_\alpha \cup \{\{\alpha, \beta\}\}, \qquad q' = q_\alpha \cup q_\beta$$

(as above), then (p', q') is still a condition, as $q'(A)$ colours neither end vertex of $\{\alpha, \beta\}$. But then (p', q') forces that $f(\alpha) = f(\beta)$ and then they are joined, which is a contradiction. □

Lemma 4 *If, in* V^{P*Q}, *either* κ *or* $E(X)$ *is decomposed into uncountably many parts, then one of them is countably chromatic.*

Proof The vertex-decomposition claim easily follows from the well-known observation that an ω_1-saturated ideal generates an ω_1-saturated ideal in a c.c.c. extension, whence one of the parts is forced to be covered by an $A \in I$ in V, which is ω-coloured by $Q(A)$.

Assume now that $1 \Vdash f \colon E(X) \to \omega_1$. For $\alpha < \beta < \kappa$ there is a $\gamma < \omega_1$ such that $1 \Vdash f(\alpha, \beta) < \gamma$ or undefined (by c.c.c.). By the Erdős–Hajnal–Rowbottom partition theorem, there is a measure-one set, $\kappa - A$, and a $\gamma < \omega_1$ with $f(\alpha, \beta) = \gamma$ for $\{\alpha, \beta\} \in [\kappa - A]^2$. Then it is forced that $f^{-1}(\{\gamma\})$ is covered by A plus an independent set, that is, $f^{-1}(\{\gamma\})$ is ω-coloured by $Q(A)$. (Of course, this gives another proof for the vertex-partition claim.) □

Now Lemmas 3 and 4 prove the theorem. □

References

[1] F. Galvin, Chromatic numbers of subgraphs, *Period Math. Hung.*, **4** (1973), 117–9

[2] A. Hajnal, On some combinatorial problems involving large cardinals, *Fundamenta Mathematica*, **69** (1970), 39–53

[3] P. Komjáth, Consistency results in infinite graphs, *Israel Journal of Mathematics*, **61** (1988), 285–94

[4] P. Komjáth & S. Shelah, Forcing constructions for uncountably chromatic graphs, *Journ. Symb. Logic*, **53** (1988), 696–707

[5] K. Prikry, *Changing Measurable into Accessible Cardinals*, Dissertationes Mathematicae, LXVIII, Warsaw, 1970

[6] R. Solovay, Real-valued measurable cardinals, in *Axiomatic Set Theory I* (ed. D. Scott), *Proc. Symp. Pure Maths.*, *XIII*, pt. 1, 397–428

Necessary conditions for mean convergence of Hermite–Fejér interpolation

Attila Máté and Paul Nevai*

Abstract

Necessary conditions are given for the Hermite–Fejér interpolation polynomials based at the zeros of orthogonal polynomials to converge in weighted L^p spaces at the Jackson rate. These conditions are known to be sufficient in the case of the generalized Jacobi polynomials.

1 Introduction

The first detailed study of weighted mean convergence of Hermite–Fejér interpolation based at the zeros of orthogonal polynomials was accomplished in [13] and [14], where it was shown that some of the most delicate problems associated with mean convergence of Hermite–Fejér interpolation can be approached through the general theory of orthogonal polynomials; in particular, a distinguished role is played by Christoffel functions. As opposed to Lagrange interpolation operators, Hermite–Fejér interpolation operators are not projectors, and thus in general the rate of convergence cannot be expected to equal the rate of the best approximation. Nevertheless, Jackson rates can be obtained.

Unaware of the general theory in [13] and [14] and of a variety of technical tools developed in [6], [9] and [10] (see [11] for a survey), A. K. Varma & J. Prasad in [22] investigated mean convergence of Hermite–Fejér interpolation in a particular case, namely in the case of

* This material is based upon work supported by the National Science Foundation under Grant Nos. DMS-8801200 (A.M.) and DMS-8814488 (P.N.), by the PSC-CUNY Research Award Program of the City University of New York under Grant No. 668337 (A.M.) and by NATO (P.N.). The first author made his contribution partly while he was visiting the Institute for Advanced Study in Princeton, New Jersey, during the summer of 1988.}

interpolation based at the zeros of the Chebyshev polynomials. Subsequently, P. Vértesi & Y. Xu [23] wrote a paper dealing with the case of generalized Jacobi polynomials. However, their results left a significant gap between the necessary and the sufficient conditions. As it turns out, the recent extension of Szegő's theory by V. Totik and us (cf. [4], [6] and [7]) can be used to fill the gap left open by P. Vértesi & Y. Xu.

Let $d\alpha$ be a positive Borel measure on the real line such that its support (denoted as $\operatorname{supp}(d\alpha)$ and defined as the smallest closed set the complement of which has measure zero) is an infinite set and all moments of $d\alpha$ are finite; such a measure will be called a moment measure (the term is motivated by Freud's [3] term m-distribution). Let

$$p_n(x) = p_n(d\alpha, x) = \gamma_n(d\alpha)x^n + \cdots \qquad (n = 0, 1, \ldots)$$

denote the orthonormal polynomials associated with $d\alpha$. These polynomials are defined by the stipulation that $\gamma_n(d\alpha) > 0$ for $n \geqslant 0$ and the requirement that

$$\int_{-\infty}^{\infty} p_m(x)p_n(x)\, d\alpha(x) = \delta_{mn} \quad (m, n \geqslant 0). \tag{1}$$

Let

$$x_{1n}(d\alpha) > x_{2n}(d\alpha) > \cdots > x_{nn}(d\alpha) \tag{2}$$

denote the zeros of $p_n(d\alpha)$. Write $H_n(d\alpha, f, x)$ for the Hermite–Fejér interpolation polynomial to f determined at the zeros of $p_n(d\alpha, x)$ as interpolation points. This is the polynomial of degree at most $2n-1$ that agrees with f at the interpolation points and has derivative zero at these points. The modulus of continuity on a set S of reals of a function f is defined as

$$\omega(f, \delta, S) = \sup\{|f(x) - f(y)| : |x - y| \leqslant \delta,\ x, y \in S\}.$$

We will consider measures $d\alpha$ satisfying one of the following two conditions. Measures satisfying the first condition are said to belong to the Szegő class, and measures with support $[-1, 1]$ such that $\alpha' > 0$ almost everywhere in this interval are said to belong to the Erdős class.

Condition A $\operatorname{supp}(d\alpha) = [-1, 1]$ *and* $\log \alpha'(\cos\theta) \in L^1[0, \pi]$.

Condition B $\operatorname{supp}(d\alpha) = [-1, 1]$, $\alpha'(x) > 0$ *for almost every* $x \in [-1, 1]$ *and there exist a positive constant* C *and an interval* $\Delta \subset [-1, 1]$ *such that* $\alpha'(x) > C$ *for* $x \in \Delta$.

The main result of this paper is the following theorem.

Theorem 1 *Let* $d\alpha$ *satisfy either Condition A or Condition B and let* $0 < p \leqslant \infty$. *Let* u *be a non-negative integrable function in* $[-1,1]$ *and let* $K > 0$. *Assume that*

$$\left(\int_{-1}^{1} |H_n(d\alpha, f, x) - f(x)|^p u(x)\, dx\right)^{1/p} \leqslant K\omega\left(f, \frac{1}{n}, [-1,1]\right) \tag{3}$$

holds for $n \geqslant 1$ *and for every function* f *continuous in* $[-1,1]$. *Then*

$$\int_{-1}^{1} \{\alpha'(x)\sqrt{1-x^2}\}^{-p} u(x)\, dx < \infty. \tag{4}$$

2 The Main Lemma

Before we can state the main lemma on which the proof of the above theorem is based, it will be useful to introduce the following notation. For a non-negative measurable function u on the real line and for $0 < p \leqslant \infty$ define the weighted p-norm of a function f as

$$\|f\|_{u,p} = \left(\int_{-\infty}^{\infty} |f(x)|^p u(x)\, dx\right)^{1/p};$$

of course, for $0 < p < 1$ this is not a norm in the usual sense. Write $l_{kn}(d\alpha, x)$ for the kth Langrange interpolating polynomial of degree $n-1$ associated with the measure $d\alpha$; that is, $l_{kn}(d\alpha)$ is the polynomial of degree $n-1$ such that $l_{kn}(d\alpha, x_{jn}(d\alpha)) = \delta_{kj}$. Furthermore, write $\lambda_{kn}(d\alpha)$ for the corresponding Christoffel number, that is, put

$$\lambda_{kn}(d\alpha) = \int_{-\infty}^{\infty} l_{kn}^2(d\alpha, x)\, dx. \tag{5}$$

Our main lemma is the following.

Lemma 3 *Suppose that* $d\alpha$ *satisfies either Condition A or Condition B. Then*

$$\liminf_{n\to\infty} n \sum_{k=1}^{n} \lambda_{kn}^2 p_{n-1}^2(x_{kn}) > 0. \tag{6}$$

In this lemma we did not indicate $d\alpha$ in the notation; that is, we wrote λ_{kn} instead of $\lambda_{kn}(d\alpha)$, $p_{n-1}(x)$ instead of $p_{n-1}(d\alpha, x)$ and x_{kn} instead of $x_{kn}(d\alpha)$. Similarly, in what follows we will not indicate $d\alpha$ in the notation if it is obvious from the context. The proof of this lemma will be given in two parts, in section 4 with Condition A and in section 6 with Condition B. Assuming this lemma, we are next going to give the proof of our main result.

3 Derivation of the Main Result

In this section we will show how Lemma 2 can be used to establish Theorem 1. To this end, we need the following lemma.

Lemma 3 *Assume that the support of* $d\alpha$ *is compact and write*

$$a_n = a_n(d\alpha) = \frac{\gamma_{n-1}(d\alpha)}{\gamma_n(d\alpha)} \qquad (n \geqslant 1). \tag{7}$$

Let u *be a non-negative integrable function on the real line, let* $K > 0$ *and let* $0 < p \leqslant \infty$. *Let* I *be a closed interval that includes the support of* $d\alpha$ *and assume that*

$$\|f(x) - H_n(d\alpha, f, x)\|_{u,p} < K\omega\left(f, \frac{1}{n}, I\right) \quad (n \geqslant 1) \tag{8}$$

holds for every function f *continuous in* I. *Then*

$$a_n^2 \|p_n^2\|_{u,p} \sum_{k=1}^{n} \lambda_{kn}^2 p_{n-1}^2(x_{kn}) = O\left(\frac{1}{n}\right). \tag{9}$$

Here the bound implied by the symbol $O(\cdot)$ *is independent of* f *and* n.

Proof As is easily seen, for every polynomial P of degree at most $2n-1$ we have

$$P(x) - H_n(d\alpha, P, x) = \sum_{k=1}^{n} P'(x_{kn})(x - x_{kn})\, l_{kn}^2(x). \tag{10}$$

Put $P_1(x) \equiv x$ and $P_2(x) \equiv \frac{1}{2}x^2$. Then (8) implies that

$$\|\{P_1^2 - P_1 H_n(P_1)\} - \{P_2 - H_n(P_2)\}\|_{u,p} = O\left(\frac{1}{n}\right).$$

According to (10), from here we can conclude that

$$\left\| \sum_{k=1}^{n} (x - x_{kn})^2 l_{kn}^2(x) \right\|_{u,p} = O\left(\frac{1}{n}\right).$$

Equation (9) follows from here, since

$$l_{kn}(x) = a_n \lambda_{kn} p_{n-1}(x_{kn}) \frac{p_n(x)}{x - x_{kn}}. \tag{11}$$

For the sake of completeness, we include a proof of this formula. First, note that

$$l_{kn}(x) = \frac{cp_n(x)}{x - x_{kn}}, \tag{12}$$

for some number c independent of x, since the polynomial on the right-hand side has exactly the same zeros as the one on the left-hand side does. Moreover,

$$\int_{-\infty}^{\infty} p_{n-1}(x) \frac{p_n(x)}{x - x_{kn}} \, \mathrm{d}\alpha(x) = \frac{\gamma_n}{\gamma_{n-1}} = \frac{1}{a_n}. \tag{13}$$

In calculating the integral on the left-hand side note that the terms other than the highest-order term of $p_n(x)/(x - x_{kn})$ have zero contribution to this integral in view of the orthogonality relations (1). Thus, noting that

$$\int_{-\infty}^{\infty} p_{n-1}(x) \gamma_{n-1} x^{n-1} \, \mathrm{d}x = \int_{-\infty}^{\infty} p_{n-1}^2(x) \, \mathrm{d}x = 1 \tag{14}$$

in view of the orthogonality relations, (13) follows by calculation the leading coefficient of the polynomial $p_n(x)/(x - x_{kn})$. Now, by the Gauss–Jacobi quadrature formula ([3, formula (3.9), p. 23] or [21, Theorem 3.4.1, p. 47]), (12) and (13) imply

$$\begin{aligned} \frac{c}{a_n} = \int_{-\infty}^{\infty} p_{n-1}(x) \, l_{kn}(x) \, \mathrm{d}\alpha(x) &= \sum_{j=1}^{n} \lambda_{jn} p_{n-1}(x_{jn}) \, l_{kn}(x_{jn}) \\ &= \lambda_{kn} p_{n-1}(x_{kn}). \end{aligned}$$

This establishes (11), completing the proof. □

In deducing Theorem 1 from Lemma 2, the following result will play an important role.

Lemma 4 *Assume that* $\mathrm{d}\alpha$ *is in the Erdős class, that is, assume that* $\mathrm{supp}(\mathrm{d}\alpha) = [-1,1]$ *and* $\alpha' > 0$ *a.e. in* $[-1,1]$. *Let* $0 < p \leqslant \infty$ *and let* u *be a non-negative Lebesgue measurable function on the real line that is zero outside the interval* $[-1,1]$. *Then*

$$\left\| \frac{1}{\alpha'(x)\sqrt{1-x^2}} \right\|_{u,p} \leqslant M_p \liminf_{n\to\infty} \| p_n^2(\mathrm{d}\alpha) \|_{u,p} \tag{15}$$

holds with some positive constant M_p *depending only on* p.

This result is a direct consequence of Máté–Nevai–Totik [6, Theorem 2, p. 317], with $2p$ replacing p. The value for M_p implied by this theorem is

$$\pi 2^{\max\{1/p-1,0\}},$$

but the best value for M_p is not known. We conjecture that the limit $\lim_{n\to\infty}\|p_n^2(d\alpha)\|_{u,p}$ exists under the assumptions of the above lemma and that relation (15) can be strengthened to

$$\left\|\frac{1}{\alpha'(x)\sqrt{1-x^2}}\right\|_{u,p} = M'_p(d\alpha) \lim_{n\to\infty}\|p_n^2(d\alpha)\|_{u,p}, \tag{16}$$

where the quantity $M'_p(d\alpha)$ may depend on p and $d\alpha$, but not on u. We can now turn to the proof of Theorem 1.

Proof of Theorem 1 We have $\alpha'(x) > 0$ for almost every $x \in [-1,1]$ according to Condition A or Condition B. Therefore we have

$$\lim_{n\to\infty} a_n(d\alpha) = \tfrac{1}{2} \tag{17}$$

(cf. (7)) according to a result of [17, p. 212, just before Remark 4] (see [18] for a correction; a simplified proof is given in [7, formula (9.11), p. 264] together with [5] or [12]). Therefore, if we extend u to the whole real line by stipulating $u = 0$ outside the interval $[-1,1]$, (9) in Lemma 3 implies with the aid of (6) that

$$\limsup_{n\to\infty}\|p_n^2(d\alpha)\|_{u,p} < \infty$$

holds for the p specified in Theorem 1 (since (8) is satisfied with this p). Therefore, the right-hand side of (15) is finite. Thus (4) follows from (15). The proof of Theorem 1 is complete. □

4 Proof of the Main Lemma with Condition A

Our next lemma allows us to estimate the expression on the left-hand side of (9); this estimate will be useful in case Condition A holds.

Lemma 5 *For an an arbitrary positive Borel measure* $d\alpha$ *with* $\operatorname{supp}(d\alpha) \subset [-1,1]$ *we have*

$$\frac{2^{2n-4}}{n\gamma_{n-1}^2} \leqslant \sum_{k=1}^{n} \lambda_{kn}^2 p_{n-1}^2(x_{kn}). \tag{18}$$

Here, as before, we used the abbreviations $\gamma_{n-1} = \gamma_{n-1}(d\alpha)$, $\lambda_{kn} = \lambda_{kn}(d\alpha)$, $p_{n-1}(x) = p_{n-1}(d\alpha, x)$ and $x_{kn} = x_{kn}(d\alpha)$.

Proof Write T_m for the m-th Chebyshev polynomial; that is, $T_0 \equiv 1$ and

$$T_m(\cos t) = \cos mt = 2^{m-1}(\cos t)^m + \cdots \qquad (m \geqslant 1).$$

We can conclude

$$\frac{2^{n-2}}{\gamma_{n-1}} = \int_{-\infty}^{\infty} p_{n-1}(x)\, T_{n-1}(x)\, d\alpha(x) \qquad (n \geq 2) \tag{19}$$

by (14) in a way similar to the proof of (13) above. The right-hand side here equals

$$\sum_{k=1}^{n} \lambda_{kn} p_{n-1}(x_{kn})\, T_{n-1}(x_{kn})$$

in view of the Gauss–Jacobi quadrature formula, quoted above after (14). As $-1 < x_{kn} < 1$ (cf. [3, Theorem I.2.2, p. 17]), we have $|T_{n-1}(x_{kn})| \leq 1$; hence this expression is

$$\leq \sum_{k=1}^{n} \lambda_{kn} |p_{n-1}(x_{kn})| \leq \sqrt{n \sum_{k=1}^{n} \lambda_{kn}^2 p_{n-1}^2(x_{kn})}\,;$$

the second inequality here follows by Schwarz's inequality. Thus (18) follows, completing the proof. □

It will be easy now to complete the proof of Lemma 2 with Condition A.

Proof of Lemma 2 with Condition A According to a well-known result of Szegő, Condition A implies that $\gamma_n(d\alpha) = O(2^n)$ as $n \to \infty$ (cf. e.g. [3, section V.6, p. 245]). Thus (6) follows from (18). □

5 The class *M*

Before we turn to the proof of Theorem 1 with Condition B, we need a result about the Erdős class of measures (cf. Corollary 9 below). Since the result we need is valid for a wider class of measures (cf. Lemma 7 below), in what follows we are going to discuss this wider class of of measures. As is well known, the orthogonal polynomials $p_n(x) = p_n(d\alpha, x)$ associated with a measure $d\alpha$ satisfy a recurrence relation

$$xp_n(x) = a_{n+1}p_{n+1}(x) + b_n p_n(x) + a_n p_{n-1}(x) \qquad (n \geq 0) \tag{20}$$

with some coefficients $a_n = a_n(d\alpha) > 0$ and $b_n = b_n(d\alpha)$; here $p_{-1} \equiv 0$, $p_0 \equiv$ constant > 0 and $a_0 = 0$ (see e.g. [3, formula (I.2.4), p. 17] or [21, formula (3.2.1), p. 42]). In fact, $a_n(d\alpha)$ here is the same as in (7).
We *define the class* ***M*** as the class of measures $d\alpha$ for which

$$\lim_{n\to\infty} a_n(d\alpha) = \tfrac{1}{2} \qquad \text{and} \qquad \lim_{n\to\infty} b_n(d\alpha) = 0. \tag{21}$$

If $d\alpha$ is in the Erdős class, that is, if $\operatorname{supp}(d\alpha) \subset [-1,1]$ and $\alpha' > 0$ a.e. in $[-1,1]$, then $d\alpha \in M$ according to the sources quoted to justify (17).

It is well known that the support of every measure $d\alpha$ in M is bounded. In fact, using the notation of (2), we have

$$|x_{kn}| < \sup_{j\geq 0}(a_j + a_{j+1} + |b_j|) \qquad \text{for all } n \geq 0 \quad (1 \leq k \leq n) \tag{22}$$

for every moment measure $d\alpha$ (see just before (1) for the definition of moment measure). Indeed, it is easy to prove by induction on n that we have

$$|p_{n+1}(x)| \geq |p_n(x)| \qquad \text{for } |x| \geq \sup_{j\geq 0}(a_j + a_{j+1} + |b_j|) \qquad (n \geq 0).$$

As $p_0(x) \neq 0$, (22) follows. Hence

$$\operatorname{supp}(d\alpha) \subset [-\sup_{j\geq 0}(a_j + a_{j+1} + |b_j|), \sup_{j\geq 0}(a_j + a_{j+1} + |b_j|)] \tag{23}$$

holds according to [21, Theorem 6.1.1, p. 111]. Note that for $d\alpha \in \boldsymbol{M}$, the right-hand side of (22) is clearly finite; hence the interval on the right-hand side of (23) is also finite. Information more precise than given in (22) about the zeros of orthogonal polynomials associated with measures $d\alpha$ in $\boldsymbol{M}$ is contained in Blumenthal's theorem (see [1, sections IV.3–4, pp. 113–124, especially after formula (4.2), p. 121}).

Denote by $\mathbb{C}$ and by $\overline{\mathbb{C}} = \mathbb{C} \cup \{\infty\}$ the complex plane and the Riemann sphere, respectively. Let $\sqrt{z^2-1}$ denote the branch of the function $\pm\sqrt{z^2-1}$ that is holomorphic in $\overline{\mathbb{C}} \setminus [-1, 1]$ and is such that

$$|z - \sqrt{z^2-1}| < 1 \qquad \text{for } z \notin [-1, 1]. \tag{24}$$

The aim of this section is to prove the following lemmas.

Lemma 6 *Assume* $d\alpha \in \boldsymbol{M}$. *Then*

$$\lim_{n\to\infty} \frac{z p_{n-1}(z)}{p_n(z)} = z(z - \sqrt{z^2-1}) \qquad (z \notin \operatorname{supp}(d\alpha)), \tag{25}$$

and the convergence is uniform on every compact subset of $\overline{\mathbb{C}} \setminus \operatorname{supp}(d\alpha)$.

Lemma 7 *Assume* $d\alpha \in \boldsymbol{M}$. *Then for every function f Riemann integrable in* $[-1, 1]$ *and bounded on a closed interval including* $\operatorname{supp}(d\alpha)$ *we have*

$$\lim_{n\to\infty} \sum_{k=1}^{n} \lambda_{kn} f(x_{kn}) p_{n-1}^2(x_{kn}) = \frac{2}{\pi} \int_{-1}^{1} f(x)\sqrt{1-x^2}\, dx. \tag{26}$$

Both of these lemmas are known. For the first one, see [9, Theorem 4.1.13, p. 33] and for the second one (with continuous f), see [9, Lemma 3.2.1, p. 16]. In [9] the first lemma is derived with the aid of the second one; here we give new proofs of these results and we will derive the second lemma from the first one. Actually, instead of Lemma 6, we will use the following, seemingly weaker, lemma to prove Lemma 7.

Lemma 8 *Assume* $d\alpha \in M$. *Then*

$$\lim_{n\to\infty} \frac{zp_{n-1}(z)}{p_n(z)} = z(z-\sqrt{z^2-1}) \qquad (z \notin \operatorname{supp}(d\alpha) \cup [-1,1]), \quad (27)$$

and the convergence is uniform on every compact subset of $\bar{\mathbb{C}} \setminus (\operatorname{supp}(d\alpha) \cup [-1,1])$.

We will show at the end of the proof of Lemma 7 that

$$[-1,1] \subset \operatorname{supp}(d\alpha); \quad (28)$$

with this additional piece of information, Lemma 8 becomes identical to Lemma 6. Much more is known about the support of $d\alpha$ than is expressed by (23) and (28); see [9, Lemma 3.3.6 and Theorem 3.3.7, p. 23].

Proof of Lemma 8 By Poincaré's theorem [16] on recurrence equations (see also [8, section 17.1, p. 526] or [15, section X.6, p. 300]), (20) and (21) imply that

$$\lim_{n\to\infty} \frac{p_{n+1}(z)}{p_n(z)} = z \pm \sqrt{z^2-1} \qquad \text{for every finite } z \notin [-1,1]. \quad (29)$$

It is important in applying Poincaré's theorem that, for fixed z, we cannot have $p_n(z) = 0$ for all large n, since p_n and p_{n+1} have no common zeros; indeed, with the notation of (2) we have

$$x_{kn} > x_{k,n-1} > x_{k+1,n} \qquad (1 \leqslant k < n) \quad (30)$$

(see e.g. [21, Theorem 3.3.2, p. 46]). For $z \in [-1,1]$ Poincaré's theorem is not applicable (and, incidentally, the conclusion in (29) is known not to be true), since then the roots $z \pm \sqrt{z^2-1}$ of the characteristic equation associated with the recurrence equation in (20) have the same absolute value. We claim that if $z \notin \operatorname{supp}(d\alpha) \cup [-1,1]$ then the + sign applies in (29). Indeed, if the − sign applies in (29) for a particular $z \notin [-1,1]$, then (24) implies that

$$\sum_{n=0}^{\infty} |p_n(z)|^2 < \infty.$$

As the solution of the Hamburger moment problem is uniquely determined for a measure with bounded support (see e.g. [3, Theorem II.2.2, p. 64]), this implies that z is real (cf. [20, Theorem II.2.9, p. 50]) and, indeed, that $z \in \operatorname{supp}(d\alpha)$ (cf. [20, Corollary 2.6, pp. 45–46]), establishing our claim. Hence (27) follows with pointwise convergence, except at the point $z = \infty$.

It follows from (27) (with $z \neq \infty$) that for any particular (finite) $z \notin \operatorname{supp}(d\alpha) \cup [-1,1]$ we have $p_n(z) \neq 0$ if n is large enough. Thus if

$$\Omega \subset \mathbb{C} \setminus (\operatorname{supp}(d\alpha) \cup [-1,1])$$

is compact, then there is a real $\eta > 0$ and an integer N such that, with ρ denoting Euclidean distance, we have

$$\rho(x_{kn}, \Omega) \geqslant \eta \qquad \text{for all } n > N \qquad (1 \leqslant k \leqslant n). \tag{31}$$

For $n > N$ and $z \in \Omega$ we have

$$\left| \frac{p_{n-1}(z)}{p_n(z)} \right| = \left| \frac{\gamma_{n-1} \prod_{k=1}^{n-1} (z - x_{k,n-1})}{\gamma_n \prod_{k=1}^{n} (z - x_{kn})} \right| \leqslant \frac{\gamma_{n-1}}{\gamma_n \min_{1 \leqslant k \leqslant n} |z - x_{kn}|}. \tag{32}$$

The inequality here holds in view of (30), since according to this we have

$$|z - x_{k,n-1}| < |z - x_{k+1,n}| \qquad \text{or} \qquad |z - x_{k,n-1}| < |z - x_{kn}|$$

according as $\operatorname{Re} z < x_{k,n-1}$ or $\operatorname{Re} z > x_{k,n-1}$. Thus (31), (32), and (7) imply that

$$\left| \frac{p_{n-1}(z)}{p_n(z)} \right| \leqslant \frac{a_n}{\eta} \qquad \text{for } n > N \text{ and } z \in \Omega.$$

Hence the sequence $\langle p_{n-1}(z)/p_n(z) : n > N \rangle$ is uniformly bounded for $z \in \Omega$ in view of (21). Thus the convergence in (27) is uniform in Ω by Vitali's theorem (see e.g. [2, section 1.3, p. 9]; a simpler proof can be given with the aid of nonstandard analysis; cf. [19, section 6.4.2, pp. 174–176]). The convergence at $z = \infty$ in (27) now follows from the maximum-modulus theorem. The proof of Lemma 8 is complete. □

Next we can turn to the proof of Lemma 7.

Proof of Lemma 7 As a first step, we are going to show that

$$\lim_{n \to \infty} \sum_{k=1}^{n} \lambda_{kn} \frac{p_{n-1}^2(z_{kn})}{z - x_{kn}} = \frac{2}{\pi} \int_{-1}^{1} \frac{\sqrt{1-x^2}}{z - x} \, dx$$

$$\text{for } z \notin \operatorname{supp}(d\alpha) \cup [-1,1] \tag{33}$$

(the square root in this formula is to be taken to be positive; however, we still retain the interpretation of the complex square root described in (24)). According to (11), the sum on the left-hand side equals

$$\lim_{n\to\infty} \frac{1}{a_n p_n(z)} \sum_{k=0}^{n} l_{kn}(z) p_{n-1}(x_{kn}) = \lim_{n\to\infty} \frac{p_{n-1}(z)}{a_n p_n(z)}$$
$$= 2(z^2 - \sqrt{z^2-1}); \tag{34}$$

the first equation here follows by the Lagrange interpolation formula, and the second equation holds according to Lemma 8 and (21). Using the substitution $x = \cos\theta$, the integral on the right-hand side of (33) can be seen to equal

$$\int_{-1}^{1} (z+x)\frac{\sqrt{1-x^2}}{z^2-x^2}\,\mathrm{d}x = \int_{-1}^{1} \frac{z\sqrt{1-x^2}}{z^2-x^2}\,\mathrm{d}x = \int_{0}^{\pi} \frac{z\sin^2\theta}{z^2-\cos^2\theta}\,\mathrm{d}\theta.$$

Now the integral on the right-hand side can be easily calculated by the calculus of residues integrating along the path composed of straight line segments connecting the points 0, $\mathrm{i}L$, $\mathrm{i}L+\pi$, and π, and noting that the integrals on the two vertical segments cancel each other, while on the horizontal segment the integrand converges to z uniformly as $L \to \infty$. Comparing the result with (34), we can see that (33) indeed holds.

Next, we can prove (26) in case f is a polynomial by multiplying both sides of (33) by $f(z)\,\mathrm{d}z$, and then integrating along a closed path containing $\operatorname{supp}(\mathrm{d}\alpha) \cup [-1,1]$ in its interior. Interchanging the order of the integration and the limit on the left-hand side (this is possible, since the convergence along the path of integration in (34) is uniform in view of the uniformness of the convergence in (27)), and interchanging the order of integration on the right-hand side, we obtain (26) by the Cauchy integral formula.

Let now A and B be such that

$$\operatorname{supp}(\mathrm{d}\alpha) \cup [-1,1] \subset [A,B], \tag{35}$$

and let f be an arbitrary function bounded in $[A,B]$ and Riemann integrable in $[-1,1]$. We are going to use the one-sided approximation theorem of Pólya & Szegő (see [21, Theorem 1.5.4, p. 11]) to establish (26) for this f. It is easy to conclude from this theorem and the Weierstrass approximation theorem that, given an arbitrary $\epsilon > 0$, there are polynomials P_1 and P_2 such that

$$P_1(x) \leqslant f(x) \leqslant P_2(x) \qquad \text{for } x \in [A,B]$$

and

$$\int_{-1}^{1} \{P_2(x) - P_1(x)\}\sqrt{1-x^2}\,\mathrm{d}x < \epsilon.$$

Since $x_{kn} \in [A,B]$ for all $n>0$ $(1 \leqslant k \leqslant n)$ (see e.g. [3, Theorem I.2.2, p. 17]), we have

$$\sum_{k=1}^{n} \lambda_{kn} P_1(x_{kn}) p_{n-1}^2(x_{kn}) \leqslant \sum_{k=1}^{n} \lambda_{kn} f(x_{kn}) p_{n-1}^2(x_{kn}) \leqslant \sum_{k=1}^{n} \lambda_{kn} P_2(x_{kn}) p_{n-1}^2(x_{kn}).$$

As we already know (26) for P_1 or P_2 replacing f, (26) follows for the above f.

We can now conclude that for every non-empty open interval $I \in [-1,1]$ and every large enough n there is a k with $1 \leqslant k \leqslant n$ such that $x_{kn} \in I$. Indeed, assuming the contrary for a given I and taking f to be the characteristic function of I, the limit inferior of the sum on the left-hand side in (26) will be zero, while the right-hand side will be positive: a contradiction. Thus it follows that supp($d\alpha$) is dense in $[-1,1]$, since an interval disjoint from supp($d\alpha$) may contain at most one zero of p_n (see e.g. [3, Theorem I.2.4, p. 18]). Since supp($d\alpha$) is also closed (by definition), (28) follows.

According to (28), formula (35) is equivalent to saying that supp($d\alpha$) $\subset [A,B]$. Thus the assumption on f made after (35) is not more restrictive than the assumption made on f in Lemma 7. (In fact, as a Riemann-integrable function is bounded, the former assumption could be more restrictive than the latter only if the set supp($d\alpha$) and the interval $[-1,1]$ were disjoint.) Hence the proof of Lemma 7 is complete. □

6 Proof of the Main Lemma with Condition B

According to the remark made after formula (21), if $d\alpha$ is in the Erdős class then (21) holds. Therefore, Lemma 7 has the following corollary.

Corollary 9 *Let* $d\alpha$ *be a measure with* supp($d\alpha$) $= [-1,1]$ *such that* $\alpha' > 0$ *a.e. in* $[-1,1]$. *Let* f *be a function Riemann integrable in* $[-1,1]$. *Then*

$$\lim_{n\to\infty} \sum_{k=1}^{n} \lambda_{kn} f(x_{kn}) p_{n-1}^2(x_{kn}) = \frac{2}{\pi} \int_{-1}^{1} f(x)\sqrt{1-x^2}\, dx. \tag{36}$$

We conjecture that the assumptions on $d\alpha$ in the above lemma guarantee the validity of the following strong convergence result, in analogy with [7, formula (10.2), p. 268] (cf. also [11, formula (4.5.32),

p. 23] for the expected size of λ_{nk}):

$$\lim_{n\to\infty} \sum_{k=1}^{n} \left| \lambda_{kn} p_{n-1}^2(x_{kn}) - \frac{2}{n}(1-x_{kn}^2) \right| = 0. \tag{37}$$

It will be easy now to accomplish the proof of Lemma 2 with Condition B.

Proof of Lemma 2 with Condition B Let Δ' be a closed subinterval of the interior of Δ. Then, according to [3, Theorem III.3.3, p. 104], there is a positive number c such that

$$n\lambda_{kn} > c \qquad \text{whenever } x_{kn} \in \Delta' \quad (n > 0) \tag{38}$$

(note that $\lambda_{nk} = \lambda(d\alpha; x_{kn})$ in the notation of [3]). Choose the function f of Lemma 7 as the characteristic function of the interval Δ'. Then (36) implies that

$$\lim_{n\to\infty} \sum_{k:\, x_{kn}\in\Delta'} \lambda_{kn} p_{n-1}^2(x_{kn}) = \frac{2}{\pi}\int_{\Delta'} \sqrt{1-x^2}\, dx > 0.$$

Therefore (38) implies

$$\liminf_{n\to\infty} n \sum_{k:\, x_{kn}\in\Delta'} \lambda_{kn}^2 p_{n-1}^2(x_{kn}) > 0.$$

As $\lambda_{nk} > 0$ for all k and n (cf. (5)), formula (6) follows from here, completing the proof. □

References

[1] T. S. Chihara, *An Introduction to Orthogonal Polynomials*, Gordon & Breach, New York–London–Paris, 1978

[2] P. L. Duren, *Univalent Functions*, Grundlehre der mathematischen Wissenschaften, 259, Springer, New York–Berlin–Heidelberg–Tokyo, 1983

[3] G. Freud, *Orthogonal Polynomials*, Pergamon Press, Oxford–New York–Toronto, 1966

[4] Máté, P. Nevai & V. Totik, What is beyond Szegő's theory of orthogonal polynomials?, in *Rational Approximation and Interpolation* (eds. P. R. Graves-Morris, E. B. Saff & R. S. Varga), Lecture Notes in Mathematics, **1105**, Springer, New York, 1984, 502–10

[5] A. Máté, P. Nevai & V. Totik, Asymptotics for the ratio of leading coefficients of orthonormal polynomials on the unit circle, *Constr. Approx.*, **1** (1985), 63–9

[6] Máté, P. Nevai & V. Totik, Necessary conditions for weighted mean convergence of Fourier series in orthogonal polynomials, *J. Approx. Theory*, **46** (1986), 314–22

[7] Máté, P. Nevai & V. Totik, Strong and weak convergence of orthogonal polynomials, *Amer. J. Math.*, **109** (1987), 239–81

[8] L. M. Milne-Thomson, *The Calculus of Finite Differences*, Macmillan, London, 1933

[9] P. Nevai, *Orthogonal polynomials*, Memoirs of the Amererican Mathematical Society, Providence, Rhode Island, 1979, 1–196

[10] P. Nevai, Mean convergence of Lagrange interpolation. III, *Trans. Amer. Math. Soc.*, **282** (1984), 669–98

[11] P. Nevai, Géza Freud, orthogonal polynomials and Christoffel functions. A case study, *J. Approx. Theory*, **48** (1986), 3–167

[12] P. Nevai, Weakly convergent sequences of functions and orthogonal polynomials, *J. Approx. Theory* (submitted)

[13] P. Nevai & P. Vértesi, Hermite–Fejér interpolation at zeros of generalized Jacobi polynomials, in *Approximation Theory, IV* (eds. C. K. Chui, L. L. Schumaker & J. D. Wards), Academic Press, New York, 1983, 629–33

[14] P. Nevai & P. Vértesi, Mean convergence of Hermite–Fejér interpolation, *J. Math. Anal. Appl.*, (1985), 26–58

[15] N. E. Nörlund, *Vorlesungen über Differenzenrechnung*, Springer, Berlin–New York, 1924

[16] H. Poincaré, Sur les équations linéaires aux différentielles et aux différences finies, *Amer. J. Math.*, **7** (1885), 203–58

[17] E. A. Rahmanov, On the asymptotics of orthogonal polynomials, *Math. USSR Sbornik*, **32** (1977), 199–213; Russian Original: *Mat. Sb.*, **103** (145) (1977), 237–52

[18] E. A. Rahmanov, On the asymptotics of orthogonal polynomials. II, *Math. USSR Sbornik*, **46** (1983), 105–17; Russian Original: *Mat. Sb.*, **118** (160) (1982), 104–17

[19] A. Robinson, Non-standard analysis, *Studies in Logic and the Foundations of Mathematics,*, revised edn., North-Holland, Amsterdam–New York–Oxford, 1974

[20] J. A. Shohat & J. D. Tamarkin, *The Problem of Moments*, Mathematical Surveys, No. I, American Mathematical Society, Providence, Rhode Island, 1943

[21] G. Szegő, *Orthogonal Polynomials*, Colloquium Publications, 23, American Mathematical Society, Providence, Rhode Island, 1967

[22] A. K. Varma & J. Prasad, An analogue of a problem of P. Erdős and E. Feldheim on L_p convergence of interpolatory processes, *J. Approx. Theory*, **56** (1989), 225–40

[23] P. Vértesi & Y. Xu, Order of mean convergence of Hermite–Fejér interpolation, Mathematical Institute of the Hungarian Academy of Sciences, Preprint 37 (1987), *Studia Sci. Math. Hungar.* (to appear)

On the Erdős–Fuchs theorems

H. L. Montgomery* and R. C. Vaughan

1 Introduction

Let $a_1, a_2, \ldots$ be an infinite sequence of integers such that $0 \leqslant a_1 \leqslant a_2 \leqslant \cdots$, let $r(n)$ denote the number of representations $n = a_i + a_j$ and let $R(N) = \sum_{n=0}^{N} r(n)$. For example, if $a_1 = 0$, $a_2 = a_3 = 1$, $a_4 = a_5 = 4$, $a_6 = a_7 = 9$, $a_8 = a_9 = 16$, …, then $r(n)$ is the number of lattice points (i, j) for which $i^2 + j^2 = n$ and $R(N)$ is the number of lattice points in the closed disc $x^2 + y^2 \leqslant N$. In this case $R(N)$ is approximately πN, but the error term $R(N) - \pi N$ oscillates substantially as N tends to infinity. Erdős & Turán [2] conjectured that in any situation of this sort there is no constant $A > 0$ for which $R(N) = AN + O(1)$. Subsequently Erdős & Fuchs [1] proved much more than this, namely that, for any constant $A > 0$,

$$R(N) - AN = \Omega(N^{1/4}(\log N)^{-1/2}) \tag{1}$$

as $N \to \infty$. For brevity we shall write $R(N) = AN + E(N)$. Here AN is the putative main term and $E(N)$ is the 'error term'. It will be convenient to put $E(N) = 0$ when $N < 0$. Erdős & Fuchs also considered other ways of counting the number of representations of n. For simplicity we restrict our attention to $r(n)$.

Theorem *Let A be a fixed positive number. Then there exist constants $c > 0$ and r_0 with $0 < r_0 < 1$ such that*

$$\sum_{n=0}^{\infty} \{E(n) - E(n-H)\}^2 r^n \gg \frac{H}{1-r} \tag{2}$$

uniformly for $r_0 \leqslant r < 1$ and integers H in the interval $1 \leqslant H \leqslant c/\sqrt{1-r}$.

* Research suppoted in part by National Foundation Grant NSF-DMS-88-05216.

Here the constants r_0 and c may depend on A. This dependence could be explicitly determined, but it is of less interest, since in any specific application only one value of A is relevant.

Corollary 1 *Let A be a fixed positive constant. Using the relation above,*

$$E(N) = \Omega(N^{1/4}). \tag{3}$$

This sharpening of the Erdős–Fuchs theorem [1] was first obtained by Jurkat and, while it appeared in the Ph.D. thesis of Hayashi [3], remains otherwise unpublished. At the end of the next section we describe the differences between the arguments of Erdős–Fuchs and Jurkat, and the one we give.

Erdős–Fuchs [1] also showed that if $A > 0$ then

$$\sum_{n=0}^{N} \{r(n) - A\}^2 = \Omega(N). \tag{4}$$

This can be derived from our theorem by taking $H = 1$.

Corollary 2 *Let A be a fixed positive constant. Then there is a number N_0 such that*

$$\max_{0 \leqslant n \leqslant N} |E(n)| \gg \frac{N^{1/4}}{(\log N)^{1/4}} \tag{5}$$

uniformly for $N \geqslant N_0$.

This is a localized form of (3), but quantitatively weaker. It would be interesting to have a proof of (5) without the power of the logarithm on the right and, moreover, it would be interesting to have proof that

$$\sum_{n=0}^{N} \{E(n) - E(n-H)\}^2 \gg HN \tag{6}$$

uniformly for $H \leqslant c\sqrt{N}$.

2 Proof of the theorem

Let $c(n)$ denote the number of positive integers i for which $a_i = n$. Then the formal power-series generating function for the a_i may be written in the form

$$g(z) = \sum_{i=1}^{\infty} z^{a_i} = \sum_{n=0}^{\infty} c(n) z^n.$$

Hence

$$g(z)^2 = \sum_{n=0}^{\infty} r(n)\, z^n$$

and

$$\frac{g(z)^2}{1-z} = \sum_{n=0}^{\infty} R(n)\, z^n.$$

Since $\sum nz^n = z/(1-z)^2$, we deduce that

$$\frac{g(z)^2}{1-z} = \frac{Az}{(1-z)^2} + \sum_{n=0}^{\infty} E(n)\, z^n.$$

On multiplying both sides by $(1-z^H)$, this gives

$$\frac{1-z^H}{1-z} g(z)^2 = \frac{Az(1-z^H)}{(1-z)^2} + \sum_{n=0}^{\infty} \{E(n)-E(n-H)\}\, z^n.$$

We multiply both sides by $h(z) = (1-z^H)/(1-z)$ to see that

$$g(z)^2 h(z)^2 = \frac{Azh(z)^2}{1-z} + h(z) \sum_{n=0}^{\infty} \{E(n)-E(n-H)\}\, z^n. \tag{7}$$

With this identity among formal power series established, we now consider the question of convergence. Suppose that the real number r is given $(0 < r < 1)$. If the series in (2) diverges then there is nothing to show. Suppose that the series converges. Then the series on the right in (7) is absolutely convergent when $|z| < \sqrt{r}$ for, by Cauchy's inequality,

$$\sum_{n=0}^{\infty} |E(n)-E(n-H)|\,|z|^n \leqslant \sqrt{\sum_{n=0}^{\infty} \{E(n)-E(n-H)\}^2 r^n \sum_{n=0}^{\infty} \frac{|z|^{2n}}{r^n}}\,. \tag{8}$$

Since the power-series expansion of the first term on the right in (7) is absolutely convergent for $|z| < 1$, it follows that the right-hand side is analytic for $|z| < \sqrt{r}$. As $h(z)$ has zeros only at certain roots of unity, we may divide by $h(z)^2$ and deduce that $g(z)^2$ is regular in this same disc. Since the coefficients of $g(z)$ are non-negative, we deduce that the power series for $g(\rho)$ converges if and only if the series for $g(\rho)^2$ converges. Here $0 < \rho < 1$. Thus we conclude that the series for $g(z)$ converges in the disc $|z| < \sqrt{r}$.

Before underaking the main argument, we dispense with some preliminary considerations. Suppose it were the case that $g(r^2) \leqslant \frac{7}{8}\sqrt{A/(1-r^2)}$. Then, by taking $z = r^2$ in (7), we deduce that the two terms on the right are of the same order of magnitude. Since $h(r^2) \approx H$, it follows that

$$\sum_{n=0}^{\infty} |E(n)-E(n-H)|r^{2n} \gg \frac{H}{1-r}.$$

Then, by (8), we deduce that the sum in (2) is $\gg H^2/(1-r)$, and we have (2) in this case. Thus we may suppose that

$$g(r^2) \geqslant \frac{7}{8}\sqrt{\frac{A}{1-r^2}}. \tag{9}$$

If it were the case that $g(r^4) \geqslant \frac{9}{8}\sqrt{A/(1-r^4)}$ then, by taking $z = r^4$ in (7) and arguing similarly, we would obtain (2). Thus we may suppose that

$$g(r^4) \leqslant \frac{9}{8}\sqrt{\frac{A}{1-r^4}}. \tag{10}$$

We are now ready to embark on the main portion of the proof. For $\rho < \sqrt{r}$ we set $z = \rho e(\theta)$ and write (7) briefly as

$$T(\rho,\theta) = U(\rho,\theta)+V(\rho,\theta).$$

We take $\rho = r$ and apply the triangle inequality to see that

$$|T(r,\theta)| \leqslant |U(r,\theta)|+|V(r,\theta)|.$$

Alternatively, we may take $\rho = r^2$ and apply the triangle inequality to see that

$$|T(r^2,\theta)| \geqslant |U(r^2,\theta)|-|V(r^2,\theta)|.$$

On subtracting this from the former inequality we deduce that

$$|T(r,\theta)|-|T(r^2,\theta)| \leqslant |U(r,\theta)|-|U(r^2,\theta)|+|V(r,\theta)|+|V(r^2,\theta)|.$$

We now integrate this with respect ot θ for $0 \leqslant \theta \leqslant 1$. Let T denote the resulting left-hand side, let U denote the resulting contribution of the first two terms on the right and V denote the resulting contribution of the last two terms on the right, so that

$$T \leqslant U+V \tag{11}$$

We deal with these three terms separately.

Consider T. Since $h(z) = 1+z+z^2+\cdots+z^{H-1}$, we see that

$$g(z)\,h(z) = \sum_{n=0}^{\infty}\left(\sum_{m=0}^{H-1} c(n-m)\right)z^n$$

for $|z| < \sqrt{r}$. Here $c(n) = 0$ for $n < 0$. Hence, by Parseval's identity,

$$\int_0^1 |T(\rho,\theta)|\ \mathrm{d}\theta = \sum_{n=0}^{\infty}\left|\sum_{m=0}^{H-1} c(n-m)\right|^2 \rho^{2n}$$

for $\rho < \sqrt{r}$. Taking first $\rho = r$ and then $\rho = r^2$ and subtracting, we find that

$$T = \sum_{n=0}^{\infty} \left| \sum_{m=0}^{H-1} c(n-m) \right|^2 r^{2n}(1-r^{2n}).$$

We note that $k^2 \geqslant k$ for any integer k. Since the sum over m above is an integer, it follows that

$$T \geqslant \sum_{n=0}^{\infty} \left(\sum_{m=0}^{H-1} c(n-m) \right) r^{2n}(1-r^{2n}) = g(r^2)\,h(r^2) - g(r^4)\,h(r^4).$$

The four functions on the right are all non-negative and $h(r^4) \leqslant h(r^2)$. Hence the above is $\geqslant h(r^2)\{g(r^2) - g(r^4)\}$. From (9) and (10) we see that the difference in braces here is $\gg (1-r)^{-1/2}$. Since $h(r^2) \asymp H$, it follows that

$$T \gg H(1-r)^{-1/2}. \tag{12}$$

We now treat U. Suppose that $r^2 \leqslant \rho \leqslant r$. Clearly $h(1) = H$ and $h'(z) \ll H^2$ uniformly for $|z| \leqslant 1$, so that

$$h(\rho e(\theta)) = H + O(H^2\{1-\rho+|\theta|\}).$$

Thus

$$\int_0^{1-r} |U(\rho,\theta)| \,\mathrm{d}\theta \ll H^2.$$

If $1 - r \leqslant \theta \leqslant \frac{1}{2}$ then

$$\begin{aligned}\frac{1}{|1-\rho e^{i\theta}|} &= \frac{1}{2\sin \pi\theta} + O\left(\frac{1-r}{\theta^2}\right) \\ &= \frac{1}{2\pi\theta} + O(\theta) + O\left(\frac{1-r}{\theta^2}\right),\end{aligned} \tag{13}$$

so that

$$\begin{aligned}&\int_{1-r}^{1/H} |U(\rho,\theta)| \,\mathrm{d}\theta \\ &\quad = A\rho \int_{1-r}^{1/H} \{H^2 + O(H^3|\theta|)\}\left\{\frac{1}{2\pi\theta} + O(\theta) + O\left(\frac{1-r}{\theta^2}\right)\right\} \mathrm{d}\theta \\ &\quad = \frac{AH^2}{2\pi} \log \frac{1}{H(1-r)} + O(H^2).\end{aligned}$$

We note that $h(z) \ll 1/|1-z|$ uniformly for $|z| \leqslant 1$. Thus from (13) we deduce that $h(\rho(\theta)) \ll 1/\theta$ for $1-r \leqslant \theta \leqslant \frac{1}{2}$ and $r^2 \leqslant \rho \leqslant r$. Thus

$$\int_{1/H}^{1/2} |U(\rho, \theta)| \, d\theta \ll \int_{1/H}^{1/2} \theta^{-3} \, d\theta \ll H^2.$$

We add these estimates to determine the contribution of the interval $[0, \frac{1}{2}]$. The interval $[\frac{1}{2}, 1]$ contributes the same amount, so we conclude that

$$\int_0^1 |U(\rho, \theta)| \, d\theta = \frac{1}{\pi} AH^2 \log \frac{1}{H(1-r)} + O(H^2)$$

uniformly for $r^2 \leqslant \rho \leqslant r$. We take $\rho = r$ and then $\rho = r^2$ and subtract, and thus find that

$$U \ll H^2. \tag{15}$$

We now consider V. By Cauchy's inequality,

$$\int_0^1 |V(\rho, \theta)| \, d\theta$$

$$\leqslant \sqrt{\int_0^1 |h(\rho e(\theta))|^2 \, d\theta \int_0^1 \left| \sum_{n=0}^{\infty} \{E(n) - E(n-H)\} \rho^n e(\theta) \right|^2 d\theta}.$$

By two applications of Parseval's identity we see that the right-hand side here is

$$\sqrt{\sum_{m=0}^{H} \rho^{2m} \sum_{n=0}^{\infty} \{E(n) - E(n-H)\}^2 \rho^{2n}}.$$

Here the first sum is $\leqslant H$. Let X denote the sum in (2). If $\rho \leqslant \sqrt{r}$ then the second sum above is $\leqslant X$. Taking $\rho = r$ and then $\rho = r^2$ and summing, we deduce that

$$V \ll \sqrt{HX}. \tag{16}$$

To complete the proof we combine (11), (12), (15) and (16). Thus we see that

$$\frac{H}{\sqrt{1-r}} \ll H^2 + \sqrt{HX}.$$

If $H \leqslant c/\sqrt{1-r}$ then the first term on the right does not majorize the left-hand side and it follows that

$$\frac{H}{\sqrt{1-r}} \ll \sqrt{HX}.$$

This gives (2). □

We have followed the method of Erdős & Fuchs, with two differences. Whereas Erdős & Fuchs integrated only over a short interval, we have introduced the 'peak function' $h(z)$ and integrated over $0 \leqslant \theta \leqslant 1$. The effect is the same, but by using $h(z)$ we are able to derive information concerning the change $E(n)-E(n-H)$ in the error term over a short interval. Secondly, we consider two radii and difference, so that the main term in (14) is cancelled. The resulting estimate (15) is thus slightly stronger than it would have been if we had considered only a single radius. Our understanding is that Jurkat differentiated the generating function and thus achieved a similar effect.

3 Proofs of the corollaries

To derive Corollary 1 we take $H = [c/\sqrt{1-r}]$ in the theorem. Since

$$\{E(n)-E(n-H)\}^2 \leqslant E(n)^2 + E(n-H)^2 \qquad \text{and} \qquad r^H \approx 1$$

in the range under consideration, it follows from (2) that

$$\sum_{n=0}^{\infty} E(n)^2 r^n \gg (1-r)^{-3/2}. \tag{17}$$

If it were the case that $E(n) = o(n^{1/4})$ then the sum on the left would be $o(\{1-r\}^{-3/2})$ as $r \to 1-$. Thus we have the result. □

We now prove Corollary 2. Let N be given and let the numbers $c(n)$ be defined as in the beginning of the proof of the theorem. Suppose that there is an $n \leqslant N-a_1$ for which $c(n) \geqslant N^2$. Then for this value of n we have $r(n+a_1) \geqslant N^2$ and, hence, $E(N) \geqslant N^2 - AN$, and we are done. Thus we may suppose that $c(n) < N^2$ for all $n \leqslant N-a_1$. We define a new sequence b_i as follows. We take $b_i = a_i$ as long as $a_i \leqslant N - a_1$ and then, for larger i, we let b_i increase rapidly to infinity. We apply the theorem with a_i replaced by b_i to obtain (17) with $r = \exp(-12(\log N)/N)$. We note that $E(n)$ is unchanged for $n \leqslant N$. Moreover, for this new sequence we have $c(n) < N^2$ for all positive integers n. Let M denote the left-hand side of (5). Then

$$\sum_{n=0}^{N} E(n)^2 r^n \leqslant \frac{M^2}{1-r} \ll M^2 \frac{N}{\log N}.$$

From our construction of the b_i it follows that $E(n) \ll n^5$ for all $n > N$, so that

$$\sum_{n>N} E(n)^2 r^n \ll \sum_{n>N} n^{10} \exp\left(-12n\frac{\log N}{N}\right) \ll \frac{1}{N}.$$

Since this is small compared with the right-hand side of (17), we deduce that

$$M^2 \frac{N}{\log N} \gg \frac{1}{(1-r)^{3/2}} \approx \frac{N^{3/2}}{(\log N)^{3/2}}.$$

This gives (5) and the proof is complete. □

References

[1] P. Erdős & W. H. J. Fuchs, On a problem of additive number theory, *J. London. Math. Soc.*, **31** (1956), 67–73

[2] P. Erdős & P. Turán, On a problem of Sidon in additive number theory, and some related problems. *J. London. Math. Soc.*, **16** (1941), 212–15; addendum (by P. Erdős), *ibid.*, **19** (1944), 208

[3] E. K. Hayashi, Omega theorems for the iterated additive convolution of a non-negative arithmetic function. Ph.D. thesis, University of Illinois at Urbana-Champaign, 1973

A tournament which is not finitely representable

Z. Nagy and Z. Szentmiklóssy*

A tournament $\langle V(T), E(T)\rangle$ is said to be representable by a collection $\{A_i : i \in I\}$ if and only if $A_i \subset V(T)$ for $i \in I$ and, whenever $B_i \subset A_i$ with $|B_i| = |A_i|$ for $i \in I$ and $x \in V(T)$, there exists an $i \in I$ such that, for some $y \in B_i$, $x = y$ or $\langle y, x\rangle \in E(T)$. It is easy to see that every countable tournament has a finite representation. Our aim is to construct a tournament T in ZFC such that T is not finitely representable. The existence of such a tournament was asked of by E. C. Milner.

Let $\{\{A_\xi^n : n \in \omega\} : \xi \in \omega_1\}$ be a system of ω-partitions of ω which is independent in the following sense: if $I \subset \omega_1$ is finite, the indices $\xi_i : i \in I$ and the indices $n_i : i \in I$ are arbitrary and

$$\left|\bigcap \{A_{\xi_i}^{n_i} : i \in I\}\right| = \omega.$$

It is easy to construct such a system from a system of independent 2-partitions ([2]) using finite intersections from a countable number of 2-partitions to construct an ω-partition.

We define the tournament T on the set $\omega_1 \times \omega$ in the following way: if $\xi \in \omega_1$ and $n \neq m$ then $\langle\langle \xi, n\rangle, \langle \xi, m\rangle\rangle \in E(T)$ iff $n < m$, and if $\xi < \eta \in \omega_1$ and $n, m \in \omega$ then $\langle\langle \xi, n\rangle, \langle \eta, m\rangle\rangle \in E(T)$ iff $n \in A_\eta^m$.

For $A \subset V(T)$ let

$$r(A) = \{x \in V(T) : |\{y \in A : \langle x, y\rangle \in E(T)\}| < |A|\}.$$

For a collection $\{A_i : i \in I\}$ the statements that $\{A_i : i \in I\}$ represents T and $\bigcup \{r(A_i) : i \in I\} = V(T)$ are equivalent. We show that for arbitrary $\{A_i : i \in I\}$ with I finite there exists a $\xi \in \omega_1$ such that $\{\xi\} \times \omega \not\subset \bigcup \{r(A_i) : i \in I\}$ and therefore T does not have a finite representation.

* Research supported by OTKA grant no. 1805.

Claim 1 *If* $A \subset V(T)$ *and* $\sup(\operatorname{dom} A) < \xi \in \omega_1$ *then* $\{\xi\}\times\omega \cap r(A)$ *is finite.*

Proof If A is finite then $r(A)$ can contain those $\langle\xi, n\rangle$ pairs for which $(\operatorname{range} A) \cap A^n_\xi$ is non-empty, and the number of these pairs is at most $|A|$. If A is infinite and $\langle\xi,n\rangle, \langle\xi,n'\rangle \in r(A)$ then there exists an $\langle\eta,m\rangle \in A$ such that

$$\langle\langle\eta,m\rangle,\langle\xi,n\rangle\rangle \in E(T) \quad \text{and} \quad \langle\langle\eta,m\rangle,\langle\xi,n'\rangle\rangle \in E(T),$$

i.e. $m \in A^n_\xi \cap A^{n'}_\xi$. Therefore $n = n'$ and $|r(A) \cap \{\xi\}\times\omega| \leqslant 1$, which proves the claim. □

Claim 2 *If* I *is finite,* $|A_i| = \omega_1$ *for* $i \in I$ *and* $\xi \in \omega_1$ *then*

$$\{\xi\}\times\omega - \bigcup\{r(A_i) : i \in I\}$$

is infinite.

Proof If $\langle\xi,n\rangle \in r(A_i)$ then there exists an $\eta_{n,i}$ such that if $\langle\eta,m\rangle \in A_i$ and $\eta > \eta_{n,i}$ then $\langle\langle\eta,m\rangle,\langle\xi,n\rangle\rangle \in E(T)$. Let $\eta_i > \sup\{\eta_{i,n} : n \in \omega\}$ and $m_i \in \omega$ such that $\langle\eta_i, m_i\rangle \in A_i$ and all the η_i−s are different. We have that for any $n \in \omega$ if $\langle\langle\eta_i,m_i\rangle,\langle\xi,n\rangle\rangle \notin E(T)$, then $\langle\xi,n\rangle \notin r(A_i)$, i.e.

$$\{\xi\}\times A^{m_i}_{\eta_i} \cap r(A_i) = \emptyset.$$

As the η_i−s are different, $\{\xi\}\times\bigcap\{A^{m_i}_{\eta_i} : i \in I\}$ is infinite and

$$\{\xi\}\times\bigcap\{A^{m_i}_{\eta_i} : i \in I\} \subset \{\xi\}\times\omega - \bigcup\{r(A_i) : i \in I\},$$

which proves the claim. □

A combination of Claim 1 and Claim 2 gives the necessary statement.

If $\lambda^k = \lambda$ then we can replace our system of partitions by a system $\{\{A^\nu_\xi : \nu \in \lambda\} : \xi \in \lambda^+\}$ of λ-partitions of λ such that, whenever $I \subset \lambda$ with $|I| < k$, $\xi_i : i \in I$ are different and $\nu_i : i \in I$ are arbitrary, then

$$\left|\bigcap\{A^{\nu_i}_{\xi_i} : i \in I\}\right| = \lambda \qquad [1, \text{Theorem } 3.1].$$

After this modification our construction gives a tournament on $\lambda^+\times\lambda$ which cannot be represented by less than λ sets.

References

[1] W. W. Comfort & S. Negrepontis, On families of large oscillations, *Fund. Math.*, **75** (1972), 275–90

[2] G. Fichtenholtz & L. Kantorovitch, Sur les opérations linaires dans l'espace des fonctions bornées, *Studia Math.*, **5** (1935), 69–98

On the volume of the spheres covered by a random walk

P. Révész

1 Introduction

Let $X_1, X_2, \ldots$ be a sequence of independent, identically distributed vectors taking values from $\mathbb{R}^d$ with distribution

$$P\{X_1 = e_i\} = P\{X_1 = -e_i\} = \frac{1}{2d} \qquad (i = 1, 2, \ldots, d).$$

where $\{e_1, e_2, \ldots, e_d\}$ is the standard orthogonal basis of $\mathbb{R}^d$. Let

$$S_0 = 0 = \{0, 0, \ldots, 0\}$$

and

$$S(n) = S_n = X_1 + X_2 + \cdots + X_n \qquad (n = 1, 2, \ldots),$$

i.e. let $\{S_n\}$ be the simple symmetric random walk in $\mathbb{R}^d$. Furthermore, let

$$\xi(x, n) = \#\{k : 0 < k \leqslant n,\ S_k = x\}$$

$$(n = 1, 2, \ldots;\ x = (x_1, x_2, \ldots, x_d);\ x_j = 0, \pm 1, \pm 2, \ldots;\ j = 1, 2, \ldots, d)$$

be the local time of the random walk. We say that the sphere

$$Q(N, u; d) = \left\{ x = (x_1, x_2, \ldots, x_d) : \|x - u\| = \sqrt{\sum_{i=1}^{d} (x_i - u_i)^2} \leqslant N \right\},$$

(where $u = (u_1, u_2, \ldots, u_d)$) is *covered by the random walk in time n* if

$$\xi(x, n) > 0 \qquad \text{for every } x \in Q(N, u; d).$$

Let $\tilde{R}_2(n)$ be the largest integer for which $Q(\tilde{R}_2(n), 0; 2)$ is covered in time n. The problem of describing the properties of $\tilde{R}_2(n)$ were proposed by Erdős & Taylor (1960). They also suggested that the order of magnitude of $\tilde{R}_2(n)$ is $\exp(\sqrt{\log n})$. In [1] and [3] we have proved the following result.

Theorem A *For any $\epsilon > 0$ we have*

$$\exp\left(\frac{\sqrt{\log n}}{(\log\log n)^{\frac{3}{4}+\epsilon}}\right) \leqslant \tilde{R}_2(n) \leqslant \exp(2\sqrt{\log n}\log\log\log n) \qquad a.s.$$

for all but finitely many n. Furthermore, for any $C > 0$, we have

$$\tilde{R}_2(n) \leqslant \exp(C\sqrt{\log n}) \qquad i.o., a.s.$$

Theorem A describes the area of the largest disc *around the origin* covered by the random walk $\{S_k : k \leqslant n\}$ in $\mathbb{R}^2$. In $\mathbb{R}^d$ $(d \geqslant 3)$ the analogous problem is clearly meaningless since the largest covered sphere around the origin is finite with probability one. However, one can ask in any dimension about the radius of the largest sphere (not necessarily around the origin) covered by the random walk in time n. In fact, let $R_d(n)$ be the largest integer for which there exists a random variable $u = u(n) \in \mathbb{R}^d$ such that $Q(R_d(n), u; d)$ is covered by the random walk in time n i.e.

$$\xi(x, n) > 0 \qquad \text{for any } x \in Q(R_d(n), u; d).$$

The properties of $R_d(n)$ were studied in [3], where the following result was proved.

Theorem B *For any n sufficiently large, $d \geqslant 3$ and $\epsilon > 0$, we have*

$$C(d)(\log n)^{1/d} \leqslant R_d(n) \leqslant (\log n)^{(2d-3)/(d-1)(d-2)+\epsilon} \qquad a.s.$$

for some $C(d) > 0$.

In the present paper we give a much stronger upper estimate than that of Theorem B.

Theorem 1 *For any $\epsilon > 0$ and $d \geqslant 3$*

$$R_d(n) \leqslant (\log n)^{1/(d-2)+\epsilon} \qquad a.s.$$

for all but finitely many n.

We also improve the lower bound of Theorem B.

Theorem 2 *For any $\epsilon > 0$ and $d \geqslant 3$*

$$R_d(n) \geqslant (\log n)^{1/(d-1)-\epsilon} \qquad a.s.$$

2 Proof of Theorem 1

In the proof the following lemmas will be used.

Lemma A ([3] Lemma 1) *For any $d \geqslant 3$ there exists a positive constant C_d such that*

$$P\{S_n = x \text{ eventually}\} = P(J(x) = 1) = \frac{C_d + o(1)}{R^{d-2}} \qquad (R \to \infty),$$

where $R = \|x\|$ and

$$J(x) = \begin{cases} 0 & \text{if } \xi(x,n) = 0 \text{ for every } n = 1,2,\ldots, \\ 1 & \text{otherwise.} \end{cases}$$

Lemma B *For any $0 < \alpha < 1$ and $L > 0$, there exists a sequence $x_1, x_2, \ldots, x_T$ in $\mathbb{R}^d$ such that*

$$L \leqslant \|x_i\| < L+1 \qquad (i = 1,2,\ldots,T),$$
$$\|x_i - x_j\| \geqslant L^\alpha \qquad (i,j = 1,2,\ldots,T;\, i \neq j),$$
$$T = KL^{(1-\alpha)(d-1)},$$

where $K = K(d)$ is a positive constant depending only on d.

Lemma 3 *Let*

$$\alpha = \alpha_1 = \alpha_1(k) = \frac{D^{k+1} - D^k}{D^{k+1} - 1} + \epsilon,$$

where

$$0 < \epsilon < \frac{D^k - 1}{D^{k+1} - 1} \quad \left(D = \frac{d-1}{d-2},\, k = 1,2,\ldots\right)$$

and define T and $x_1, x_2, \ldots, x_T$ as in Lemma B. Then, for any L large enough, we have

$$P\{Q(L,0;d) \text{ is covered eventually}\}$$
$$\leqslant P\{x_1, x_2, \ldots, x_t \text{ are covered eventually}\} \leqslant e^{-(T-1)}.$$

Consequently, for any $n > 0$ and L large enough, we have

$$P\{\exists\, u = u(n) \in Z^d \text{ such that } S_n \in Q(L,u;d)$$
$$\text{and } Q(L,u;d) \text{ is covered eventually}\} \leqslant C^* L^d e^{-(T-1)}.$$

Proof Define the sequence $\alpha_1, \alpha_2, \ldots, \alpha_k$ $(k = 1,2,\ldots)$ by

$$\alpha_i = \frac{D^{k+1} - D^{k+1-i}}{D^{k+1} - 1} \qquad (i = 2,3,\ldots,k)$$

and

$$\alpha_1 = \frac{D^{k+1}-D^k}{D^{k+1}-1}+\epsilon.$$

Assuming that

$$0<\epsilon<\frac{D^k-D^{k-1}}{D^{k+1}-1}<\frac{D^k-1}{D^{k+1}-1},$$

we have

$$0<\alpha_1<\alpha_2<\cdots<\alpha_k<1 \qquad (k=1,2,\ldots)$$

and

$$(\alpha_{i+1}-\alpha_1)(d-1)<\alpha_i(d-2) \qquad (i=1,2,\ldots,k),$$

where $\alpha_{k+1}=1$.

Let $x_{i_1},x_{i_2},\ldots,x_{i_T}$ be an arbitrary permutation of $x_1,x_2,\ldots,x_T$. Consider the consecutive distances

$$\|x_{i_2}-x_{i_1}\|, \qquad \|x_{i_3}-x_{i_2}\|, \qquad \ldots, \qquad \|x_{i_T}-x_{i_{T-1}}\|.$$

Let the number of these distances between L^{α_1} and L^{α_2} be l_i ($i=1,\ldots,k-1$) and let the number between L^{α_k} and $2L$ be l_k. Then, by Lemma A, the probability that the random walk visits the points $x_{i_1},x_{i_2},\ldots,x_{i_T}$ in this given order is less than or equal to

$$\left(\frac{C_d+o(1)}{L^{\alpha_1(d-2)}}\right)^{l_1}\left(\frac{C_d+o(1)}{L^{\alpha_2(d-2)}}\right)^{l_2}\cdots\left(\frac{C_d+o(1)}{L^{\alpha_k(d-2)}}\right)^{l_k}.$$

Noting the fact that the number of those j ($j=1,2,\ldots,T$) for which

$$L^{\alpha_i}\leqslant\|X_j-X_s\|<L^{\alpha_{i+1}} \qquad (s=1,2,\ldots,T)$$

is less than or equal to

$$KL^{(\alpha_{i+1}-\alpha_1)(d-1)} \qquad (i=1,2,\ldots,k),$$

we have

$P\{x_1,x_2,\ldots,x_T$ are covered eventually$\}$

$$\leqslant \sum_{l_1+l_2+\cdots+l_{k=T-1}}\frac{(T-1)!}{l_1!\,l_2!\cdots l_k!}\prod_{i=1}^{k}\left(\frac{[C_d+o(1)]KL^{(\alpha_{i+1}-\alpha_1)(d-1)}}{L^{\alpha_i(d-2)}}\right)^{l_i}$$

$$=\left(\sum_{i=1}^{k}\frac{[C_d+o(1)]KL^{(\alpha_{i+1}-\alpha_1)(d-1)}}{L^{\alpha_i(d-2)}}\right)^{T-1}\leqslant \mathrm{e}^{-(T-1)},$$

if L is large enough. This comletes the proof of Lemma 3. □

Proof of Theorem 1 Let $L = [(\log n)^{\theta}]$ with

$$\theta = \theta_k = \frac{1}{(d-1)(1-\alpha)} + \epsilon = \frac{1}{d-1} \frac{1}{\dfrac{D^{k-1}}{D^{k+1}-1} - \epsilon} + \epsilon.$$

Then $T \geqslant (\log n)^{\psi}$ with some $\psi > 1$ and so we obtain Theorem 1, noting that

$$\lim_{k\to\infty} \theta_k = \frac{1}{(d-2)-\epsilon(d-1)} + \epsilon. \qquad \square$$

3 Proof of Theorem 2

Lemma 4 *There exists a constant $K > 0$ such that*

$$P\{\xi(x,n) > 0\} \geqslant (\tfrac{1}{2}C_d)R^{2-d} \qquad (R = \|x\|)$$

if $n \geqslant KR^2$, where C_d is the constant in Lemma A.

Proof It is easy to see that

$$\sum_{n=KR^2}^{\infty} P\{S_n = x\} \leqslant \frac{C_d}{2R^{d-2}}$$

if K is large enough. This implies Lemma 4. $\square$

Let

$$L = L(n) = [(\log n)^{(d-2)/(d-1)}]$$

and define

$$\tau_1 = \tau_1(n) < \psi_1 = \psi_1(n) < \tau_2 = \tau_2(n) < \psi_2 = \psi_2(n) < \cdots < \tau_L = \tau_L(n).$$

by

$$\tau_1 = n + [(\log n)^{2/(d-1)}], \qquad \psi_1 = \inf\{k : k > \tau_1,\ S_n = S_k\},$$
$$\tau_2 = \psi_1 + [(\log n)^{2/(d-1)}], \qquad \psi_2 = \inf\{k : k > \tau_2,\ S_n = S_k\}, \qquad \ldots.$$

Clearly, with a positive probability (depending on n) ψ_1 is not defined. However, we have the following lemma.

Lemma 5 *There exists a constant $C > 0$ such that*

$$P\{\max_{1\leqslant i\leqslant L} \psi_i \leqslant C(\log n)^{(d-1)^{-1}}\} \geqslant n^{-1/2}.$$

Proof Clearly, for any $C > 0$ there exists a constant $0 < p = p(C) < 1$ such that

$$P\{\|S_{\tau_1} - S_n\| \leqslant C(\log n)^{1/(d-1)}\} \geqslant p > 0.$$

Since

$$
\begin{aligned}
P\{\psi_1 \leqslant C(\log n)^{1/(d-1)}\} \\
&\geqslant P\{\psi_1 \leqslant C(\log n)^{1/(d-1)} \mid \|S_{\tau_1} - S_n\| \leqslant C(\log n)^{1/(d-1)}\} \\
&\qquad \times P\{\|S_{\tau_1} - S_n\| \leqslant C(\log n)^{1/(d-1)}\} \\
&\geqslant \left(\frac{1}{2d}\right)^{C(\log n)^{1/(d-1)}} p
\end{aligned}
$$

and $\psi_1, \psi_2, \ldots$ are independent random variables, we have

$$
P\{\max_{1\leqslant i\leqslant L} \psi_i \leqslant C(\log n)^{1/(d-1)}\} \geqslant \left(\left(\frac{1}{2d}\right)^{C(\log n)^{1/(d-1)}} p\right)^L \geqslant n^{-1/2}
$$

for a suitable $C > 0$. Hence we have Lemma 5. □

Proof of Theorem 2 Let

$$
A = A(n) = \{\max_{1\leqslant i\leqslant L} \psi_i \leqslant C(\log n)^{1/(d-1)}\}.
$$

and let $x \in \mathbb{R}^d$ be an arbitrary latice point for which

$$
\|x - S_n\| \leqslant (\log n)^{1/(d-1)-\epsilon}.
$$

Then

$$
\begin{aligned}
P\{\xi(x, n + [(\log n)^{2/(d-1)}]) - \xi(x, n) = 0 \mid A(n)\} \\
\leqslant 1 - \frac{C_d}{2(\log n)^{\{(d-1)^{-1}-\epsilon\}(d-2)}}.
\end{aligned}
$$

Hence the conditional probability (given $A(n)$) that x is not covered is less than or equal to

$$
(1 - \tfrac{1}{2}C_d(\log n)^{\{\epsilon-(d-1)^{-1}\}(d-2)})^L \leqslant \exp(-\tfrac{1}{2}C_d(\log n)^{\epsilon(d-2)}).
$$

Consequently the conditional probability that there exists a point $x \in Q((\log n)^{(d-1)^{-1}-\epsilon}, S_n; d)$ being not covered is less than or equal to

$$
C(\log n)^{d\{(d-1)^{-1}-\epsilon\}} \exp(-\tfrac{1}{2}C_d(\log n)^{\epsilon(d-2)}).
$$

This fact together with Lemma 5 proves Theorem 2. □

References

[1] P. Erdős & P. Révész, On the area of the circles covered by a random walk. *J. Mult. Anal.*, **27** (1988), 169–80

[2] P. Erdős & S. J. Taylor, Some problems concerning the structure of random walk paths, *Acta Math. Acad. Sci. Hung.*, **11** (1960), 137–62

[3] P. Révész, Simple symmetric random walk in $\mathbb{R}^d$ (1989), in *Almost Everywhere Convergence*, Proceedings of the International Conference on Almost Everywhere Convergence in Probability and Ergodic Theory (ed. G. A. Edgar-L. Sucheston). Academic Press, Boston

Special Lucas sequences, including the Fibonacci sequence, modulo a prime

A. Schinzel

We shall consider Lucas sequences defined by the conditions

$$u_0 = 0, \qquad u_1 = 1, \qquad u_n = au_{n-1} + u_{n-2} \quad (a \in \mathbb{Z})$$

and for a fixed prime p the sequence $\bar{u}_n$ is defined by the same condition in $\mathbb{F}_p$, the finite field with p elements. Let $k = k(p)$ be the length of the shortest period of the latter sequence. We shall prove the following result.

Theorem *Let $S = S(p)$ be the set of frequencies with which different residues occur in the sequence $\bar{u}_n$ $(0 \leqslant n < k(p))$. For $p > 7$ and $p \nmid a(a^2+4)$ we have*

$$S = \begin{cases} \{0,1,2\} \text{ or } \{0,1,2,3\} & \text{if } k(p) \not\equiv 0 \pmod 4, \\ \{0,2,4\} & \text{if } k(p) \equiv 4 \pmod 8, \\ \{0,1,2\} \text{ or } \{0,2,3\} \\ \quad \text{or } \{0,1,2,4\} \text{ or } \{0,2,3,4\} & \text{if } k(p) \equiv 0 \pmod 8. \end{cases}$$

Corollary 1 *If $p > 7$ and $p \nmid a^2+4$ then at least one residue modulo p does not occur in the sequence $\bar{u}_n$.*

Corollary 2 *If $p \nmid a(a^2+4)$ then at least one residue modulo p occurs exactly twice in the shortest period of the sequence $\bar{u}_n$.*

Corollary 3 *If $a = 1$ and $p > 7$*

$$S = \begin{cases} \{0,1,2,3\} & \text{if } k(p) \not\equiv 0 \pmod 4, \\ \{0,2,4\} & \text{if } k(p) \equiv 4 \pmod 8, \\ \{0,1 2,4\} \text{ or } \{0,2,3,4\} & \text{if } k(p) \equiv 0 \pmod 8. \end{cases}$$

In the case $a = 1$ corresponding to the Fibonacci sequence, Corollary 1 has been proved by Shah [2] for $p \equiv 1 \pmod 4$ and by Brückner [1] for $p \equiv 3 \pmod 4$; Corollary 3 has been conjectured, in a somewhat weaker form, by E. Jacobson. It has been this last conjecture, proposed during the First Conference of the Canadian Number Theory Association in April 1988, which prompted the present work.

L. Somer, in the paper [4] presented at the Third International Conference on Fibonacci Numbers in July 1988, has proved Corollary 1 except for the case where simultaneously $k(p) \equiv 4 \pmod 8$, $a \not\equiv \pm 1 \pmod p$ and $p \equiv 1$ or $9 \pmod{20}$. The results of [4] also cover the case $k(p) \equiv 0 \pmod 8$ of the theorem and furnish the following conclusions:

$$\{1\} \subset S \subset \{0,1,2,3\} \qquad \text{if } k(p) \not\equiv 0 \pmod 4,$$

$$\{4\} \subset S \subset \{0,2,4\} \qquad \text{if } k(p) \equiv 4 \pmod 8.$$

Finally, one can extract from [4] the following result: if in the definition of u_n the condition $u_n = au_{n-1}+u_{n-2}$ is replaced by $u_n = au_{n-1}-u_{n-2}$ then, for $p > 3$ and $p \nmid a^2-4$,

$$S = \begin{cases} \{0,1\} & \text{if } k(p) \equiv 1 \pmod 2, \\ \{0,2\} & \text{if } k(p) \equiv 2 \pmod 4, \\ \{0,12\} & \text{if } k(p) \equiv 0 \pmod 4. \end{cases}$$

It does not seem possible to obtain similar theorems for Lucas sequences defined by the conditions

$$u_0 = 0, \qquad u_1 = 1, \qquad u_n = au_{n-1}+bu_{n-2} \quad (b \neq 0, \pm 1).$$

Let $D = a^2+4$, ξ be a zero of x^2-ax-1 in the finite field $\mathbb{F}_q$, where $q = p$ if $\left(\frac{D}{p}\right) = 1$ and $q = p^2$ if $\left(\frac{D}{p}\right) = -1$ (we exclude the case $p|D$). For $\bar{u}_n$ we have the formula

$$\bar{u}_n = \frac{\xi^n-(-\xi^{-1})^n}{\xi+\xi^{-1}}. \tag{1}$$

Let δ be the least positive exponent such that

$$\xi^\delta = 1.$$

Of the following seven lemmata at least three (Lemma 1, Lemma 4 and Lemma 5) can be deduced from published results, (e.g. Theorem 13 of [3]). However, the deductions would not be much shorter than the following proofs.

Lemma 1 *For $p \nmid 2D$ we have $k(p) = [\delta, 2]$.*

Proof We shall show first that $k \equiv 0 \pmod 2$. Indeed, otherwise we infer from $\bar{u}_k \equiv 0$, $\bar{u}_{k+1} \equiv 1$ and (1) that

$$\xi^k+\xi^{-k} = 0, \qquad \xi^{k+1}-\xi^{-k-1} = \xi+\xi^{-1}.$$

Hence

$$(\xi+\xi^{-1})(\xi^{-k}+1) = -\xi^{1-k}-\xi^{-1-k}-\xi-\xi^{-1} = 0,$$

from which we deduce in turn that

$$\xi^{-k}+1 = 0; \quad 2 \equiv 0 \pmod p; \quad p = 2.$$

For $k \equiv 0 \pmod 2$ the conditions $\bar{u}_k \equiv 0$ and $\bar{u}_{k+1} \equiv 1$ are equivalent to

$$\xi^k - \xi^{-k} = 0; \quad \xi^{k+1} - \xi^{-k-1} = \xi - \xi^{-1},$$

i.e.

$$(\xi^k - 1)(\xi^k + 1) = 0, \quad (\xi^k + \xi^{-1})(\xi^k - 1) = 0,$$

whence, in turn,

$$\xi^k = 1; \quad \delta | k.$$

Since k is the length of the shortest period, the lemma follows. □

Lemma 2 *Let $p \nmid 2D$. The conditions*

$$n \equiv m \pmod 2 \quad \text{and} \quad \bar{u}_n = \bar{u}_m \tag{2}$$

hold if and only if either $n \equiv m \pmod k$ or $n \equiv m \equiv 1 \pmod 2$ and $n+m \equiv 0 \pmod k$ or $k \equiv 0 \pmod 4$, $n \equiv m \equiv 0 \pmod 2$ and $n+m \equiv \frac{1}{2}k \pmod k$.

Proof We obtain from (1) and (2)

$$\xi^n - (-\xi^{-1})^n = \xi^m - (-\xi^{-1})^m.$$

Hence

$$(\xi^n - \xi^m)(1+[-1]^n \xi^{-n-m}) = 0.$$

Thus either $\xi^{n-m} = 1$, i.e. $n-m \equiv 0 \pmod \delta$, or $\xi^{-n-m} = (-1)^{n+1}$, i.e., $\delta(n+1) \equiv 0 \pmod 2$, whence $n+m \equiv \frac{1}{2}\delta(n+1) \pmod \delta$.

Since $n-m \equiv n+m \equiv 0 \pmod 2$, in view of Lemma 1 the first possibility gives $n \equiv m \pmod k$ and the second possibility gives either $n \equiv 1 \pmod 2$ and $n+m \equiv 0 \pmod k$ or $k = \delta \equiv 0 \pmod 4$, $n \equiv 0 \pmod 2$ and $n+m \equiv \frac{1}{2}k \pmod k$. The converse implication is trivial. □

Lemma 3 *Let $p \nmid 2D$. The conditions*

$$n \equiv m \pmod 2 \quad \text{and} \quad \bar{u}_n = -\bar{u}_m \tag{3}$$

are equivalent to

$$n \equiv m \equiv 0 \pmod 2 \quad \text{and} \quad n+m \equiv 0 \pmod k \quad \text{if } k \equiv 2 \pmod 4$$

and

$$n \equiv m+\tfrac{1}{2}k \pmod k \qquad \text{and} \qquad \bar{u}_n = \bar{u}_{m+\frac{1}{2}k} \qquad \text{if } k \equiv 0 \pmod 4.$$

Proof If $k \equiv 0 \pmod 4$, we have by Lemma 1 that $\delta = k$ and $\xi^{\frac{1}{2}k} = -1$, and by (1) that $\bar{u}_{m+\frac{1}{2}k} = -\bar{u}_m$.

If $k \equiv 2 \pmod 4$, we obtain from (1) and (3)

$$\xi^n-(-\xi^{-1})^n = -\xi^m+(-\xi^{-1})^m,$$

whence

$$(\xi^n+\xi^m)(1-[-1]^n\xi^{-n-m}) = 0.$$

Thus either $\delta \equiv 0 \pmod 2$ and $n-m \equiv \frac{1}{2}\delta \pmod \delta$, or $\delta n \equiv 0 \pmod 2$ and $n+m \equiv \frac{1}{2}\delta n \pmod \delta$. The first possibility contradicts $\delta = k \equiv 2 \pmod 4$ and the second gives $n \equiv 0 \pmod 2$ and $n+m \equiv 0 \pmod k$. The converse implication is trivial. □

Lemma 4 *Let $p \nmid 2D$. We have $\bar{u}_n \equiv 0$ if and only if either $k \equiv 2 \pmod 2$ and $n \equiv 0 \pmod k$ or $k \equiv 4 \pmod 8$ and $n \equiv 0 \pmod{\frac{1}{4}k}$ or $k \equiv 0 \pmod 8$ and $n \equiv 0 \pmod{\frac{1}{2}k}$.*

Proof By Lemma 3 with $n = m$, the condition $\bar{u}_n = 0$ is equivalent to $n \equiv 0 \pmod 2$ and $2n \equiv 0 \pmod k$ if $k \equiv 2 \pmod 4$ or to $\bar{u}_n = \bar{u}_{n+\frac{1}{2}k}$ if $k \equiv 0 \pmod 4$. Since in the latter case $n \equiv n+\frac{1}{2}k \pmod 2$, an application of Lemma 2 gives the assertion. □

Lemma 5 *Let $p \nmid 2D$. We have*

$$k\,|\,p-1 \quad \text{if} \left(\frac{D}{p}\right) = 1; \qquad k\,|\,2(p+1) \quad \text{if} \left(\frac{D}{p}\right) = -1.$$

Proof If $\left(\frac{D}{p}\right) = 1$, we have $\xi^{p-1} = 1$. Hence $\delta\,|\,p-1$ and, by Lemma 1, $k\,|\,p-1$. If $\left(\frac{D}{p}\right) = -1$, we have $-\xi^{-1} = \xi^p$. Hence $\xi^{2(p+1)} = 1$, i.e. $\delta\,|\,2(p+1)$ and, by Lemma 1, $k\,|\,2(p+1)$. □

Lemma 6 *If $k = 2(p+1) \equiv 0 \pmod 8$ then for every integer e there is an n such that*

$$\bar{u}_{n+e} = \bar{u}_n. \tag{4}$$

Proof If $\bar{u}_e = 0$ it suffices to take $n = 0$. If $\bar{u}_e \neq 0$ we use that identity

$$u_n u_{m+e} - u_m u_{n+e} = (-1)^m u_e u_{n-m}$$

and find, by virtue of Lemma 4, that the quotients

$$\frac{\bar{u}_{n+e}}{\bar{u}_n} \qquad \text{for } 0 < n < \tfrac{1}{2}k$$

are all distinct. Since $\frac{1}{2}k = p+1$, we have p distinct elements of $\mathbb{F}_p$. One of them must be 1, which gives (4). □

Remark For the case $a = 1$ the lemma has been proved by Brückner [1]. The above proof follows the same pattern. There is also a similar argument in [4].

Lemma 7 *Let* $p \nmid 2aD$. *We have*

$$D\sum_{j=0}^{\frac{1}{2}k-1} \bar{u}_{2j}^2 = -k, \qquad D\sum_{j=0}^{\frac{1}{2}k-1} \bar{u}_{2j+1}^2 = k, \qquad D^2\sum_{j=0}^{k-1} \bar{u}_j^4 = 6k.$$

Proof Since $p \nmid D$ the formula (1) applies and we find that

$$D\sum_{j=0}^{\frac{1}{2}k-1} \bar{u}_{2j}^2 = \sum_{j=0}^{\frac{1}{2}k-1} (\xi^{2j}-\xi^{-2j})^2 = \sum_{j=0}^{\frac{1}{2}k-1} \xi^{4j} - 2\times\tfrac{1}{2}k + \sum_{j=0}^{\frac{1}{2}k-1} \xi^{-4j}.$$

We have $\xi^4 \neq 1$, since otherwise $\xi^2 = \pm 1$ which for $p \nmid aD$ is incompatible with $\xi^2 - a\xi - 1 = 0$. Hence

$$\sum_{j=0}^{\frac{1}{2}k-1} \xi^{4j} = \frac{\xi^{2k}-1}{\xi^4-1} = 0 = \frac{\xi^{-2k}-1}{\xi^{-4}-1} = \sum_{j=0}^{\frac{1}{2}k-1} \xi^{-4j}$$

and so

$$D\sum_{j=0}^{\frac{1}{2}k-1} \bar{u}_{2j}^2 = -k.$$

Similarly

$$\begin{aligned} D\sum_{j=0}^{\frac{1}{2}k-1} \bar{u}_{2j+1}^2 &= \sum_{j=0}^{\frac{1}{2}k-1} (\xi^{2j+1}+\xi^{-2j-1})^2 \\ &= \sum_{j=0}^{\frac{1}{2}k-1} \xi^{4j+2} + 2\times\tfrac{1}{2}k + \sum_{j=0}^{\frac{1}{2}k-1} \xi^{-4j-2} \\ &= \xi^2\frac{\xi^{2k}-1}{\xi^4-1} + k + \xi^{-2}\frac{\xi^{-2k}-1}{\xi^{-4}-1} = k \end{aligned}$$

and

$$\begin{aligned} D^2\sum_{j=0}^{k-1} \bar{u}_j^4 &= \sum_{j=0}^{k-1} (\xi^j - [-\xi^{-1}]^j)^4 \\ &= \sum_{j=0}^{k-1} \xi^{4j} - 4\sum_{j=0}^{k-1} (-\xi^2)^j + 6k - 4\sum_{j=0}^{k-1} (-\xi^{-2})^j + \sum_{j=0}^{k-1} \xi^{4j} \\ &= \frac{\xi^{4k}-1}{\xi^4-1} - 4\frac{\xi^{2k}-1}{-\xi^2-1} + 6k - 4\frac{\xi^{-2k}-1}{-\xi^{-2}-1} + \frac{\xi^{-4k}-1}{\xi^{-4}-1} = 6k. \end{aligned}$$ □

Proof of the theorem We shall consider successively the cases $k \equiv 2 \pmod 4$, $k \equiv 4 \pmod 8$ and $k \equiv 0 \pmod 8$.

If $k \equiv 2 \pmod 4$ we have, by Lemma 5, $k \mid p \pm 1$: thus $k \leqslant p+1$. Moreover, by Lemma 2 for all odd n,

$$\bar{u}_n = \bar{u}_{k-n}. \tag{5}$$

It follows that the number of distinct $\bar{u}_n$ for $n = 0, 1, \dots, k-1$ does not exceed

$$\tfrac{1}{2}k + \tfrac{1}{2}(\tfrac{1}{2}k+1) \leqslant \tfrac{1}{4}(3k+2) \leqslant \tfrac{1}{4}(3p+5) < p \qquad (p \geqslant 7),$$

which shows that $0 \in S$.

By Lemma 4, 0 occurs exactly once among the $\bar{u}_k$ $(0 \leqslant n < k)$: thus $1 \in S$.

By Lemma 2 the conditions

$$n \equiv 1 \pmod 2, \qquad \bar{u}_n = \bar{u}_{\frac{1}{2}k}$$

imply $n \equiv \frac{1}{2}k \pmod k$ and the conditions

$$n \equiv 0 \pmod 2, \qquad \bar{u}_n = \bar{u}_{\frac{1}{2}k} \quad (0 \leqslant n < k) \tag{6}$$

have at most one solution. Thus if (6) is soluble, $\bar{u}_{\frac{1}{2}k}$ occurs exactly twice in the sequence $\bar{u}_n$ $(0 \leqslant n < k)$ and so $2 \in S$.

On the other hand, if for $n_1 \equiv 1 \pmod 2$ $(0 < n_1 < k,\ n_1 \neq \frac{1}{2}k)$ the system

$$n \equiv 0 \pmod 2, \qquad \bar{u}_n = \bar{u}_{n_1}$$

is insoluble, $\bar{u}_{n_1}$ occurs exactly twice in the sequence $\bar{u}_n$ $(0 \leqslant n < k)$, namely for $n = n_1$ and $n = k - n_1$. Thus $2 \in S$. Therefore, assuming that $2 \notin S$ we infer that $\bar{u}_{\frac{1}{2}k}$ does not occur among the $\bar{u}_{2j}$, and $\bar{u}_{2i+1}$ occurs among the $\bar{u}_{2j}$ $(0 < 2j < k)$ for all $2i+1 \not\equiv \frac{1}{2}k \pmod k$.

By Lemmas 2 and 3, every residue which occurs in the sequence $\bar{u}_{2j+1}^2$ $(0 < 2j+1 < k,\ 2j+1 \neq \frac{1}{2}k)$ occurs in the sequence $\bar{u}_{2j}^2$ $(0 < 2j < k)$ and in both sequences exactly twice. Hence one sequence is a permutation of the other. Thus

$$\sum_{j=0}^{\frac{1}{2}k-1} \bar{u}_{2j+1}^2 - \bar{u}_{\frac{1}{2}k}^2 = \sum_{j=0}^{\frac{1}{2}k-1} \bar{u}_{2j}^2$$

and, by Lemma 7,

$$k - D\bar{u}_{\frac{1}{2}k}^2 = -k, \qquad \text{whence} \qquad D\bar{u}_{\frac{1}{2}k}^2 = 2k.$$

However, by (1),

$$D\bar{u}^2_{\frac{1}{2}k} = (\xi^{\frac{1}{2}k}+\xi^{-\frac{1}{2}k})^2 = \xi^k+2+\xi^{-k} = 4.$$

Thus $2k \equiv 4 \pmod p$, i.e. $k \equiv 2 \pmod p$, and, since $k \leqslant p+1$,

$$k = 2, \qquad \text{whence} \qquad a = u_2 \equiv 0 \pmod p,$$

contrary to the assumption.

The obtained contradiction shows that $2 \in S$. By virtue of Lemma 2, every residue occurs at most once among the $\bar{u}_{2j}$ ((0 $\leqslant 2j < k$) and at most twice among the $\bar{u}_{2j+1}$ $(0 < 2j+1 < k)$. Hence

$$S = \{0,1,2\} \quad \text{or} \quad \{0,1,2,3\}.$$

Consider now the case $k(p) \equiv 4 \pmod 8$. Then by Lemma 4

$$\bar{u}_n = 0, \qquad \text{for } n = 0, \tfrac{1}{4}k, \tfrac{1}{2}k, \tfrac{3}{4}k \pmod k \text{ exclusively.}$$

Thus $4 \in S$. Further, by Lemma 2, $\bar{u}_n = \bar{u}_m$ and $n \equiv m \pmod 2$ holds if and only if either $n \equiv m \pmod k$ or $n \equiv 0 \pmod 2$ and $m \equiv \frac{1}{2}k-n \pmod k$ or $n \equiv 1 \pmod 2$ and $m \equiv k-n \pmod k$.

Note that the last two options are impossible with $m = n$. Thus $S \subset \{0,2,4\}$. If $2 \notin S$ then every residue which occurs among the $\bar{u}_{2j+1}$ $(0 < 2j+1 < k)$ occurs among the $\bar{u}_{2j}$ $(0 < 2j < k)$ and in both sequences exactly twice. Hence one sequence is a permutation of the other and

$$\sum_{j=0}^{\frac{1}{2}k-1} \bar{u}^2_{2j} = \sum_{j=0}^{\frac{1}{2}k-1} \bar{u}^2_{2j+1}.$$

By Lemma 7 this gives

$$-k \equiv k \pmod p, \qquad \text{whence} \qquad 2k \equiv 0 \pmod p,$$

which contradicts Lemma 5.

If $0 \in S$ then, since every residue occurs in the sequence $\bar{u}_j$ $(0 \leqslant j < k)$ at least twice and 0 occurs 4 times, we have by Lemma 5, that $2p+2 \geqslant k \geqslant 2(p-1)+4$, whence $k = 2p+2$ and every non-zero residue occurs in the sequence $\bar{u}_j$ exactly twice. Thus $\bar{u}_j$ $(0 \leqslant j < k,\ \bar{u}_j \neq 0)$ is a permutation of $(1,1,2,2,\dots,p-1,p-1)$ and we obtain

$$\sum_{j=0}^{k-1} \bar{u}_j^4 = 2 \sum_{j\in\mathbb{F}_p} j^4 = 0.$$

By Lemma 7 this gives $6k \equiv 0 \pmod p$, contrary to Lemma 5. The obtained contradictions show that $2 \in S$ and $0 \in S$: thus

$$S = \{0,2,4\}.$$

It remains to consider the case $k(p) \equiv 0 \pmod 8$. In this case the conditions $n \equiv 1 \pmod 2$ and $n \equiv k-n \pmod k$ are impossible, but the conditions $n \equiv 0 \pmod 2$ and $n \equiv \frac{1}{2}k-n \pmod k$ are possible, namely for $n = \frac{1}{4}k$ and $n = \frac{3}{4}k$. Thus every residue which occurs among the $\bar{u}_j$ $(0 \leqslant j < k)$ occurs either twice or four times except for $\bar{u}_{\frac{1}{4}k}$ and $\bar{u}_{\frac{3}{4}k}$, which occur either twice or once, depending on whether the conditions

$$n \equiv 1 \pmod 2, \quad \bar{u}_n = \bar{u}_{\frac{1}{4}k}$$

are soluble or not. (Note that $\bar{u}_n = \bar{u}_{\frac{1}{4}k}$ if and only if $\bar{u}_{\frac{1}{2}k+n} = \bar{u}_{\frac{3}{4}k}$.) Since, by Lemma 4, 0 occurs among the $\bar{u}_j$ $(0 \leqslant j < k)$ exactly twice, we have $2 \in S$ and it remains to show that $0 \in S$. Supposing the contrary, we infer that $k \geqslant 2+2(p-2) = 2p-2$, whence, by Lemma 5, $k = 2p+2$ and, by Lemma 6, the equation

$$\bar{u}_{n+1} = \bar{u}_n$$

is soluble. If $n_i \equiv i \pmod 2$ $(i = 0$ or $1)$ is a solution we have, by Lemma 2,

$$\bar{u}_n = \bar{u}_{n_i} \quad \text{if and only if} \quad n \equiv n_i, n_{i+1}, \tfrac{1}{2}k-n_i-i, k-n_i+i-1 \pmod k. \tag{8}$$

Let us choose $e \equiv 1 \pmod 2$ such that

$$e \not\equiv \pm 1, \pm(2n_i+1) \pmod{\tfrac{1}{2}k}. \tag{9}$$

Such choice is possible since $\frac{1}{2}k = p+1 > 8$. By Lemma 6 there exists an n_e such that

$$\bar{u}_{n_e+e} = \bar{u}_{n_e}. \tag{10}$$

In view of (8), (9) and (10), we have

$$\bar{u}_{n_e+e} \neq \bar{u}_{n_i} \tag{11}$$

and, since by (10) and Lemma 3

$$\bar{u}_{\frac{1}{2}k+n_e+e} = -\bar{u}_{n_e+e} = -\bar{u}_{n_e} = \bar{u}_{\frac{1}{2}k+n_e},$$

also

$$\bar{u}_{\frac{1}{2}k+n_e} \neq \bar{u}_{n_i}. \tag{12}$$

In view of (11) and (12), we have either $\bar{u}_{n_i} \neq \pm\bar{u}_{\frac{1}{4}k}$ or $\bar{u}_{n_e} \neq \pm\bar{u}_{\frac{1}{4}k}$. Thus among the residues $\bar{u}_j$ $(0 \leqslant j < k)$ we have either two residues $(\pm\bar{u}_{\frac{1}{4}k})$ occurring thrice, two residues $(\pm\bar{u}_{n_i}$ or $\pm\bar{u}_{n_e})$ occurring four times and the remaining residues occurring at least twice or two residues

($\pm\bar{u}_{\frac{1}{4}k}$) occurring once, four residues ($\pm\bar{u}_{n_i}$ and $\pm\bar{u}_{n_e}$) occurring four times and the remaining residues occurring at least twice. This gives

$$2(p+1) = k \geqslant \min\{3\times2+4\times2+2(p-4), 2+4\times4+2(p-6)\} = 2(p+3),$$

a contradiction. Therefore $0 \in S$ and the proof is complete. □

Proof of Corollary 1 For $p \nmid a$ the corollary follows directly from the theorem. For $p \mid a$ we have

$$\bar{u}_n = 0 \text{ or } 1.$$

Hence, for $p > 2$, $0 \in S$. □

Proof of Corollary 2 For $p > 7$ the corollary follows directly from the theorem. For $p = 3$, 5 or 7, the proof given above for $2 \in S$ applies, provided that $p \nmid a(a^2+4)$. Finally, for $p = 2$ and a odd, the residue 1 has the desired property. □

Proof of Corollary 3 For $a = 1$ we have $u_2 = 1$ and hence, by Lemma 2,

$$\bar{u}_n = 1 \quad \text{for } n \equiv 1, 2, k-1 \pmod{k} \quad \text{if } k \equiv 2 \pmod{4};$$

$$\bar{u}_n = 1 \quad \text{for } n \equiv 1, 2, \tfrac{1}{2}k-2, k-1 \pmod{k} \quad \text{if } k \equiv 0 \pmod{4}.$$

This gives $3 \in S$ if $k \equiv 2 \pmod{4}$ and $4 \in S$ if $k \equiv 0 \pmod{4}$. □

References

[1] G. Brückner, Fibonacci sequence modulo a prime $p \equiv 3 \pmod{4}$, *Fibonacci Quart.*, **8** (1970), 217–20
[2] A. P. Shah, Fibonacci sequence modulo m, *Fibonacci Quart.*, **6** (1968), 139–41
[3] L. Somer, The divisibility properties of primary Lucas recurrences with respect to primes, *Fibonacci Quart.*, **18** (1980), 316–34
[4] L. Somer, Distribution of residues of certain second-order linear recurrences modulo p (to appear)

A remark on the heights of subspaces

Wolfgang M. Schmidt

We define the height of an algebraic subspace as in [3] or [1]. Let K be an algebraic number field, $M(K)$ the set of absolute values of K which extend the standard or a p-adic absolute value $\mathbb{Q}$ and, for $v \in M(K)$, let n_v be the local degree, so that we have the product formula $\prod_v |c|_v^{n_v} = 1$ for $c \neq 0$ in K. Here and below, the product is over $v \in M(K)$. Given a vector $\boldsymbol{A} = (A_1, \ldots, A_l)$ in K^l, we put

$$|\boldsymbol{A}|_v = \begin{cases} \max(|A_1|_v, \ldots, |A_l|_v) & \text{when } v \text{ is Archimedean,} \\ \sqrt{|A_1|_v^2 + \cdots + |A_l|_v^2} & \text{when } v \text{ is non-Archimedean.} \end{cases}$$

When $\boldsymbol{A} \neq \boldsymbol{0}$, set

$$H(\boldsymbol{A}) = \prod_v |\boldsymbol{A}|_v^{n_v}.$$

By the product formula, $H(c\boldsymbol{A}) = H(\boldsymbol{A})$ for $c \neq 0$ in K. When S is a subspace of K^n of some dimension $p > 0$, spanned by $\boldsymbol{a}_1, \ldots, \boldsymbol{a}_p$, then the exterior product $\boldsymbol{A} = \boldsymbol{a}_1 \wedge \cdots \wedge \boldsymbol{a}_p$ lies in K^l with $l = \binom{n}{p}$. Moreover, $\boldsymbol{A} \neq \boldsymbol{0}$ and $\boldsymbol{A}$ is determined by S up to a factor of $c \neq 0$. The height of S is defined by

$$H(S) = H(\boldsymbol{A}).$$

When S is of dimension 0, set $H(S) = 1$.

The following first appeared in hand-written and photo-copied lecture notes in the fall of 1987.

Proposition *Let S and T be subspaces of K^n. Then*

$$H(S+T)\,H(S \cap T) \leqslant H(S)\,H(T).$$

Proof Let $\boldsymbol{u}_1, \ldots, \boldsymbol{u}_r$ be a basis of $S \cap T$, further $\boldsymbol{u}_1, \ldots, \boldsymbol{u}_r, \boldsymbol{x}_1, \ldots, \boldsymbol{x}_s$ a basis of S and $\boldsymbol{u}_1, \ldots, \boldsymbol{u}_r, \boldsymbol{y}_1, \ldots, \boldsymbol{y}_t$ a basis of T. It will suffice to show that for each $v \in M(K)$ we have a local inequality

$$|\boldsymbol{u}_1\wedge\cdots\wedge\boldsymbol{u}_r\wedge\boldsymbol{x}_1\wedge\cdots\wedge\boldsymbol{x}_s\wedge\boldsymbol{y}_1\wedge\cdots\wedge\boldsymbol{y}_t|_v\,|\boldsymbol{u}_1\wedge\cdots\wedge\boldsymbol{u}_r|_v \leqslant |\boldsymbol{u}_1\wedge\cdots\wedge\boldsymbol{u}_r\wedge\boldsymbol{x}_1\wedge\cdots\wedge\boldsymbol{x}_s|_v\,|\boldsymbol{u}_1\wedge\cdots\wedge\boldsymbol{u}_r\wedge\boldsymbol{y}_1\wedge\cdots\wedge\boldsymbol{y}_t|_v. \quad (1)$$

In the case when $r = 0$, i.e. $S \cap T = \{\boldsymbol{0}\}$, this is to be interpreted as

$$|\boldsymbol{x}_1\wedge\cdots\wedge\boldsymbol{x}_s\wedge\boldsymbol{y}_1\wedge\cdots\wedge\boldsymbol{y}_t|_v \leqslant |\boldsymbol{x}_1\wedge\cdots\wedge\boldsymbol{x}_s|_v\,|\boldsymbol{y}_1\wedge\cdots\wedge\boldsymbol{y}_t|_v. \quad (2)$$

This is well known when v is Archimedean (attributed to Fischer [2] by Bombieri & Vaaler [1]) and trivial when v is non-Archimedean. In the Archimedean case we have equality in (2) if each $\boldsymbol{x}_i$ is orthogonal to each $\boldsymbol{y}_j$. To prove (1), we remark that both sides are unchanged if we add linear combinations of $\boldsymbol{u}_1,\ldots,\boldsymbol{u}_r$ to the $\boldsymbol{x}_i$'s and $\boldsymbol{y}_j$'s. In the Archimedean case, after adding suitable such combinations, we may suppose that all the $\boldsymbol{x}_i$'s and $\boldsymbol{y}_j$'s are orthogonal to the $\boldsymbol{u}_k$'s. But then (1) reduces to

$$|\boldsymbol{u}_1\wedge\cdots\wedge\boldsymbol{u}_r|_v^2\,|\boldsymbol{x}_1\wedge\cdots\wedge\boldsymbol{x}_s\wedge\boldsymbol{y}_1\wedge\cdots\wedge\boldsymbol{y}_t|_v \leqslant |\boldsymbol{u}_1\wedge\cdots\wedge\boldsymbol{u}_r|_v^2\,|\boldsymbol{x}_1\wedge\cdots\wedge\boldsymbol{x}_s|_v\,|\boldsymbol{y}_1\wedge\cdots\wedge\boldsymbol{y}_t|_v, \quad (3)$$

which follows from (2). In the non-Archimedean case, writing $\boldsymbol{u}_i = (u_{i1},\ldots,u_{in})$ $(1 \leqslant i \leqslant r)$, we may suppose that $|\boldsymbol{u}_1\wedge\cdots\wedge\boldsymbol{u}_r|_v = |\det U|_v$, where U is the matrix (u_{ij}) with $1 \leqslant i, j \leqslant r$. By adding suitable linear combinations of $\boldsymbol{u}_1,\ldots,\boldsymbol{u}_r$, we may suppose that the first r coordinates of each $\boldsymbol{x}_i$ and $\boldsymbol{y}_j$ are zero. But then (1) reduces to (3) and hence is correct. □

The proposition can be used in the theory of Siegel's lemma. For instance, a certain assumption in Theorem 12 of Bombieri & Vaaler [1], namely that the conjugates of $S^{(1)},\ldots,S^{(d)}$ of an algebraic subspace S have $\dim(S^{(1)}+\cdots+S^{(d)}) = d\dim S$, turns out to be unnecessary. (Actually, they defined the height of matrices, rather than of subspaces.)

Added in proof Our proposition has also been proved in a paper by T. Struppeck & J. D. Vaaler, Inequalities for heights of algebraic subspaces and the Thue–Siegel principle (to appear).

References

[1] E. Bombieri & J. Vaaler, On Siegel's lemma, *Invent. Math.*, **73** (1983), 11–32
[2] E. Fischer, Über den Hadamardschen Determinantensatz, *Arch. Math.*, Basel, **13** (1908), 32–40
[3] W. M. Schmidt, On the heights of algebraic subspaces and diophantine approximations, *Annals of Math.*, (1967), 430–72

Incompactness for chromatic numbers of graphs

Saharon Shelah*

0 Introduction

We have proved the singular cardinal compactness theorem ([12, 13]). A special case of it is that if G is a graph of size a singular cardinal λ such that every subgraph of power less than λ has colouring number less than or equal to ω, then G has countable colouring number. We asked in [12] if this held for the chromatic number. Komjáth showed in [10] that it is consistent that there exists a counterexample of size $\aleph_{\omega_1}$. In this model the continuum is $\aleph_{\omega_1+1}$. Answering his question, we show that such a counterexample is consistent even with GCH (Section 1) and show that similar examples exist in $V = L$ (Section 1).

P. Erdős and A. Hajnal showed that under GCH there is a graph G of size $\aleph_2$ with $\mathrm{Chr}(G) = \aleph_1$ such that every subgraph of size $\aleph_1$ is countably chromatic. They asked in [5] if a similar example which is $\aleph_2$-chromatic exists. The consistency and independence of this statement were shown by Baumgartner and Foreman & Laver, respectively ([2], [6]). Whether or not similar examples exist under $V = L$ was an old problem. We show that this is the case and much more (Section 3): for every regular non-weakly compact κ there is a graph G on κ, with $\mathrm{Chr}(G) = \kappa$, such that every smaller subgraph is countably chromatic. We notice that our earlier proof with just $\mathrm{Chr}(G) = \omega_1$ was published in [4].

Galvin [9] observed that it is not obvious whether or not an $\aleph_2$-chromatic graph should contain an $\aleph_1$-chromatic subgraph. Komjáth showed that this is in fact independent ([10]). Here we show that, e.g. under $V = L$, no counterexample of size $\aleph_2$ exists.

* **Research partially supported by BSF. Number 347 in Shelah's publications.**

Under GCH, P. Erdős & A. Hajnal showed that for $0 < k \leqslant n < \omega$ there is a graph of size $\aleph_n$ with chromatic number with $\aleph_k$ all subgraphs of size $\aleph_{n-1}$ being $\leqslant \aleph_{k-1}$-chromatic. In Section 5 we show that it is consistent (relative to a supercompact) that every $\aleph_n$-chromatic graph $(0 < n < \omega)$ contains an $\aleph_n$-chromatic subgraph of power less than $\aleph_\omega$.*

Notation

A graph is a pair $G = (V, E)$, where $E \subseteq [V^2] = \{x \subseteq V : |x| = 2\}$. E and G are sometimes confused. $\mathrm{Chr}(G)$ is the chromatic number of G. $\mathrm{tp}(A)$ is the order type of A.

1 Incompactness in singular cardinals via forcing

Theorem 1 (GCH) *If $\lambda > \mathrm{cf}(\lambda) = \omega_1$, then there exists a cardinality, cofinality and GCH-preserving partial ordering which adds an $\aleph_1$-chromatic graph on λ such that every subgraph of power less than λ is countably chromatic.*

We can, of course, replace ω_1 by any regular cardinal.

Proof The proof is broken into a series of definitions and lemmas. Let $\{\kappa_\alpha : \alpha < \omega_1\}$ be an increasing, continuous sequence of singular cardinals converging to λ and let $\lambda_\alpha = \kappa_\alpha^+$. Fix a sequence $\{D_\alpha : \alpha < \omega_1\}$ of disjoint sets with $|D_\alpha| = \lambda_\alpha$, $D = \bigcup\{D_\alpha : \alpha < \omega_1\}$ and $E_\alpha = \bigcup\{D_\beta : \beta < \alpha\}$. For $A, B \subseteq D$ we use the following convention: $A_\alpha = A \cap D_\alpha$ and $B_\alpha = B \cap D_\alpha$.

Definition $p = (A, X) \in P$ if $A \subseteq D$, $|A_\alpha| < \lambda_\alpha$ for $\alpha < \omega_1$ and X is a graph on D with

(a) $X \cap [D_\alpha]^2 = \emptyset$;
(b) if $\{x, y\} \in X$, $y \in E_\alpha$ and $x \in D_\alpha$, then $x \in A$;
(c) for $x \in A$, the set $\{y \in D - A : \{x, y\} \in X\}$ is finite and is included in E_α;
(d) for $x \in A_\alpha$ and $\beta < \alpha$, the set $\{y \in E_\beta : \{y, x\} \in X\}$ is finite;
(e) $\mathrm{Chr}(X) \leqslant \omega$.

Next we define extension.

Definition $q = (B, Y) \geqslant p = (A, X)$, that is, (B, Y) extends (A, X), if

(f) $B \supseteq A$ and $Y \supseteq X$;
(g) if $x \in A_\alpha$, then $\{y \in E_\alpha : \{y, x\} \in Y\} = \{y \in E_\alpha : \{y, x\} \in X\}$;

* We thank Peter Komjáth for rewriting the paper.

(h) if $x \in B_\alpha - A_\alpha$, then $\{y \in A : \{y,x\} \in Y - X\}$ is finite and is included in E_α.

Notice that the second clause of (h) follows from (g). It is a trivial calculation to check that the partial order is transitive.

Definition $(B,Y) \geqslant_\alpha (A,X)$ if $(B,Y) \geqslant (A,X)$, $B \cap E_{\alpha+1} = A \cap E_{\alpha+1}$ and $Y \cap [E_{\alpha+1}]^2 = X \cap [E_{\alpha+1}]^2$. Similarly $(B,Y) \geqslant^\alpha (A,X)$ denotes that $(B,Y) \geqslant (A,X)$, $B \cap (D - E_{\alpha+1}) = A \cap (D - E_{\alpha+1})$ and

$$Y \cap [D - E_{\alpha+1}]^2 = X \cap [D - E_{\alpha+1}]^2.$$

Obviously $\leqslant_\alpha$ and $\leqslant^\alpha$ are transitive suborderings.

Lemma 1.2 *If $\theta \leqslant \kappa_{\alpha+1}$ and $\{p_\xi : \xi < \theta\}$ form a continuous $\leqslant_\alpha$-increasing sequence, then they have a common $\leqslant_\alpha$-extension.*

Proof Put $p_\xi = (A^\xi, X^\xi)$. We take $A = \bigcup \{A^\xi : \xi < \theta\}$ and $X = \bigcup \{X^\xi : \xi < \theta\}$. We show that (A,X) is a condition and that $(A,X) \geqslant_\alpha (A^\xi, X^\xi)$ $(\xi < \theta)$.

Everything is trivial except that $\mathrm{Chr}(X) \leqslant \omega$. As every D_β $(\beta \leqslant \alpha)$ is independent in X (i.e. there is no edge of X joining two vertices of D_β), $E_{\alpha+1}$ is certainly countably chromatic. The vertex set of X on $D - E_{\alpha+1}$ is the union of an independent set, $D - A$, and $A - E_{\alpha+1} = \bigcup \{A^{\xi+1} - A^\xi - E_{\alpha+1} : \xi < \theta\}$. X on $A^{\xi+1}$ is countably chromatic, and from every vertex in $A^{\xi+1} - A^\xi$ only finitely many edges go to A^ξ. This implies that $\mathrm{Chr}(X) \leqslant \omega$. □

Lemma 1.3 *If $q \geqslant p$ and $\alpha < \omega_1$, then there are r and s with $p \leqslant_\alpha r \leqslant^\alpha q$ and $p \leqslant^\alpha s \leqslant_\alpha q$.*

Proof If $q = (B,Y)$ and $p = (A,X)$, put $r = (C,Z)$, where $C_\beta = A_\beta$ for $\beta \leqslant \alpha$, $C_\beta = B_\beta$ for $\beta > \alpha$ and

$$Z = X \cup \{\{x,y\} \in Y : x \in D_\beta,\ y \in B_\gamma\ (\beta < \gamma,\ \alpha < \gamma)\}.$$

Similarly for s. □

Lemma 1.4 *Assume that $\alpha < \omega_1$, $p \in P$ and $p \Vdash$ 'τ is a name for an ordinal'. Then there exists a $q \geqslant_\alpha p$ and a set A $(|A| \leqslant \lambda_\alpha)$ such that*

(i) $q \Vdash$ '$\tau \in A$';

(j) if $q \leqslant q^*$, q^* decides a value for τ and $q \leqslant^\alpha r \leqslant_\alpha q^*$, then r decides a value for τ.

Proof We let $\{r_\xi : \xi < \lambda_\alpha\}$ enumerate the possible restrictions $(A \cap E_{\alpha+1}, X \cap [E_{\alpha+1}]^2)$ for $(A,X) \in P$. By transfinite recursion on ξ

we construct an $\leqslant_\alpha$-ascending sequence p_ξ with $p_0 = p$ such that $p_{\xi+1} \cup r_\xi$ decides a value for τ if there exists a $q \geqslant_\alpha p$ with $q \cup r_\xi$ deciding τ. □

Lemma 1.5 *Cardinals and cofinalities remain.*

Proof As usual, it suffices to show that if κ is a regular cardinal in the ground model, then $\theta = \text{cf}(\kappa) < \kappa$ is impossible in the enlarged model. As $|P| = \lambda^+$, no problem arises with $\kappa \geqslant \lambda^{++}$.

Assume first that $\lambda_\alpha < \kappa < \lambda_{\alpha+1}$, $p \Vdash$ '$S \subseteq \kappa$ is cofinal and $|S| = \theta$' and $\theta < \kappa$. By Lemma 1.4 there is a $q \geqslant p$ and a T with $|T| \leqslant \theta + \lambda_\alpha < \kappa$ such that $q \Vdash$ 'T is cofinal' with T in the ground model: a contradiction.

If $\alpha < \omega_1$, α a limit and $\kappa = \kappa_\alpha^+$, then (as κ_α is singular) $\theta < \kappa_\alpha$, so that $\theta < \kappa_\beta$ for some $\beta < \alpha$. Again, we get that $\text{cf}(\kappa) \leqslant \lambda_\beta < \kappa$ in the ground model. Assume, finally, that $\kappa = \lambda^+$. Then $\theta \leqslant \lambda_\alpha$ for some $\alpha < \omega_1$ and we may proceed as in the previous case. □

Lemma 1.6 *GCH survives.*

Proof As P is ω_1-closed, it suffices (by Silver's theorem) to show that $2^\theta = \theta^+$ holds in an enlarged model for every regular cardinal θ. There is no problem for $\theta > \lambda$; so assume that $\kappa_\alpha^+ \leqslant \theta < \kappa_{\alpha+1}$ and that $p \Vdash$ '$T_\xi \subseteq \theta$ are different ($\xi < \theta^{++}$)'. By Lemma 1.4, there is a q, as there, and a partial function $F(r, \xi, \zeta)$ such that if $r = (A, X)$ with $A \subseteq E_{\alpha+1}$ and $X \subseteq [E_{\alpha+1}]^2$, then $r \cup q$ forces either that $\zeta \in T_\xi$ or that $\zeta \notin T_\xi$ according to whether $F(r, \xi, \zeta)$ is 0 or 1. As the number of different r's is λ_α, the number of $F(r, \cdot, \zeta)$ functions is $\leqslant (\lambda_\alpha^+)^\theta = \theta^+$, and so there are $\xi_1 \neq \xi_2$ with $F(r, \xi_1, \zeta) = F(r, \xi_2, \zeta)$, that is, $q \Vdash$ '$T_{\xi_1} \doteq T_{\xi_2}$': a contradiction. □

Lemma 1.7 *P forces that the generic graph is countably chromatic on every set of size less than λ.*

Proof Assume that $p \Vdash$ '$\tau \subseteq D$ with $|\tau| \leqslant \lambda_\alpha$'. There is a $q \geqslant_\alpha p$ such that $q \Vdash$ '$\tau \subseteq F$', where $|F| \leqslant \lambda_\alpha$, by Lemma 1.4. Extend q to an $r = (X, A)$ with $F \subseteq A$; then we are done by (e). □

Lemma 1.8 *The generic graph is $\aleph_1$-chromatic.*

Assume that $p \Vdash$ '$f \colon \lambda \to \omega$ is a good colouring'. We let $p_0 = p$ and, by induction, define p_n, x_n and α_n with $p_n = (A^n, X^n)$, $p_n \leqslant p_{n+1}$ and $\alpha_n < \alpha_{n+1}$ for $n < \omega$ and such that either $p_{n+1} \Vdash$ '$f(x_n) = n$ and

$x_n \in D_{\alpha_{n+1}} - A^n$', or else $p_{n+1} \Vdash$ '$f(x_n) \neq n$ and $x_n \in D_{\alpha_{n+1}} - A^n - E_{\alpha_n}$ and for every $x \in D - A^n - E_{\alpha_n}$ we have $h(x) \neq n$'. This can easily be done. Put $q = (B, Y)$, where $\alpha = \sup(\alpha_n)$, $y \in D_\alpha - \bigcup\{A^n_{\alpha_n} : n < \omega\}$, $B = \bigcup A^n \cup \{y\}$ and $Y = \bigcup\{X^n : n < \omega\} \cup \{\{x_n, y\} : n < \omega\}$. Obviously, q is a condition and $q \geqslant p_n$ for $n < \omega$. If $r \geqslant q$ forces $f(y) = n$, then r forces a contradiction. □

That completes the proof of Theorem 1.1. □

2 Incompactness in singular cardinals under $V = L$.

Theorem 2.1 ($V = L$) *If $\kappa = \mathrm{cf}(\kappa)$ is not weakly compact, $\omega \leqslant \theta < \kappa$ and $\lambda > \mathrm{cf}(\lambda) = \kappa$, then there is a θ^+-chromatic graph of power λ in which every subgraph of power less than λ is $\leqslant \theta$-chromatic.*

Definition If f and g are functions on a common domain, a set of ordinals of limit type, then $f <^* g$ denotes that there is a $\beta \in \mathrm{Dom} f$ such that $f(\beta') < g(\beta')$ holds for every $\beta' > \beta$.

Lemma 2.2 ([16]) ($V = L$) *Assume that λ_i $(i \leqslant \mu)$ is an increasing continuous sequence of singular cardinals. Put*

$$\Gamma = \{f : \mathrm{Dom} f = \mu,\ f(i) < \lambda_i^+\ (i < \mu)\}.$$

Then there is a $<^*$-increasing, $<^*$-cofinal sequence $\{f_\xi : \xi < \lambda_\mu^+\}$ in Γ such that for every $\xi < \lambda_\mu^+$ the system $\{f_\zeta : \zeta < \xi\}$ can be disjointed, that is, there is a function $g : \xi \to \mu$ such that if $\zeta_0 < \zeta_1 < \xi$ and $i > g(\zeta_0), g(\zeta_1)$, then $f_{\zeta_0}(i) < f_{\zeta_1}(i)$ holds.

By the result of in Section 3, there is a graph G on κ with $\mathrm{Chr}(G) = \theta$ and $\mathrm{Chr}(G \restriction \alpha) \leqslant \theta^+$ for $\alpha < \kappa$, and if, for $i < \kappa$, $G(i) := \{j < i : \{j, i\} \in G\}$, then $G(i)$ is either empty or of type θ.

Let $\{\lambda_i : i < \kappa\}$ be a continuous, increasing sequence of singular cardinals, converging to λ, with $\lambda_0 > \kappa$. Put $A_i = \{i\} \times \lambda_i^+ \times \kappa$. We are going to build a graph H on $\bigcup\{A_i : i < \kappa\}$ such that, for every $x \in A_i$, there are g_x and h_x defined on $G(i)$, with $g_x(j) < \lambda_j^+$ and $h_x(j) < \kappa$, and the vertices in $\bigcup\{A_j : j < i\}$ joined to x are $H(x) = \{\langle j, g_x(j), h_x(j)\rangle : j \in G(i)\}$. As there is a natural projection of H onto G, mapping A_i onto i, $\mathrm{Chr}(H) \leqslant \theta^+$ is obvious. We stipulate that $h_x(j) > i$ holds for $x \in A_i$ and $j \in G(i)$.

Definition $X \subseteq A_i$ is *large* if, for every $\xi < \lambda_i^+$ and $\nu < \kappa$, there is an $\langle i, \xi', \nu'\rangle \in X$ with $\xi' > \xi$ and $\nu' > \nu$.

We add the following stipulation on H. Let $\{f^i_\xi : \xi < \lambda_i^+\}$ be a <*-cofinal sequence, as in Lemma 2.2, for $G(i) \neq \emptyset$. So $\operatorname{Dom} f^i_\xi = G(i)$ and $f^i_\xi(j) < \lambda_j^+$.

(a) For $x \in A_i$ there are $\gamma_x < \delta_x < \lambda_i^+$ such that $f^i_{\gamma_x} <^* g_x <^* f^i_{\delta_x}$ and the intervals $[\gamma_x, \delta_x]$ $(x \in A_i)$ are pairwise disjoint;

(b) if, for $j \in G(i)$, $B_j \subseteq A_j$ is large, then

$$\{x \in A_i : H(x) \subseteq \bigcup \{B_j : j \in G(i)\}\}$$

is large.

This selection can be made by an obvious transfinite recursion. The graph H is already constructed: we first show that $\operatorname{Chr}(\theta) = \theta^+$. If $F: \bigcup\{A_i : i < \kappa\} \to \theta$ is a good colouring, by recursion on $i < \kappa$ we can choose a large $X_i \subseteq A_i$ such that

(c) F on X_i is constant;

(d) if $x \in X_i$, $H(x) \subseteq \{X_j : j \in G(i)\}$.

One only needs to notice that the union of θ non-large sets is not large, either. By (c), we have a θ-colouring on G and so we are finished by $\operatorname{Chr}(G) = \theta^+$.

We finally show that every $B \subseteq \bigcup\{A_i : i < \kappa\}$ with $|B| < \lambda$ spans a subgraph which is θ-chromatic. Let $|B| \leqslant \lambda_i$. The graph on $B \cap \bigcup\{A_j : j \leqslant i\} \subseteq \bigcup\{A_j : j \leqslant i\}$ is θ-chromatic by our assumptions on G (using the projection). Assume now that $B \subseteq \bigcup\{A_j : j > i\}$ (and that $|B| \leqslant \lambda_i$). For every $j > i$, there is, by Lemma 2.2, a disjointing function ξ_x $(x \in B \cap A_j)$ for g_x. Decompose the edges of $H \restriction B$ into two classes: $\{y, x\} \in H_1$ if $y = \langle j, g_x(j), h_x(j)\rangle$ if $j \leqslant \xi_x$ and $\{y, x\} \in H_2$ otherwise. Now H_1 has the property that there is a well-ordering (the ordered sum of $A_j \cap B$ $(j > i)$) such that every vertex is joined to less than θ smaller vertices. As is well known, this implies that $\operatorname{Chr}(H_1) \leqslant \theta$. It suffices to show that $\operatorname{Chr}(H_2) \leqslant \theta$. If $\{y, x\} \in H_2$, $y \in A_j$, $x \in A_i$, $j < i$ and if $y = \langle j, g_x(j), h_x(j)\rangle$, then, given y and i, there is at most one x, and $i < h_x(j)$. Therefore, every vertex has not more than θ edges 'going down' and less than κ edges 'going up'. So every connected component is of size less than κ. By the properties of G, every component is $\leqslant \theta$-chromatic. Thus so is H_2. □

Note Even $\bigcap_i |B \cap A_i| \leqslant \lambda_i$ implies that $\operatorname{Chr}(G \restriction B) \leqslant \theta$.

3 Large gaps in regular cardinals under $V = L$

Theorem 3.1 ($V = L$) *If κ is a cardinal then there is a graph G on κ^+ such that* $\operatorname{Chr}(G) = \kappa^+$ *but, for every $\alpha < \kappa$,* $\operatorname{Chr}(G \restriction \alpha) \leqslant \omega$ *holds.*

Proof We use the following principle deduced from $V = L$ in [1].

(⊠) There is a sequence $\langle C_\delta, M_\delta : \delta < \kappa^+, \text{limit}\rangle$ such that

(a) $C_\delta \subseteq \delta$ is a club;
(b) if $\alpha \subseteq C'_\delta$ then $C_\alpha = C_\delta \cap \alpha$;
(c) M_δ is a model on δ;
(d) if $\alpha \in C'_\delta$, then $M_\alpha \prec M_\delta$;
(e) if M is a model on κ^+ with vocabulary $\leqslant \kappa$, then

$$\{\delta < \kappa^+ : \text{tp}(C_\delta) = \kappa,\ M_\delta \prec M\}$$

is stationary.

We assume that for every limit δ, $M_\delta = \langle \delta, f_\delta \rangle$, where f_δ is a function from δ into κ. We define for every $\delta < \kappa^+$ (δ a limit) $g_\delta : C'_\delta \to \kappa^+$ as follows. Let $B = \{\delta < \kappa^+ : \text{tp}(C_\delta) = \kappa\}$ and, for $\delta \in B$, let $h^*(\delta) = \min C'_\delta$. Then

(f) if $\alpha \in C'_\delta$, then $g_\alpha \subseteq g_\delta$;
(g) if $\text{tp}(C'_\delta) = \xi + 1$ and $\epsilon = \max(C'_\delta)$, then

$$g_\delta(\epsilon) = \min\{\tau : \tau \in B,\ h^*(\tau) \geqslant \epsilon,\ f_\delta(\tau) = \xi\}$$

if such a τ exists and is undefined otherwise.

To define G we join every $\delta < \kappa^+$ with $\text{tp}(C_\delta) = \kappa$ into the vertex set $\{g_\delta(\xi) : \xi \in C'_\delta\}$.

We show that $\text{Chr}(G) = \kappa^+$. Assume that $f : \kappa^+ \to \kappa$ is a good colouring. Select a δ as in (e). Then, for every $\xi < \kappa$ with $E_\xi = \{h^*(\delta) : \delta \in B, f(\delta) = \xi\}$ unbounded (in κ^+), $g_\delta(\xi)$ is defined and so $f(\delta) = \xi$ is ruled out by construction. If E_ξ is bounded, then this bound is less than $h^*(\delta)$, and so $f(\delta) = \xi$ is impossible again.

We now turn to the proof of the other property.

Definition $F : \alpha \to \omega$ is *suitable* if it is a good colouring and, for every limit $\beta \leqslant \alpha$, $|\omega - \{F(g_\beta(\xi)) : \xi \in C'_\beta\}| = \omega$.

The following claim clearly suffices for the proof.

Claim *If $\beta < \alpha$, $\text{tp}(C_\beta) \neq \kappa$, F is a suitable colouring of β and F' is a colouring of a finite subset of $[\beta, \kappa^+)$ such that $F \cup F'$ is a good colouring, then there is a good colouring on α, compatible with $F \cup F'$.*

Proof of the claim (by transfinite induction on α) If $\alpha = \alpha' + 1$, add α' to the domain of F' and apply the claim.

Assume that α is a limit. Enumerate C_α as $\{\gamma_\xi : \xi < \text{tp}(C_\alpha)\}$ and suppose that $\gamma_\zeta \leqslant \beta < \gamma_{\zeta+1}$ ($\gamma_0 = 0$ is assumed). As F is suitable on β, $A = \omega - F(g_\alpha(\gamma_{\omega\xi})) : \omega\xi < \zeta\}$ is infinite. Select $k^* \in A$. Applying the claim we can extend F from β to $\gamma_{\zeta+1}$, from $\gamma_{\zeta+1}$ to γ_{ζ_2}, and so on, but colouring vertices $g_\alpha(\gamma_{\omega\epsilon})$ ($\gamma_\zeta \leqslant \epsilon \in C'_\alpha$) only with the colour k^*. For a limit ordinal $\xi \leqslant \mathrm{tp}(C_\alpha)$, $\{F(g_\alpha(\gamma_{\omega\tau})) : \omega\tau < \xi\}$ contains only one element of A. The inductive step is possible as $g_\alpha(\gamma_{\omega\xi})$ is connected by an edge to no ordinal less than or equal to $h^*(g_\alpha(\gamma_{\omega\xi}))$ which is greater or equal to $\gamma_{\omega\xi}$. □

Theorem 3.2 ($V = L$) *If κ is an inaccessible, not weakly compact cardinal, then there is a graph G on κ with* $\mathrm{Chr}(G) = \kappa$, *but for $\alpha < \kappa$,* $\mathrm{Chr}(G \restriction \alpha) \leqslant \omega$.

Proof Similar to the proof of the previous theorem, only we use the appropriate principle with
(e*) if M is a model on κ with vocabulary $\leqslant \kappa$ and $\mu < \kappa$, then

$$\{\delta < \kappa : \mathrm{tp}(C_\delta) = \mu,\ M_\delta \prec M\}$$

is stationary. □

Remark It is easy to modify the construction to get graphs as in Theorems 3.1 and 3.2 with arbitrary chromatic number less than $|G|$.

4 Non-spanned subgraphs

Theorem 4.1 ($V = L$) *If G is a graph on $\lambda = \mathrm{cf}(\lambda) > \omega$ with* $\mathrm{Chr}(G) \geqslant \theta \geqslant \omega$ *and, for every $\alpha < \lambda$ we have* $\mathrm{Chr}(G \restriction \alpha) < \theta$, *then there exists a subgraph G' of G with* $\mathrm{Chr}(G') = \theta$.

Proof We are going to use the following consequence of $V = L$, proved like the proof of $\diamondsuit$ by R. L. Jensen. Let $L_\mathrm{a} \subseteq L_\mathrm{b} \subseteq L_\mathrm{c}$ be extensions of ZF vocabulary by finitely many new symbols. $M^\mathrm{a}(\delta)$ denotes a model of L_a and similarly for $M^\mathrm{b}(\delta)$, etc.

Lemma 4.2 ($V = L$) *If $\lambda = \mathrm{cf}(\lambda) > \omega$, M^a is a model on λ and φ is a first-order sentence in L_c, then there exist models*

$$\langle M^\mathrm{c}_\xi(\delta) : \xi < \epsilon_\delta,\ \delta < \lambda \text{ limit}\rangle$$

such that

(a) $M^\mathrm{c}_\xi(\delta)$ expands $M^\mathrm{a} \restriction \delta$;
(b) for $\xi \neq \zeta$, $M^\mathrm{c}_\xi(\delta) \restriction L_\mathrm{b} \neq M^\mathrm{c}_\zeta(\delta) \restriction L_\mathrm{b}$;
(c) if M^c expand M^a satisfies φ, then there is an $N^\mathrm{c} \supseteq M^\mathrm{a}$ satisfying φ, such that for a closed unbounded set of δ there is a $\xi < \epsilon_\delta$ with $M^\mathrm{c}_\xi(\delta) = N^\mathrm{c} \restriction \delta$.

If $\diamondsuit^*_\lambda$ holds, we can take $\epsilon_\delta = \delta$.

Definition Let C and D be closed, unbounded sets in $\lambda = \mathrm{cf}(\lambda)$ and $\tau(\alpha) = \min(C-(\alpha+1))$ for $\alpha < \lambda$. Then

$$\Delta(C,D) = \tau(0) \cup \bigcup \{[\alpha, \tau(\alpha)) : \alpha \in D\}.$$

Lemma 4.3 *If I has the property that for every club C there is a club D such that $\Delta(C,D) \in I$, then λ is the union of countably many elements of I.*

Proof Let $C_0 = \lambda$ and let C_{n+1} satisfy $\Delta(C_n, C_{n+1}) \in I$. If $\alpha \notin \bigcup \Delta(C_n, C_{n+1})$ and $\alpha_n = \max(\alpha \cap C_n)$, then $\alpha = \alpha_0 > \alpha_1 > \cdots$: a contradiction. □

In order to prove the result, we try to formulate the fact that no subgroup G' of G has $\mathrm{Chr}(G') = \theta$. $\mathrm{Chr}(G') \leqslant \theta$ means that there is an $F: \lambda \to \theta$ good colouring of G'. On the other hand, given F, we may assume that G' consists of the edges $\{\alpha, \beta\}$ with $\{\alpha, \beta\} \in G$ *and* $F(\alpha) \neq F(\beta)$. So the property can be translated as follows. For every $F: \lambda \to \theta$ there is a $\sigma < \theta$ and an $H: \lambda \to \sigma$ with

(d) if $\{\alpha, \beta\}$ is in G and $F(\alpha) \neq F(\beta)$, then $H(\alpha) \neq H(\beta)$.

We now let M^{a} be G, $L_{\mathrm{b}} = L_{\mathrm{a}} \cup \{F, \theta\}$, $L_{\mathrm{c}} = L_{\mathrm{b}} \cup \{H\}$ and φ the sentence in (d).

If I is the collection of subsets of λ spanning subgraphs with chromatic number less than θ, then by Lemma 4.3 and the fact that $\mathrm{Chr}(G) > \theta$ (otherwise we are done), there is a club C such that, for no club D, $\Delta(C,D) \in I$ holds. Enumerate $C \cup \{0\}$ as $\{\gamma_\alpha : \alpha < \lambda\}$. We construct an $F: \lambda \to \theta$ by recursively defining $F \restriction [\gamma_\alpha, \gamma_{\alpha+1})$.

If there exists a ξ such that $M^{\mathrm{c}}_\xi(\gamma_\alpha) \restriction L_{\mathrm{b}} = (M^{\mathrm{a}} \restriction \delta, F \restriction \gamma_\alpha, \theta)$, then the range of H in $M^{\mathrm{c}}_\xi(\gamma_\alpha)$ is bounded (in θ) and ξ is unique by (b). For $\gamma_\alpha \leqslant \tau < \gamma_{\alpha+1}$ we then put

$$B(\tau) = \{\beta < \gamma_\alpha : \{\beta, \tau\} \text{ is in } G, \text{ and no } \beta' < \beta \text{ has } \{\beta', \tau\} \in G \;\; \textit{and } M^{\mathrm{c}}_\xi(\delta_\alpha) \models H(\beta') = H(\beta).\}$$

$|B(\tau)| < \theta$ as $\mathrm{Rang}\, H$ is bounded (in θ). Now define

$$F(\tau) = \min\{\theta - F''B(\tau)\}.$$

If no such ξ exists, any extension works.

Having constructed $F: \lambda \to \theta$, by our indirect assumption there is $H: \lambda \to \sigma < \theta$, a 'better colouring of G', determined by F. So, by Lemma 4.2, there is (a possibly different) H such that, for a closed unbounded D, if $\delta \in D$ there is a ξ with $M^{\mathrm{c}}_\xi(\delta) = M_{\mathrm{c}} \restriction \delta$. We assume that $D \subseteq C$.

Claim $\mathrm{Chr}(G \restriction \Delta(C,D)) < \theta$.

Clearly this claim gives the desired contradiction.

Proof of the claim As $\mathrm{Chr}(G \restriction \alpha) < \theta$ for every $\alpha < \lambda$ and $\mathrm{cf}(\lambda) > \theta$, there exists a $\sigma < \theta$ such that $\mathrm{Chr}(G \restriction \alpha) \leqslant \sigma$ $(\alpha < \lambda)$. From this, those edges joining vertices in the *same* interval of $\Delta(C,D)$ can get a good colouring by not more than σ colours. It suffices to show that H is a good colouring for the edges between different intervals. Otherwise there is $\{\tau', \tau\} \in G$ with $\tau' < \gamma_\alpha \leqslant \tau < \gamma_{\alpha+1}$ $(\gamma_\alpha \in D)$ and $H(\tau') = H(\tau)$. Fix τ and take τ' *minimal.* Then $F(\tau') \neq F(\tau)$ and so $H(\tau') \neq H(\tau)$: a contradiction. □

5 Compactness is consistent

We mention that Foreman & Laver showed, from an almost huge cardinal, the consistency of GCH and the statement that every graph with power and chromatic number $\aleph_2$ contains a subgraph of power and chromatic number $\aleph_1$. See also [7, 8, 15]. We use the following result.

Lemma 5.1 (Ben-David & Magidor [3]) *If the existence of a supercompact cardinal is consistent, then so is ZFC+GCH and that for every regular $\lambda > \omega_\omega$ there is an ultrafilter D on λ such that $\aleph_n^\lambda/D = \aleph_n$ for $0 < n < \omega$, and there are $A_\xi \in D$ $(\xi < \lambda)$ such that, for $\alpha < \lambda$, the set $w_\alpha = \{\xi : \alpha \in A_\xi\}$ has power less than $\aleph_\omega$.*

Theorem 5.2 *In the model of Lemma 5.1, if $0 < n < \omega$, G is a graph such that every subgraph of G of power less than $\aleph_\omega$ has chromatic number not exceeding $\aleph_n$, then* $\mathrm{Chr}(G) \leqslant \aleph_n$.

Proof We may assume that the vertex set of G is λ, a regular cardinal (otherwise we may use λ^+). By assumption, there is a good colouring, $f_\alpha : w_\alpha \to \omega_n$. For $\xi < \lambda$, let $g_\xi(\alpha) = f_\alpha(\xi)$ and, finally, define $h: \lambda \to \omega_n^\lambda/D$ by $h(\xi) = g_\xi/D$. This h is a good colouring of G by ω_n colours. [It is a good colouring because, if we assume that $\xi, \zeta < \lambda$, $\{\xi, \zeta\} \in G$ and $h(\xi) = h(\zeta)$, then

$$\begin{aligned}\{\alpha : g_\xi(\alpha) = g_\zeta(\alpha)\} &= \{\alpha : f_\alpha(\xi) = f_\alpha(\zeta)\} \\ &\subseteq \{\alpha : \{\xi,\zeta\} \not\subseteq W_\alpha\} \\ &= \{\alpha : \alpha \notin A_\xi \text{ or } \alpha \notin A_\zeta\} \\ &\subseteq \{\alpha : \alpha \notin A_\xi \cap \in A_\zeta\} = \varnothing \quad (\mathrm{mod}\, D).\end{aligned}$$

Hence $g_\xi/D \neq g_\zeta/D$. The number of colours is $\aleph_n$ because, as $|\omega_n^\lambda/D| = \aleph_n$, the function h has the right number of colours.]

References

[1] U. Abraham, S. Shelah & R. M. Solovay, Squares with diamonds and Souslin trees with special squares, *Fundamenta Math.*, **127** (1986), 133–62

[2] J. E. Baumgartner, Generic graph construction, *J. Symb. Logic*, **49** (1984), 234–40

[3] Sh. Ben-David & M. Magidor, The weak □ is really weaker than the full □, *J. Symb. Logic*, **51** (1986), 1029–33

[4] K. Devlin, Constructibility, in *Handbook of Mathematical Logic* (ed. J. Barwise), North-Holland, 1977, 453–89

[5] P. Erdős & A. Hajnal, On chromatic number of graphs and set systems, *Acta Math. Acad. Sci. Hung.*, **17** (1966), 61–99

[6] M. Foreman & R. Laver, A graph reflection property (to appear)

[7] M. Foreman, M. Magidor & S. Shelah, Martin's maximum, saturated ideals and non-regular ultrafilters, Part I, *Annals of Math.*, **127** (1988), 1–47

[8] M. Foreman, M. Magidor & S. Shelah, Martin's maximum, saturated ideals and non-regular ultrafilters, Part II, *Annals of Math.*, **127** (1988), 521–45

[9] F. Galvin, Chromatic numbers of subgraphs, *Period. Math. Hung.*, **4** (1973), 117–19

[10] P. Komjáth, Consistency results in finite graphs, *Israel J. of Math.*, **61** (1988), 285–94

[11] P. Komjáth & S. Shelah, Forcing constructions for uncountably chromatic graphs, *J. Symb. Logic* (to appear)

[12] S. Shelah, A compactness theorem for singular cardinals, free algebras, Whitehead problem, and transversals, *Israel J. of Math.*, **21** (1975), 319–49

[13] S. Shelah, Incompactness in regular cardinals, *Notre Dame J. of Formal Logic*, **26** (1985), 195–228

[14] S. Shelah, Remarks on squares, *Around Classification Theory of Models*, Lecture Notes **1182**, Springer, Heidelberg, 276–9

[15] S. Shelah, Iterated forcing and normal ideals on ω_1, *Israel J. of Math.*, **60** (1987), 345–80

[16] S. Shelah, Gap 1 two cardinal principles and omitting type theorem for $L(Q)$, *Israel J. of Math.*, **65** (1989), 133–52

[17] S. Shelah, $UP_1(I)$, large ideal on ω_1, general preservation, in *Proper and Improper Forcing*, Springer Verlag Prospective in Mathematics (to appear)

Graphs with no unfriendly partitions

Saharon Shelah* and E. C. Milner†

Abstract

An unfriendly n-partition of a graph $G = (V, E)$ is a map $c: V \to \{0, 1, \ldots, n-1\}$ such that, for every vertex x, there holds

$$|\{y \in E(x) : c(x) = c(y)\}| \leqslant |\{y \in E(x) : c(x) \neq c(y)\}|,$$

where $E(x)$ is the set of vertices joined to x by an edge of G. We disprove a conjecture of Cowen & Emerson by showing that there is a graph which has no unfriendly 2-partition. However, we also show that every graph has an unfriendly 3-partition.

1 Introduction

Let $G = (V, E)$ be a simple graph. A map $c: V \to \{0, 1, \ldots, n-1\}$ is called an unfriendly n-partition of G (see [1]) if, for every vertex x, there holds

$$|\{y \in E(x) : c(x)=c(y)\}| \leqslant |\{y \in E(x) : c(x) \neq c(y)\}|,$$

where $E(x)$ is the set of vertices joined to x by an edge of G.

It is easily seen that any finite graph has an unfriendly 2-partition and hence, by compactness, so does every locally finite graph. Cowan & Emerson [2] conjectured that every graph has an unfriendly 2-partition and Aharoni, Milner & Prikry [1] proved this for graphs satisfying either (1) *there are only finitely many vertices with infinite degrees,* or (2) *there are a finite number of infinite cardinals $m_0 < m_1 < \cdots < m_k$ such that*

* No. 356. The first author wishes to express his thanks to the Niedersächsches Ministerium für Wissenshaft und Kunst for partially supporting this research.
† Research supported by NSERC grant #A5198.

m_i is regular for $0 < i \leqslant k$, every vertex of infinite degree has degree m_i for some $i \leqslant k$ and the number of vertices of finite degree is less than m_0.

The following result disproves the conjecture of [2]. For a cardinal $\lambda = \omega_\alpha$ and an ordinal β, we use the notation $\lambda^{(+\beta)}$ to denote the cardinal $\omega_{\alpha+\beta}$.

Theorem 1 *There is a graph $G = (V, E)$, of size $|V| = (2^\omega)^{(+\omega)}$, which has no unfriendly 2-partition and in which every vertex has infinite degree.*

A similar argument also proves the following more general version of Theorem 1.

Theorem 2 *For any infinite cardinal λ, there is a graph $G = (V, E)$, of size $|V| = \kappa = (2^\lambda)^{(+\omega)}$, which has no unfriendly 2-partition and in which every vertex has infinite degree.*

Before giving proofs of these results, we shall prove the following consistency result which, although weaker, illustrates the main idea in a simpler setting.

Theorem 3 *It is consistent that there is a graph $G = (V, E)$ of size $|V| = \omega_\omega$ which has no unfriendly 2-partition and the degree of each vertex is either ω, or ω_1, or ω_ω.*

We conclude the paper with a proof of the following positive result.

Theorem 4 *Every graph has an unfriendly 3-partition.*

2 Proof of Theorem 3

For subsets A and B of ω, we write $A > B$ if $|A \backslash B| = \omega$ and $|B \backslash A| < \omega$. It is well known that the following statement $(*)$ is independent of the axioms of set theory. (For example, CH $\Rightarrow (*)$ and $(\mathrm{MA} + 2^\omega > \omega_1) \Rightarrow \neg(*)$.)

$(*)$ There is a uniform, non-principal ultrafilter $\mathfrak{U}$ on ω which is generated by ω_1 sets A_ξ $(\xi < \omega_1)$ such that $A_\xi > A_\zeta$ for $\xi < \zeta < \omega_1$ so that, for any set $A \in \mathfrak{U}$, there is some $\xi < \omega_1$ such that $|A_\zeta \backslash A| < \omega$ for $\xi \leqslant \zeta < \omega_1$.

We show that $(*)$ implies there is a graph with the properties stated in Theorem 3.

We construct the desired graph $G = (V, E)$ as follows. Let $V = X \cup Y \cup Z$, where $X = \{x_n : n < \omega\}$, $Y = \{y_{\alpha,\xi} : \alpha < \omega_\omega, \xi < \omega_1\}$ and

$Z = \{z_\alpha : \alpha < \omega_\omega\}$ and let $E = E_1 \cup E_2 \cup E_3$, where

$$E_1 = \{(x_n, y_{\alpha,\xi}) : \alpha \leqslant \omega,\ \xi < \omega_1,\ n \in A_\xi\},$$

$$E_2 = \{(y_{\alpha,\xi}, z_\alpha) : \alpha < \omega_\omega,\ \xi < \omega_1\},$$

$$E_3 = \{(x_n, z_\alpha) : \alpha < \omega_\omega, n < \omega\}.$$

Note that each vertex of X has degree ω_ω, each vetex of Y has degree ω and each vertex of Z has degree ω_1.

We want to show that G has no unfriendly 2-partition. Suppose for a contradiction that $c\colon V \to \{0,1\}$ is an unfriendly partition of G. Since $\mathfrak{U}$ is an ultrafilter on ω, there are $\epsilon < 2$ and $A \in \mathfrak{U}$ such that $c(x_n) = \epsilon$ if and only if $n \in A$. There is $\xi < \omega_1$ such that $|A_\zeta \backslash A| < \omega$ for $\xi \leqslant \zeta < \omega_1$. Since, by assumption, c is an unfriendly partition, since

$$E(y_{\alpha,\zeta}) = \{z_\alpha\} \cup \{x_n : n \in A_\zeta, \alpha \leqslant \omega_n\} \qquad (\alpha < \omega_\omega, \zeta < \omega_1)$$

and since $c(x_n) = \epsilon$ for $n \in A$, it follows that $c(y_{\alpha,\zeta}) = 1-\epsilon$ for $\alpha < \omega_\omega$ and $\xi \leqslant \zeta < \omega_1$. Further, since $E(z_\alpha) = X \cup \{y_{\alpha,\zeta} : \zeta < \omega_1\}$ for $\alpha < \omega_\omega$, we must also have $c(z_\alpha) = \epsilon$. But, for $n \in A$,

$$E(x_n) = \{y_{\alpha,\zeta} : n \in A_\zeta, \alpha \leqslant \omega_n\} \cup Z,$$

and this contradicts the assumption that c is an unfriendly partition since $c(x_n) = c(z)$ $(z \in Z)$ and $|E(x_n) \backslash Z| < |Z|$. □

3 Proof of Theorem 1

We will use the following notation. For an ordinal we α define $\|\alpha\|$ to be $|\alpha|$ if α is infinite and 0 if α is finite. If $\mathfrak{u} = (u_0, u_1, \ldots, u_{l-1})$ is a sequence of ordinals, the length of $\mathfrak{u}$ is $l(\mathfrak{u}) = l$, and the *last term* of $\mathfrak{u}$ is

$$\operatorname{lt}(\mathfrak{u}) = \begin{cases} u_{l-1} & \text{if } l \geqslant 1, \\ 2^\omega & \text{if } l = 0. \end{cases}$$

If $\mathfrak{v} = (v_0, v_1, \ldots, v_l)$ has length $l+1$ and $v_i = u_i$ $(i < l)$, then we write $\mathfrak{v} = \mathfrak{u}^\wedge v_l$ and we also write $\mathfrak{u} = \mathfrak{v}^*$ to indicate that $\mathfrak{u}$ is obtained from $\mathfrak{v}$ by omitting the last term v_l. Put

$$\mathcal{I} = \{(u_0, u_1, \ldots, u_{l-1}) : 2^\omega > \|u_0\| > \|u_1\| > \ldots > \|u_{l-1}\|\},$$

$$\mathcal{J} = \{(v_0, v_1, \ldots, v_{l-1}) : v_i < \omega_1\ (i < l)\}.$$

Let $\mathfrak{U}$ be a uniform, non-principal ultrafilter on ω. We shall define sets $A_{\mathfrak{i},\rho} \in \mathfrak{U}$ for $\mathfrak{i} \in \mathcal{I}$ and $\rho < |\operatorname{lt}(\mathfrak{i})|$ by induction on $l(\mathfrak{i})$ as follows. Let $A_{\square,\rho}$ $(\rho < 2^\omega)$ be any enumeration of the members of $\mathfrak{U}$, where □

denotes the empty sequence. Now suppose that $A_{\mathrm{i},\rho}$ has been defined for $\mathrm{i} \in \mathcal{I}$, $l(\mathrm{i}) \leqslant l$ and $\rho < |\mathrm{lt}(\mathrm{i})|$. For $\mathrm{i} \in \mathcal{I}$, $l(\mathrm{i}) = l+1$ and $\rho < |\mathrm{lt}(\mathrm{i})|$, put

$$A_{\mathrm{i},\rho} = A_{\mathrm{i}^*, h(\theta,\rho)},$$

where $\theta = \mathrm{lt}(\mathrm{i})$ and $h(\theta,\cdot)$ is any one–one map from $|\theta|$ onto θ.

Put $\kappa_n = (2^\omega)^{(+n)}$ and $\kappa = \sum\{\kappa_n : n < \omega\}$. We define the graph $G = (V, E)$ of size κ as follows. Put $V = X \cup Y \cup Z$, where

$$X = \{x_n : n < \omega\}, \qquad Y = \{y^\alpha_{\mathrm{i},\mathrm{j}} : \alpha < \kappa, \mathrm{i} \in \mathcal{I}, \mathrm{j} \in \mathcal{I}, l(\mathrm{j}) = l(\mathrm{i})+1\},$$
$$Z = \{z^\alpha_{\mathrm{i},\mathrm{j}} : \alpha < \kappa, \mathrm{i} \in \mathcal{I}, \mathrm{j} \in \mathcal{I}, l(\mathrm{j}) = l(\mathrm{i})\}.$$

The edge set of G is $E = E_1 \cup E_2 \cup E_3$, where

$$E_1 = \{\{x_n, y^\alpha_{\mathrm{i},\mathrm{j}}\} : y^\alpha_{\mathrm{i},\mathrm{j}} \in Y, k = \mathrm{lt}(\mathrm{i}) < \omega, n \in \bigcap\{A_{\mathrm{i},\rho} : \rho < k\} \text{ and } \alpha \leqslant \kappa_n\},$$

$$E_2 = \{\{y^\alpha_{\mathrm{i},\mathrm{j}}, z^\alpha_{\mathrm{i}_1,\mathrm{j}_1}\} : y^\alpha_{\mathrm{i},\mathrm{j}} \in Y, z^\alpha_{\mathrm{i}_1,\mathrm{j}_1} \in Z, \alpha < \kappa \text{ and either } \mathrm{i} = \mathrm{i}_1^*, \mathrm{j}_1 = \mathrm{j} \text{ or } \mathrm{i} = \mathrm{i}_1, \mathrm{j}_1 = \mathrm{j}^*\},$$

$$E_3 = \{\{x_n, z^\alpha_{\square,\square}\} : n < \omega, \alpha < \kappa\}.$$

Note that every vertex has infinite degree.

We will assume that there is an unfriendly partition $c : V \to \{0,1\}$ of G and derive a contradiction.

Since $\mathfrak{U}$ is an utrafilter, there are $A \in \mathfrak{U}$ and $\epsilon \in \{0,1\}$ such that $c(x_n) = \epsilon$ if and only if $n \in A$. We will prove that, whenever

$$\alpha < \kappa, \quad \mathrm{i} \in \mathcal{I}, \quad \mathrm{j} \in \mathcal{I}, \quad l(\mathrm{j}) = l(\mathrm{i})+1, \quad \gamma = \mathrm{lt}(\mathrm{i}), \tag{1}$$

and there is a $\rho < \gamma$ such that $A_{\mathrm{i},\rho} = A$ holds, then

$$c(y^\alpha_{\mathrm{i},\mathrm{j}}) = 1-\epsilon \tag{2}$$

and

$$c(z^\alpha_{\mathrm{i},\mathrm{j}^*}) = \epsilon. \tag{3}$$

Note first that (3) follows from (2). For $E(z^\alpha_{\mathrm{i},\mathrm{j}^*}) = C_1 \cup C_2$, where

$$C_1 = \{y^\alpha_{\mathrm{i},\mathrm{j}^*\frown\zeta} : \zeta < \omega_1\} \quad \text{and} \quad C_2 = \begin{cases} \{y^\alpha_{\mathrm{i}^*,\mathrm{j}^*}\} & \text{if } \mathrm{i} \neq \square, \\ X & \text{if } \mathrm{i} = \mathrm{j}^* = \square. \end{cases}$$

Since $|C_1| = \omega_1 > |C_2|$ and since, by (2), $c(y_{\mathrm{i},\mathrm{j}^*\frown\zeta}) = 1-\epsilon$, (3) follows.

We will prove (2) by induction on $\gamma = \mathrm{lt}(\mathrm{i})$.

Consider first the case when $\gamma < \omega$. In this case there is no $\mathrm{i}_1 \in \mathcal{I}$ such that $\mathrm{i} = \mathrm{i}_1^*$. Therefore,

$$E(y^{\alpha}_{\mathrm{i},\mathrm{j}}) = \{x_n : \kappa_n > \alpha \text{ and } n \in \bigcap\{A_{\mathrm{i},\sigma} : \sigma < \gamma\} \cup \{z^{\alpha}_{\mathrm{i},\mathrm{j}^*}\}.$$

But $\bigcap\{A_{\mathrm{i},\sigma} : \sigma < \gamma\}$ is an infinite subset of $A_{\mathrm{i},\rho} = A$ and only finitely many $n < \omega$ fail to satisfy the condition $\kappa_n > \alpha$. Therefore, since c is an unfriendly partition of G, it follows that $c(y^{\alpha}_{\mathrm{i},\mathrm{j}}) = 1-\epsilon$.

Now suppose that $\gamma \geqslant \omega$. In this case,

$$E(y^{\alpha}_{\mathrm{i},\mathrm{j}}) = \{z^{\alpha}_{\mathrm{i}^\frown\tau,\mathrm{j}} : \tau < |\gamma|\} \cup \{z^{\alpha}_{\mathrm{i},\mathrm{j}^*}\}.$$

By the hypothesis (1), there is some $\rho < |\gamma|$ such that $A = A_{\mathrm{i},\rho}$. Also, for any τ such that $\rho < \tau < |\gamma|$, there is some $\sigma < |\tau|$ such that $h(\tau,\sigma) = \rho$, and so

$$A_{\mathrm{i}^\frown\tau,\sigma} = A_{\mathrm{i},\rho} = A.$$

Thus, by the inductive hypothesis, $c(z^{\alpha}_{\mathrm{i}^\frown\tau,\mathrm{j}}) = \epsilon$. It follows that $c(y^{\alpha}_{\mathrm{i},\mathrm{j}}) = 1-\epsilon$, and this completes the proof of (2) and (3) under the hypothesis (1).

In particular, by (3), $c(z^{\alpha}_{\square,\square}) = \epsilon$ for every $\alpha < \kappa$.

For $n \in A$, we have that

$$E(x_n) = D_1 \cup D_2,$$

where

$$D_1 = \{y^{\alpha}_{\mathrm{i},\mathrm{j}} \in Y : \gamma = \mathrm{lt}(\mathrm{i}) < \omega,\ n \in \bigcap\{A_{\mathrm{i},\rho} : \rho < \gamma\} \text{ and } \alpha \leqslant \kappa_n\},$$
$$D_2 = \{z^{\alpha}_{\square,\square} : \alpha < \kappa\}.$$

Since $|D_1| \leqslant |\mathcal{I}|\kappa_n < \kappa = |D_2|$ and $c(x_n) = c(z)$ for all $z \in D_2$, this contradicts the assumption that c is an unfriendly partition. □

4 Sketch of the proof of Theorem 2

The proof is similar to the proof of Theorem 1. First we choose an ultrafilter $\mathfrak{U}$ on λ such that

$$B_n = \{\omega\alpha + n : \alpha < \lambda\} \notin \mathfrak{U} \qquad (n < \omega).$$

Now continue as in the proof of Theorem 1 using this ultrafilter and replacing 2^ω by 2^λ, ω_1 by λ^+, the cardinal successor of λ, X by $\{x_\xi : \xi < \lambda\}$ and replacing E_1 by

$$\{\{x_\xi, y^{\alpha}_{\mathrm{i},\mathrm{j}}\} : \gamma = \mathrm{lt}(\mathrm{i}) < \omega,\ \xi \in \bigcap\{A_{\mathrm{i},\rho} : \rho < \gamma\}$$
$$\text{and } (\exists\, n)(\xi \in B_n \text{ and } \alpha \leqslant \kappa_n)\}. \quad \square$$

5 Unfriendly 3-partitions

The following Bernstein-type lemma is probably known.

Lemma 1 *Let $\mathcal{A} = \langle A_i : i \in I \rangle$ be a family of sets such that $|A_i| \geqslant |I| \geqslant \omega$. Then there are pairwise disjoint sets $B_i \subseteq A_i$ $(i \in I)$ such that $|B_i| = |A_i|$.*

Proof Let

$$D = \{|A_i| : i \in I\}, \qquad R = \{\kappa \in D : \kappa > \sum\{\mu : \mu < \kappa, \mu \in D\}\}$$

and, for $\kappa \in R$, let $I(\kappa) = \{i \in I : |A_i| \geqslant \kappa\}$. We can inductively choose subsets $A_i(\kappa) \subseteq A_i$ for $\kappa \in R$ and $i \in I(\kappa)$ so that $|A_i(\kappa)| = \kappa$ and so that $A_i(\kappa) \cap A_j(\mu) = \varnothing$ if $(\kappa, i) \neq (\mu, j)$. The sets

$$B_i = \bigcup\{A_i(\kappa) : i \in I(\kappa), \kappa \in R\} \qquad (i \in I)$$

satisfy the conditions of the lemma. □

Let $G = (V, E)$ be a graph. For a subset $A \subseteq V$, we define

$$\text{nbly}(A) = \{x \in V : |E(x) \cap A| = |E(x)|\}.$$

The set A is *closed* if $\text{nbly}(A) \subseteq A$, and the *closure* of A is $\bar{A}$, the smallest closed set containing A. Note that, if we write $A^* = A \cup \text{nbly}(A)$, then $\bar{A} = A_\alpha$, where $\langle A_\xi : \xi \leqslant \alpha \rangle$ is a continuous increasing sequence of sets such that $A_0 = A$, $A_{\xi+1} = A_\xi^*$ and $A_\alpha^* = A_\alpha$. Thus we may write $\bar{A} \backslash A = \{a_i : i < \lambda\}$, where

$$|E(a_i)| = |E(a_i) \cap (A \cup \{a_j : j < i\}| \qquad (i < \lambda).$$

If h is a function defined on a subset $A \subseteq V$, then we say that h is *satisfactory* for the element $a \in A$ if

$$|\{y \in A \cap E(a) : h(y) = h(a)\}| \leqslant |\{y \in A \cap E(a) : h(y) \neq h(a)\}|,$$

and h is *completely satisfactory* for a if

$$|\{y \in E(a) : y \notin A \text{ or } h(y) = h(a)\}| \leqslant |\{y \in A \cap E(a) : h(y) \neq h(a)\}|.$$

Of course, if h is satisfactory on the set $B \subseteq A$, then it is completely satisfactory on $B \cap \text{nbly}(A)$. It is also clear that, if h is completely satisfactory on $B \subseteq A$, then so also is any extension of h. In particular, if the domain of h is V, the terms satisfactory and completely satisfactory coincide. An *unfriendly 3-partition* of the graph G is a function $h: V \to \{0, 1, 2\}$ which is satisfactory for every vertex.

Lemma 2 *Let $A, B \subseteq V$, B infinite and $A \cap B = \varnothing$, and suppose that, for $z \in B$,*

$$|E(z)\backslash A| \leqslant |B| \Rightarrow E(z)\backslash A \subseteq B,$$

$$|E(z)\backslash A| > |B| \Rightarrow |[E(z)\backslash A] \cap B| = |B|.$$

If $h\colon A \cup B \to \{0,1,2\}$, then there is $g\colon \overline{A \cup B} \to \{0,1,2\}$ extending h which is satisfactory for every element of $[\overline{A \cup B}\backslash(A \cup B)] \cup B'$, where $B' = \{b \in B : |E(b) \cap [\overline{A \cup B}\backslash(A \cup B)]| > |E(b) \cap (A \cup B)|\}$.

Proof Let $b \in B'$. If $|E(b)\backslash A| \leqslant |B|$, then $E(b)\backslash A \subseteq B$ and so

$$E(b) \cap \overline{A \cup B} = E(b) \cap (A \cup B),$$

which is a contradiction. Therefore, $|E(b)\backslash A| > |B|$ and hence

$$|[E(b)\backslash A] \cap B| = |B|.$$

It follows that

$$|E(b) \cap [\overline{A \cup B}\backslash(A \cup B)]| > |B|$$

for $b \in B'$ and, hence, by Lemma 1, there are pairwise disjoint sets

$$F(b) \subseteq E(b) \cap [\overline{A \cup B}\backslash(A \cup B)] \qquad (b \in B')$$

such that $|F(b)| = |E(b) \cap \overline{A \cup B}|$.

Let $\{z_i : i < \lambda\}$ be an enumeration of the elements of $\overline{A \cup B}\backslash(A \cup B)$ such that

$$|E(z_i)| = |E(z_i) \cap (A \cup B \cup \{z_j : j < i\})| \qquad (i < \lambda).$$

We extend h to the function $g\colon \overline{A \cup B} \to \{0,1,2\}$ by choosing $g(z_i) \in \{0,1,2\}$ inductively for $i < \lambda$. At the i-th step there are two possible choices for $g(z_i)$ that will ensure that g is satisfactory for z_i; consequently, if $z_i \in F(b)$ for some $b \in B'$, then we may also choose $g(z_i)$ different from $g(b)$. The function g so constructed satisfies the requirements of the lemma. □

We now prove Theorem 4 that *every graph has an unfriendly 3-partition.*

Proof We will prove by induction on the infinite cardinal μ that the following assertion holds.

$\mathcal{P}_\mu$: *Let $G = (V, E)$ be a graph and let $A, B \subseteq V$ be subsets such that*

$$A = \bar{A}, \qquad A \cup B = V, \qquad A \cap B = \varnothing, \qquad |B| = \mu.$$

If $x \in B$, $c < 3$ and $h: A \to \{0,1,2\}$, then there is $g: V \to \{0,1,2\}$ extending h such that $g(x) \neq c$ and g is satisfactory for every element of B.

The theorem follows from this since every finite graph has an unfriendly 2-partition and $\mathcal{P}_\mu$ (with $A = \varnothing$ and $B = V$) implies that every graph of cardinality μ has an unfriendly 3-partition.

Case $\mu = \omega$.

Since A is closed and B is denumerable, it follows that $0 < |E(y)| \leqslant \omega$ for $y \in B$. We define an ordinal $\alpha < \omega_1$ and subsets B_β $(\beta \leqslant \alpha)$ of B so that

$$B_0 = \{y \in B : |E(y)| < \omega\},$$

$$B_\beta = \left\{y \in B \Big\backslash \bigcup_{\gamma<\beta} B_\gamma : \left|E(y) \cap \bigcup_{\gamma<\beta} B\right| = \omega\right\} \qquad (0 < \beta < \alpha)$$

and $E(y) \cap \bigcup_{\beta<\alpha} B_\beta$ is finite for all $y \in B_\alpha = B \Big\backslash \bigcup_{\beta<\alpha} B_\beta$. Let $\{\epsilon_1, \epsilon_2, \epsilon_3\} = \{0,1,2\}$ be such that

$$c \notin \begin{cases} \{\epsilon_0, \epsilon_2\} & \text{if } x \in B_0, \\ \{\epsilon_0, \epsilon_1\} & \text{if } x \notin B_0. \end{cases}$$

We will construct the extension g of h so that range$(g \restriction B_0) \subseteq \{\epsilon_0, \epsilon_2\}$ and range$(g \restriction B \backslash B_0) \subseteq \{\epsilon_0, \epsilon_1\}$. This will ensure that $g(x) \neq c$.

First define $g_1 = \{(y, \epsilon_1) : y \in B_1\}$. Now inductively define $g_\beta: B_\beta \to \{\epsilon_0, \epsilon_1\}$ for $1 < \beta < \alpha$ in such a way that, for each $y \in B_\beta$,

$$|\{z : z \in B_\gamma \ (1 \leqslant \gamma < \beta),\ g_\gamma(z) \neq g_\beta(y)\}| = \omega.$$

The set B_α is either empty or denumerable and every vertex of $G \restriction B_\alpha$ has infinite degree; so there is a map $g_\alpha: B_\alpha \to \{\epsilon_0, \epsilon_1\}$ that is satisfactory for every element of B_α. The function $g' = h \cup \bigcup_{1 \leqslant \beta \leqslant \alpha} g_\beta$, defined on $V \backslash B_0$, is completely satisfactory for the elements of $\bigcup_{1<\beta\leqslant\alpha} B_\beta$.

We now imitate the proof that any locally finite graph has an unfriendly 2-partition to define $g'': B_0 \to \{\epsilon_0, \epsilon_1\}$. For each finite set $K \subseteq B_0$, we can choose a map $g_K: K \to \{\epsilon_0, \epsilon_2\}$ so that $g_K \cup g'$ is satisfactory for elements of K. Since every vertex of B_0 has finite degree, it follows by compactness that there is $g: V \to \{0,1,2\}$ extending g' which is satisfactory for elements of B_0 and satisfies range$(g \restriction B_0) \subseteq \{\epsilon_0, \epsilon_2\}$. Since g is constantly ϵ_1 on B_1, it follows that g is satisfactory for all elements of B.

Case $\mu > \omega$.

We may assume without loss of generality that

(a) $\overline{A \cup B'} \neq V$ for $B' \subseteq B$ with $|B'| < |B|$.

For, suppose that $\overline{A \cup B'} = V$, where $\omega \leqslant \kappa = |B'| < \mu$. We can assume that $x \in B'$ and also that, for all $y \in B'$,

$$|E(y)\backslash A| \leqslant \kappa \Rightarrow E(y)\backslash A \subseteq B',$$

$$|E(y)\backslash A| > \kappa \Rightarrow |[E(y)\backslash A] \cap B'| = \kappa.$$

By the inductive hypothesis, $\mathcal{P}_\kappa$ holds and so there is an extension $h' \colon A \cup B' \to \{0, 1, 2\}$ of h which is satisfactory for elements of B'. Now it follows from Lemma 2 that there is $g \colon V = \overline{A \cup B'} \to \{0, 1, 2\}$ extending h' which is satisfactory for all elements of $[V\backslash(A \cup B')] \cup B''$, where

$$B'' = \{y \in B' : |E(y) \cap [V\backslash(A \cup B')]| > |E(y) \cap (A \cup B')|\}.$$

But if $y \in B' \backslash B''$, then $|E(y)| = |E(y) \cap (A \cup B')|$ and so h' is completely satisfactory for y and, hence, so also is g. Thus g is satisfactory for all the elements of $B = V\backslash A$.

By the assumption (a) it follows that there are subsets A_α $(\alpha \leqslant \mu)$ and B_α $(\alpha < \mu)$ of V such that

(b) $A_0 = A$, $A_{\alpha+1} = \overline{A_\alpha \cup B_\alpha}$, $A_\alpha = \bigcup_{\beta < \alpha} A_\beta$ (α a limit) and $A_\mu = V$;

(c) $x \in B_0$ and $B_\alpha \subseteq B\backslash A_\alpha$ $(\alpha < \mu)$;

(d) $B_\alpha = \varnothing$ if α is a limit;

(e) if α is a non-limit then

$$|B_\alpha| = |\alpha| + \omega$$

and, for every $y \in \bigcup_{\beta \leqslant \alpha} B_\beta$,

$$|E(y)\backslash A_\alpha| \leqslant |\alpha| + \omega \Rightarrow E(y)\backslash A_\alpha \subseteq B_\alpha,$$

$$|E(y)\backslash A_\alpha| > |\alpha| + \omega \Rightarrow |[E(y)\backslash A_\alpha] \cap B_\alpha| = |B_\alpha|.$$

If $\alpha < \mu$ and B_β has been defined for $\beta < \alpha$, then A_α is defined by (b); and it follows by (a) and the fact that (e) holds for $\beta < \alpha$ that $|B\backslash A_\alpha| = |B|$, and so we can choose B_α satisfying (c), (d) and (e). At the same time, at non-limit stages, we can also choose the set B_α so that it contains the first element of $B\backslash A_\alpha$ in some well ordering of B (in type μ); this will ensure that the construction stops with $A_\mu = V$.

For an infinite cardinal $\kappa < \mu$, denote by Y_κ the set of all elements $y \in \bigcup_{\beta < \kappa^+} B_\beta$ such that $|E(y) \cap \bigcup_{\beta < \kappa^+} B_\beta| = \kappa^+$. Since $|Y_\kappa| \leqslant \kappa^+$, it follows that there are pairwise disjoint sets $I_\kappa(y) \subseteq \{\alpha : \kappa \leqslant \alpha < \kappa^+\}$

($y \in Y_\kappa$) each of cardinality κ^+ such that $E(y) \cap B_\alpha \neq \emptyset$ for $\alpha \in I(y)$. Now choose elements $x_\alpha \in B_\alpha$ for non-limit $\alpha < \mu$ so that $x_0 = x$ and $x_\alpha \in E_\kappa(y)$ if $\alpha \in I_\kappa(y)$ for some $\kappa < \mu$ and $y \in Y_\kappa \cap \bigcup_{\beta<\alpha} B_\beta$ (and x_α is chosen arbitrarily in B_α if there is no such y).

We shall define inductively a continuously increasing sequence of functions $g_\alpha : A_\alpha \to \{0,1,2\}$ for $\alpha < \mu$ so that, at non-limit stages, the following conditions hold:

(f) $g_{\alpha+1}(x_{\alpha+1}) \neq g(y)$ if there are $\kappa < \mu$ and $y \in Y_\kappa$ such that $\alpha+1 \in I_\kappa(y)$;

(g) $g_{\alpha+1}$ is satisfactory for every element of
$$[A_{\alpha+1} \backslash (A_\alpha \cup B_\alpha)] \cup B'_\alpha,$$
where
$$B'_\alpha = \left\{ y \in \bigcup_{\beta \leqslant \alpha} B_\beta : |E(y) \cap [A_{\alpha+1} \backslash (A_\alpha \cup B_\alpha)]| > |E(y) \cap (A_\alpha \cup B_\alpha)| \right\}.$$

Put $g_0 = h$. At limit stages we define $g_\alpha = \bigcup_{\beta<\alpha} g_\beta$. Suppose that $\alpha < \mu$ and that $g_\alpha : A_\alpha \to \{0,1,2\}$ has already been defined. We want to define $g_{\alpha+1}$ so that (f) and (g) hold. If α is a non-limit, then A_α is closed and so, by the inductive hypothesis $\mathcal{P}_{|\alpha|+\omega}$ applied to the subgraph $G_\alpha = G \restriction A_\alpha \cup B_\alpha$, there is $g'_\alpha : A_\alpha \cup B_\alpha \to \{0,1,2\}$ which extends g and which is satisfactory for every element of B_α. Further, we may assume that $g'_\alpha(x_\alpha) \neq c_\alpha$, where $c_\alpha = g_\alpha(y)$ if $\alpha \in I_\kappa(y)$ for some $\kappa < \mu$ and $y \in Y_\kappa$, and $c_\alpha = c$ otherwise. If α is a limit ordinal, we simply put $g'_\alpha = g_\alpha$.

We want to apply Lemma 2, with
$$A = A_\alpha \Big\backslash \bigcup_{\beta \leqslant \alpha} B_\beta \quad \text{and} \quad B = \bigcup_{\beta \leqslant \alpha} B_\beta.$$

Let $z \in \bigcup_{\beta \leqslant \alpha} B_\beta$. If
$$\left| E(z) \Big\backslash \Big(A_\alpha \Big\backslash \bigcup_{\beta \leqslant \alpha} B_\beta \Big) \right| \leqslant \left| \bigcup_{\beta \leqslant \alpha} B_\beta \right| = |\alpha| + \omega,$$
then
$$E(z) \Big\backslash \Big(A_\alpha \Big\backslash \bigcup_{\beta \leqslant \alpha} B_\beta \Big) \subseteq \bigcup_{\beta \leqslant \alpha} B_\beta \quad \text{by (e);}$$
if

$$\left|E(z)\setminus\left(A_\alpha\setminus\bigcup_{\beta\leqslant\alpha}B_\beta\right)\right| > \left|\bigcup_{\beta\leqslant\alpha}B_\beta\right|$$

then

$$\left|\left[E(z)\setminus\left(A_\alpha\setminus\bigcup_{\beta\leqslant\alpha}B_\beta\right)\right]\cap B_\alpha\right| = \left|\bigcup_{\beta\leqslant\alpha}B_\beta\right|.$$

Thus the conditions of the lemma are satisfied. Therefore, there is a function $g_{\alpha+1}\colon A_{\alpha+1}\to\{0,1,2\}$ extending g'_α which satisfies both (f) and (g).

This defines the g_α for $\alpha\leqslant\mu$. It remains to show that $g=g_\mu$ is satisfactory for every element of B. Let $z\in B$. If $z\notin\bigcup_{\alpha<\mu}B_\alpha$, then $z\in A_{\alpha+1}\setminus(A_\alpha\cup B_\alpha)$ for some $\alpha<\mu$. Since $A_{\alpha+1}$ is the closure of $A_\alpha\cup B_\alpha$, it follows that $|E(z)|=|E(z)\cap A_{\alpha+1}|$. Since $g_{\alpha+1}$ is satisfactory for z, it is completely satisfactory and, hence, g is also satisfactory for z. Suppose now that $z\in B_\alpha$ for some non-limit $\alpha<\mu$. Let $\beta\leqslant\mu$ be minimal such that $|E(z)|=|E(z)\cap A_\beta|$. Then $\beta>\alpha$ since A_α is closed and $z\notin A_\alpha$. In order to show that g is satisfactory for z we shall consider separately the following cases.

Case 1 $\beta=\gamma+1$ is a successor ordinal.

Case 1(i) $|E(z)|=|E(z)\cap B_\gamma|$.

For non-limit ξ $(\alpha\leqslant\xi<\gamma)$, there holds $|E(z)\setminus A_\xi|>|\xi|+\omega$, otherwise $E(z)\subseteq A_\xi\cup B_\xi\subseteq A_\gamma$, which contradicts the choice of β. If $\gamma>\alpha$, then

$$|E(z)\cap A_\gamma|\geqslant\sum\{|\xi|+\omega:\alpha\leqslant\xi<\gamma\}=|\gamma|+\omega=|B_\gamma|\geqslant|E(z)|.$$

This again is a contradiction, and so $\gamma=\alpha$. Therefore,

$$|E(z)|=|E(z)\cap B_\alpha|.$$

Since g'_α is satisfactory for z, it is completely satisfactory and, hence, so is g.

Case 1(ii) $|E(z)|>|E(z)\cap B_\gamma|$.

Since $|E(z)|>|E(z)\cap A_\gamma|$, it follows that

$$\begin{aligned}|E(z)|=|E(z)\cap A_{\gamma+1}|&=|E(z)\cap[A_{\gamma+1}\setminus(A_\gamma\cup B_\gamma)]|\\&>|E(z)\cap(A_\gamma\cup B_\gamma)|.\end{aligned}$$

Therefore, $g_{\gamma+1}$ is completly satisfactory for z, and so is g.

Case 2 β is a limit ordinal.

Case 2(i) $|E(z)| \leqslant |\beta|$.

For non-limit ξ $(\alpha \leqslant \xi < \beta)$, we have $|E(z) \backslash A_\xi| > |\xi| + \omega$ (else $E(z) \subseteq A_{\xi+1}$) and, hence, $E(z) \cap B_\xi \neq \emptyset$. It follows that $|\beta| = \beta$ and

$$|\{\xi : \alpha \leqslant \xi < \beta, \xi \in I_\kappa(z) \text{ for some } \kappa < \beta\}| = |\beta|.$$

Since $g_\xi(z) \neq g_\xi(x_\xi)$ if $\xi \in I_\kappa(z)$, it follows that g_β is completely satisfactory for z, and therefore so is g.

Case 2(ii) $|E(z)| > |\beta|$.

In this case $|E(z)| = \lambda$ is singular and there are an increasing sequence of cardinals λ_ι $(\iota < \mathrm{cf}(\lambda))$ and an increasing sequence of ordinals β_ι $(\iota < \mathrm{cf}(\lambda))$ such that $\lambda = \sup \lambda_\iota$, $\beta = \sup \beta_\iota$ and

$$|E(z) \cap [A_{\beta_\iota+1} \backslash (A_{\beta_\iota} \cup B_{\beta_\iota})]| = \lambda_\iota > |E(z) \cap (A_{\beta_\iota} \cup B_{\beta_\iota})|.$$

But this implies that $g_{\beta_\iota+1}$ $(\iota < \mathrm{cf}(\lambda))$ is satisfactory for z, i.e.

$$|\{y \in E(z) \cap A_{\beta_\iota+1} : g_{\beta_\iota}(y) \neq g_{\beta_\iota}(z)\}| = \lambda_\iota \qquad (\iota < \mathrm{cf}(\lambda)).$$

From this it follows that g is satisfactory for z, and this completes the proof. □

References

[1] R. Aharoni, E. C. Milner & K. Prikry, Unfriendly partitions of a graph (to appear in *J. Combin. Theory*)

[2] R. Cowen & W. Emerson, Proportional colorings of graphs (unpublished)

On the greatest prime factor of an arithmetical progression

T. N. Shorey* and R. Tijdeman

Abstract

We prove that the greatest prime factor of the terms of a finite arithmetical progression of positive integers with difference greater than one is greater than the number of terms, with one explicitly given exception.

Let a, d and k be positive integers with $\gcd(a,d) = 1$. A classical theorem of Sylvester [14] states that the product $a(a+d)\cdots(a+(k-1)d)$ is divisible by some prime number greater than k provided that $a \geqslant d+k$. Sylvester added that his proof is wearisome. Langevin [6] indicated how a simple proof can be derived under the weaker condition $a > k$. The latter result has been applied by Marszałek [9] to a problem of Erdős. In view of further application we want to get rid of the condition. In order to avoid trivialities we assume $k > 2$. Let $P(x)$ denote the greatest prime factor of x.

Theorem *Let a, d and k be positive integers with $d > 1$, $k > 2$ and $\gcd(a,d) = 1$. Then*

$$P\big(a(a+d)\cdots(a+(k-1)d)\big) \leqslant k \tag{1}$$

implies $a = 2$, $d = 7$ and $k = 3$.

The complementary case $d = 1$ is different and admits infinitely many exceptions. If $a = k+1$, Sylvester's result is equivalent to the Postulate of Bertrand which was first proved by Chebyshev [2]. The result for any $a > k$ was rediscovered and again proved by Schur [13] in 1929. The proofs were simplified by Erdős [3]. It follows from Sylvester's theorem

* Research supported by the Netherlands Organization for the Advancement of Pure Research (Z.W.O.).

that the greatest prime factor of $a(a+1)\cdots(a+k-1)$ is at most k if and only if $\pi(k) = \pi(a+k-1)$. The problem is therefore equivalent to the well-known problem on gaps between consecutive prime numbers. Using results on this gap problem we see that the condition $a > k$ can be replaced by $a > \frac{1}{13}k$ for $k > 118$ by a result of Rohrbach & Weis [11] and by $a > k^{23/42}$ for k sufficiently large by improvements of the theorem of Hoheisel & Ingham (cf. Iwaniec & Pintz [4]).

If $d = 2$, then the condition $a > k$ can be dropped by Bertrand's Postulate. For $d = 3$, 4, 6 and 12, Breusch [1] and Molsen [10] obtained extensions of Bertrand's Postulate. Breusch proved that for $n \geqslant 7$ there is a prime number between n and $2n$ in each of the arithmetical progressions $3j+1$, $3j+2$, $4j+1$ and $4j+3$ and Molsen showed that for $n \geqslant 118$ the interval $n < x < \frac{4}{3}n$ contains a prime number from each of the arithmetical progressions $12j+1$, $12j+5$, $12j+7$ and $12j+11$. It follows from these results that in Sylvester's theorem the condition $a \geqslant d+k$ can be dropped if $d = 3$, 4 and 6 and further if $d = 12$ and $k \geqslant 14$.

There are various other improvements of the theorems of Sylvester and Langevin in the literature, in particular improvements of the bound k for the greatest prime factor (at the cost of restrictions on a and d). For such results we refer the reader to Langevin [5, 6, 7, 8], where further references can also be found. We are greatful to Dr M. Langevin and Dr J. Pintz for some helpful comments.

Proof Suppose that we have integers a, d and k as in the theorem such that (1) holds. By Breusch's result we may assume that $d \geqslant 5$ and $d \neq 6$.

If $k = 3$ then

$$P(a(a+d)(a+2d)) \leqslant 3.$$

Hence either $a = 1$, $a+d = 2^l$ and $a+2d = 3^m$ or $a = 2^l$, $a+d = 3^m$ and $a+2d = 2^n$ for certain positive integers l, m and n. In the first instance we have $2^{l+1} = 3^m+1$ and $l > 1$. Since $3^m \equiv 7 \pmod 8$, there are no solutions. In the second instance we have $2\times 3^m = 2^l+2^n$ with $l < n$. This implies $l = 1$, $n > 2$ and $3^m = 2^{n-1}+1$. Since $3^m \equiv 1 \pmod 4$, m is even. Hence $2^{n-1} = (3^{m/2}-1)(3^{m/2}+1)$. Since 2 and 2^2 are the only powers of 2 which differ by 2, we find that $m = 2$ and $n = 4$ and that the only case with $k = 3$ is $a = 2$ and $d = 7$.

If $k = 4$ then

$$P(a(a+d)(a+2d)) \leqslant 3 \quad \text{and} \quad P(a+3d) \leqslant 3.$$

The only values satisfying the first inequality are $a = 2$ and $d = 7$. Since $P(23) > 3$, we obtain $k \neq 4$.

If $k = 5$ or 6, then $P(a(a+d)(a+2d)(a+3d)(a+4d)) \leqslant 5$. At most one of the numbers a, $a+d$, $a+2d$, $a+3d$ and $a+4d$ is divisible by 5. Applying the result in the case $k = 3$ to a, $a+d$ and $a+2d$, and to $a+2d$, $a+3d$ and $a+4d$, we obtain that $5|a(a+d)(a+2d)$ and $5|(a+2d)(a+3d)(a+4d)$, whence $5|a+2d$. If a is even, then $a+d$ and $a+3d$ are odd and at most one of them is divisible by 3. The other number has to be 1, which is impossible. If a is odd, then $a+4d$ must be a power of 3 and $a = 1$. Furthermore, $a+3d$ must be a power of 2. Since $7 \leqslant a+3d < a+4d$, we obtain $2\|a+d$ and $3\|a+d$, whence $a+d = 6$. It follows that $a = 1$ and $d = 5$, but the sequence 1, 6, 11, 16, 21 does not meet the requirements.

If $k = 7$, 8, 9 or 10, then

$$P(a(a+d)(a+2d)(a+3d)(a+4d)(a+5d)(a+6d)) \leqslant 7.$$

If $a+d$ is even, then a, $a+2d$, $a+4d$ and $a+6d$ are all odd. If, moreover, $3|a$, then $5|a+2d$ and $7|a+4d$ or $7|a+2d$ and $5|a+4d$. It follows that $P((a+d)(a+3d)(a+5d)) \leqslant 3$, which contradicts the result in the case $k = 3$. If, on the other hand, $3 \nmid a$, then there is at most one multiple of 3, at most one multiple of 5 and at most one multiple of 7 among a, $a+2d$, $a+4d$ and $a+6d$. It follows that $a = 1$ and that $a+3d$ is a perfect power of 2. Furthermore, $2\|a+5d$, $5 \nmid a+5d$, $7 \nmid a+5d$ and $a+5d > 2$, whence $3|a+5d$. Also $2|a+d$, $3 \nmid a+d$ and $7 \nmid a+d$, whence $5|a+d$. Since $a+6d$ is a power of 5 larger than 5, we have that $5\|a+d$. We conclude that $a+d = 10$, but the sequence with $a = 1$ and $d = 9$ does not meet the requirements. Thus $a+d$ cannot be even, and $a+d$, $a+3d$ and $a+5d$ are odd. It follows that one among them is a power of 3 greater than 3, one among them is a power of 5 and one among them is a power of 7. We infer that $5 \nmid (a+2d)(a+4d)$, $7 \nmid (a+2d)(a+4d)$ and $9 \nmid (a+2d)(a+4d)$. Thus one among $a+2d$ and $a+4d$ is a power of 2 greater than 4 and the other has only one factor 2 and one factor 3. Thus $a+2d \leqslant 6$, which is a contradiction.

In the sequel we may assume without loss of generality that $d \geqslant 5$, $d \neq 6$ and $k \geqslant 11$. Denote the number of primes p with $p \leqslant k$ and $\gcd(p,d) = 1$ by $\pi_{d,k}$. For each such prime we choose a term from $a, a+d, \ldots, a+(k-1)d$ such that none of the others is divisible by a higher power of p. Taking away the chosen terms we are left with at least $k - \pi_{d,k}$ numbers. By counting prime factors p for each p, we find that the product of the remaining terms is a divisor of

$$\prod_{p<k} p^{[(k-1)/p]+[(k-1)/p^2]+\cdots} = (k-1)!\,.$$

We infer, by (1) and $\gcd(a,d)=1$, that

$$\prod_{j=0}^{k-\pi_{d,k}-1} (a+jd) \leqslant (k-1)!\,.$$

On the other hand

$$\prod_{j=0}^{k-\pi_{d,k}-1} (a+jd) \geqslant (k-\pi_{d,k}-1)!\, d^{k-\pi_{d,k}-1}$$

and, if $a \geqslant d$, we have the sharper inequality

$$\prod_{j=0}^{k-\pi_{d,k}-1} (a+jd) \geqslant (k-\pi_{d,k})!\, d^{k-\pi_{d,k}}.$$

We conclude that

$$d^{k-\pi_{d,k}-1} \leqslant (k-1)(k-2)\cdots(k-\pi_{d,k}), \tag{2}$$

and

$$d^{k-\pi_{d,k}} \leqslant (k-1)(k-2)\cdots(k-\pi_{d,k}+1) \qquad \text{when } a \geqslant d. \tag{3}$$

Since $\pi_{d,k} \leqslant \pi(k) \leqslant \frac{1}{2}(k-1)$ for $k \geqslant 11$, we obtain $d \leqslant k-1$. Hence there is a prime p with $p\,|\,d$, $p \leqslant k$ and $\pi_{d,k} \leqslant \pi(k)-1$. We substitute this inequality in (2) and (3) and derive upper bounds for d when $k = 11, 12, \ldots, 275$. From the calculations it follows that $d < 6$ for all these k and that $d < 5$ unless $a < d$ and $k \in \{13, 19, 20, 23, 31\}$. Therefore at least one of the prime numbers 41, 43, 47 and 59 must belong to the arithmetical progression a, $a+d$, $a+2d$, $\ldots$, $a+(k-1)d$. Then (1) cannot hold. Thus $k \geqslant 276$.

By formula (3.2) of Rosser & Schoenfeld [12] we have

$$\pi(x) \leqslant \frac{x}{\log x}\left(1+\frac{3}{2\log x}\right) \qquad \text{for } x > 1. \tag{4}$$

We substitute $\pi_{d,k} \leqslant \pi(k)-1$ in (2) and use estimate (4) for $\pi(k)$. Hence

$$d \leqslant k^{\{\pi(k)-1\}/\{k-\pi(k)\}} \leqslant \exp\left(\frac{1+\dfrac{3}{2\log k}-\dfrac{\log k}{k}}{1-\dfrac{1}{\log k}-\dfrac{3}{2\log^2 k}}\right).$$

Observe that the right-hand side is monotone decreasing in $k \geqslant 276$. We infer that $d < 5$ for $k \geqslant 276$. This completes the proof of the theorem. □

References

[1] R. Breusch, Zur Verallgemeinerung des Bertrandschen Postulates, dass zwischen x und $2x$ stets Primzahlen liegen, *Math. Z.*, **34** (1932), 505–26

[2] P. L. Chebyshev, Mémoire sur les nombres premiers, in *Oeuvres de P. L. Tchebychef*, Tome 1, St Pétersbourg, 1899, 51–70

[3] P. Erdős, A theorem of Sylvester and Schur, *J. London Math. Soc.*, **9** (1934), 282–8

[4] H. Iwaniec & J. Pintz, Primes in short intervals. *Monatsh. Math.*, **98** (1984), 115–43

[5] M. Langevin, Méthodes élémentaires en vue du théorème de Sylvester, *Séminaire Délange–Pisot–Poitou*, 17e année, 1975/76, no. G2, 9 p.

[6] M. Langevin, Plus grand facteur premier d'entiers en progression arithmétiques, *Séminaire Délange–Pisot–Poitou*, 18e année, 1976/77, no. 3, 7 p.

[7] M. Langevin, Facteurs premiers d'entiers en progression arithmétique, *Séminaire Délange–Pisot–Poitou*, 19e année, 1977/78, no. 4, 7 p.

[8] M. Langevin, Facteurs premiers d'entiers en progression arithmétique, *Acta Arith.*, **39** (1981), 241–9

[9] R. Marszałek, On the product of consecutive elements of an arithmetic progression, *Monatsh. Math.*, **100** (1985), 215–22

[10] K. Molsen, Zur Verallgemeinerung des Bertrandschen Postulates, *Deutsche Math.*, **6** (1941), 248–56

[11] H. Rohrbach & J. Weis, Zum finiten Fall des Bertrandschen Postulates, *J. reine angew. Math.*, **214–5** (1964), 432–40

[12] B. Rosser & L. Schoenfeld, Approximate formulas for some functions of prime numbers, *Illinois J. Math.*, **6** (1962), 64–94

[13] I. Schur, Einige Sätze über Primzahlen mit Anwendungen auf Irreduzibilitätsfragen I, *Sitzungsber. preuss. Akad. d. Wissensch., phys.-math. Klasse*, 1929, p. 128

[14] J. J. Sylvester, On arithmetic series, *Messenger Math.*, **21** (1892), 1–19 and 87–120

References

[1] [illegible] allgemeine [illegible] (1931) [illegible]

[2] [illegible] sur les nombres premiers [illegible] Oeuvres de P. L. [illegible]

[3] [illegible] 407 [illegible] Math. [illegible]

[4] H. [illegible] Math. [illegible] (19[illegible])

[5] [illegible]

[6] [illegible] nombres premiers [illegible] arithmétiques [illegible]

[7] [illegible] progression arithmétique [illegible]

[8] [illegible] progression [illegible]

[9] [illegible] On the product of consecutive elements of an arithmetic progression [illegible] Math. [illegible]

[10] [illegible]

[11] [illegible] Zahlen [illegible] (1951) [illegible]

[12] [illegible] Approximate formulas for some functions of prime numbers [illegible]

[13] [illegible] Primzahlen [illegible] arithmetischen [illegible]

[14] [illegible] 27 (1953) [illegible] and [illegible]

The probabilistic lens: Sperner, Turan and Bregman revisited

Joel Spencer

Over the past half-century Paul Erdős has been the prime developer of the probabilistic method – a methodology in which random structures are studied to prove combinatorial theorems. The probabilistic method is now well recognized as a vital weapon in the combinatorialist's arsenal. In this revisionist work we examine three classical results: Sperner's theorem on maximal antichains, Turan's theorem on maximal graphs of given independence number and Bregman's theorem on maximal permanents of 0-1 matrices. We prove these theorems using the language and technique of the probabilistic method. The proofs indicate the versatility of the method and – we believe – cast fresh light on the results themselves. Viewed through the probabilistic lens these three results exhibit surprising similarities.

1 Sperner

Let $\mathcal{P}_n$ denote the power set of $[n] = \{1, \ldots, n\}$. Sperner's theorem states that the maximal antichain $\mathcal{F} \subseteq \mathcal{P}_n$ consits of all sets of size $i = \lfloor \frac{1}{2}n \rfloor$ or $\lceil \frac{1}{2}n \rceil$. Let $\mathcal{F} \subseteq \mathcal{P}_n$ be an antichain. Let $\mathcal{C} \subseteq \mathcal{P}_n$ be a random chain; that is, select $\sigma \in S_n$ uniformly and let

$$\mathcal{C} = \{\{\sigma(j) : 1 \leqslant j \leqslant i\}, 0 \leqslant i \leqslant n\}.$$

Let $A \subseteq [n]$ with $|A| = i$. $\mathcal{C}$ contains precisely one i-set and all are equally likely. So

$$\Pr[A \in \mathcal{C}] = 1 \Big/ \binom{n}{i}.$$

Let X_A be the indicator random variable for $A \in \mathcal{C}$ and let $X = |\mathcal{C} \cap \mathcal{F}|$ so that

$$X = \sum_{A \in \mathcal{F}} X_A.$$

Linearity of expectation gives

$$E[X] = \sum_{A \in \mathcal{F}} E[X_A] = \sum_{A \in \mathcal{F}} \Pr[X \in \mathcal{C}]$$
$$= \sum_{A \in \mathcal{F}} 1 \Big/ \binom{n}{|A|}.$$

Since $\mathcal{F}$ is an antichain, $|\mathcal{C} \cap \mathcal{F}| \leqslant 1$ for all chains $\mathcal{C}$. So $X \leqslant 1$ always and thus $E[X] \leqslant 1$; that is

$$\sum_{A \in \mathcal{F}} 1 \Big/ \binom{n}{|A|} \leqslant 1. \qquad (*)$$

This result, known as the LYM inequality, implies Sperner's theorem since $1/\binom{n}{i}$ is minimized at $i = \lfloor \frac{1}{2} n \rfloor$.

Moreover, we may characterize those $\mathcal{F}$ for which equality holds in $(*)$. When equality holds, $E[X] = 1$; but $X \leqslant 1$ always and therefore $X = 1$ always, that is, $|\mathcal{C} \cap \mathcal{F}| = 1$ for all chains $\mathcal{C}$. Now let A, x and y be such that $x, y \notin A$ and suppose $A \cup \{x\} \in \mathcal{F}$. There is a chain $\mathcal{C}$ that contains A, $A \cup \{x\}$ and $A \cup \{x, y\}$, and that $\mathcal{C}$ intersects $\mathcal{F}$ exactly at $A \cup \{x\}$ (as $|\mathcal{C} \cap \mathcal{F}| = 1$). Let $\mathcal{C}'$ be $\mathcal{C}$ with $A \cup \{x\}$ replaced by $A \cup \{y\}$. As $|\mathcal{C}' \cap \mathcal{F}| = 1$, we deduce that $A \cup \{y\} \in \mathcal{F}$. From this implication we deduce that if $\mathcal{F}$ contains one i-set then it contains all i-sets: but therefore it must consist precisely of all i-sets. Thus equality holds in the LYM inequality $(*)$ if and only if $\mathcal{F}$ consists of all sets of a given cardinality. □

2 Turan

Let $\alpha(G)$ denote the size of the largest independent set in a graph G. For any integers $m \leqslant n$ let $n = qm + r$ $(0 < r < m)$. Let $T(n, m)$ denote the graph on n vertices consisting of r cliques of size $q+1$ and $m-r$ cliques of size q, all vertex disjoint. Let

$$e = e(n, m) = r\binom{q+1}{2} + (m-r)\binom{q}{2}$$

denote the number of edges of $T(n, m)$. Turan's theorem states that any graph G with n vertices and at most e edges has $\alpha(G) > m$, with equality if and only if $G \cong T(n, m)$.

Let G be a graph on vertex set $\{1, \ldots, n\}$ with degrees $d_1, \ldots, d_n$. Let $<$ be a random order of $\{1, \ldots, n\}$ (i.e., chosen uniformly from the $n!$

posibilities) and define

$$C = \{i : j < i \Rightarrow \{j, i\} \notin G\};$$

that is, $i \in C$ if and only if all edges from i go to the 'right' under $<$. As i and its d_i neighbours are ranked randomly,

$$\Pr[i \in C] = \frac{1}{d_i + 1}.$$

Let X_i be the indicator random variable for $i \in C$ and let $X = |C|$ so that

$$X = \sum_{i=1}^{n} X_i.$$

Linearity of expectation gives

$$E[X] = \sum_{i=1}^{n} E[X_i] = \sum_{i=1}^{n} \Pr[i \in C] = \sum_{i=1}^{n} \frac{1}{d_i + 1}.$$

Thus, for some ordering $<$,

$$|C| = X \geqslant \sum_{i=1}^{n} \frac{1}{d_i + 1}.$$

C is an independent set since, if $i, j \in C$, either $i < j$ or $j < i$, but in either case $\{i, j\} \notin G$. Thus

$$\alpha(G) \geqslant |C| \geqslant \sum_{i=1}^{n} \frac{1}{d_i + 1}.$$

Now suppose G has $e' \leqslant e(n, m)$ vertices. Then $d_1 + \cdots + d_n = 2e'$ is bounded from above. The function $f(x) = 1/(x+1)$ is concave and decreasing. Hence the value $f(d_1) + \cdots + f(d_n)$ is minimized when the d_i are as nearly equal as possible and $d_1 + \cdots + d_n$ is as large as possible. This occurs when the d_i are as in $T(n, m)$. For those d_i,

$$\sum_{i=1}^{n} \frac{1}{d_i + 1} = m,$$

as each clique contributes one to the summation. Hence, for any given G,

$$\alpha(G) \geqslant \sum_{i=1}^{n} \frac{1}{d_i + 1} \geqslant m.$$

For $\alpha(G) = m$ we must have $e' = e$ and the d_i as in $T(n,m)$. Then $E[X] = m$ and so we must further have $X = m$ for all orderings $<$. (Note that this is the case for $T(n,m)$, as for any ordering $<$ the set C consists of the 'first' element of each clique.) Suppose G contains three vertices x, y and z with $\{x,y\},\{x,z\} \in G$ but $\{y,z\} \notin G$. Let $<$ be an ordering with $x < y < z$ the first three elements. Let $<'$ be the same ordering except that $y <' z <' x$. The corresponding sets C and C' would then be the same except that $x \in C$ and $y,z \notin C$, whereas $x \notin C'$ and $y,z \in C'$. But then X would not be constant: a contradiction. Thus no such x, y and z can exist and thus G must be a union of cliques. The values of d_i further force G to be isomorphic to $T(n,m)$. □

3 Bregman

Let $A = [a_{ij}]$ be an $n \times n$ 0-1 matrix with row sums $r_1,\ldots,r_n$ all positive. The permanent per(A) is the number of permutaions $\sigma \in S_n$ with $a_{i,\sigma(i)} = 1$ for all i. Bregman's theorem is a bound

$$\text{per}(A) \leqslant \prod_{i=1}^{n} (r_i!)^{1/r_i}$$

with equality if and only if A is decomposible into 'blocks' of all-1 submatrices.

Let S be the set of all σ with $a_{i,\sigma(i)} = 1$ for all i so that $|S| = \text{per}(A)$. Pick $\sigma \in S$ and $\tau \in S_n$ uniformly and independently. Let $A^1 = A$. Let $R_{\tau(i)}$ be the $\tau(1)$-row sum in A^1 (so that $R_{\tau(1)} = r_{\tau(1)}$). Delete row $\tau(1)$ and column $\sigma\tau(1)$ from A^1 to give A^2. In general, let A^i be A with rows $\tau(1),\ldots,\tau(i-1)$ and columns $\sigma\tau(1),\ldots,\sigma\tau(i-1)$ deleted and let $R_{\tau(i)}$ be the $\tau(i)$-row sum in A^i. (This is non-zero as $a_{\tau(i),\sigma\tau(i)} = 1$.) Set

$$L = L(\sigma,\tau) = \prod_{i=1}^{n} R_{\tau(i)} = \prod_{i=1}^{n} R_i.$$

We think, roughly, of L as Lazyman's permanent calculation. There are $r_{\tau(1)}$ choices for a 1 in row $\tau(1)$, each of which leads to a different subpermanent calculation. Lazyman takes the factor $r_{\tau(1)}$, picks the 1 from permutation σ and examines only A^2. As $\sigma \in S$ is chosen uniformly, Lazyman tends towards high subpermanents and so it should not be surprising that he tends to overestimate the permanent. To make this precise we define the geometric mean $G[Y]$. If $Y > 0$ takes values $a_1,\ldots,a_s$ with probabilities $p_1,\ldots,p_s$, respectively, then

$$G[Y] = \prod_{i=1}^{s} a_i^{p_i}.$$

Equivalently, $G[Y] = \exp(E[\ln Y])$. Linearity of expectation translates into the geometric mean of a product being the product of the geometric means.

Claim $\text{per}(A) \leqslant G[L]$.

In fact, this holds for any fixed τ. Let us set $\tau(i) = i$ for convenience of notation. We use induction on the size n of the matrix. Reorder so that the first row has ones in the first $r = r_1$ columns. For $1 \leqslant j \leqslant r$ let t_j be the number of $\sigma \in S$ with $\sigma = j$. Let $\bar{t} = (t_1, \ldots, t_r)/r$, so that $r\bar{t} = \text{per}(A)$. Conditioning on $\sigma(1) = j$, $R_2 \cdots R_n$ becomes Lazyman's calculation of the permanent of A with the first row and j-th column removed. So, by induction,

$$G[R_2 \cdots R_n \mid \sigma(1) = j] \geqslant t_j.$$

As $\Pr[\sigma(1) = j] = t_j/\text{per}(A)$,

$$G[L] = G[R_1 R_2 \cdots R_n] \geqslant r \prod_{j=1}^{r} t_j^{t_j/\text{per}(A)}.$$

Lemma $[t_1^{t_1} \cdots t_r^{t_r}]^{1/r} \geqslant \bar{t}^{\bar{t}}$.

Proof Taking logarithms, this is equivalent to

$$\frac{1}{r} \sum_{i=1}^{r} t_i \ln t_i \geqslant \bar{t} \ln \bar{t},$$

which follows from the convexity of $f(x) = x \ln x$. □

Applying the lemma

$$G[L] \geqslant r(\bar{t}^{\bar{t}})^{r/\text{per}(A)} = r\bar{t} = \text{per}(A),$$

completing the claim. □

Now fix σ and calculate $G[L]$. For convenience of notation reorder so that $\sigma(i) = i$ for all i and so that the first row has ones in the first $r = r_1$ columns. With τ selected uniformly, the columns $1, \ldots, r$ are selected in random, uniform order. R_1 is the number of those columns remaining when column 1 is to be deleted and so R_1 is uniformly distributed from 1 to $r = r_1$. Thus $G[R_1] = (r_1!)^{1/r_1}$; in general $G[R_i] = (r_i!)^{1/r_i}$ and 'linearity' gives

$$G[L] = \prod_{i=1}^{n} G[R_i] = \prod_{i=1}^{n} (r_i!)^{1/r_i}.$$

This gives Bregman's inequality

$$\mathrm{per}(A) \leqslant G[L] = \prod_{i=1}^{n} (r_i!)^{1/r_i}.$$

For equality to hold it must hold in the claim for any τ. As $f(x) = x \ln x$ is strictly convex it must be that $t_1 = \cdots = t_r$. Thus Lazyman may delete any column with a 1 in row $\tau(1)$ to give A^2. By induction Lazyman may at each stage delete any column with a 1 in row $\tau(i)$ to go from A^{i-1} to A^i and, regardless of his choice, $L = \mathrm{per}(A)$. (Note that this is the case when A is decomposible into blocks.) Now suppose that A is not decomposible into blocks so that, reordering, $a_{11} = a_{12} = a_{22} = 1$, but $a_{21} = 0$. Lazyman might begin by deleting row 1 and column 1, and then row 2 and column 2; so there must be a $\sigma \in S$ with $\sigma(1) = 1$ and $\sigma(2) = 2$. Now imagine Lazyman first takes rows 3 to n, deleting row i and column $\sigma(i)$, and then deletes row 1 and column 2. As $a_{21} = 0$ this would leave $L = 0$: a contradiction. □

On mean convergence of derivatives of Lagrange interpolation

J. Szabados and A. K. Varma*

1 Results

Let

$$(-1 \leqslant)\ x_1 < x_2 < \cdots < x_n\ (\leqslant 1) \tag{1}$$

be a system of nodes of interpolation, $L_n(f,x)$ the corresponding Lagrange interpolation polynomial of degree at most $n-1$ of a continuous function $f \in C[-1,1]$ and

$$w(x) = \prod_{k=1}^{N} |x-y_k|^{\Gamma_k} \qquad (|x| \leqslant 1;\ -1 = y_1 < y_2 < \cdots < y_N = 1;\ \Gamma_k > -1;\ k = 1,\ldots,N)$$

a so-called generalized Jacobi weight ($w \in GJ$).

Let $r \geqslant 0$ be an arbitrary integer; we will be concerned with the weighted mean convergence of $L_n^{(r)}(f,x)$. Our first result will reduce this problem to the weighted convergence of $L_n(f,x)$.

Theorem 1 *Let (1) be the roots of the n-th orthogonal polynomial associated with the weight-function $w(x) \in GJ$. Then, for any $f^{(r)} \in C[-1,1]$, we have*

$$\int_{-1}^{1} |f^{(r)}(x) - L_n^{(r)}(f,x)|^p (1-x^2)^{rp/2} w(x)\,\mathrm{d}x \leqslant c\left(n^{rp}\int_{-1}^{1} |f(x) - L_n(f,x)|^p w(x)\,\mathrm{d}x + E_{n-r-1}(f^{(r)})^p\right)$$

$$(n \geqslant r+1;\ r = 0,1,\ldots;\ 0 < p < \infty), \tag{2}$$

* This paper was completed while the first-named author visited the University of Florida in Gainesville. Research partially supported by Hungarian National Foundation for Scientific Research Grant No. 1801.

where $c > 0$ depends only on p and r and $E_n(\cdot)$ is the best polynomial approximation of degree at most n of the corresponding function.

In some particular cases, Theorem 1 yields the mean convergence of $L_n^{(r)}(f,x)$ (see also the classical papers of Erdős & Feldheim [1] about the case $0 < p < \infty$ and $w(x) = (1-x^2)^{-1/2}$ and Erdős & Turán [2] for the case $p = 2$ and $w(x) \in L^1$).

Corollary *If*

$$w(x) \in GJ, \qquad \frac{w(x)^{1/p-1/2}}{(1-x^2)^{1/4}} \in L^p \tag{3}$$

then, for $f^{(r)}(x) \in C[-1,1]$ *we have*

$$\int_{-1}^{1} |f^{(r)}(x) - L_n^{(r)}(f,x)|^p (1-x^2)^{rp/2} w(x)\, dx \leq cE_{n-r-1}(f^{(r)})^p \qquad (n \geq r+1;\ r = 0,\ldots). \tag{4}$$

Namely, according to a theorem of Nevai [6, Theorem 1], under the conditions (3) we have

$$\int_{-1}^{1} |L_n(f,x)|^p w(x)\, dx \leq c(\max_{|x|\leq 1} |f(x)|)^p.$$

Applying this for $f - p_{n-1}$ instead of f (p_{n-1} being the best approximating polynomial of f of degree at most n), we obtain from Theorem 1,

$$\int_{-1}^{1} |f^{(r)}(x) - L_n^{(r)}(f,x)|^p (1-x^2)^{rp/2} w(x)\, dx \leq c\{n^{rp} E_{n-1}(f)^p + E_{n-r-1}(f^{(r)})^p\}.$$

Hence, and from the well-known inequality

$$E_n(f) \leq \frac{cE_{n-1}(f')}{n}, \tag{5}$$

(4) follows.

In Theorem 1 the weight $(1-x^2)^{rp/2} w(x)$ on the left-hand side of (2) depends on r. In the particular case $p = 2$ and $w(x) = 1/\sqrt{1-x^2}$ we succeeded in getting an estimate with weight independent of r.

Theorem 2 *For any integer* $r \geq 0$, *there exists a system of nodes (1) such that for the corresponding Lagrange interpolation* $L_n(f,x)$ *we have, for all* $f^{(r)} \in C[-1,1]$,

$$\int_{-1}^{1} \frac{\{f^{(r)}(x) - L_n^{(r)}(f,x)\}^2}{\sqrt{1-x^2}} \, dx \leqslant c_r \omega\left(f^{(r)}, \frac{1}{n}\right)^2,$$

where ω is the modulus of continuity of the corresponding function.

Remark Theorem 2 probably remains true for the L^p metric instead of L^2 (for $0 < p < 2$ this is trivial).

2 Proofs

Let

$$\Delta_n(x) = \frac{\sqrt{1-x^2}}{n} + \frac{1}{n^2} \qquad (|x| \leqslant 1; n = 1,2,\ldots)$$

and let Π_n denote the set of algebraic polynomials of degree at most n. For the proof of Theorems 1 and 2 we need the following lemma.

Lemma *If $w(x) \in GJ$ and $P(x) \in \Pi_n$ then*

$$\int_{-1}^{1} |P^{(r)}(x)\Delta_n(x)^r|^p w(x) \, dx \leqslant c \int_{-1}^{1} |P(x)|^p w(x) \, dx \tag{7}$$

and

$$\int_{-1}^{1} |P^{(r)}(x)|^p w(x) \, dx \leqslant c \int_{-1}^{1} |P(x)\Delta_n(x)^{-r}|^p w(x) \, dx \tag{8}$$

for all $n, r = 0, 1, \ldots$ and $0 < p < \infty$.

Proof We make use of the following special case of a theorem of Lubinsky & Nevai [5, Theorem 3]:

$$\int_{-1}^{1} |P'(x)|^p \Delta_n(x)^{p+q} w(x) \, dx \leqslant c \int_{-1}^{1} |P(x)|^p \Delta_n(x)^q w(x) \, dx,$$

where $0 < p < \infty$ and q is an arbitrary real number. Iterating this inequality, we easily obtain

$$\int_{-1}^{1} |P^{(r)}(x)|^p \Delta_n(x)^{p+q} w(x) \, dx \leqslant c \int_{-1}^{1} |P(x)|^p \Delta_n(x)^{q-(r-1)p} w(x) \, dx$$

$$(r = 1,2,\ldots).$$

Choosing $q = (r-1)p$ and $q = -p$, we obtain (7) and (8), respectively. □

Proof of Theorem 1 Let $p_{n-1} \in \Pi_{n-1}$ be the best approximating polynomial of $f^{(r)}(x) \in C[-1,1]$. According to a theorem of Leviatan [4,

Theorem 1], we have

$$|f^{(j)}(x)-p_{n-1}^{(j)}(x)| \leqslant \frac{c}{\{n\Delta_n(x)\}^j}E_{n-j}(f^{(j)}) \qquad (j=0,\ldots,r;\ n\geqslant r).$$

Applying this for $j=r$, we obtain

$$\int_{-1}^{1} |f^{(r)}(x)-p_{n-1}^{(r)}(x)|^p(1-x^2)^{rp/2}w(x)\,\mathrm{d}x \leqslant cE_{n-r-1}(f^{(r)}). \tag{9}$$

On the other hand, from (7)

$$\begin{aligned}\int_{-1}^{1} &|L_n^{(r)}(p_{n-1}-f,x)|^p(1-x^2)^{rp/2}w(x)\,\mathrm{d}x\\ &\leqslant cn^{rp}\int_{-1}^{1} |L_n(p_{n-1}-f,x)|^p w(x)\,\mathrm{d}x\\ &\leqslant cn^{rp}\left(\int_{-1}^{1} |f(x)-L_n(f,x)|^p w(x)\,\mathrm{d}x + E_{n-1}(f)^p\right).\end{aligned} \tag{10}$$

Using (5) again, the inequalities (9) and (10) yield (2). □

Proof of Theorem 2 Define the nodes of interpolation (1) as follows (see also Szabados & Vértesi [7]). Let

$$a[\tfrac{1}{2}(r+1)], \qquad N=n-2a$$

and let the x_k's be the roots of the polynomial

$$\Omega_n(x) = T_N(x)\prod_{m=1}^{a}\left(x^2-\cos^2\frac{m-1}{N}\pi\right),$$

where $T_N(x) = \cos N\arccos x$ is the Chebyshev polynomial of degree N. Let

$$y_i = \cos\frac{2i-1}{2N}\pi \qquad (i=1,\ldots,N) \tag{11}$$

and let

$$l_i(x) = \frac{T_N(x)}{T_N'(y_i)(x-y_i)} \qquad (i=1,\ldots,N)$$

be the fundamental polynomials corresponding to the roots (11). Thus, if we write

$$z_{\pm m} = \pm\cos\frac{m-1}{N}\pi \qquad (m=1,\ldots,a) \tag{12}$$

then the fundamental polynomials corresponding to the roots (11) *in the system* (1) will be

$$\bar{l}_i(x) = l_i(x) \prod_{j=1}^{a} \frac{z_j^2 - x^2}{z_j^2 - y_i^2} \qquad (i = 1, \ldots, N), \tag{13}$$

while those corresponding to (12) in the system (1) are

$$\bar{\bar{l}}_m(x) = (-1)^{m-1} \frac{x + z_m}{2z_m} T_N(x) \prod_{\substack{j=1 \\ j \neq m}}^{a} \frac{x^2 - z_j^2}{z_m^2 - z_j^2} \qquad (m = \pm 1, \ldots, \pm a). \tag{14}$$

Since by (6) and (12)

$$\frac{z_j^2 - x^2}{\Delta_n(x)^2} = \frac{1 - x^2 + O(n^{-2})}{\Delta_n(x)^2}$$
$$= n^2 + \frac{1 - x^2 - n^2 \Delta_n(x)^2 + O(n^{-2})}{\Delta_n(x)^2} = n^2 + O\left(\frac{1}{\Delta_n(x)}\right),$$

i.e.

$$\Delta_n(x)^{-2a} \prod_{j=1}^{a} (z_j^2 - x^2) = n^{2a} + O\left(\frac{n^{2a-2}}{\Delta_n(x)}\right),$$

with the notation

$$c_i = \frac{N^{2a}}{\prod_{j=1}^{a} (z_j^2 - y_i^2)} = O\left(\frac{N^{2a}}{(1 - y_i^2)^a}\right) \qquad (i = 1, \ldots, N), \tag{15}$$

we obtain from (13)

$$\bar{l}_i(x) = \Delta_n(x)^{2a} c_i l_i(x) + O(\Delta_n(x)^{2a-1} n^{-2} c_i l_i(x)) \qquad (i = 1, \ldots, N). \tag{16}$$

As for the other type of fundamental functions (14), we get from (12) that

$$|\bar{\bar{l}}_m(x)| = O\left(\prod_{\substack{j=1 \\ j \neq m}} \left|\frac{x^2 - z_j^2}{z_m^2 - z_j^2}\right|\right) = O(\{n^2 \Delta_n(x)\}^{2a-2}) \qquad (m = \pm 1, \ldots, \pm a). \tag{17}$$

Now let $f^{(r)}(x) \in C[-1, 1]$; then according to a theorem of Gopengauz [3], there exist polynomials $P(x) \in \Pi_{n-1}$ such that

$$|f^{(j)}(x) - P^{(j)}(x)| = O\left(\left\{\frac{\sqrt{1 - x^2}}{n}\right\}^{r-j} \omega\left(f^{(r)}, \frac{1}{n}\right)\right) \quad (|x| \leq 1,\ j = 0, \ldots, r). \tag{18}$$

Hence, with the notations

$$\epsilon_i = f(y_i) - P(y_i) = O\left(\frac{(1-y_i^2)^{r/2}}{n^r}\right)\omega\left(f^{(r)}, \frac{1}{n}\right) \qquad (i = 1,\ldots,N), \quad (19)$$

$$\delta_m = f(z_m) - P(z_m) = O(n^{-2r})\omega\left(f^{(r)}, \frac{1}{n}\right) \qquad (m = \pm 1,\ldots,\pm a), \quad (20)$$

we obtain from (16), (17) and (20) that

$$\begin{aligned}
&\Delta_n(x)^{-2r} L_n(P-f,x)^2 \\
&\qquad = \Delta_n(x)^{-2r}\left(\sum_{i=1}^{N} \epsilon_i \bar{l}_i(x) + \sum_{m=-a}^{a} \delta_m \bar{\bar{l}}_m(x)\right)^2 \\
&\qquad \leqslant 2\Delta_n(x)^{-2r}\left\{\left(\sum_{i=1}^{N} \epsilon_i \bar{l}_i(x)\right)^2 + \left(\sum_{m=-a}^{a} \delta_m \bar{\bar{l}}_m(x)\right)^2\right\} \\
&\qquad \leqslant 4\Delta_n(x)^{4a-2r}\left\{\left(\sum_{i=1}^{N} c_i\epsilon_i l_i(x)\right)^2 + O\left(\frac{1}{n^4\Delta_n(x)^2}\right)\right. \\
&\qquad\qquad \left.\times\left(\sum_{i=1}^{N} c_i\epsilon_i |l_i(x)|\right)^2\right\} \\
&\qquad\qquad + O([n^2\Delta_n(x)]^{4a-2r-4})\omega\left(f^{(r)}, \frac{1}{n}\right)^2. \quad (21)
\end{aligned}$$

Here, by (15) and (19),

$$c_i\epsilon_i = \begin{cases} \omega\left(f^{(r)}, \dfrac{1}{n}\right) & (r \text{ even}) \\[2ex] \dfrac{n}{\sqrt{1-y_i^2}}\omega\left(f^{(r)}, \dfrac{1}{n}\right) & (r \text{ odd}) \end{cases} \qquad (i = 1,\ldots,N).$$

Thus, in case r even, (21) yields

$$\begin{aligned}
&\Delta_n(x)^{-2r} L_n(P-f,x)^2 \\
&\qquad \leqslant O\left(\omega\left(f^{(r)}, \frac{1}{n}\right)^2\right)\left\{\sum_{i=1}^{N} l_i(x)^2 + \frac{1}{n^4\Delta_n(x)^2}\left(\sum_{i=1}^{N} |l_i(x)|\right)^2 + 1\right\} \\
&\qquad\qquad + 4\sum_{j\neq k} c_i\epsilon_i c_k\epsilon_k l_i(x) l_k(x). \quad (22)
\end{aligned}$$

Using the well-known facts

$$\sum_{i=1}^{N} l_i(x)^2 = O(1), \qquad \sum_{i=1}^{N} |l_i(x)| = O(\log n) \qquad (|x| \leqslant 1),$$

the estimate

$$\int_{-1}^{1} \frac{1}{n^4 \Delta_n(x)^2 \sqrt{1-x^2}}\,dx = O\left(\frac{1}{n}\right),$$

as well as the orthogonality of the fundamental functions $l_i(x)$ with respect to the weight $(1-x^2)^{-1/2}$, we get from (22) that

$$\int_{-1}^{1} \frac{\Delta_n(x)^{-2r} L_n(P-f,x)^2}{\sqrt{1-x^2}}\,dx = O\left(\omega\left(f^{(r)}, \frac{1}{n}\right)^2\right). \tag{23}$$

On the other hand, if r is odd, then (21) yields

$$\begin{aligned}\Delta_n(x)^{-2r} L_n(P-f,x)^2 \leqslant O\left(\frac{1-x^2}{n^2}\right) \sum_{i,k} c_i \epsilon_i c_k \epsilon_k l_i(x) l_k(x) \\ + O\left(n^{-2}\omega\left(f^{(r)}, \frac{1}{n}\right)^2\right) \sum_{i,k} \frac{|l_i(x) l_k(x)|}{\sqrt{1-x_i^2}\sqrt{1-x_k^2}} \\ + O\left(\omega\left(f^{(r)}, \frac{1}{n}\right)^2\right).\end{aligned}$$

Here, using the obvious relations,

$$\begin{aligned}\int_{-1}^{1} \frac{(1-x^2) l_i(x) l_k(x)}{\sqrt{1-x^2}}\,dx &= -\frac{\pi}{2T_N'(x_i) T_N'(x_k)} \\ &= O\left(\frac{\sqrt{1-x_i^2}\sqrt{1-x_k^2}}{n^2}\right) \qquad (i \neq k),\end{aligned}$$

and

$$\sum_{i=1}^{N} \frac{1-x^2}{1-y_i^2} l_i(x)^2 = O(1), \qquad \sum_{i=1}^{N} \frac{|l_i(x)|}{1-y_i^2} = O(n) \qquad (|x| \leqslant 1),$$

we obtain (23) for odd r.

Finally, the theorem follows from

$$f^{(r)}(x) - L_n^{(r)}(f,x) = f^{(r)}(x) - P^{(r)}(x) + L_n^{(r)}(P-f,x)$$

on using (23), (18) with $j = r$ as well as (8) with $p = 2$, $w(x) = (1-x^2)^{-1/2}$ and $L_n(P-f,x)$ instead of P. □

References

[1] P. Erdős & E. Feldheim, Sur le mode de convergence pour l'interpolation de Lagrange, *C. R. Acad. Sci. Paris Ser. A-B*, **203** (1936) 913–15

[2] P. Erdős & P. Turán, On interpolation, I, *Annals of Maths.*, **38** (1937) 142–55

[3] I. E. Gopengauz, On a theorem of A. F. Timan on approximation of functions by polynomials on a finite interval, *Mat. Zam.*, **1** (1967) 163–72

[4] D. Leviatan, The behaviour of the derivatives of the algebraic polynomial of best approximation, *J. Appr. Theory*, **35** (1982) 169–76

[5] D. S. Lubinsky & P. Nevai, Markov–Bernstein inequalities revisited, *Appr. Theory and its Appl.*, **3** (1987) 98–119

[6] P. Nevai, Mean convergence of Lagrange interpolation, III, *Trans Amer. Math. Soc.*, **282** (1984) 669–98

[7] J. Szabados & P. Vértesi, On simultaneous optimization of norms of derivatives of Lagrange interpolation polynomials *Bull. London Math. Soc.*, **21** (1989), 457–81

Sur une question d'Erdős et Schinzel

Gérald Tenenbaum

A Paul Erdős, qui ouvre la voie.

1 Introduction

Désignons par $P^+(m)$ le plus grand facteur premier d'un entier générique m, avec la convention $P^+(1) = 1$. Un théorème de Tchébychev, publié en 1895, après sa mort, par Markov, énonce que

$$x^{-1}P^+\left(\prod_{n\leqslant x}(n^2+1)\right) \to \infty \qquad (x \to \infty).$$

Il est naturel de conjecturer beaucoup plus: si la suite polynomiale $\{n^2+1 : n = 1, 2, \ldots\}$ contient effectivement son quota heuristique de nombres premiers, on doit avoir pour x assez grand

$$P^+\left(\prod_{n\leqslant x}(n^2+1)\right) \gg x^2 \tag{1.1}$$

Ce problème a suscité, par le passé, la curiosité de nombreux mathématiciens et fait encore régulièrement, de nos jours, l'objet d'intéressantes publications. En 1952, Erdős [5] montré que, pour tout polynôme irréductible $F(X)$ à coefficients entiers, on a

$$P^+\left(\prod_{n\leqslant x}F(n)\right) > x\exp\{c\log_2 x\log_3 x\} \qquad (x > x_0(F)) \tag{1.2}$$

(Ici et dans toute la suite de cet article, nous notons $\log_k$ la k-ième itérée de la fonction logarithme.)

Dans le cas des polynômes quadratiques, un progrès important a été accompli par Hooley [9] qui, en 1967, a obtenu dans (1.1) la minoration $x^{1+\mu}$ avec $\mu = \frac{1}{10}$ – cf. également le chapitre 2 de [10]. Le meilleur résultat actuellement connu dans cette direction est dû à Deshouillers & Iwanieč [2], avec une constante μ excédant légèrement $\frac{1}{5}$.

Toutefois, en l'absence de majoration suffisamment fine pour certaines sommes de Kloosterman incomplètes, ces dernières estimations n'ont pas d'équivalent en degré plus grand que 2. Récemment, Erdős & Schinzel se sont réattaqués à cette question et ont montré [6] que l'on peut remplacer la borne inférieure de (1.2) par

$$x \exp\exp\{c(\log_2 x)^{1/3}\},$$

où c est une constante absolue positive. Cette évaluation a pu être obtenue grâce à une amélioration du *Lemma* 1 d'Erdős [5] que l'on peut formuler ainsi: *Désignons, pour $x \geqslant z \geqslant y \geqslant 2$, par $H_F(x,y,z)$ le nombre des entiers n n'excédant pas x et pour lesquels $F(n)$ possède au moins un diviseur d tel que $y < d \leqslant z$. Alors*

$$H_F(x, \tfrac{1}{2}x, x) > \frac{x}{\log x} \exp\{c'(\log_2 x)^{1/3}\}. \tag{1.3}$$

De plus, tout renforcement de (1.3) conduit par la même méthode à une amélioration correspondante dans le problème de Tchébychev.

Dans cette situation, Erdős & Schinzel ont formulé le problème, d'intérêt propre, de l'évaluation asymptotique de $H_F(x,y,z)$ sans restriction de primalité sur le polynôme F et pour toutes valeurs relatives des variables x, y et z. Ils posent en particulier la question de déterminer pour quels polynômes F la densité

$$D_F(y) := \lim_{x\to\infty} x^{-1} H_F(x,y,2y) \tag{1.4}$$

tend vers 0 lorsque $y \to \infty$.

Nous nous proposons ici d'aborder l'étude asymptotique de $H_F(x,y,z)$. Pour souci de simplicité, nous n'envisageons que le cas $z \leqslant 2y$. Nous définissons alors implicitement la quantité $\beta = \beta(y,z)$ par la formule

$$z = y\{1 + (\log y)^{-\beta}\}. \tag{1.5}$$

Notre approximation pour $H_F(x,y,z)$ est de la forme

$$x(\log y)^{-\delta(\beta,F)},$$

où l'exposant $\delta(\beta,F)$ est optimal, en un sens que nous préciserons ultérieurement. Quelques notations sont nécessaires pour définir explicitement cette quantité.

Considérons un polynôme $F(X)$ à coefficients entiers, prenant des valeurs positives sur les entiers positifs, et dont la décomposition canonique en produit de facteurs irréductibles dans $\mathbb{Z}[X]$ est

$$F(X) = \prod_{j=1}^{r} F_j(X)^{\alpha_j}. \tag{1.6}$$

Par analogie avec les fonctions arithmétiques classiques, nous posons

$$\hat{\omega}(F) := r, \qquad \hat{\Omega}(F) := \sum_{j=1}^{r} \alpha_j, \qquad \hat{\tau}(F) := \prod_{j=1}^{r} (\alpha_j+1). \tag{1.7}$$

Un rôle particulier est joué par la fonction de variable réelle

$$\gamma(v) = \gamma_F(v) := \sum_{j=1}^{r} \{(\alpha_j+1)^v - 1\} \tag{1.8}$$

(cf. en particulier le lemme 4.1). Nous désignons par $u = u(\beta, F)$ l'unique solution de l'équation

$$\gamma_F'(u) = \sum_{j=1}^{r} (\alpha_j+1)^u \log(\alpha_j+1) = \max\{\beta+1, \gamma_F'(0)\} \tag{1.9}$$

(noter que $\gamma_F'(0) = \log \hat{\tau}(F)$) et nous définissons $\delta(\beta, F)$ par la formule

$$\delta(\beta, F) = \begin{cases} u\gamma_F'(u) - \gamma_F(u) & (0 \leqslant \beta \leqslant \gamma_F'(1)-1), \\ \beta+1-\gamma_F(1) & (\beta > \gamma_F'(1)-1). \end{cases} \tag{1.10}$$

Il est peut-être plus agréable de transformer cette expression en introduisant la fonction

$$Q(t) := t\log t - t + 1 \qquad (t > 0), \tag{1.11}$$

qui est non-négative et possède un unique zéro (double) au point $t = 1$. On peut alors écrire

$$\delta(\beta, F) = \begin{cases} \sum_{j=1}^{r} Q((\alpha_j+1)^u) & (0 \leqslant u \leqslant 1), \\ \beta+1-\hat{\Omega}(F) & (u > 1). \end{cases} \tag{1.12}$$

Il est à noter que $\delta(\beta, F)$ est continue en $\beta = \gamma_F'(1) - 1$. On a $\delta(\beta, F) \geqslant 0$ pour tout $\beta \geqslant 0$, avec égalité si et seulement si $\beta + 1 \leqslant \log \hat{\tau}(F)$.

Théorème *Soient β_0 et B deux nombres réels positifs. Il existe des constantes positives y_0, c_j $(0 \leqslant j \leqslant 5)$, ne dépendant que de β_0, B et F, telles que les propositions suivantes soient vérifiées pour $y_0 < x < x^{c_0}$ et $0 \leqslant \beta(y, z) \leqslant B$.*

(a) *Si $\beta + 1 \leqslant \log \hat{\tau}(F) - c_1/\sqrt{\log_2 y}$, alors*

$$c_2 x \leqslant H_F(x, y, z) \leqslant (1 - c_2)x. \tag{1.13}$$

(b) *Si $\beta + 1 > \log \hat{\tau}(F) - c_1/\sqrt{\log_2 y}$, alors*

$$c_3 x(\log y)^{-\delta(\beta, F)} e^{-c_4\sqrt{\log_2 y \log_3 y}} \leqslant H_F(x, y, z) \leqslant c_5 x (\log y)^{-\delta(\beta, F)}. \tag{1.14}$$

De plus, le terme $\log_3 y$ *peut être omis dans le membre de gauche lorsque* $\beta \geqslant \beta_0$.

Complément *Quitte à altérer les valeurs de y_0 et c_j $(1 \leqslant j \leqslant 5)$, la minoration de (1.13) et l'encadrement (1.14) sont valables pour tout* $c_0 < 1$.

Remarque On a $\hat{\tau}(F) \geqslant 3$ dès que F est réductible. Quitte à choisir $\beta_0 < \log 3 - 1$, l'hypothèse du point (b) du théorème implique alors $\beta \geqslant \beta_0$.

Lorsque F est irréductible, on a en particulier

$$H_F(x, y, 2y) = x(\log y)^{-\delta + o(1)} \tag{1.15}$$

pour x et y tendant vers $+\infty$ dans tout domaine $y \leqslant x^{c_0}$ $(c_0 < 1)$, avec

$$\delta := \delta(0, F) = 1 - \frac{1 + \log\log 2}{\log 2} = 0.086071\ldots. \tag{1.16}$$

Cela conduirait à une amélioration considérable de (1.3) si l'on pouvait choisir $c_0 = 1$. Malheureusement, la méthode du présent travail repose de manière essentielle sur l'estimation par le crible de quantités du type

$$\operatorname{card}\{n \leqslant x : d \,|\, F(n),\, p \,|\, F(n) \Rightarrow p \,|\, d \text{ ou } p > t\}$$

où $d > y$ et où t est un paramètre 'grand'. Cela interdit *a priori* de considérer des valeurs de y de l'ordre de x. Le complément au théorème représente donc pratiquement la limite naturelle de notre méthode. En l'absence d'une profonde altération du raisonnement suivi dans [5], un tel résultat semble inutilisable dans le problème de Tchébychev – voir cependant la note p. 442.

Dans [5], Erdős avait suggéré qu'il était possible d'obtenir une borne inférieure du type

$$P^+\Bigl(\prod_{n \leqslant x} F(n)\Bigr) > x \exp\{(\log x)^c\}. \tag{1.17}$$

Cette inégalité découlerait, pour tout $c < 1 - \delta$, de la minoration contenue dans (1.15) si la valeur $y = x$ était admissible.

Dans le cas des 'petites' valeurs de y, nos résultats permettent de répondre complètement à la question d'Erdős & Schinzel concernant la densité (1.4).

Corollaire *$D_F(y)$ tend vers 0 à l'infini si et seulement si $F(X)$ est irréductible dans $\mathbb{Z}[X]$. Dans ce cas, on a de plus*

$$D_F(y) = (\log y)^{-\delta + o(1)} \qquad (y \to \infty),$$

où δ est défini par (1.16).

Lorsque F est réductible, on a $c_2(F) \leqslant D_F(y) \leqslant 1-c_2(F)$, et Erdős conjecture que $D_F(y)$ tend vers une limite $l(F)$, nécessairement dans $]0,1[$. Nos méthodes ne semblent pas permettre l'obtention d'un tel résultat.

Le point (b) du théorème généralise essentiellement l'étude faite dans l'article [18] et dans le chapitre 2 de [8], qui correspond au cas $F(X) = X$. Nous avions alors également considéré les 'grandes' valeurs de z, i.e. $z > 2y$. Dans cette circonstance, les nombres premiers $\leqslant z/y$ jouent un rôle particulier engendrant des difficultés techniques supplémentaires que nous avons préféré éviter ici. Leur présence ne modifie cependant pas la méthode dans son principe et il serait tout-à-fait possible, le cas échéant, d'étendre dans cette direction le champ de validité de nos estimations. De même, on pourrait s'affranchir de la condition $\beta(y,z) \leqslant B$ en adaptant convenablement la méthode élémentaire présentée au début de la section 2.5 de [8]. Il est d'ailleurs vraisemblable que l'on obtienne ainsi un équivalent asymptotique de $H_F(x,y,z)$ si $\beta(y,z)$ ne tend pas trop vite vers l'infini.

Le changement de comportement asymptotique de $H_F(x,y,z)$ au passage par la valeur critique

$$\beta = \gamma_F'(1) - 1 = \sum_{j=1}^{r} (\alpha_j + 1)\log(\alpha_j + 1) - 1$$

est susceptible d'une interprétation probabiliste. Lorsque z est suffisamment proche de y, les différentes conditions $F(n) \equiv 0 \pmod d$ $(y < d \leqslant z)$ sont essentiellement indépendantes et l'on a

$$H_F(x,y,z) \approx \sum_{n \leqslant x} \tau(F(n);y,z), \tag{1.18}$$

où $\tau(m;y,z)$ désigne le nombre de diviseurs de m dans l'intervalle $]y,z[$. En revanche, lorsque z/y dépasse un certain seuil, l'effet de la dépendance est perceptible et $H_F(x,y,z)$ devient notablement moindre que le membre de droite de (1.18). Le lecteur trouvera dans la section 2.2 de [8] une discussion plus approfondie sur ce point, contenant notamment, dans le cas $F(X) = X$, une justification heuristique de la valeur $\beta = 2\log 2 - 1$ du seuil de dépendance. Le raisonnement s'étend sans difficulté au cas général.

La première partie du théorème appelle encore une remarque. Considérons, pour fixer les idées, le cas $F(X) = X^2$, $\beta = 0$, y fixé. Nous obtenons que, pour une densité positive (minorée indépendamment de y) d'entiers n, le nombre n^2 possède un diviseur dans $]y,2y]$.

Cette propriété équivaut clairement à l'existence de deux diviseurs d et d' de n tels que $y < dd' < 2y$. Le problème apparaît donc comme une version duale de la conjecture d'Erdős, résolue dans [13], selon laquelle il est 'presque toujours' (i.e.: dans une suite de densité 1) possible de trouver d et d' tels que $1 < d'/d \leqslant 2$. Dans les deux cas, le résultat est suggéré par l'hypothèse heuristique d'équiprobabilité des quantités $\log(dd')$ ou $\log(d'/d)$ – puisque le nombre des diviseurs de n^2, comme celui des rapports distincts d'/d, est *normalement* $(\log n)^{\log 3+o(1)}$. Or, la majoration de (1.13) indique les limites de cette analogie: alors que la conjecture d'Erdős est effectivement satisfaite 'presque partout', la propriété relative aux diviseurs de n^2 n'a pas lieu sur un ensemble de densité 1 ou même proche de 1. Il faut voir une explication de ce phénomène dans le principe, souvent observé, qu'un ensemble-différence est mieux réparti qu'un ensemble-somme.

L'auteur a le plaisir de remercier ici Paul Erdős, Andrej Schinzel, Léo Murata et Jean-Marie De Koninck pour leur aide durant la préparation de cet article.

2 Notations et conventions – rappels sur les congruences algébriques

Les lettres p et q sont réservées pour désigner exclusivement des nombres premiers. On note $P^-(n)$ (resp. $P^+(n)$) le plus petit (resp. le plus grand) facteur premier de l'entier $n > 1$. Par convention, $P^-(1) = +\infty$, $P^+(1) = 1$.

Nous désignons par $\tau(n)$ (resp. $\omega(n)$) le nombre des diviseurs (resp. des facteurs premiers) de l'entier n, et nous définissons, pour chaque valeur du paramètre réel θ, les fonctions arithmétiques multiplicatives

$$\tau(n,\theta) := \sum_{d|n} d^{i\theta}, \tag{2.1}$$

$$\tau_\theta(n) := \prod_{\substack{p^\nu \| n \\ p \leqslant \exp(1/|\theta|)}} (\nu+1) \tag{2.2}$$

Pour tous y et z tels que $1 \leqslant y \leqslant z$, nous posons également

$$\tau(n;y,z) = \sum\{1 : d|n,\ y < d \leqslant z\}.$$

Etant donné un entier $n \geqslant 1$, nous notons $d|n^\infty$ la propriété: $p|d \Rightarrow p|n$.

Nous employons indifféremment les symboles O de Landau et $\ll$ de Vinogradov. L'écriture $A \asymp B$ signifie: $A \ll B$ et $B \ll A$.

Dans tout l'article, nous supposons fixé un polynôme $F(X)$ à coefficients entiers, de degré $g \geqslant 1$, prenant des valeurs positives sur les entiers positifs, et dont la décomposition canonique dans $\mathbb{Z}[X]$ est fournie par (1.6). Nous posons également

$$F^*(X) := \prod_{j=1}^{r} F_j(X). \tag{2.3}$$

Pour chaque polynôme $G(X)$ de $\mathbb{Z}[X]$, nous désignons par $\rho(n;G)$ le nombre des solutions modulo n de la congruence

$$G(x) \equiv 0 \pmod n.$$

Il découle du théorème chinois que $\rho(n;G)$ est une fonction multiplicative de n. Pour la concision, nous notons

$$\rho(n) := \rho(n;F), \qquad \rho_j(n) := \rho(n;F_j) \quad (1 \leqslant j \leqslant r), \qquad \rho^*(n) := \rho(n;F^*). \tag{2.4}$$

Soit D^* le discriminant de $F^*(X)$. On sait que

$$\rho^*(p^\nu) = \rho^*(p) \qquad (p \nmid D^*, \nu \geqslant 1) \tag{2.5}$$

– cf. par exemple Nagell [15], *Theorem* 52. Comme l'égalité $\rho^*(p) = p$ n'est possible que si $p \mid F(1)$, on peut écrire

$$\rho^*(p^\nu) \leqslant \min\{g, p-1\} \qquad (p \nmid D^*F(1), \nu \geqslant 1). \tag{2.6}$$

Pour la commodité du lecteur, nous rappelons maintenant quelques faits simples qui sont pour la plupart établis dans [3].

Lorsque $1 \leqslant i < j \leqslant r$, les polynômes $F_i(X)^{\alpha_i}$ et $F_j(X)^{\alpha_j}$ sont premiers entre eux dans $\mathbb{Z}[X]$. Nous désignons par $m_{ij} \in \mathbb{Z}$ leur pgcd, c'est-à-dire le plus petit entier positif de la forme

$$m_{ij} = U(X)F_i(X)^{\alpha_i} + V(X)F_j(X)^{\alpha_j},$$

avec $U(X), V(X) \in \mathbb{Z}[X]$. Nous posons alors

$$M := \prod_{1 \leqslant i < j \leqslant r} m_{ij} \tag{2.7}$$

avec la convention $M = 1$ si $r = 1$. Il est clair que les congruences $F_i(x) \equiv 0 \pmod p$ et $F_j(x) \equiv 0 \pmod p$ sont incompatibles lorsque $p \nmid m_{ij}$. D'où

$$\rho(p^\nu) = \sum_{j=1}^{r} \rho(p^\nu; F_j^{\alpha_j}) \qquad (p \nmid M). \tag{2.8}$$

Maintenant, on vérifie facilement que l'on a pour tous p, $\nu \geqslant 1$ et $1 \leqslant j \leqslant r$,

$$\rho(p^\nu; F_j^{\alpha_j}) = p^{\nu - \lceil \nu/\alpha_j \rceil} \rho_j(p^{\lceil \nu/\alpha_j \rceil}), \tag{2.9}$$

où, ici et dans tout l'article, nous notons $\lceil x \rceil$ le plus petit entier au moins égal à $x \in \mathbb{R}$.

Nous posons une fois pour toutes

$$D := D^* F(1) M. \tag{2.10}$$

Ainsi l'ensemble des facteurs premiers de D contient tous les nombres premiers pour lesquels la fonction $\rho(p^\nu)$ est susceptible de posséder un comportement exceptionnel. Il découle en particulier de (2.5), (2.8) et (2.9) que l'on a

$$\rho(p^\nu) = \sum_{j=1}^{r} \rho_j(p) p^{\nu - \lceil \nu/\alpha_j \rceil} \qquad (p \nmid D,\ \nu \geqslant 1). \tag{2.11}$$

Nous associons également à $F^*(X)$ une 'fonction d'Euler'

$$\varphi^*(n) := n \prod_{p|n} (1 - \rho^*(p) p^{-1}) \tag{2.12}$$

Par (2.6), on a

$$\varphi^*(n) \geqslant 1, \qquad (n \geqslant 1, (n, D) = 1). \tag{2.13}$$

Il nous sera utile dans la suite de disposer d'une majoration uniforme pour $\rho(p^\nu)$. Nous la déduirons, comme dans [3], du résultat de Nagell & Ore – cf. [15], *Theorem* 54 –

$$\rho^*(p^\nu) \leqslant g \cdot (D^*)^2 \qquad (p \geqslant 2,\ \nu \geqslant 1). \tag{2.14}$$

Cela implique

$$\rho(p^\nu) \ll \sum_{j=1}^{r} p^{\nu - \lceil \nu/\alpha_j \rceil} \tag{2.15}$$

lorsque $p \nmid M$ (grâce à (2.8) et (2.9)). Nous allons voir que cette estimation persiste en fait même lorsque $p | M$. Supposons en effet que $p^\mu \| M$. Nous pouvons certainement nous restreindre au cas $\nu > \mu r$. Alors la relation $p^\nu | F(n)$ implique $p^{\mu+1} | F_j(n)^{\alpha_j}$ pour au moins un j. La plus grande puissance de p divisant le produit de tous les autres $F_i(n)^{\alpha_i}$ est nécessairement $\leqslant \mu$, donc j est unique. On en déduit que $p^{\nu-\mu} | F_j(n)^{\alpha_j}$, et il suit

$$\rho(p^\nu) \leqslant p^\mu \sum_{j=1}^{r} \rho(p^{\nu-\mu}; F_j^{\alpha_j}) = \sum_{j=1}^{r} p^{\nu - \lceil (\nu-\mu)/\alpha_j \rceil} \rho_j(p^{\lceil (\nu-\mu)/\alpha_j \rceil}).$$

Au vu de (2.14), cela implique bien (2.15).

3 Arithmétique des suites polynomiales – lemmes généraux

Nous nous proposons dans cette section d'établir les principaux résultats auxiliaires nécessaires à la démonstration de notre théorème, et qui concernent le comportement asymptotique de fonctions multiplicatives restreintes à des suites polynomiales. Bien entendu, les fonctions $\rho_j(p)$ s'avèrent fondamentales, et les trois premiers lemmes leur sont dévolus.

Lemme 3.1 *Soit $G(X)$ un polynôme irréductible de $\mathbb{Z}[X]$. On a pour x infini*

$$\sum_{p\leqslant x} \rho(p\,;G) = \mathrm{li}(x)+O\big(x\exp(-c\sqrt{\log x})\big), \tag{3.1}$$

où $c = c(G)$ est une constante positive.

Démonstration Nous nous contentons de brèves indications pour cette estimation classique. Soit $K = \mathbb{Q}(\xi)$ le corps de nombres engendré par une racine de l'équation $G(\xi) = 0$. Désignons par $\mathbb{O}$ l'anneau des entiers de K, par N la norme de K sur $\mathbb{Q}$ et par $\mathfrak{p}$ un idéal premier générique de $\mathbb{O}$. Le théorème des idéaux premiers sous une forme forte (cf. Landau [11], *Satz* 191) s'écrit

$$\sum_{N\mathfrak{p}\leqslant x} 1 = \mathrm{li}(x)+O\big(x\exp(-c\sqrt{\log x})\big). \tag{3.2}$$

Or on a, pour chaque $\mathfrak{p}$, $N\mathfrak{p} = p^f$, où p est un nombre premier rationnel. Comme l'a remarqué Erdős dans [4], la contribution à (3.2) des $\mathfrak{p}$ tels que p divise le discriminant D_G de G est $O(1)$, et celle des $\mathfrak{p}$ tels que $f > 1$ est $O(\sqrt{x})$ puisque le nombre des $\mathfrak{p}$ associés à un p fixé ne pas dépasse pas le degré de G. Lorsque $f = 1$ et $p \nmid D_G$, on a

$$\mathrm{card}\{\mathfrak{p} : N\mathfrak{p} = p\} = \rho(p\,;G)$$

puisque ce nombre est exactement celui des facteurs linéaires distincts dans la réduction de $G \bmod p$ – cf. par exemple Lang [12], chapitre I, section 8. Cela suffit pleinement à établir (3.1). □

Lemme 3.2 *On a pour x infini*

$$\sum_{p\leqslant x} \rho^*(p)p^{-1} = r\log_2 x+O(1). \tag{3.3}$$

Démonstration Cela découle immédiatement de (2.11) et (3.1), par sommation d'Abel. □

Lemme 3.3 *Soit f une fonction périodique de période 2π, à variation bornée sur l'intervalle $[0,2\pi[$, et de valeur moyenne*

$$\bar{f} := \frac{1}{2\pi}\int_0^{2\pi} f(t)\,\mathrm{d}t.$$

Pour tout polynôme irréductible G de $\mathbb{Z}[X]$, il existe une constante positive $c = c(G)$ telle que l'on ait pour $\theta \in \mathbb{R}$ $(\theta \neq 0)$ et $2 \leqslant y \leqslant x$,

$$\sum_{y<p\leqslant x} \rho(p;G)\frac{f(\theta\log p)}{p}$$
$$= \bar{f}\log\frac{\log x}{\log y} + O\left(\frac{V(f)}{|\theta|\log y} + \{M(f) + (1+|\theta|)V(f)\}\exp(-c\sqrt{\log y})\right), \tag{3.4}$$

où l'on a posé

$$V(f) := \int_0^{2\pi} |\mathrm{d}f(t)|, \qquad M(f) := \sup_{0\leqslant t\leqslant 2\pi} |f(t)|.$$

Démonstration La démonstration est identique à celle du *Lemma* 30.1 de [8], à ceci près que $\pi(x)$ doit être remplacé par la fonction sommatoire de $\rho(p;G)$. Le lemme 3.1 remplaçant le théorème des nombres premiers, les calculs sont inchangés. □

Le résultat suivant est une application simple de la théorie du crible.

Lemme 3.4 *Désignons par $d_0, d_1, \ldots, d_r$ des nombres entiers deux à deux premiers entre eux et tels que*

$$d_0 | D^\infty, \qquad (d_j, D) = 1 \qquad (1 \leqslant j \leqslant r).$$

Soient K et K′ des nombres réels satisfaisant à $0 < K < 1 \leqslant K'$. Il existe une constante positive $c_6 = c_6(K,F) < 1$ telle que sous la condition

$$2 \leqslant y \leqslant x^{c_6}, \qquad d_0 d_1 \ldots d_r \leqslant x^{1-K}, \qquad P^+(d_0 d_1 \ldots d_r) \leqslant y^{K'},$$

on ait uniformément

$$\sum_{\substack{n\leqslant x \\ d_0|F(n),\, d_j|F_j(n)\ (1\leqslant j\leqslant r) \\ p|F(n) \Rightarrow p|D\prod_{j=1}^r d_j \text{ ou } p>y}} 1 \asymp \frac{x}{(\log y)^r}\frac{\rho(d_0)}{d_0}\prod_{j=1}^r \frac{\rho_j(d_j)}{\varphi^*(d_j)}. \tag{3.5}$$

De plus, la relation (3.5) persiste, en remplaçant le signe $\asymp$ par $\leqslant$, lorsque $P^+(d_0 d_1 \ldots d_r) > y^{K'}$.

Démonstration On peut supposer

$$\rho(d_0)\prod_{j=1}^r \rho_j(d_j) \geqslant 1,$$

car le résultat est trivial dans le cas contraire. On applique alors le crible de Selberg à l'ensemble d'entiers

$$\mathcal{A} := \{F^*(n) : n \leqslant x,\ d_0 | F(n),\ d_j | F_j(n)\ (1 \leqslant j \leqslant r)\}$$

pour l'ensemble de nombres premiers

$$\mathcal{P} := \{p : p \leqslant y,\ p \nmid D \prod_{j=1}^{r} d_j\}.$$

Par le théorème chinois, on a

$$|\mathcal{A}| = X + O(R)$$

avec

$$X := x\frac{\rho(d_0)}{d_0}\prod_{j=1}^{r}\frac{\rho_j(d_j)}{d_j} \geqslant x^K\rho(d_0), \quad R := \rho(d_0)\prod_{j=1}^{r}\rho_j(d_j) \ll x^{K/3}\rho(d_0), \tag{3.6}$$

où la seconde estimation découle de (2.6). Lorsque $d | \prod_{p\in\mathcal{P}} p$, on a similairement (notant, comme c'est l'usage, $\mathcal{A}_d$ l'ensemble des multiples de d qui appartiennent à $\mathcal{A}$)

$$|\mathcal{A}_d| = X\frac{\rho^*(d)}{d} + O(R\rho^*(d)),$$

avec

$$\rho^*(d) \leqslant g^{\omega(d)} \qquad \Big(d \,\Big|\, \prod_{p\in\mathcal{P}} p\Big). \tag{3.7}$$

Le lemme fondamental du crible de Selberg, tel qu'il est énoncé, par exemple, par Halberstam & Richert dans [7] (*Theorem* 7.1), implique alors que, pour tout choix du paramètre $v \geqslant 1$, le membre de gauche de (3.5) vaut

$$X\prod_{p\in\mathcal{P}}\left(1 - \frac{\rho^*(p)}{p}\right)\{1 + O(\mathrm{e}^{-v\log v})\} + O\Big(R\sum_{d\leqslant y^{2v}}(3g)^{\omega(d)}\Big). \tag{3.8}$$

(La condition traditionnelle '$\Omega_2(\kappa)$' est ici trivialement impliquée par (3.7), avec $\kappa = g$.) Choisissons v assez grand pour que le terme entre accolades soit dans l'intervalle $[\frac{1}{2}, \frac{3}{2}]$, puis c_6 assez petite pour que le dernier terme d'erreur soit $\ll x^{K/2}\rho(d_0)$. Cela est possible, compte tenu de (3.6), puisque l'on a classiquement

$$\sum_{d\leqslant z} b^{\omega(d)} \ll_b z(\log 2z)^{b-1} \qquad (z \geqslant 1,\ b > 0).$$

Observons ensuite que, puisque $\rho^*(p) \leqslant \min(p-1, g)$ lorsque $p \in \mathcal{P}$, le terme principal de (3.8) est alors

$$\geqslant \tfrac{1}{2}X \prod_{\substack{p\leqslant y\\ p\nmid D}}\left(1-\frac{\rho^*(p)}{p}\right) \gg \frac{x^K\rho(d_0)}{(\log x)^r}.$$

Cela implique que l'expression (3.8) est

$$\asymp X \prod_{p\in\mathcal{P}}\left(1-\frac{\rho^*(p)}{p}\right).$$

Le résultat annoncé découle donc de (3.3). □

Nous allons maintenant utiliser le résultat précédent pour majorer des sommes de fonctions multiplicatives d'arguments polynomiaux. Nous avons en fait besoin d'une estimation uniforme lorsque la sommation est restreinte aux entiers n tels que $F(n) \equiv 0 \pmod d$. Il serait théoriquement possible d'étendre *stricto sensu* les travaux d'Ennola [3] ou de Wolke [19], tous deux reposant sur la méthode développée par Erdős [4] pour évaluer $\sum_{n\leqslant x} \tau(G(n))$ lorsque $G(X)$ est irréductible. Un tel résultat, qui pourrait être l'analogue du théorème de Shiu [17], serait fort utile et mériterait certainement un article pour lui-même. En tout état de cause, la généralisation n'est pas de pure routine et induit de notables difficultés techniques. Nous nous sommes, pour ces raisons, cantonnés ici à une version plus simple, mais cependant suffisante pour l'application que nous envisageons. Les calculs sont grandement facilités par l'hypothèse de forte multiplicativité imposée aux fonctions arithmétiques considérées.

Lemme 3.5 *Soit A une constante positive arbitraire et $h_1, h_2, \ldots, h_r$ des fonctions arithmétiques fortement multiplicatives satisfaisant pour tout nombre premier p à*

$$0 \leqslant h_j(p) \leqslant A \qquad (1 \leqslant j \leqslant r).$$

Soit K un nombre réel $(0 < K < 1)$. On a uniformément pour $x \geqslant 1$ et $1 \leqslant d \leqslant x^{1-K}$,

$$\sum_{\substack{n\leqslant x\\ d\mid F(n)}} \prod_{j=1}^{r} h_j(F_j(n)) \ll x\frac{g(d)}{d}\exp\left(\sum_{j=1}^{r}\sum_{p\leqslant x}\frac{\rho_j(p)(h_j(p)-1)}{p}\right), \tag{3.9}$$

où g est la fonction multiplicative définie par

$$g(p^\nu) = \begin{cases} \rho(p^\nu) & \text{si } p\mid D,\\ \left(1-\dfrac{\rho^*(p)}{p}\right)^{-1} \displaystyle\sum_{j=1}^{r} \rho_j(p)h_j(p)p^{\nu-\lceil \nu/\alpha_j\rceil} & \text{si } p\nmid D. \end{cases} \tag{3.10}$$

Démonstration Pour chaque entier n tel que $d|F(n)$, on définit les ensembles de nombres premiers

$$\mathcal{P}_j(n) := \{p : p\,|\,F_j(n),\ p \nmid D\} \qquad (1 \leqslant j \leqslant r).$$

Ainsi que nous l'avons remarqué à la section 2, les $\mathcal{P}_j(n)$ sont deux à deux disjoints. Notant $\mathcal{P}_0$ l'ensemble des facteurs premiers de D, on voit que $\{\mathcal{P}_0, \mathcal{P}_1(n), \ldots, \mathcal{P}_r(n)\}$ induit une partition de l'ensemble des facteurs premiers de d. Nous lui associons la décomposition

$$d = d_0 \prod_{j=1}^{r} d_j(n) \qquad \left(d_0 = \prod_{\substack{p^\nu \| d \\ p | D}} p^\nu, \quad d_j(n) = \prod_{\substack{p^\nu \| d \\ p \in \mathcal{P}_j(n)}} p^\nu \ (0 \leqslant j \leqslant r) \right).$$

Considérons maintenant un r-uple $\{d_1, d_2, \ldots, d_r\}$ d'entiers deux à deux premiers entre deux tels que $d_1 d_2 \ldots d_r = d d_0^{-1}$ et un entier n tel que $d_j(n) = d_j$ $(1 \leqslant j \leqslant r)$. L'hypothèse $d|F(n)$ implique

$$m_j\,|\,F_j(n) \qquad (0 \leqslant j \leqslant r)$$

avec

$$m_j := \prod_{p^\nu \| d_j} p^{\lceil \nu/\alpha_j \rceil}. \tag{3.11}$$

Le membre de gauche de (3.9), disons S, satisfait donc à l'inégalité

$$S \leqslant \sum_{\substack{d_1 \ldots d_r = d/d_0 \\ (d_i, Dd_j) = 1\ (1 \leqslant i \neq j \leqslant r)}} S(d_1, \ldots, d_r) \tag{3.12}$$

avec

$$S(d_1, \ldots, d_r) := \sum_{\substack{n \leqslant x \\ m_j | F_j(n)\ (1 \leqslant j \leqslant r), \\ d_0 | F(n)}} \prod_{j=1}^{r} h_j(F_j(n)).$$

Ici et dans toute la suite de cette démonstration, m_j est défini en fonction de d_j par (3.11)

Pour chaque n apportant une contribution positive à $S(d_1, \ldots, d_r)$, on peut décomposer $F_j(n)$ $(1 \leqslant j \leqslant r)$ de manière unique sous la forme

$$F_j(n) = D_{nj} M_{nj} A_{nj} B_{nj} \tag{3.13}$$

avec les conditions

$$D_{nj}\,|\,D^\infty, \qquad m_j\,|\,M_{nj}\,|\,m_j^\infty, \qquad (A_{nj}B_{nj}, Dm_j) = 1,$$
$$P^+(A_{nj}) \leqslant \xi \leqslant P^-(B_{nj}),$$

où $\xi = \xi(n)$ est un paramètre que nous préciserons plus loin. Notons d_{nj}, a_{nj} et b_{nj} les noyaux sans facteur carré respectifs de D_{nj}, A_{nj} et B_{nj}. On a alors

$$h_j(F_j(n)) = h_j(d_{nj})h_j(m_j)h_j(a_{nj})h_j(b_{nj}), \tag{3.14}$$

et il sera utile dans la suite de garder à l'esprit la majoration

$$h_j(d_{nj}) \leqslant A^{\omega(D)} \ll 1 \qquad (1 \leqslant j \leqslant r).$$

Posons maintenant

$$\tilde{a}_n := \prod_{j=1}^{r} a_{nj}, \qquad \tilde{b}_n := \prod_{j=1}^{r} b_{nj}, \qquad q_n := P^-(\tilde{b}_n)$$

et

$$X = x^{c_7}$$

avec $c_7 = \min\{K, c_6(\frac{1}{2}K, F)\}$. Nous choisissons, pour chaque n, $\xi = \xi(n)$ aussi grand que possible sous la contrainte

$$\tilde{a}_n \leqslant \sqrt{X}.$$

On a donc nécessairement

$$\tilde{a}_n q_n > \sqrt{X}. \tag{3.15}$$

Nous désignons respectivement par $S_1(d_1,\ldots,d_r)$ et $S_2(d_1,\ldots,d_r)$ les contributions à $S(d_1,\ldots,d_r)$ des entiers $n \leqslant x$ tels que $\tilde{a}_n \leqslant X^{1/3}$ et $\tilde{a}_n > X^{1/3}$.

Pour majorer $S_1(d_1,\ldots,d_r)$ nous utilisons (3.15) sous la forme $q_n > \sqrt{X}/\tilde{a}_n \geqslant X^{1/6}$. En remarquant que $h_j(b_{nj}) \ll A^{6g/c_7} \ll 1$, il suit

$$S_1(d_1,\ldots,d_r) \ll \prod_{j=1}^{r} h_j(m_j) \sideset{}{'}\sum_{\substack{a_1\ldots a_r \leqslant X \\ (a_i, Dm_i a_j)=1 \\ 1\leqslant i \neq j \leqslant r}} \prod_{j=1}^{r} h_j(a_j) \sum_{n\leqslant x}^{(\dagger)} 1$$

avec les conditions de sommation

$$(\dagger)\begin{cases} a_j m_j \,|\, F_j(n) & (1 \leqslant j \leqslant r), \\ d_0 \,|\, F(n), \\ p \,|\, F(n) \Rightarrow p \,|\, D \prod_{j=1}^{r} a_j m_j \text{ ou } p > X^{1/6}. \end{cases}$$

Ici et dans la suite de cette section, l'apostrophe indique que la sommation ainsi désignée est restreinte à des entiers sans facteur carré. Puisque

$$d_0 \prod_{j=1}^{r} a_j m_j \leqslant x^{1-K/2},$$

on peut appliquer le lemme 3.4 pour estimer la somme intérieure. On a ainsi

$$\sum_{n\leqslant x}^{(\dagger)} 1 \ll \frac{x}{(\log x)^r}\,\frac{\rho(d_0)}{d_0}\prod_{j=1}^{r}\frac{\rho_j(a_jm_j)}{\varphi^*(a_jm_j)}\,. \tag{3.16}$$

En remarquant que

$$\begin{aligned}\sum_{\substack{a_i\ldots a_r\leqslant X\\(a_i\ldots a_r,D)=1}}' \prod_{j=1}^{r}\frac{h_j(a_j)\,\rho_j(a_j)}{\varphi^*(a_j)} &\leqslant \prod_{j=1}^{r}\sum_{\substack{a\leqslant X\\(a,D)=1}}'\frac{\rho_j(a)\,h_j(a)}{\varphi^*(a)}\\ &\leqslant \exp\left(\sum_{j=1}^{r}\sum_{p\leqslant x}\left\{\frac{\rho_j(p)\,h_j(p)}{p}+O\left(\frac{1}{p^2}\right)\right\}\right),\end{aligned}$$

on obtient finalement

$$S_1(d_1,\ldots,d_r) \ll x\frac{\rho(d_0)}{d_0}\prod_{j=1}^{r}\frac{\rho_j(m_j)\,h_j(m_j)}{\varphi^*(m_j)}\exp\left(\sum_{j=1}^{r}\sum_{p\leqslant x}\frac{\rho_j(p)(h_j(p)-1)}{p}\right) \tag{3.17}$$

où nous avons fait appel au lemme 3.2 pour transformer le facteur $(\log x)^{-r}$ dans (3.16).

Désignons le facteur initial d'indice j dans (3.17) par $G_j(d_j)$. On a

$$G_j(d_j) = \prod_{\substack{p\mid d_j\\ p^\nu\| d}} G_j(p^\nu)$$

d'où

$$\sum_{\substack{d_1\ldots d_r=d/d_0\\(d_i,d_j)=1\,(1\leqslant i<j\leqslant r)}}\prod_{j=1}^{r} G_j(d_j) = \prod_{p^\nu\| d/d_0}\sum_{j=1}^{r} G_j(p^\nu) = \frac{d_0}{d}g\left(\frac{d}{d_0}\right). \tag{3.18}$$

Compte tenu de (3.17), cela montre que la contribution à S des $S_1(d_1,\ldots,d_r)$ est compatible avec l'estimation annoncée (3.9).

Pour évaluer la contribution des $S_2(d_1,\ldots,d_r)$, nous utilisons toujours l'identité (3.14), mais nous majorons $h_j(b_{nj})$ en fonction de $q := P^+(\tilde{a}_n)$. Puisque $b_{nj} \ll x^g$, on a pour x assez grand

$$\omega(b_{nj}) = \sum_{\substack{p\mid b_{nj}\\ p>q}} 1 \leqslant (g+1)\frac{\log x}{\log q}\,.$$

D'où

$$\prod_{j=1}^{r} h_j(b_{nj}) \leqslant x^{C/\log q}, \tag{3.19}$$

avec

$$C := r(g+1)\max\{0, \log A\}.$$

Il vient ainsi

$$S_2(d_i, \ldots, d_r) \ll \prod_{j=1}^{r} h_j(m_j) \sum_{\substack{q \leqslant \sqrt{X} \\ q \nmid D}} x^{C/\log q} \sum_{a_1, \ldots, a_r}{}'^{(\ddagger)} \prod_{j=1}^{r} h_j(a_j) \sum_{n \leqslant x}{}^{(*)} 1 \tag{3.20}$$

avec les conditions de sommation

$$(\ddagger) \begin{cases} X^{1/3} < a_1 \ldots a_r \leqslant X^{1/2} \\ (a_i, Dm_i a_j) = 1 \qquad (1 \leqslant i, j \leqslant r, i \neq j) \\ P^+(a_i \ldots a_r) = q \end{cases}$$

et

$$(*) \begin{cases} a_j m_j \,|\, F_j(n) \qquad (1 \leqslant j \leqslant r), \\ d_0 \,|\, F(n), \\ p \,|\, F(n) \Rightarrow p \,|\, D \prod_{j=1}^{r} a_j m_j \text{ ou } p > q. \end{cases}$$

Par le lemme 3.4, on a

$$\sum_{n \leqslant x}{}^{(*)} 1 \ll \frac{x}{(\log q)^r} \frac{\rho(d_0)}{d_0} \prod_{j=1}^{r} \frac{\rho_j(a_j m_j)}{\varphi^*(a_j m_j)}.$$

En reportant dans (3.20) et en faisant appel à (3.18), il suit

$$S_2 := \sum_{\substack{d_1 \ldots d_r = d/d_0 \\ (d_i, d_j) = 1\, (1 \leqslant i \neq j \leqslant r)}} S_2(d_1, \ldots, d_r) \ll x \frac{g(d)}{d} \sum_{q \leqslant \sqrt{X}} x^{C/\log q} \frac{T_q}{(\log q)^r} \tag{3.21}$$

avec

$$T_q := \sum_{a_1, \ldots, a_r}{}'^{(\ddagger)} \prod_{j=1}^{r} \frac{\rho_j(a_j)\, h_j(a_j)}{\varphi^*(a_j)}.$$

En regroupant la somme r-uple suivant les valeurs de $a = a_1 \ldots a_r$, on peut écrire

$$T_q \leqslant \sum_{\substack{X^{1/3} < a \leqslant X^{1/2} \\ P^+(a) = q \\ (a, D) = 1}}{}' \frac{\theta(a)}{a},$$

où θ est la fonction multiplicative définie sur les entiers sans facteur carré par

$$\theta(p) = \{1-\rho^*(p)p^{-1}\}^{-1}\sum_{j=1}^{r}\rho_j(p)h_j(p).$$

On a en particulier pour tout $\alpha > 0$

$$\begin{aligned} T_q &\leqslant \frac{\theta(q)}{q}\sum_{\substack{X^{1/3}q^{-1}<b\leqslant X^{1/2}q^{-1}\\ P^+(b)=q\\ (b,D)=1}}{}' \frac{\theta(b)}{b}\left(\frac{bq}{X^{1/3}}\right)^{\alpha} \\ &\ll q^{\alpha-1}X^{-\alpha/3}\prod_{\substack{p\leqslant q\\ p\nmid D}}\{1+\theta(p)p^{\alpha-1}\} \\ &\ll q^{\alpha-1}X^{-\alpha/3}\exp\left(\sum_{j=1}^{r}\sum_{p\leqslant q}\frac{\rho_j(p)h_j(p)}{p}p^{\alpha}\{1+O(p^{-1})\}\right). \end{aligned}$$

Choisissons $\alpha = 6C/c_7\log q$. On a, pour chaque j $(1\leqslant j\leqslant r)$,

$$\sum_{p\leqslant q}\frac{\rho_j(p)h_j(p)}{p}(p^{\alpha}\{1+O(p^{-1})\}-1) \ll \sum_{p\leqslant q}\frac{\rho_j(p)\{\alpha\log p+p^{-1}\}}{p}\ll 1,$$

où nous avons fait appel au lemme 3.1, après sommation d'Abel. Nous pouvons donc écrire

$$T_q \ll x^{-2C/\log q}q^{-1}\exp\left(\sum_{j=1}^{r}\sum_{p\leqslant x}\frac{\rho_j(p)h_j(p)}{p}\right).$$

En reportant dans (3.21), on constate donc que la somme en q est

$$\ll T\exp\left(\sum_{j=1}^{r}\sum_{p\leqslant x}\frac{\rho_j(p)\{h_j(p)-1\}}{p}\right)$$

avec

$$T := \sum_{q\leqslant x}x^{-C/\log q}\left(\frac{\log x}{\log q}\right)^{r}q^{-1}\ll 1.$$

Cela achève la démonstration. □

Le résultat suivant est une conséquence facile du lemme 3.5. Il montre, dans le cas $d = 1$, que le membre de gauche de (3.9) est dominé par des entiers n tels que le produit des 'petits' facteurs premiers de $F(n)$ est lui-même 'petit'.

Lemme 3.6 *Dans les hypothèses du lemme 3.5, on a uniformément pour* $w\geqslant v\geqslant 2$ *et* $x\geqslant 2$

$$\sum\left\{\prod_{j=1}^{r} h_j(F_j(n)) : n \leqslant x,\ \prod_{j=1}^{r}\prod_{\substack{p|F_j(n)\\ p\leqslant v}} p > w\right\}$$
$$\ll x\exp\left(-\frac{\log w}{\log v} + \sum_{j=1}^{r}\sum_{p\leqslant x}\frac{\rho_j(p)(h_j(p)-1)}{p}\right).$$

Démonstration On applique le lemme 3.5 aux fonctions fortement multiplicatives g_j $(1 \leqslant j \leqslant r)$ définies par

$$g_j(p) = \begin{cases} h_j(p)p^{\alpha} & \text{si } p \leqslant v \\ h_j(p) & \text{si } p > v \end{cases}$$

avec $\alpha := 1/\log v$. On a ainsi $g_j(p) \leqslant eA$ pour tout p et tout j. La somme à majorer n'excède pas

$$w^{-\alpha}\sum_{n\leqslant x}\prod_{j=1}^{r} g_j(F_j(n))$$

et le résultat découle de (3.9) – avec g_j à place de h_j – puisque l'on a pour chaque j

$$\sum_{p\leqslant v}\frac{\rho_j(p)(h_j(p)-1)(p^{\alpha}-1)}{p} \ll \sum_{p\leqslant v}\rho_j(p)\frac{\alpha\log p}{p} \ll 1. \qquad \square$$

Nous aurons également besoin de la variante suivante du lemme 3.6, dans laquelle $h_j \equiv 1$ mais où les petits facteurs premiers sont comptés avec multiplicité. Nous n'avons pas recherché la forme optimale du résultat – le point essentiel étant ici la complète uniformité. Il est certainement possible d'améliorer l'estimation dans un domaine convenable en v et w – cf. par exemple [20].

Lemme 3.7 *On a uniformement pour $w \geqslant v \geqslant 2$ et $x \geqslant 2$*

$$\operatorname{card}\left\{n \leqslant x : \prod_{\substack{p^{\nu}\|F(n)\\ p\leqslant v}} p^{\nu} > w\right\} \ll \exp\left(-c_8\frac{\log w}{\log v}\right) \tag{3.22}$$

où $c_8 = c_8(F)$ est une constante positive.

Démonstration Posons $\alpha = \max_{1\leqslant j\leqslant r}\alpha_j$ et $\kappa = 1/(3g\alpha+1)$. Nous répartissons les entiers n comptés dans (3.22) en deux classes, selon que l'on a ou non

$$\prod_{\substack{p|F(n)\\ p\leqslant v}} p \geqslant w^{\kappa}. \tag{3.23}$$

Le cas de la première classe relève du lemme 3.6 avec $h_j \equiv 1$

$(1 \leqslant j \leqslant r)$. Son cardinal est

$$\ll x \exp\left(-\kappa \frac{\log w}{\log v}\right),$$

l'estimation étant trivialement satisfaite si $v \geqslant w^{\kappa}$.

Lorsque (3.23) n'a pas lieu, on peut écrire

$$\sum_{\substack{p^{\nu} \| F(n) \\ p \leqslant v}} (\nu - 3g\alpha) \log p > \log w - 3g\alpha\kappa \log w = \kappa \log w.$$

Cela implique

$$\prod_{\substack{p^{\nu} \| F(n) \\ \nu \geqslant 3g\alpha + 1}} p^{\nu} > w^{\kappa}. \tag{3.24}$$

Notons t un entier générique tel que $p^{\nu} \| t \Rightarrow \nu \geqslant 3g$. L'inégalité (3.24) implique l'existence d'un diviseur t de $F^*(n)$ satisfaisant à $t > w^{\kappa/\alpha}$. Le cardinal de la seconde classe est donc au plus égal à

$$\sum_{w^{\kappa/\alpha} < t \leqslant c_9 x^g} \rho^*(t)\left(\left[\frac{x}{t}\right] + 1\right) \ll \sum_{t \leqslant c_9 x^g} c_{10}^{\omega(t)} \left\{\frac{x}{t}(tw^{-\kappa/\alpha})^{\eta} + \left(\frac{x^g}{t}\right)^{1-\eta}\right\}$$

où η est un paramètre arbitraire $(0 \leqslant \eta < 1)$ et où nous avons utilisé la majoration $\rho^*(t) \leqslant c_{10}^{\omega(t)}$ qui découle de (2.14). La série

$$\sum_t c_{10}^{\omega(t)} t^{\eta - 1}$$

est convergente pour tout $\eta < 1 - 1/3g$. En choisissant par exemple $\eta = 1 - 1/2g$, nous obtenons la majoration

$$\ll x w^{-\kappa/2\alpha g} + \sqrt{x}.$$

Cela suffit pleinement à compléter la démonstration. □

La lemme suivant montre que, lorsque $d = 1$, l'estimation (3.9) du lemme 3.5 fournit l'ordre de grandeur exact de la somme considérée si les $h_j(p)$ sont minorés par une constante positive. Ce résultat est analogue au *Satz* 2 de Wolke [19].

Lemme 3.8 *Soient A' et A'' des constantes positives et $h_1, h_2, \ldots, h_r$ des fonctions arithmétiques fortement multiplicatives satisfaisant pour tout nombre premier p à*

$$A' \leqslant h_j(p) \leqslant A'' \qquad (1 \leqslant j \leqslant r).$$

Alors on a pour $x \geqslant 1$

$$\sum_{n \leqslant x} \prod_{j=1}^{r} h_j(F(n)) \asymp x \exp\left(\sum_{j=1}^{r} \sum_{p \leqslant x} \frac{\rho_j(p)(h_j(p) - 1)}{p}\right). \tag{3.25}$$

Démonstration La majoration découle de (3.9) avec $d = 1$. Pour établir la minoration, nous restreignons la somme en n aux entiers qui satisfont les conditions suivantes:

$$\begin{cases} F_j(n) = D_jA_jB_j \qquad (1 \leqslant j \leqslant r), \\ D_j \mid D^\infty, \\ (A_j, D) = 1,\ A_j \leqslant X := x^{c_6/2r},\ P^-(B_j) > X, \end{cases}$$

où $c_6 = c_6(\frac{1}{2}, F)$ est la constante du lemme 3.4. Notant d_j, a_j et b_j les noyaux sans facteur carré respectifs de D_j, A_j et B_j, on a alors

$$h_j(F(n)) = h_j(d_j)\,h_j(a_j)\,h_j(b_j) \gg h_j(a_j)$$

puisque

$$h_j(d_j) \geqslant \min(1, A')^{\omega(D)}$$

et

$$h_j(b_j) \geqslant \min(1, A')^{(2rg/c_6)+1} \qquad (x \geqslant x_0).$$

La somme en n de (3.25) est donc

$$\gg \sum_{\substack{a_1,\ldots,a_r \leqslant X \\ (a_j, D)=1\,(1\leqslant j\leqslant r)}}' \prod_{j=1}^{r} h_j(a_j) \sum_{\substack{n\leqslant x \\ a_j \mid F_j(n)\ (1\leqslant j\leqslant r) \\ p\mid F(n) \Rightarrow p \mid D\prod_{j=1}^{r} a_j \text{ ou } p>X}} 1$$

$$\gg \sum_{\substack{a_1,\ldots,a_r \leqslant X \\ (a_j, D)=1\,(1\leqslant j\leqslant r)}}' \prod_{j=1}^{r} \frac{\rho_j(a_j)\,h_j(a_j)}{\varphi^*(a_j)} \frac{x}{(\log x)^r},$$

d'après le lemme 3.4. Il nous suffit donc maintenant de montrer que l'on a pour chaque j $(1 \leqslant j \leqslant r)$

$$\sum_{\substack{a\leqslant X \\ (a, D)=1}}' \frac{\rho_j(a)\,h_j(a)}{\varphi^*(a)} \gg \exp\left(\sum_{p\leqslant x} \frac{\rho_j(p)\,h_j(p)}{p}\right). \tag{3.26}$$

Soit $\epsilon > 0$ et $Z := X^\epsilon$. Le membre de gauche de (3.26) est au moins égal à

$$\sum_{\substack{P^+(a)\leqslant Z \\ (a, D)=1}}' \frac{\rho_j(a)\,h_j(a)}{a} - \sum_{\substack{P^+(a)\leqslant Z \\ a>X \\ (a, D)=1}}' \frac{\rho_j(a)\,h_j(a)}{a}.$$

La première de ces deux sommes vaut

$$\prod_{\substack{p\leqslant Z\\ p\nmid D}}\left(1+\frac{\rho_j(p)h_j(p)}{p}\right)\geqslant \exp\left(\sum_{p\leqslant Z}\frac{\rho_j(p)h_j(p)}{p}\right) \tag{3.27}$$

puisque $\rho_j(p)h_j(p)\leqslant 1$. La seconde est majorée pour tout $\eta>0$ par

$$X^{-\eta}\prod_{p\leqslant Z}\left(1+\frac{\rho_j(p)h_j(p)}{p^{1-\eta}}\right)\leqslant \exp\left(-\frac{\eta}{\epsilon}\log Z+\sum_{p\leqslant Z}\frac{\rho_j(p)h_j(p)}{p^{1-\eta}}\right).$$

Pour le choix $\eta = 1/\log Z$, cette expression ne dépasse pas

$$\exp\left\{-\frac{1}{\epsilon}+\sum_{p\leqslant Z}\frac{\rho_j(p)h_j(p)}{p}+O\left(\frac{1}{\log Z}\sum_{p\leqslant Z}\frac{\log p}{p}\right)\right\} \tag{3.28}$$

où le terme d'erreur est borné indépendamment de ϵ. Lorsque ϵ est assez petit, (3.28) est donc au plus égal à la moitié de (3.27). Cela établit bien (3.26) et achève la démonstration. □

Lemme 3.9 *Soit $g(d)$ la fonction arithmétique multiplicative définie au lemme 3.5. On suppose de plus l'existence de nombres réels $\mu_j\geqslant 0$ $(1\leqslant j\leqslant r)$ tels que l'on ait*

$$h_j(p)=\mu_j \qquad (1\leqslant j\leqslant r)$$

sauf au plus pour un nombre fini de nombres premiers p. Alors, pour chaque entier $N\geqslant 1$, il existe un polynôme $P_N(t)$, de degré N et dont les coefficients ne dépendent que des h_j et de F, tel que l'on ait pour x infini

$$\sum_{d\leqslant x} g(d) = x(\log x)^{\mu-1}\left\{P_N\left(\frac{1}{\log x}\right)+O_N\left(\left(\frac{1}{\log x}\right)^{N+1}\right)\right\} \tag{3.29}$$

avec

$$\mu := \sum_{j=1}^{r}\alpha_j\mu_j.$$

Démonstration Soit $s=\sigma+i\tau$ un nombre complexe de partie réelle $\sigma>1$. Pour tous les p sauf un nombre fini, on a, par (3.10),

$$\sum_{\nu=0}^{\infty} g(p^\nu)p^{-\nu s}=1+\left(1-\frac{\rho^*(p)}{p}\right)\sum_{j=1}^{r}\mu_j\rho_j(p)\sum_{\nu=0}^{\infty}p^{\nu(1-s)-\lceil\nu/\alpha_j\rceil}.$$

La somme en ν vaut

$$\sum_{m=1}^{\infty}\sum_{(m-1)\alpha_j<\nu\leqslant m\alpha_j}p^{\nu(1-s)-m}=\sum_{\nu=1}^{\alpha_j}p^{-1-\nu(s-1)}+R_p(s), \tag{3.30}$$

où $R_p(s)$ est une fonction holomorphe de s pour $\sigma>1-1/2\alpha_j$ et satisfait dans ce domaine à la majoration

$$R_p(s) \ll p^{2\alpha_j(1-\sigma)-2}(1+p^{(\alpha_j-1)(\sigma-1)}).$$

Pour $\sigma \geqslant \sigma_0 := 1-1/(4\max_{1\leqslant j\leqslant r}\alpha_j)$ et p assez grand, on peut donc écrire

$$\sum_{\nu=0}^{\infty} g(p^\nu)p^{-\nu s} = \prod_{j=1}^{r}\prod_{\nu=1}^{\alpha_j}(1-p^{-1-\nu(s-1)})^{-\mu_j\rho_j(p)}\exp V_p(s),$$

où $V_p(s)$ est une fonction holomorphe en s dans le demi-plan $\sigma \geqslant \sigma_0$ qui satisfait dans la même région à

$$V_p(s) \ll p^{-3/2}.$$

De plus, lorsque p est borné les facteurs locaux

$$\sum_{\nu=0}^{\infty} g(p^\nu)p^{-\nu s} \tag{3.31}$$

définissent des fonctions holomorphes pour $\sigma \geqslant \sigma_0$. Cela découle immédiatement de (3.10) et (2.15) sous la forme

$$g(p^\nu) \ll \sum_{j=1}^{r} p^{\nu-\lceil \nu/\alpha_j\rceil}.$$

Nous pouvons donc écrire pour $\sigma > 1$

$$\sum_{d=1}^{\infty} g(d)\,d^{-s} = \prod_{j=1}^{r}\prod_{\nu=1}^{\alpha_j}\prod_{p}(1-p^{-1-\nu(s-1)})^{-\mu_j\rho_j(p)}\Phi(s), \tag{3.32}$$

où Φ est une fonction holomorphe et bornée pour $\sigma \geqslant \sigma_0$, avec $\Phi(1) \neq 0$.

Considérons maintenant la fonction zêta de Dedekind, $\zeta_j(s)$, du corps de nombres $\mathbb{Q}(\theta_j)$ engendré par un zéro θ_j de F_j. On peut écrire, pour $\sigma > 1$, cette fonction comme un produit eulérien, soit

$$\zeta_j(s) = \prod_{\mathfrak{p}}\{1-(N\mathfrak{p})^{-s}\}^{-1}, \tag{3.33}$$

où $\mathfrak{p}$ parcourt l'ensemble des idéaux premiers de l'anneau des entiers de $\mathbb{Q}(\theta_j)$ et N désigne la norme de $\mathbb{Q}(\theta_j)$ sur $\mathbb{Q}$. Pour chaque $\mathfrak{p}$ on a $N\mathfrak{p} = p^f$, où p est un nombre rationnel. Comme l'a montré Erdős dans [4], la partie du produit (3.33) correspondant aux $\mathfrak{p}$ tels que $f \geqslant 2$ ou $p\,|\,D$ définit une fonction holomorphe, bornée et sans zéro pour $\sigma \geqslant \sigma_0 > \frac{1}{2}$. Pour tous les autres $\mathfrak{p}$, la valeur $p = N\mathfrak{p}$ apparaît exactement $\rho_j(p)$ fois, de sorte que l'on peut écrire

$$\prod_{p}(1-p^{-s})^{-\rho_j(p)} = \zeta_j(s)\,\phi_j(s) \qquad (\sigma > 1), \tag{3.34}$$

où $\phi_j(s)$ est holomorphe, bornée et sans zéro pour $\sigma \geqslant \sigma_0$. En reportant dans (3.32), il suit

$$\sum_{d=1}^{\infty} g(d)\, d^{-s} = \prod_{j=1}^{r} \prod_{\nu=1}^{\alpha_j} \zeta_j(1+\nu(s-1))^{\mu_j} \Psi(s) \qquad (\sigma > 1), \tag{3.35}$$

où $\Psi(s)$ est holomorphe et bornée pour $\sigma \geqslant \sigma_0$, avec $\Psi(1) \neq 0$. La série (3.35) possède donc en $s = 1$ une singularité de type $(s-1)^{-\mu}$ et est prolongeable holomorphiquement dans une région

$$\left\{ s : \sigma \geqslant 1 - \frac{c}{1+\log^+ |\tau|} \right\} \Big\backslash [1-c, 1],$$

où c est une constante positive convenable. Cela découle des propriétés classiques des fonctions zêta de Dedekind - cf. Landau [11] *Satz* 185, p. 105. De plus, le fait que les $\zeta_j(s)$ soient d'ordre fini dans toute bande verticale (cf. [11] *Satz* 171, p. 87) implique, grâce à un théorème général concernant les séries de Dirichlet, que le prolongement $\mathscr{G}$ de (3.35) satisfait, pour tout $\epsilon > 0$, à

$$\mathscr{G}(s) \ll_\epsilon |\tau|^\epsilon \qquad \left(|\tau| \geqslant 1, \ \sigma \geqslant 1 - \frac{c}{1+\log|\tau|} \right). \tag{3.36}$$

On peut alors évaluer le membre de gauche de (3.29) par la technique usuelle d'intégration complexe, faisant apparaître un contour de Hankel autour de $s = 1$. Cette méthode a été employée par Selberg [16] dans le cas des puissances complexes de la fonction zêta de Riemann et fut développée par Delange dans les années soixante (voir en particulier [1]). Les calculs étant identiques *mutatis mutandis*, nous omettons les détails.

4 Répartition globale des diviseurs dans les suites polynomiales

Cette section est dévolue à l'étude de diverses moyennes pondérées de fonctions liées à la répartition globale des diviseurs de $F(n)$.

Nous nous donnons un paramètre L satisfaisant à

$$0 \leqslant L \leqslant \log_2 y$$

et nous posons

$$Y := \exp\exp L \leqslant y.$$

Pour chaque entier n et tout j $(1 \leqslant j \leqslant r)$, nous introduisons les fonctions

$$n_{jL} := \prod_{\substack{p \mid F_j(n) \\ p \nmid D,\, p \leqslant Y}} p^{\alpha_j}$$

et

$$n_L := \prod_{j=1}^{r} n_{jL}.$$

D'après le choix de D à la section 2, les n_{jL} sont deux à deux premiers entre eux. On a identiquement

$$\tau(n_L) = \prod_{j=1}^{r} H_j(F_j(n)), \tag{4.1}$$

où H_j est la fonction arithmétique fortement multiplicative définie par

$$H_j(p) = \begin{cases} \alpha_j + 1 & \text{si } p \nmid D,\ p \leqslant Y, \\ 1 & \text{si } p \mid D \text{ ou } p > Y. \end{cases} \tag{4.2}$$

Dans toute cette section, nous supposons que $\beta(y,z) \in [0,B]$. Nous nous donnons une constante positive arbitraire v_0 et considérons un nombre réel v $(0 \leqslant v \leqslant v_0)$. Toutes les constantes, implicites ou explicites, peuvent dépendre de B, v_0 et F, mais pas de L.

Lemme 4.1 *Pour $x \geqslant 2$, $2 \leqslant y \leqslant x$ et $0 \leqslant v \leqslant v_0$, on a*

$$\sum_{n \leqslant x} \tau(n_L)^v \asymp x\mathrm{e}^{\gamma(v)L}. \tag{4.3}$$

Démonstration Cela découle du lemme 3.8 et de l'identité (4.1), en posant $h_j = H_j^v$ $(1 \leqslant j \leqslant r)$. □

Lemme 4.2 *Posons $W(t) := \exp(t - 2\sqrt{t \log t})$ $(t > 1)$. Désignons, pour tous $v > 0$, $\epsilon > 0$, par $\mathscr{A}(v,\epsilon)$ l'ensemble des entiers n tels que*

$$\min_{\epsilon^{-1}\mathrm{e}^{-L} \leqslant |\theta| \leqslant 1} \tau_\theta(n_L)^{-1} W(L + \log|\theta|)^{-\gamma'(v)} \leqslant \tau(n_L)^{-1}. \tag{4.4}$$

Pour chaque $\epsilon > 0$, il existe une constante $y_0(\epsilon)$ telle que l'on ait, lorsque $y_0(\epsilon) \leqslant y \leqslant x$ et $0 \leqslant v \leqslant v_0$,

$$\sum_{\substack{n \leqslant x \\ n \in \mathscr{A}(v,\epsilon)}} \tau(n_L)^v \ll \left(\log \frac{1}{\epsilon}\right)^{-1} x\mathrm{e}^{\gamma(v)L}. \tag{4.5}$$

Démonstration Pour chaque θ $(\epsilon^{-1}\mathrm{e}^{-L} \leqslant |\theta| \leqslant 1)$, posons

$$m(\theta) = L + [\log|\theta|],$$

de sorte que $\log(1/\epsilon) - 1 < m(\theta) \leqslant L$. Si n est compté dans $\mathscr{A}(v,\epsilon)$ et si θ réalise le minimum (4.4), on a

$$\begin{aligned} 1 &\leqslant \mathrm{e}^{\gamma'(v)} W(m(\theta))^{\gamma'(v)} \tau_\theta(n_L)\tau(n_L)^{-1} \\ &\leqslant \mathrm{e}^{\gamma'(v)} W(m(\theta))^{\gamma'(v)} \prod_{\substack{p^\nu \| n_L \\ L - m(\theta) \leqslant \log_2 p \leqslant L}} (\nu+1)^{-1}. \end{aligned}$$

Soit $w = w(m(\theta))$ un paramètre tel que $0 < w \leqslant 1$. Elevons l'inégalité précédente à la puissance w, multiplions les deux membres par $\tau(n_L)^v$ et sommons sur toutes les valeurs possibles de $m = m(\theta)$. Nous obtenons, pour n dans $\mathscr{A}(v, \epsilon)$,

$$\tau(n_L)^v \leqslant \sum_{\log(1/\epsilon)-1<m\leqslant L} W(m)^{w(m)\gamma'(v)} \prod_{j=1}^{r} h_j(F_j(n); m), \tag{4.6}$$

où $h_j(n; m)$ est la fonction fortement multiplicative de n définie par

$$h_j(p; m) = \begin{cases} (\alpha_j+1)^v & \text{si } p \nmid D, \log_2 p \leqslant L-m, \\ (\alpha_j+1)^{v-w(m)} & \text{si } p \nmid D, L-m < \log_2 p \leqslant L, \\ 1 & \text{si } p \mid D \text{ ou } \log_2 p > L. \end{cases}$$

Sommons l'inégalité (4.6) pour $n \leqslant x$ et $n \in \mathscr{A}(v, \epsilon)$. Après interversion de sommations, nous pouvons faire appel au lemme 3.5 pour évaluer la somme en n dans le membre de droite. Nous obtenons que le membre de gauche de (4.5) est

$$\ll x\mathrm{e}^{\gamma(v)L} \sum_{\log(1/\epsilon)-1\leqslant m\leqslant L} \mathrm{e}^{-m\Xi(m)} \tag{4.7}$$

avec

$$\Xi(m) := \sum_{j=1}^{r} (\alpha_j+1)^v \left\{1-(\alpha_j+1)^{-w(m)} - w(m)\left(1-2\sqrt{\frac{\log m}{m}}\right)\log(\alpha_j+1)\right\}.$$

L'expression entre accolades est certainement positive lorsque

$$1-2\sqrt{\frac{\log m}{m}} \leqslant (\alpha_j+1)^{-w(m)} < 1.$$

Nous choisissons

$$w(m) = -\frac{\log(1-2\sqrt{(\log m)/m})}{\log \max\limits_{1\leqslant j\leqslant r}(\alpha_j+1)}.$$

On a alors

$$\Xi(m) \geqslant Q\left(1-2\sqrt{\frac{\log m}{m}}\right) \geqslant 2\frac{\log m}{m}.$$

En reportant dans (4.7), on obtient que la somme en m ne dépasse pas

$$\sum_{m\geqslant\log(1/\epsilon)-1} m^{-2} \ll \left(\log\frac{1}{\epsilon}\right)^{-1}.$$

Cela achève la démonstration. □

Lemme 4.3 *Pour tous v $(0 \leqslant v \leqslant v_0)$ et $\psi > 0$, soit $\mathcal{B}(v,\psi)$ l'ensemble des entiers n tel que*

$$|\log \tau(n_L) - \gamma'(v)\,L| > \psi\gamma'(v)\sqrt{L}. \tag{4.8}$$

On a pour $x \geqslant 2$, $2 \leqslant y \leqslant x$ et $0 \leqslant \psi \leqslant \frac{1}{3}\sqrt{L}$,

$$\sum_{\substack{n\leqslant x\\ n\in\mathcal{B}(v,\psi)}} \tau(n_L)^v \ll \mathrm{e}^{-\frac{1}{3}\psi^2} x \mathrm{e}^{\gamma(v)L}. \tag{4.9}$$

Démonstration Nous procédons de manière analogue à la preuve du lemme précédent, mais les détails sont ici plus simples car la situation correspond à $\theta = 1$. Contentons-nous d'estimer par exemple la contribution R au membre de gauche de (4.9) des entiers n tels que

$$\tau(n_L) > \exp\{\gamma'(v)(L+\psi\sqrt{L})\}. \tag{4.10}$$

Le cas des autres exceptions relève d'une manipulation symétrique.

Pour chaque valeur du paramètre $w \geqslant 0$, on a

$$\begin{aligned} R &\leqslant \sum_{n\leqslant x} \tau(n_L)^{v+w} \exp\{-w\gamma'(v)(L+\psi\sqrt{L})\} \\ &\ll x\exp\{(\gamma(v)-X)L\}, \end{aligned}$$

d'après le lemme 4.1, avec

$$X := \sum_{j=1}^{r} (\alpha_j+1)^v \left\{1-(\alpha_j+1)^w + \left(1+\frac{\psi}{\sqrt{L}}\right) w\log(\alpha_j+1)\right\}.$$

Comme précédemment, on a

$$X \geqslant Q\left(1+\frac{\psi}{\sqrt{L}}\right) \geqslant \frac{\psi^2}{3L},$$

pour le choix

$$w = \frac{\log(1+\psi/\sqrt{L})}{\log\max(\alpha_j+1)}.$$

Cela implique bien l'estimation annoncée. □

Lemme 4.4 *Posons*

$$\kappa = \kappa(v,F) = 2\sum_{j=1}^{r} \alpha_j(\alpha_j+1)^v.$$

On a pour $x \geqslant 2$, $2 \leqslant y \leqslant x$, $0 \leqslant v \leqslant v_0$ et $\theta \in \mathbb{R}$,

$$\sum_{n\leqslant x} \tau(n_L)^{v-1}\tau_\theta(n_L)^{-1}|\tau(n_L,\theta)|^2 \ll x\{\log(3+|\theta|)\}^\kappa \mathrm{e}^{\gamma(v)L}. \tag{4.11}$$

Démonstration Lorsque $|\theta| \leqslant e^{-L}$, on a $\tau_\theta(n_L) = \tau(n_L)$ pour tout n et la majoration triviale

$$|\tau(n_L, \theta)| \leqslant \tau(n_L)$$

fournit l'estimation souhaitée, compte tenu de (4.3).

Supposons désormais $|\theta| > e^{-L}$. La somme à majorer, disons $S(\theta)$, peut encore s'écrire

$$S(\theta) = \sum_{n \leqslant x} \prod_{j=1}^{r} g_j(F_j(n)),$$

où g_j $(1 \leqslant j \leqslant r)$ est la fonction arithmétique fortement multiplicative définie sur les nombres premiers par

$$g_j(p) = \begin{cases} (\alpha_j+1)^{v-2}\left|\sum_{\nu=0}^{\alpha_j} p^{i\nu\theta}\right|^2 & \text{si } p \nmid D,\ \log p \leqslant |\theta|^{-1}, \\ (\alpha_j+1)^{v-1}\left|\sum_{\nu=0}^{\alpha_j} p^{i\nu\theta}\right|^2 & \text{si } p \nmid D,\ |\theta|^{-1} < \log p \leqslant e^L, \\ 1 & \text{si } p \mid D \text{ ou } \log p > e^L. \end{cases}$$

Nous pouvons donc faire appel au lemme 3.5.

Lorsque $|\theta| \leqslant 2$, nous majorons trivialement $g_j(p)$ par $(\alpha_j+1)^v$ lorsque $\log p \leqslant 1/|\theta|$. Nous obtenons

$$S(\theta) \ll x \exp\left\{\sum_{j=1}^{r}\left(\sum_{|\theta|\log p \leqslant 1} (\alpha_j+1)^v \rho_j(p) p^{-1} + \sum_{1/|\theta| < \log p \leqslant e^L} (\alpha_j+1)^{v-1}\left|\sum_{\nu=0}^{\alpha_j} p^{i\nu\theta}\right|^2 \rho_j(p) p^{-1} - L\right)\right\}.$$

Estimons la première somme en p par le lemme 3.1, et la seconde par le lemme 3.3 avec

$$f(t) = (\alpha_j+1)^{v-1}\left|\sum_{0 \leqslant \nu \leqslant \alpha_j} e^{i\nu t}\right|^2,$$

de moyenne $\bar{f} = (\alpha_j+1)^v$. Nous obtenons

$$S(\theta) \ll x e^{\gamma(v) L},$$

ce qui est bien en accord avec (4.11).

Lorsque $|\theta| > 2$, on a $\tau_\theta(n_L) = 1$ pour tout n, et le lemme 3.5 fournit la borne

$$S(\theta) \ll x \exp\left\{\sum_{j=1}^{r} \sum_{p \leqslant Y} \left((\alpha_j+1)^{\nu-1} \left|\sum_{\nu=0}^{\alpha_j} p^{i\nu\theta}\right|^2 - 1\right) \rho_j(p) p^{-1}\right\}.$$

On applique le lemme 3.3 à la sous-somme en p correspondant à l'intervalle de sommation (éventuellement vide)

$$c^{-2}\{\log(3+|\theta|)\}^2 < \log p \leqslant e^L,$$

où c est une constante positive minorant tout les $c(F_j)$ tels qu'ils sont définis au lemme 3.3. On obtient une contribution ne dépassant pas

$$\{(\alpha_j+1)^{\nu}-1\}\{L-2\log_2(3+|\theta|)\}+O(1).$$

La sous-somme complémentaire est majorée trivialement, en estimant simplement $|\sum p^{i\nu\theta}|$ par α_j+1. Elle est au plus égale à

$$2\{(\alpha_j+1)^{\nu+1}-1\}\log_2(3+|\theta|)+O(1).$$

En regroupant ces évaluations, on obtient bien le résultat indiqué. □

5 Majorations de $H_F(x,y,z)$

Nous nous proposons dans cette section d'établir les bornes supérieures des encadrements (1.13) et (1.14) de notre théorème.

Commençons par (1.13). Nous allons montrer que, pour une proportion positive des entiers $n \leqslant x$, minorée indépendamment de y, le nombre $F(n)$ ne possède aucun diviseur dans $]y,2y]$. Comme $H_F(x,y,z)$ est une fonction croissante de z (et donc décroissante de β), cela suffit pleinement à prouver l'estimation supérieure de (1.13).

Considérons la décomposition canonique

$$F(n) = a_n b_n \qquad (P^+(a_n) \leqslant 2y < P^-(b_n)).$$

On a certainement $\tau(F(n);y,2y) = 0$ lorsque $a_n \leqslant y$. Pour chaque $\epsilon > 0$, on peut donc écrire

$$x - H_F(x,y,2y) \geqslant \sum_{\substack{n\leqslant x\\ a_n\leqslant y}} 1 \geqslant \sum_{\substack{n\leqslant x\\ P^+(a_n)\leqslant y^\epsilon}} 1 - \sum_{\substack{n\leqslant x\\ P^+(a_n)\leqslant y^\epsilon\\ a_n>y}} 1.$$

Le premier terme de cette minoration est égal au nombre des entiers $n \leqslant x$ tels que $F^*(n)$ ne possède aucun facteur premier dans $]y^\epsilon,2y]$, D'après le lemme fondamental de la théorie du crible (cf. par exemple [7], *Theorem 7.1*), il est

$$\asymp \epsilon^r x$$

sous l'hypothèse $y_0 \leqslant y \leqslant x^{c_0}$, avec $y_0 = y_0(\epsilon, F)$ et $c_0 = c_0(F)$ convenables. Or, le lemme 3.7 nous permet de majorer le second cardinal par

$$\ll e^{-c_8/\epsilon}x.$$

Pour ϵ assez petit mais fixé, nous obtenons bien l'estimation souhaitée.

Etablissons maintenant la majoration de (1.14), c'est-à-dire l'estimation

$$H_F(x,y,z) \ll x(\log y)^{-\delta(\beta,F)} \tag{5.1}$$

sous les conditions

$$y_0 \leqslant y \leqslant x^{1-K}, \qquad \max\left(0, \log \hat{\tau}(F) - 1 - \frac{c_1}{\sqrt{\log_2 y}}\right) \leqslant \beta \leqslant B.$$

Nous pouvons en fait supposer que $\beta \geqslant \log \hat{\tau}(F) - 1$ puisque $\delta(\beta, F) = 0$ dans le cas contraire. En particulier, on a donc

$$\gamma'(u) = \beta + 1. \tag{5.2}$$

Nous distinguons deux cas, selon que $u \leqslant 1$ ou non. Dans le premier cas, soit $\beta \leqslant \gamma'(1) - 1$, nous écrivons

$$H_F(x,y,z) \leqslant H_1 + H_2$$

avec

$$H_1 := \text{card}\{n \leqslant x : \tau(n_L) > (\log y)^{1+\beta}\}$$

$$H_2 := \sum \{\tau(F(n); y, z) : n \leqslant x,\ \tau(n_L) > (\log y)^{1+\beta}\}$$

pour le choix (valable dans cette section uniquement)

$$L = \log_2 y.$$

Le lemme 4.1 avec $v = u$ fournit

$$H_1 \leqslant \sum_{n \leqslant x} \tau(n_L)^u (\log y)^{-u(1+\beta)} \ll x(\log y)^{\gamma(u) - u(1+\beta)}.$$

Grâce à (5.2), cela équivant bien à (5.1). Pour estimer H_2, donnons-nous un paramètre $v \geqslant 0$. On a

$$\begin{aligned} H_2 &\leqslant (\log y)^{v\gamma'(u)} \sum_{n \leqslant x} \tau(F(n); y, z)\tau(n_L)^{-v} \\ &= (\log y)^{v\gamma'(u)} \sum_{y < d \leqslant z} \sum_{\substack{n \leqslant x \\ d | F(n)}} \tau(n_L)^{-v}. \end{aligned}$$

Puisque $d \leqslant 2y \leqslant 2x^{1-K}$, on peut appliquer le lemme 3.5 pour majorer la somme intérieure, compte tenu de l'identité (4.1). On obtient la majoration

$$\ll x\frac{g(d)}{d}(\log y)^{\gamma(-v)},$$

où g est définie par (3.10) avec

$$h_j(p) = (\alpha_j+1)^{-v} \qquad (p\nmid D, p \leqslant y).$$

Par sommation d'Abel, le lemme 3.9 nous permet donc de montrer que

$$\sum_{y<d\leqslant x}\frac{g(d)}{d} \ll \frac{z-y}{y}(\log y)^{\mu-1}$$

avec

$$\mu = \sum_{j=1}^{r}\alpha_j(\alpha_j+1)^{-v} = \gamma(1-v)-\gamma(-v).$$

Nous obtenons ainsi

$$H_2 \ll x(\log y)^{v\gamma'(u)+\gamma(1-v)-\beta-1}.$$

Pour le choix $v = 1-u$ (qui est licite car $u \leqslant 1$) l'exposant de $\log y$ vaut $-u\gamma'(u)+\gamma(u) = -\delta(\beta,F)$. Cela établit encore (5.1) lorsque $\gamma(0)-1 \leqslant \beta \leqslant \gamma(1)-1$.

Lorsque $\beta > \gamma(1)-1$, nous nous contentons de la majoration triviale

$$H_F(x,y,z) \leqslant \sum_{n\leqslant x}\tau(F(n);y,z) = \sum_{y<d\leqslant z}\sum_{\substack{n\leqslant x\\ d|F(n)}} 1 \ll x\sum_{y<d\leqslant z}\frac{\rho(d)}{d}.$$

Nous pouvons de nouveau faire appel au lemme 3.9 pour estimer la somme en d puisque (2.11) et (3.10) montrent que $\rho(d)$ est majorée par la fonction $g(d)$ correspondant au cas $h_j \equiv 1$ $(1 \leqslant j \leqslant r)$. On obtient comme annoncé

$$H_F(x,y,z) \ll x\frac{z-y}{y}(\log y)^{\hat{\Omega}(F)-1} = x(\log y)^{\hat{\Omega}(F)-\beta-1}.$$

Cela achève la preuve de (5.1). □

6 Minorations de $H_F(x,y,z)$

Nous allons maintenant établir la validité des bornes inférieures de notre théorème. A cet effet, nous adoptons une méthode analogue à celle du chapitre 5 de [8], reposant sur l'idée que les quantités $\log d$ $(d|F(n))$ sont bien réparties dans $[0,\log F(n)]$ dès qu'elles sont suffisamment nombreuses.

Pour chaque entier $m \geqslant 1$, nous introduisons la fonction de répartition

$$\mathcal{F}(m;t) := \tau(m;0,\mathrm{e}^t) = \mathrm{card}\{d : d|m, d \leqslant \mathrm{e}^t\}.$$

La transformée de Fourier–Stieltjes est évidemment

$$\tau(m,\theta) = \int_{-\infty}^{+\infty} \mathrm{e}^{i\theta t}\, \mathrm{d}\mathcal{F}(m;t) = \sum_{d|m} d^{i\theta}.$$

Soit $\eta := \log(z/y)$. Nous allons montrer que, relativement à la mesure discrète sur $\{n : n \leqslant x\}$ associée au poids $\tau(n_L)^v$, la fonction d'accroissement

$$\Delta(n_L;t) := \mathcal{F}(n_L;t+\eta) - \mathcal{F}(n_L;t)$$

est 'souvent' positive lorsque v est sensiblement plus grand que u. (Nous choisirons $L \approx \log_2 y - T$, où T est une constante suffisamment grande.) Cela permet de mettre en évidence l'existence d'un ensemble assez riche de nombres premiers $p > \exp\exp L$ tels que pn_L possède un diviseur dans $]y,z]$. Un argument de crible permet ensuite de conclure.

Soit $\lambda(m)$ la mesure de Lebesgue de l'ensemble $\mathcal{L}(m)$ des nombres réels t tel que $\Delta(m;t) > 0$. Commençons par établir une minoration pour $\lambda(m)$.

Lemme 6.1 *Pour chaque entier* $m \geqslant 1$, *on a*

$$\tau(m)^2 \leqslant 3\pi\lambda(m) \int_{|\theta\eta|\leqslant 1} |\tau(m,\theta)|^2\, \mathrm{d}\theta. \tag{6.1}$$

Démonstration Par Cauchy–Schwarz, on peut écrire

$$\eta^2\tau(m)^2 = \left(\int_{-\infty}^{+\infty} \Delta(m;t)\, \mathrm{d}t\right)^2 \leqslant \lambda(m) \int_{-\infty}^{+\infty} \Delta(m;t)^2\, \mathrm{d}t.$$

Posons

$$w(t) := \left(\frac{\sin \frac{1}{2}t}{\frac{1}{2}t}\right)^2 = \int_{-1}^{1} (1-|\theta|)\mathrm{e}^{i\theta t}\, \mathrm{d}\theta.$$

Alors pour tout $t \in \mathbb{R}$, on a

$$\begin{aligned}\Delta(m;t) &\leqslant w(1)^{-1} \sum_{d|m} w(\eta^{-1}(\log d - t)) \\ &= w(1)^{-1}\eta \int_{|\theta\eta|\leqslant 1} (1-|\theta\eta|)\mathrm{e}^{i\theta t}\tau(m,\theta)\, \mathrm{d}\theta.\end{aligned}$$

La formule de Parseval implique donc

$$\int_{-\infty}^{+\infty} \Delta(m;t)^2 \, dt \leqslant 2\pi w(1)^{-2}\eta^2 \int_{|\theta\eta|\leqslant 1} (1-|\theta\eta|)^2 |\tau(m,\theta)|^2 \, d\theta,$$

d'où le résultat indiqué puisque $w(1)^2 > \frac{2}{3}$. □

Le résultat suivant constitue le point-clef de la démonstration.

Lemme 6.2 *Soient ϵ et β_0 des nombres réels positifs. Il existe des constantes $c_{11} = c_{11}(\epsilon, B, F) > 0$ et $y_0 = y_0(\epsilon, \beta_0, B, F)$ telles que l'on ait pour $y_0 \leqslant y \leqslant x$, $\log_2 y - \sqrt{\log_2 y} \leqslant L \leqslant \log_2 y$, $0 \leqslant v \leqslant 1$, et $\gamma'(v) \geqslant 1$,*

$$\sum_{\substack{n\leqslant x \\ \lambda(n_L)\leqslant \epsilon e^{\sigma L}}} \tau(n_L)^v \ll \left(\log \frac{1}{\epsilon}\right)^{-1} x e^{\gamma(v)L}, \tag{6.2}$$

où l'on a posé

$$\sigma = \sigma(\beta, v) = \begin{cases} \min\left\{1 - c_{11}\sqrt{\dfrac{\log L}{L}}, \gamma'(v) - \beta - \dfrac{c_{11}}{\sqrt{L}}\right\} & \text{si } 0 \leqslant \beta < \beta_0, \\ \min\left\{1, \gamma'(v) - \beta - \dfrac{c_{11}}{\sqrt{L}}\right\} & \text{si } \beta_0 \leqslant \beta \leqslant B. \end{cases} \tag{6.3}$$

Remarque On a $\gamma'(0) = \log \hat{\tau}(F)$. Donc $\gamma'(v) \geqslant \log 3 > 1$ dès que F est réductible.

Démonstration D'après (6.1), la condition $\lambda(n_L) < \epsilon e^{\sigma L}$ implique

$$3\pi\epsilon e^{\sigma L} \int_{|\theta\eta|\leqslant 1} \frac{|\tau(n_L, \theta)|^2}{\tau(n_L)^2} \, d\theta \geqslant 1.$$

Puisque la contribution au membre de gauche des réels θ tel que $12\pi\epsilon e^{\sigma L}|\theta| \leqslant 1$ est trivialement $\leqslant \frac{1}{2}$, on voit que les entiers n comptés dans (6.2) satisfont nécessairement

$$\epsilon_1 e^{\sigma L} \int_{1/(\epsilon_1 e^{\sigma L})}^{1/\eta} \frac{|\tau(n_L, \theta)|^2}{\tau(n_L)^2} \, d\theta \geqslant 1, \tag{6.4}$$

où, pour simplifier l'écriture, nous avons posé $\epsilon_1 = 12\pi\epsilon$. Appliquons maintenant les lemmes 4.2 et 4.3. Quitte à négliger un ensemble d'entiers exceptionnels dont la contribution à (6.2) est acceptable, cela nous permet de supposer que l'on a

$$\tau(n_L)^{-1} \leqslant \tau_\theta(n_L)^{-1} W(L + \log\theta)^{-\gamma'(v)} \qquad (\epsilon_1^{-1} e^{-\sigma L} \leqslant \theta \leqslant 1) \tag{6.5}$$

et

$$\tau(n_L)^{-1} \leqslant \exp\{-\gamma'(v)(L - \psi\sqrt{L})\}, \tag{6.6}$$

où $W(t)$ est la fonction définie au lemme 4.2 et où nous avons posé

$$\psi = \psi(\epsilon) = \sqrt{3\log_2(1/\epsilon)}.$$

Reportons ces estimations dans (6.4) en utilisant (6.5) lorsque $\theta \leqslant 1$. Nous obtenons

$$\epsilon_1 e^{\sigma L}\left\{\int_{1/(\epsilon_1 e^{\sigma L})}^{1} \frac{|\tau(n_L,\theta)|^2}{\tau(n_L)\tau_\theta(n_L)} W(L+\log\theta)^{-\gamma'(v)}\, d\theta \right.$$
$$\left. + \exp\{-\gamma'(v)(L-\psi\sqrt{L})\} \int_1^{1/\eta} \frac{|\tau(n_L,\theta)|^2}{\tau(n_L)}\, d\theta\right\} \geqslant 1. \quad (6.7)$$

Multiplions les deux membres de cette inégalité par $\tau(n_L)^v$ et sommons sur tous les entiers $n \leqslant x$ satisfaisant (6.4), (6.5) et (6.6). Grâce au lemme 4.4, nous obtenons ainsi que la contribution à (6.2) des entiers non exceptionnels est

$$\ll \epsilon x e^{\gamma(v)L}(I_1+I_2)$$

avec

$$I_1 := e^{\sigma L}\int_{1/(\epsilon_1 e^{\sigma L})}^{1} W(L+\log\theta)^{-\gamma'(v)}\, d\theta \leqslant e^{-(1-\sigma)L}\int_1^L W(t)^{-\gamma'(v)} e^t\, dt$$

et

$$I_2 := \exp\{(\sigma-\gamma'(v))L+\psi\gamma'(v)\sqrt{L}\}\int_1^{1/\eta}\{\log(3+\theta)\}^\kappa\, d\theta.$$

Nous alons montrer que I_1 et I_2 sont $O(1)$ pour le choix

$$c_{11} := 2(\psi+2)\gamma'(1)+12B.$$

Cela suffit pleinement à impliquer (6.2).

Majorons I_1. Lorsque $\beta < \beta_0$, on a $(1-\sigma) \geqslant c_{11}\sqrt{(\log L)/L}$, d'où

$$I_1 \leqslant \exp(-c_{11}\sqrt{L\log L})\int_1^L \exp(\tfrac{1}{2}c_{11}\sqrt{t\log t})\, dt \ll 1.$$

Lorsque $\beta \geqslant \beta_0$, nous distinguons deux cas. Si $\gamma'(v) > 1+\frac{1}{2}\beta_0$, on a

$$I_1 \leqslant \int_1^L \exp\{-\tfrac{1}{2}\beta_0 t + 2\gamma'(1)\sqrt{t\log t}\}\, dt \ll 1.$$

Si $\gamma'(v) \leqslant 1+\frac{1}{2}\beta_0$, alors $1-\sigma \geqslant 1-\gamma'(v)+\beta \geqslant \frac{1}{2}\beta_0$, et il suit

$$I_1 \leqslant \exp(-\tfrac{1}{2}\beta_0 L)\int_1^L \exp(2\gamma'(1)\sqrt{t\log t})\, dt \ll 1.$$

Majorons I_2. Puisque $\eta^{-1} \leqslant \exp(\beta L + 2B\sqrt{L}) \leqslant \exp(\beta L + \frac{1}{6}c_{11}\sqrt{L})$, on peut écrire

$$I_2 \leqslant \exp\{-(\gamma'(v) - \sigma - \beta)L + \tfrac{2}{3}c_{11}\sqrt{L} + O(\log L)\}.$$

Or, on a en toute circonstance

$$(\gamma'(v) - \sigma - \beta)L \geqslant c_{11}\sqrt{L}.$$

On voit donc que $I_2 = O(1)$ lorsque y_0, et donc L, est assez grand. Cela achève la preuve du lemme 6.2. □

Nous déduirons facilement les minorations annoncées pour $H_F(x, y, z)$ de la proposition suivante.

Proposition 6.3 *Soient K et β_0 des nombres réels de $]0, 1[$. Il existe des constantes positives T, ϵ, c_{12} et y_0, ne dépendant que de K, β_0, B et F, telles que l'on ait pour $y_0 \leqslant y \leqslant x^{1-K}$, $0 \leqslant \beta(y, z) \leqslant B$, $0 \leqslant v \leqslant 1$, $\gamma'(v) \geqslant 1$ et $\log_2 y - 2T \leqslant L \leqslant \log_2 y - T$,*

$$\sum_{n \leqslant x}^{(*)} \tau(n_L)^v \gg x(\log y)^{\gamma(v)+\sigma-1}, \tag{6.8}$$

où $\sigma = \sigma(\beta, v; \beta_0, \epsilon)$ est défini par (6.3) et où l'astérisque indique que la sommation est restreinte aux entiers n tel que

$$(*)\begin{cases} \tau(F(n); y, z) \geqslant 1, \\ |\tau(n_L) - \gamma'(v)L| \leqslant c_{12}\sqrt{L}. \end{cases}$$

Démonstration Posons $R = \log(1/\epsilon)$, $T = \log(4R/K)$ et $\theta = \frac{1}{2}c_6(\frac{1}{2}K, F)$, où $c_6(K, F)$ est la quantité définie au lemme 3.4. Nous pouvons sans restreindre la généralité supposer que $T > \log(1/\theta)$, d'où

$$\exp\exp L \leqslant y^{\theta}. \tag{6.9}$$

La premiere étape consiste à remarquer l'on a pour ϵ suffisamment petit mais fixé

$$\sum_{n \leqslant x}^{(**)} \tau(n_L)^v \asymp x e^{\gamma(v)L}, \tag{6.10}$$

où la double astérisque indique que la variable de sommation n est astreinte aux conditions

$$(**)\begin{cases} \log(n_L) \leqslant R e^L, \\ |\log(n_L) - \gamma'(v)L| \leqslant R\sqrt{L}, \\ \lambda(n_L) > \epsilon e^{\sigma L}. \end{cases}$$

On obtient (6.10) en appliquant le lemme 4.1 et en majorant les contributions respectives des entiers n contrevenant à chacune des conditions (**) par les lemmes 3.6, 4.3 et 6.2.

Ensuite, désignons par $\mathcal{M}$ l'ensemble des entiers m qui sont de la forme $m = n_L$ pour au moins un n compté dans (6.10). Alors le choix de T implique pour tout m de $\mathcal{M}$ et y_0 assez grand

$$m \leqslant \exp(Re^{-T}\log y) \leqslant \tfrac{1}{2}y^{K/2} \leqslant \sqrt{y}. \tag{6.11}$$

De plus, chaque m de $\mathcal{M}$ possède une ou plusieurs représentations de la forme

$$m = \prod_{j=1}^{r} m_j^{\alpha_j}, \tag{6.12}$$

où les m_j sont sans facteur carré, deux à deux premiers entre eux, et premiers à D. Nous posons

$$\chi(m) := \sum \prod_{j=1}^{r} \frac{\rho_j(m_j)}{\sigma^*(m_j)},$$

où la sommation est étendue à toutes les représentations (6.12). Compte tenu de (6.9) et (6.11), il découle du lemme 3.4 que l'on a

$$\sum_{n\leqslant x}^{(**)} \tau(n_L)^v \asymp x e^{-rL} \sum_{m\in\mathcal{M}} \chi(m)\tau(m)^v,$$

d'où, d'après (6.10),

$$\sum_{m\in\mathcal{M}} \chi(m)\tau(m)^v \asymp \exp\{(\gamma(v)+r)L\}. \tag{6.13}$$

Considérons maintenant la sous-somme S de (6.10) restreinte aux entiers n pour lesquels $F(n)$ possède un facteur premier p tel que $\log y - \log p \in \mathscr{L}(n_L)$. Cette condition équivaut à l'existence d'un diviseur d de n_L tel que

$$\log y - \log p < \log d \leqslant \log\frac{z}{y} + \log y - \log p,$$

c'est-à-dire

$$y < pd \leqslant z.$$

De plus, puisque $n_L \leqslant \sqrt{y}$, d'après le choix de T, on a $\sqrt{y} < p \leqslant 2y$, donc $p \nmid d$ et $pd \mid F(n)$. Ainsi S minore le membre de gauche de (6.8) lorsque $c_{12} = R$.

Soit $V(n)$ le nombre de représentations de $F(n)$ sous la forme $F(n) = mph$ avec les conditions

$$\begin{cases} M \in \mathcal{M}, \\ \log p \in \log y - \mathcal{L}(m), \\ p'|h \Rightarrow p'|mp \text{ ou } p' > X := \min(2y, x^{\theta}). \end{cases}$$

Alors n est certainement compté dans S si $V(n) \neq 0$. De plus on a $V(n) \ll 1$ pour tout n. En effet, ou bien $2y \leqslant x^{\theta}$, et le nombre p et les facteurs premiers respectifs de m et h varient dans des intervalles disjoints – d'où $V(n) \leqslant 1$ – ou bien $2y > x^{\theta}$, et $V(n)$ n'excède pas le nombre total des facteurs premiers $> \sqrt{y}$ de $F(n)$ – d'où $V(n) \ll g/\theta \ll 1$. On peut donc écrire

$$S \gg \sum_{n \leqslant x}{}^{(**)} V(n)\tau(n_L)^{v} \gg \sum_{m \in \mathcal{M}} \tau(m)^{v} \sum_{\log p \in \log y - \mathcal{L}(m)} \sum_{\substack{n \leqslant x \\ pm|F(n)}}{}'' 1,$$

où la double apostrophe signifie que: $p'|F(n) \Rightarrow p'|pm$ ou $p' > X$. La somme intérieure relève du lemme 3.4 puisque

$$pm \leqslant 2ym \leqslant y^{1-\frac{1}{2}K} \qquad \text{et} \qquad P^{+}(pm) \leqslant 2y \leqslant X^{1/\theta}.$$

On obtient ainsi

$$S \gg \frac{x}{(\log y)^{r}} \sum_{m \in \mathcal{M}} \chi(m)\tau(m) \sum_{\log p \in \log y - \mathcal{L}(m)} \frac{\rho^{*}(p)}{\sigma^{*}(p)}. \tag{6.14}$$

Dans la somme intérieure, p parcourt une réunion d'au plus $\tau(m)$ $(\leqslant \exp\{2\gamma'(v)L\})$ intervalles de longueur logarithmique totale $\lambda(m)$ et dont les bornes ont des logarithmes de l'ordre de $\log y$ (puisque $m \leqslant \sqrt{y}$). La somme en p est donc

$$\gg \frac{1}{\log y} \sum_{p} \frac{\rho^{*}(p)\log p}{p}$$
$$\gg \frac{\lambda(m)}{\log y} + O(\exp\{2\gamma'(v)L - c\sqrt{\log y}\}),$$

d'après le lemme 3.1. Le terme d'erreur ci-dessus est négligeable compte tenu de la troisième condition (**) et l'on obtient

$$S \gg x(\log y)^{\sigma-1-r} \sum_{m \in \mathcal{M}} \chi(m)\tau(m)^{v}.$$

Le résultat souhaité découle donc de (6.13). □

Fin de la démonstration du théorème et du complément

Nous sommes maintenant en mesure d'établir les minorations de (1.13) et (1.14) avec $c_0 = 1 - K$.

Soit c_{11} la constante du lemme 6.2 pour le choix de ϵ défini à la proposition 6.3. Lorsque

$$0 \leqslant \beta \leqslant \gamma'(0) - 1 - \frac{2c_{11}}{\sqrt{\log_2 y}}, \tag{6.15}$$

on peut choisir $v = 0$ dans (6.8). De plus la condition (6.15) n'est non vide que si $\gamma'(0) > 1$, et donc $\gamma'(0) \geqslant \log 3$. Choisissons alors, dans la proposition 6.3, $\beta_0 < \log 3 - 1$. Nous pouvons minorer $H_F(x, y, z)$ par $H_F(x, y, Z)$ avec

$$\beta_1 := \beta(y, Z) = \max(\beta, \beta_0).$$

La formule (6.3) montre que l'on a alors $\sigma(\beta_1, 0) = 1$. Puisque $\gamma(0) = 0$, la formule (6.8) implique

$$H_F(x, y, Z) \geqslant x,$$

d'où la minoration du point (a) de notre théorème.

Supposons maintenant que l'on a

$$\max\left(0, \gamma'(0) - 1 - \frac{2c_{11}}{\sqrt{\log_2 y}}\right) \leqslant \beta \leqslant \gamma'(1) - 1 - \frac{2c_{11}}{\sqrt{\log_2 y}}. \tag{6.16}$$

Considérons d'abord le cas où F est réductible. Alors $\gamma'(0) \geqslant \log 3$ et l'on a $\beta \geqslant \beta_0$ pour une constante positive convenable. Nous déterminons, dans (6.8), le paramètre v par l'équation

$$\gamma'(v) = \beta + 1 + \frac{2c_{11}}{\sqrt{\log_2 y}}.$$

On a alors $0 \leqslant v \leqslant 1$ et

$$v \leqslant u + \frac{c_{13}}{\sqrt{\log_2 y}}. \tag{6.17}$$

La formule (6.3) donne encore $\sigma = 1$, d'où

$$\sum_{\substack{n \leqslant x \\ \tau(F(n); y, z) \geqslant 1 \\ \log \tau(n_L) \leqslant \gamma'(v) L + c_{12}\sqrt{L}}} \tau(n_L)^v \geqslant x(\log y)^{\gamma(v)}.$$

On en déduit immédiatement

$$H_F(x, y, z) \geqslant x(\log y)^{\gamma(v) - v\gamma'(v)} \exp(-c_{12} v \sqrt{\log_2 y}).$$

La relation (6.17) implique alors l'estimation inférieure de (1.14), i.e.

$$H_F(x, y, z) \geqslant x(\log y)^{-\delta(\beta, F)} \exp(-c_4 \sqrt{\log_2 y})$$

puisque $\delta(\beta, F) = u\gamma'(u) - \gamma(u)$.

Lorsque F est irréductible, le même raisonnement s'applique si $\beta \geqslant \beta_0 > 0$. Dans le cas contraire, nous pouvons encore opérer le même choix de v, mais la formule (6.3) donne seulement cette fois

$$\sigma \geqslant 1 - 2c_{11}\sqrt{\frac{\log_3 y}{\log_2 y}}.$$

Nous obtenons donc comme annoncé

$$H_F(x, y, z) \gg x(\log y)^{-\delta(\beta, F)}\exp(-c_4\sqrt{\log_2 y \log_3 y}).$$

Il reste à examiner le cas où

$$\gamma'(1) - 1 - \frac{2c_{11}}{\sqrt{\log_2 y}} \leqslant \beta \leqslant B.$$

On choisit alors, dans (6.8), $v = 1$. Par (6.3), on a

$$\sigma \geqslant \gamma'(1) - \beta - \frac{2c_{11}}{\sqrt{\log_2 y}}.$$

En majorant, dans (6.8), $\tau(n_L)$ par

$$(\log y)^{\gamma'(1)} \exp(2c_{12}\sqrt{\log_2 y}),$$

il vient

$$H_F(x, y, z) \gg x(\log y)^{\gamma(1) - \beta - 1} \exp(-c_{14}\sqrt{\log_2 y}),$$

d'où l'estimation requise puisque $\gamma(1) = \hat{\Omega}(F)$.

Ajouté aux épreuves Nous avons très récemment démontré, par une méthode complètement différente, l'inégalité

$$H_F(x, \tfrac{1}{2}x, x) > x(\log x)^{-\eta} \qquad (x \to \infty)$$

pour tout $\eta > \log 4 - 1$ et tout polynôme $F(X)$ irréductible dans $\mathbb{Z}[X]$ – cf. G. Tenenbaum, Sur une question d'Erdős et Schinzel, II, *Inventiones Math.*, **99** (1990), 215–24. Cela implique en particulier la validité de (1.17) pour tout constante c satisfaisant à $c < 2 - \log 4 = 0.613705\ldots$.

References

[1] H. Delange, Sur des formules de Atle Selberg, *Acta Arith.*, **19** (1971), 105–46
[2] J.-M. Deshouillers & H. Iwanieč, On the greatest prime factor of n^2+1, *Ann. Inst. Fourier (Grenoble)*, **32**, 4 (1982), 1–11
[3] V. Ennola, A note on a divisor problem, *Ann. Univ. Turku*, Ser. AI, **118–2**, (1968), 3–11
[4] P. Erdős, On the sum $\sum d\{f(n)\}$, *J. London Math. Soc.*, **27** (1952), 7–15

[5] P. Erdős, On the greatest prime factor of $\prod f(k)$, *J. London Math. Soc.*, **27** (1952), 379–84

[6] P. Erdős & A. Schinzel, On the greatest prime factor of $\prod_{k=1}^{x} f(k)$, *Acta Arith.* (à paraître)

[7] H. Halberstam & H.-E. Richert, *Sieve Methods*, Academic Press, London–New York–San Francisco (1974)

[8] R. R. Hall & G. Tenenbaum, *Divisors*, Cambridge Tracts No. 90, Cambridge University Press, Cambridge, 1988

[9] C. Hooley, On the greatest prime factor of a quadratic polynomial, *Acta Math.*, **117** (1967), 2–16

[10] C. Hooley, *Applications of Sieve Methods to the Theory of Numbers.* Cambridge Tracts No. 70, Cambridge University Press, Cambridge, 1976

[11] E. Landau, *Einfürung in die elementäre und analytische Theorie der algebraischen Zahlen und der Idealen*, Teubner, Leipzig (1927); réimpression: Chelsea, New York (1949)

[12] S. Lang, *Algebraic Number Theory*, Addison-Wesley, Reading–Menlo Park–London–Don Mills (1970)

[13] H. Maier & G. Tenenbaum, On a set of divisors of an integer, *Invent. Math.*, **76** (1984), 121–28

[14] A. A. Markov, Über die Primteiler der Zahlen von der Form $1+4x^2$, *Bull. Acad. Sci. St Petersburg*, **3** (1895), 55–9

[15] T. Nagell, *Introduction to Number Theory*, Chelsea, New York (1964)

[16] A. Selberg, Note on a paper by L. G. Sathe, *J. Indian Math. Soc.*, **18** (1954), 83–7

[17] P. Shiu, A Brun–Titchmarsh theorem for multiplicative functions, *J. reine angew. Math.*, **313** (1980), 161–70

[18] G. Tenenbaum, Sur la probabilité qu'un entier possède un diviseur dans un intervalle donné, *Compositio Math.*, **51** (1984), 243–63

[19] D. Wolke, Multiplikative Funktionen auf schnell wachsenden Folgen, *J. reine angew. Math.*, **251** (1971), 54–67

[20] D. Wolke, Polynomial values with small prime divisors, *Acta Arith.*, **19** (1971), 327–33

[5] [illegible] (1966).

[6] [illegible] On the [illegible] [illegible].

[7] [illegible] San Francisco (1967).

[8] [illegible] Cambridge Tracts No. [illegible] Cambridge University Press, Cambridge, 1958.

[9] [illegible]

[10] [illegible] Theory of [illegible] Cambridge University Press, Cambridge, 1970.

[11] [illegible]

[12] [illegible] Theory, 2nd [illegible] Reading [illegible] (1974).

[13] [illegible]

[14] [illegible]

[15] [illegible] New York [illegible]

[16] [illegible]

[17] [illegible]

[18] [illegible]

[19] [illegible]

[20] [illegible] 327–[illegible].

Large α-preserving sets in infinite α-connected graphs

Carsten Thomassen*

Abstract

We prove the conjecture of Mader that every α-connected graph G (where α is an infinite cardinal) contains a set S of α vertices such that $G-S'$ is α-connected for every subset S' of S.

1 Introduction

A graph G is *α-connected* if, for every set S of fewer than α vertices of G, $G-S$ is connected and has at least two vertices. In this paper, α will always be an infinite cardinal. Following Mader [1], a vertex set S of G is *α-preserving* if, for every subset $S' \subseteq S$, $G-S'$ is α-connected. If G is α-connected, then every vertex set of cardinality less than α is α-preserving. The same holds for directed graphs when 'α-connected' is replaced by 'strongly α-connected' (for the precise definition , see [1]). Mader [1] gave examples of strongly α-connected directed graphs D such that, for each set S of α vertices in D, $D-S$ has a vertex of outdegree 0. In particular, D has no α-preserving set of cardinality α. He conjectured that this phenomenon does not occur for undirected graphs and proved the conjecture for α-connected graphs with α vertices. We shall here verify the conjecture for all graphs.

2 Terminology and basic observations on connectivity

A *graph* G is a pair (V, E), where V is a set and E is a set of unordered pairs xy of elements of V. We call V the *vertex set* of G and denote it

* AMS classification: 05C40.

by $V(G)$. E is the *edge set* and is denoted $E(G)$. If the edge xy is present we say that x and y are *joined* by xy and that x and y are *neighbours*. We say that xy is *incident* with x and y and that x and y are the *ends* of xy. Two edges with no common end are *independent*. The *degree* of a vertex x in G is the cardinality of the set of neighbours of x in G. If $A \subseteq V(G) \cup E(G)$, then $G-A$ is obtained from G by deleting A and all edges incident with $A \cap V(G)$. For $A \subseteq V(G)$, we put $G(A) = G-[V(G) \setminus A]$. A *path* of length n from x to y is a graph with vertices $x_0, x_1, \ldots, x_n$ and edges $x_0x_1, x_1x_2, \ldots, x_{n-1}x_n$ such that $x_0 = x$ and $x_n = y$. Two paths from x to y are *internally disjoint* if they have only x and y in common. If G is a graph, $x \in V(G)$ and $A \subseteq V(G)$, then an α-*fan* from x to A is a collection of α paths from x to A such that they have only x in common pair by pair and they have only their ends in common with $\{x\} \cup A$. A *component* is a maximal connected subgraph. A component with only one vertex is an *isolated* vertex.

We shall assume Zorn's lemma that every set which is inductively ordered has a maximal element. We shall also use transfinite induction on ordinals. If β is an ordinal then the *cardinality of* β is the cardinality of the set of ordinals less than β. If A is a set, then $|A|$ denotes the cardinality of A.

The following version of Menger's theorem is an easy consequence of Zorn's lemma (see e.g. [2] for (a) and note that (b) follows from (a)).

Lemma 2.1 *Let G be an α-connected graph.*

(a) *If $x, y \in V(G)$, then G has a collection of α internally disjoint paths from x to y.*

(b) *If $x \in V(G)$ and $A \subseteq V(G)$ has cardinality α, then G has an α-fan from x to A.*

Lemma 2.2 *Let G be an α-connected graph. If $A \subseteq V(G)$ is a set of cardinality α, then G has an α-connected subgraph H of cardinality α containing A. If $B \subseteq A$ such that $G(B)$ is α-connected, then H can be chosen such that it intersects only those components of $G-B$ which intersect A.*

Proof We define a sequence $G_0, G_1, \ldots$ of subgraphs of G as follows. First $G_0 = G(A)$. Having defined G_k we let G_{k+1} be obtained from G_k as follows: for each pair of vertices x, y of G_k we let J_{xy} be a collection of α internally disjoint $x-y$ paths, which exist by Lemma 2.1(a). We let G_{k+1} be the union of G_k and all collections J_{xy}, where x, y is a pair of vertices of G_k. Since $\alpha^2 = \alpha$, G_{k+1} has cardinality α. Now, $H = G_0 \cup G_1 \cup \cdots$ is α-conected and has cardinality $\alpha + \alpha + \cdots = \alpha$. If

$G(B)$ is α-connected, then we define the sequence $G_0, G_1, \ldots$ slightly differently. Again $G_0 = G(A)$. For each vertex x in G_k but not in B we let I_x be an α-fan from x to B (which exists by Lemma 2.1 (b)). We let G_{k+1} be the union of G_k and all the α-fans I_x. Then $H = G_0 \cup G_1 \cup \cdots$ satisfies the conclusion of Lemma 2.2. □

Lemma 2.3 *Let G be an α-connected graph and $A \subseteq V(G)$ a vertex set such that $|A| < \alpha$. Let W be a set of pairs of A. Then G has a collection $\mathcal{P}$ of paths of length greater or equal to 3 such that each path in $\mathcal{P}$ joins two vertices forming a pair in W and has no other vertices in common with A such that two paths in $\mathcal{P}$ have at most an end in common. Moreover, for each pair in W, some path in $\mathcal{P}$ joins that pair.*

Proof By Zorn's lemma we consider a maximal collection $\mathcal{P}$ of paths satisfying the conclusion of Lemma 2.3 (except possibly the last statement). Since A and the union of paths in $\mathcal{P}$ have cardinality less than α, the maximality of $\mathcal{P}$ implies that $\mathcal{P}$ also satisfies the last statement of Lemma 2.3. □

3 α-preserving sets

The key idea is the following decomposition result.

Theorem 3.1 *Let G be an α-connected graph and let $V \subseteq V(G)$ be a set of α vertices. Then G has a collection of α pairwise disjoint α-connected subgraphs whose union includes V and is such that any two of the subgraphs are joined by α independent edges.*

Proof Let β be the smallest ordinal of cardinality α. Let $V = \{v_\gamma : \gamma < \beta\}$. We now define, by transfinite induction, a collection of subgraphs $G_{\gamma,\delta}$ $(\delta < \gamma < \beta)$ such that each $G_{\gamma,\delta}$ is countable or has cardinality at most the cardinality of γ and $G_{\gamma,\delta} \cap G_{\gamma,\delta'} = \emptyset$ whenever $\delta \neq \delta'$. Suppose we have already defined $G_{\gamma,\delta}$ for a fixed $\gamma < \beta$ and for all $\delta < \gamma$. We shall then define $G_{\gamma+1,\delta}$ for all $\delta \leqslant \gamma$. First we let H be the union of the disjoint subgraphs $G_{\gamma,\delta}$ taken over all δ $(\delta < \gamma)$. Then H has cardinality less than α, and we let ρ be the smallest ordinal such that $v_\rho \notin V(H)$. Put $A = V(H) \cup \{v_\rho\}$. Let $u_\delta \in V(G_{\gamma,\delta})$ for each $\delta < \gamma$ and put $u_\gamma = v_\rho$

Now $|A| < \alpha$ and so we can apply Lemma 2.3 to G and A. The collection W of pairs is the following: any pair in some $V(G_{\gamma,\delta})$ belongs to W and any pair in some $U = \{u_\delta : \delta \leqslant \gamma\}$ belongs to W. Let P be one of the pairs guaranteed by Lemma 2.3. If P joins two vertices of the same $G_{\gamma,\delta}$ we put $P' = P$. If P joins two vertices of U, then P' is obtained from P by deleting an edge of P joining two non-ends of P. Let us assign a colour γ to that edge. Now we let H' denote the union of H and all the subgraphs P'. We let $G_{\gamma+1,\delta}$ be the component of H' containing $G_{\gamma,\delta}$ when $\delta < \gamma$, and $G_{\gamma+1,\gamma}$ is the component of H' containing u_γ. If γ is a limit ordinal not greater than β and $G_{\rho,\delta}$ is defined for each ρ and δ satisfying $\delta < \rho < \gamma$, then we let $G_{\gamma,\delta}$ be the union of $G_{\rho,\delta}$ taken over all ρ between δ and γ. Now put $H_\delta = G_{\beta,\delta}$. It is easy to see that the subgraphs H_δ $(\delta < \beta)$ satisfy the conclusion of Theorem 3.1. (Note that any two of H_δ and $H_{\delta'}$ $(\delta < \delta')$ are joined by independent edges of all colours between δ' and β. There are α such colours.) □

The following result extends the aforementioned result of Mader.

Corollary 3.2 *If G is an α-connected graph of cardinality α, then $V(G)$ can be partitioned into α-preserving sets each of cardinality α.*

Proof We apply Theorem 3.1 with $V = V(G)$. For each $\delta \leqslant \beta$ (in the proof of Theorem 3.1) we write $V(H_\delta) = \{v_{\delta,\gamma} : 1 \leqslant \gamma < \beta\}$. Now each set $U_\gamma = \{v_{\delta,\gamma} : 1 \leqslant \delta < \beta\}$ is α-preserving. To see this let $U \subseteq U_\gamma$ and let S be a vertex set in $G-U$, where $|S| < \alpha$. We shall show that $(G-U)-S$ is connected. Let x and y be distinct vertices in $(G-U)-S$. The $x \in V(H_\delta)$ and $y \in V(H_{\delta'})$ $(\delta \leqslant \delta' < \beta)$. Since $|S| < \alpha$ and $|U \cap V(H_\delta)| = 1$, $(H_\delta - U) - S$ is connected. Similarly, $(H_{\delta'} - U) - S$ is connected. Moreover, there is an edge between $(H_\delta - U) - S$ and $(H_{\delta'} - U) - S$. Hence $(G-U)-S$ has a path from x to y. □

We are now ready for our main theorem.

Theorem 3.3 *Every α-connected graph G contains an α-preserving vertex set of cardinality α.*

Proof By Corollary 3.2 we can assume that $|V(G)| > \alpha$. By Lemma 2.2, G has an α-connected subgraph H of cardinality α. It is easy to see that every set of isolated vertices of $G-V(H)$ is α-preserving. So, if $G-V(H)$ has α (or more) isolated vertices we are done. Assume therefore that $G-V(H)$ has less than α isolated vertices. If $G-V(H)$ has a non-isolated vertex of degree not greater than α (in $G-V(H)$), then we extend H to an α-connected subgraph H_1 of $G-x$ such that H_1 contains all neighbours of x. By Lemma 2.2, we can assume that H_1 intersects only the component of $G-V(H)$ which contains neighbours of x. So, a vertex which is isolated in $G-V(H)$ is isolated in $G-V(H_1)$.

Moreover, x is isolated in $G-V(H_1)$ but not in $G-V(H)$. If $G-V(H_1)$ has a non-isolated vertex x_1 of degree not greater than α (in $G-V(H_1)$), then we let H_2 be an α-connected subgraph of G containing H_1 such that the isolated vertices of $G-V(H_1)$ form a proper subset of the isolated vertices of $G-V(H_2)$. By transfinite induction, we define α-connected subgraphs H_γ $(\gamma < \beta)$ such that, for all γ and δ (where $\gamma < \delta$), $H_\gamma \subseteq H_\delta$ and the set of isolated vertices of $G-H_\gamma$ is a proper subset of the isolated vertices of $G-H_\delta$. Put $M = \bigcup_\gamma H_\gamma$. Then M is α-connected and of cardinality α. If H_γ is defined for all $\gamma < \beta$, then $G-V(M)$ has at least α isolated vertices and we are done. So assume that $M = H_\rho$, where $\rho < \beta$ (that is, the above transfinite induction stops before β). Since $H_{\rho+1}$ does not exist, all vertices of $G-V(M)$ which are not isolated have degree greater than α (in $G-V(M)$). We can assume that $G-V(M)$ has less than α isolated vertices.

By Corollary 3.2, $V(M)$ can be partitioned into sets U_δ $(\delta < \beta)$ of cardinality α which are α-preserving in M. If some U_δ is α-preserving in G we are done. So we assume that, for each ordinal $\delta < \beta$, there is a set S_δ in $V(G)$ such that $|S_\delta| < \alpha$ and $G-(U_\delta \cup S_\delta)$ is disconnected. Since $M-U_\delta$ is connected, $G-(U_\delta \cup S_\delta)$ has a component N_δ which does not intersect M. Either N_δ is an isolated vertex or N_δ has more than α vertices. Since $G-S_\delta$ is α-connected, each vertex y in N_δ is joined to M by an α-fan. All ends (other than y) in that fan are in U_δ. This implies that $N_\delta \cap N_{\delta'} = \emptyset$ whenever $\delta \neq \delta'$. Put $S = \bigcup_{\delta<\beta} S_\delta$. Then $|S| \leqslant \alpha$. If y is a vertex of S such that y has at most α neighbours in N_δ, then these neighbours in N_δ are called *bad*. A graph N_δ has at most α bad vertices. Since N_δ has more than α vertices (unless N_δ is an isolated vertex), N_δ has a vertex u_δ which is not bad. We claim that the set $U = \{u_\delta : \delta < \beta\}$ is α-preserving. Clearly $|U| = \alpha$. For any subset $U' \subseteq U$ and for any set $T \subseteq V(G)$ such that $|T| < \alpha$, we shall show that $G-(T \cup U')$ is connected. Since $M-T$ is connected, it suffices to show that, for each vertex x in $G-(T \cup U')$, there is a path (in $G-(T \cup U')$) from x to M. Let $P = x_0x_1\ldots x_n$ be a path in $G-T$ from x to M. Let i be the smallest number such that x_i is in some N_δ. (If i does not exist then P is in $G-(T \cup U')$). We define z in H_δ as follows. If $x_i \neq u_\delta$ then $z = x_i$. If $x_i = u_\delta$ then $i \geqslant 1$ and x_{i-1} is in S_δ. Since u_δ is not bad, x_{i-1} is joined to more than α vertices of N_δ. In particular, x_{i-1} is joined to a vertex z which is in $N_\delta-(S \cup T)$. Since $|T \cup S_\delta| < \alpha$, $G-(T \cup S_\delta \cup \{u_\delta\})$ has a path P' from z to M. All vertices in P' are in $N_\delta \cup M$. Hence P' does not intersect U. Now $(x_0x_1\ldots x_{i-1}z) \cup P$ contains a path from x_0 to M in $G-(T \cup U')$. $\square$

References

[1] W. Mader, On infinite n-connected graphs, 1987, preprint

[2] C. Thomassen, Infinite graphs, *Selected Topics in Graph Theory II* (eds. L. W. Beineke & R. J. Wilson), Academic Press, London, 1983, 129–60

Some recent results on interpolation

P. Vértesi

We survey some problems and developments connected with interpolation with which P. Erdős has been concerned for several decades.

1 Lebesgue functions, Lebesgue constants and optimal Lebesgue constants

1.1 Let $X = \{x_{kn}\}$ $(k = 1, 2, \ldots, n;\ n = 1, 2, \ldots)$ be a fixed triangular interpolatory matrix in $[-1, 1]$, i.e., let

$$-1 \equiv x_{n+1,n} \leqslant x_{nn} < x_{n-1,n} < \cdots < x_{2n} < x_{1n} \leqslant x_{0n} \equiv 1 \qquad (n = 1, 2, \ldots).$$

In the study of the Lagrange interpolatory polynomials

$$L_n(f, X, x) := \sum_{k=1}^{n} f(x_{kn})\, l_{kn}(X, x)$$

the estimation

$$|L_n(f, X, x) - f(x)| \leqslant [\lambda_n(X, x) + 1]\,\omega\!\left(f, \frac{1}{n}\right), \qquad (n = 1, 2, \ldots), \quad (1.1)$$

– due to Lebesgue – shows the fundamental importance of the expressions

$$\lambda_n(X, x) := \sum_{k=1}^{n} |l_{kn}(X, x)|, \qquad \Lambda_n(X, x) := \|\lambda_n(X, x)\| \qquad (n = 1, 2, \ldots), \quad (1.2)$$

(Lebesgue functions and Lebesgue constants). Here, as usual, the functions

$$l_{kn}(X,x) = \omega_n(X,x)[\omega_n'(X,x_{kn})(x-x_{kn})]^{-1}$$

are the fundamental polynomials of Lagrange interpolation, where

$$\omega_n(X,x) = c_n \prod_{k=1}^{n} (x-x_{kn}),$$

$\omega(f,\delta)$ is the usual modulus of continuity of $f \in C[-1,1]$ and $\|\cdot\|$ is the maximum norm on $[-1,1]$.

In 1914 G. Faber [1] proved that *for an arbitrary fixed matrix* $X \subset [-1,1]$ we have

$$\Lambda_n(X) > c\log n \qquad (n = 1,2,\ldots).$$

In 1961 this was significantly strengthend by P. Erdős [2] to the inequality

$$\Lambda_n(X) \geqslant \frac{2}{\pi}\log n - c_1 \qquad (n = 1,2,\ldots). \tag{1.3}$$

The estimate

$$\Lambda_n(T) = \frac{2}{\pi}\log n + c_2 \tag{1.4}$$

(where $T = \{\cos(2k-1)\pi/2n\}$ is the Chebyshev matrix) clearly shows that the relation (1.3) is sharp.

The conjecture that $\lambda_n(X,x)$ is not less than $c\log n$ on a 'big set' of $[-1,1]$ (which certainly holds for the matrix T) was raised by P. Erdős. The proof of the following theorem, which answers this conjecture, is in P. Erdős & P. Vértesi [3].

Theorem 1.1 *Let* $\epsilon > 0$ *be any given number. Then for any fixed* $X \subset [-1,1]$ *there exist sets* $H_n = H_n(\epsilon, X)$ *with* $|H_n| \leqslant \epsilon$ *and* $\eta = \eta(\epsilon)$ *with* $\eta > 0$ *such that*

$$\lambda_n(X,x) > \eta\log n \qquad \textit{if } x \in [-1,1]\backslash H_n.$$

Other developments are in [4].

1.2 Nearly 60 years ago S. Bernstein suspected that

$$\Lambda_n^* := \min_X \Lambda_n(X)$$

is attained when the $n+1$ local maxima

$$\mu_{kn}(X) := \max_{x_{kn}\leqslant x\leqslant x_{k-1,n}} \lambda_n(X,x) \qquad (k = 1,2,\ldots,n+1;\ n \geqslant 3)$$

are the same. Furthermore Erdős conjectured that the smallest of these $n+1$ maxima is largest when they are equal.

These conjectures were proved by T. Kilgore [5] and C. DeBoor & A. Pinkus [6] in 1978. To formulate the result, let us first make a simple observation: to obtain Λ_n^* we may, without loss of generality, restrict our attention to those nodal configurations where $-1 \equiv x_{nn}$ and $1 \equiv x_{1n}$ (see [5, p. 274]). We call these X *canonical* matrices.

Now the statement is as follows.

Theorem 1.2 *Let the matrix X be canonical. Then $\lambda_n(X,x)$ equioscillates, i.e.,*

$$\mu_{2n}(X) = \mu_{3n} = \cdots = \mu_{nn}(X), \tag{1.5}$$

if and only if $\Lambda_n(X) = \Lambda_n^*$. Moreover, for arbitrary canonical X,

$$\min_{2\leqslant k\leqslant n} \mu_{kn}(X) \leqslant \Lambda_n^* \leqslant \max_{2\leqslant k\leqslant n} \mu_{kn} \qquad (n \geqslant 3). \tag{1.6}$$

There is a unique matrix X^* (the so-called optimal matrix) satisfying (1.5).

Using Theorem 1.2 and the works of R. Günttner [7] and L. Brutman [8], the following theorem was recently proved in P. Vértesi [9].

Theorem 1.3

$$\Lambda_n^* = \frac{2}{\pi}\log n + \chi + O\left(\left[\frac{\log\log n}{\log n}\right]^2\right),$$

where (if $\gamma = 0.577215\ldots$ denotes the Euler constant)

$$\chi = \frac{2}{\pi}\left(\gamma + \log\frac{4}{\pi}\right) = 0.521251\ldots.$$

Analogous results concerning trigonometric and complex cases when, contrary to the algebraic case, the corresponding optimal matrices and optimal constants are known, can be found in [6], [10], [11] and H. Ehlich & K. Zeller [12].

2 The divergence of the Lagrange interpolation

2.1 Fifty years ago, G. Grünwald [13] and J. Marcinkiewicz [14] simultaneously and independently proved that, for some $f \in C[-1,1]$, we have

$$\overline{\lim_{n\to\infty}} |L_n(f,T,x)| = \infty \qquad \textit{for every } x \in [-1,1]. \tag{2.1}$$

A conjecture of Erdős, extending (2.1) to an *arbitrary* interpolatory matrix X was proved in 1979 by P. Erdős & P. Vértesi [15].

Theorem 2.1 *For any interpolatory matrix $X \subset [-1,1]$ there is a function $f \in C[-1,1]$ such that*

$$\overline{\lim_{n\to\infty}} |L_n(f,X,x)| = \infty \qquad a.e.\ in\ [-1,1].$$

There are many developments of Theorem 2.1 which are listed in paper [16].

2.2 Formula (2.1) can be proved in a stronger form: *For any $\{\epsilon_n\}$ with $\lim_{n\to\infty} \epsilon_n = 0$ and $\epsilon > 0$, there is an $f \in C[-1,1]$ such that*

$$\overline{\lim_{n\to\infty}} \frac{L_n(f,T,x)}{\epsilon_n \log n} > 1 \qquad for\ every\ x \in [-1,1].$$

On the other hand, in 1950 Erdős [17] proved that *for every $f \in C[-1,1]$*

$$\frac{1}{n}\sum_{k=1}^{n} |L_k(f,T,x)| = o(\log\log n) \qquad a.e.\ in\ [-1,1].$$

i.e., taking arithmetic means has a smoothing effect (cf. [17]).

In the opposite direction, in 1938 P. Erdős & G. Grünwald [18] believed that they had proved that there exists an $f \in C[-1,1]$ so that

$$\overline{\lim_{n\to\infty}} \left| \frac{1}{n}\sum_{k=1}^{n} L_k(f,T,x)\right| = \infty \qquad \text{for every } x \in [-1,1].$$

Later Erdős discovered that the proof was erroneous (c.f. last formula of p. 92); what [18] actually gave was that *there exists an $f \in C[-1,1]$ such that*

$$\overline{\lim_{n\to\infty}} \frac{1}{n}\sum_{k=1}^{n} |L_k(f,T,x)| = \infty \qquad for\ every\ x \in [-1,1].$$

Their earlier statement remained open until very recently when, in a remarkable paper, G. Halász [19] succeeded in proving the following result.

Theorem 2.2 *For any $\{\epsilon_n\}$ with $\lim_{n\to\infty}\epsilon_n = 0$ and $\epsilon_n > 0$, there exists an $f \in C[-1,1]$ so that*

$$\overline{\lim_{n\to\infty}} \frac{1}{\epsilon_n \log\log n}\left| \frac{1}{n}\sum_{k=1}^{n} L_k(f,T,x)\right| > 1 \qquad a.e.\ in\ [-1,1]. \tag{2.2}$$

On the other hand, as G. Halász remarked, relation (2.2) probably does not hold for every x in $[-1,1]$.

3 The fine and rough theory of the Lagrange interpolation

In their paper [20], P. Erdős & P. Turán proved the following interesting result.

Theorem 3.1 *Let* $\Lambda_n(X) \sim n^\beta$ $(0 < \beta < 1)$.

(a) *If* $0 < \alpha < \beta/(\beta+2)$ *then* $\overline{\lim}_{n\to\infty}\|L_n(f,X,x)\| = \infty$ *for some* $f \in \mathrm{Lip}\,\alpha$.

(b) *If* $\beta < \alpha \leqslant 1$, *then* $\lim_{n\to\infty}\|L_n(f,X,x)-f(x)\| = 0$ *for any* $f \in \mathrm{Lip}\,\alpha$ *(cf. (1.1)).*

(c) *If* $\beta/(\beta+2) < \alpha < 1$, *there is an interpolatory matrix* Z *with* $\Lambda(Z) \sim n^\beta$ *such that* $\lim_{n\to\infty}\|L_n(f,Z,x)-f(x)\| = \infty$ *for any* $f \in \mathrm{Lip}\,\alpha$ *and, further, one can define another interpolatory matrix* Y *with* $\Lambda_n(Y) \sim n^\beta$ *such that* $\overline{\lim}_{n\to\infty}\|L_n(f,Y,x)\| = \infty$ *for some* $f \in \mathrm{Lip}\,\alpha$.

In other words, in the first two cases, using only the order of the Lebesgue constants $\lambda_n(X)$ one can characterize the behaviour of the class Lip α (rough theory), while in the third case we have to investigate the structure of the interpolatory matrix too (fine theory).

During the last 35 years there have been many, many papers generalizing the above results and its methods, but certainly one of the most interesting developments is due to the recent work [21] of G. Halász & J. Szabados.

To investigate the expression

$$|L_n^{(r)}(f,X,x)-f^{(r)}(x)| \qquad (f^{(r)} \in C[-1,1];\ r=0,1,2,\ldots,\ \text{fixed}),$$

they introduce a complicated-looking but very sensitive expression, Λ_{nr} to handle the fine and rough theory of $L_n^{(r)}(F,X,x)$. Instead of the exact definition, we mention some basic properties of $\Lambda_{nr}(X)$:

$$\Lambda_{n0}(X) = \Lambda_n(X); \qquad \Lambda_{nr}(X) \geqslant c_r \log n \qquad (\text{for any } X);$$
$$\Lambda_{nr}(X_r) = O(\log n) \qquad (\text{for special } X_r).$$

Further, with

$$\Lambda_n^{[r]} := n^{-r}\left\|\sum_{k=1}^{n}(1-x_{kn}^2)^{r/2}|l_{kn}^{(r)}(X,x)|\right\|$$

we have

$$\Lambda_{nr} = O(1)\Lambda_n^{[r]}(X), \qquad \text{where } X \text{ is arbitrary,} \tag{3.1}$$

and, with a proper interpolatory matrix Y_r,

$$\lim_{n\to\infty} \frac{\Lambda_n^{[r]}(Y_r)}{\Lambda_{nr}(Y_r)} = \infty. \tag{3.2}$$

($\Lambda_n^{[r]}(X)$ is another generalization of $\Lambda_n(X)$ which was very useful in many cases (cf. J. Szabados [22], say), but (3.1) and (3.2) show the advantage of $\Lambda_{nr}(X)$ over $\Lambda_n^{[r]}$.)

The results corresponding to Theorem 3.1 parts (a) and (b) are as follows.

Theorem 3.2 *Let* $\Lambda_{nr}(X) \sim n^\beta$.

(a) *If* $0 < \alpha < \beta/[\beta+2(r+1)]$ *then* $\overline{\lim}_{n\to\infty}\|L_n^{(r)}(f,X,x)\| = \infty$ *for some* f *with* $f^{(r)} \in \operatorname{Lip}\alpha$.

(b) *If* $\beta < \alpha \leqslant 1$, *then* $\lim_{n\to\infty}\|L_n^{(r)}(f,X,x) - f^{(r)}(x)\| = 0$ *for any* f *with* $f^{(r)} \in \operatorname{Lip}\alpha$.

Some statements in [21] are more general, but the complete analogue of part (c) of Theorem 3.1 has not yet been proved.

4 Convergent interpolatory processes and the Erdős conditions

As we have seen, the Lagrange interpolation often exhibits divergent phenomena. It is natural to ask how to *construct interpolatory processes which converge uniformly for all* $f \in C[-1,1]$, say. It turned out that one possibility was to loosen the strict condition on the degree of the interpolating polynomials, as was first shown for special matrices X by L. Fejér in 1912 (Hermite–Fejér step-parabolas $H_n(f,T,x)$ of degree $2n-1$ based on T).

The next natural requirement is to lower the degree $2n-1$. It was shown by J. Szabados [23] and B. Shekhtman [24} that *the degree cannot be less than* $n(1+\epsilon)$, *where* $\epsilon > 0$. On the other hand, the next statement shows that in many cases this condition is sufficient as well.

Let $X = \{x_{kn} = \cos\theta_{kn}\}$ and denote by $N_n(t)$ the number of θ_{kn} in the open interval $I \subseteq (0,\pi)$.

Theorem 4.1 *For every* $f \in C[-1,1]$ *and* $\epsilon > 0$, *there exists a sequence of polynomials* $P_n(f,x)$ *of degree not greater than* $n(1+\epsilon)$ *such that*

$$P_n(f,x_{kn}) = f(x_{kn}) \qquad (1 \leqslant k \leqslant n;\ n = 1,2,\ldots)$$

and

$$\lim_{n\to\infty}\|f(x) - P_n(f,x)\| = 0$$

if and only if the interpolatory matrix satisfies the so-called Erdős conditions:

$$\begin{cases} \overline{\lim_{n\to\infty}} \dfrac{N_n(I_n)}{n|I_n|} \leqslant \dfrac{1}{\pi} & \textit{whenever } \lim_{n\to\infty} n|I_n| = \infty; \\ \underline{\lim}_{n\to\infty}\, n(\theta_{i+1,n}-\theta_{in}) > 0 & \textit{for arbitrary } 1 \leqslant i \leqslant n-1. \end{cases}$$

Theorem 4.1 was originally stated in 1943 by P. Erdős [25], with only a sketchy indication of the proof. The complete argument was only recently found by P. Erdős, A. Kroó & J. Szabados [26].

Other connected results, problems and references may be found in P. Erdős [27].

References

[1] G. Faber, Über die interpolatorische Darstellung stetiger Funktionen, *Jahresber. der Deutschen Math. Ver.*, **23** (1914), 191–200

[2] P. Erdős, Problems and results on the theory of interpolation. II, *Acta Mat. Acad. Sci. Hungar.*, **12** (1961), 235–44

[3] P. Erdős & P. Vértesi, On the Lebesgue function of interpolation, Functional Analysis and Approximations, *ISNM* Vol 60 (1981), Birkhäuser, Basel, 299–309

[4] P. Vértesi, On sums of Lebesgue function type, *Acta Mat. Acad. Sci. Hungar.*, **40** (1982), 217–27

[5] T. A. Kilgore, A characterization of the Lebesgue interpolation projections with minimal Tchebycheff norm, *J. Approximation Theory*, **24** (1978), 273–88

[6] C. DeBoor & A. Pinkus, Proof of the conjectures of Bernstein and Erdős concerning the optimal nodes for polynomial interpolation, *J. Approximation Theory*, **24** (1978), 289–303

[7] R. Günttner, Approximation durch trigonometrische Interpolation, Diplomarbeit, Technische Universität, Hannover, 1970

[8] L. Brutman, On the Lebesgue function of polynomial interpolation, *SIAM J. Numer. Anal.*, **15** (1978), 694–704

[9] P. Vértesi, Optimal Lebesgue constant for Lagrange interpolation *SIAM J. Numer. Anal.*, **27** (1990)

[10] L. Brutman, On the polynomial and rational projections in the complex plane, *SIAM J. Numer. Anal.*, **17** (1980), 366–72

[11] L. Brutman & A. Pinkus, On the Erdős conjecture concerning minimal norm interpolation on the unit circle, *SIAM J. Numer. Anal.*, **17** (1980), 373–5

[12] H. Ehlich & K. Zeller, Auswertung der Normen von Interpolationsoperatoren, *Math. Ann.*, **164** (1966), 105–12

[13] G. Grünwald, Über die Divergenzerscheinungen der Lagrangeschen Interpolationspolynome stetiger Funktionen, *Annals of Math.*, **37** (1936), 908–18

[14] J. Marcinkiewicz, Sur la divergence des polynoms d'interpolation, *Acta Sci. Mat.* (Szeged), **8** (1937), 131–35

[15] P. Erdős & P. Vértesi, On the almost everywhere divergence of Lagrange interpolatory polynomials for arbitrary system of nodes, *Acta Mat. Acad. Sci. Hungar.*, **36** (1980), 71–89 and **38** (1981), 263

[16] P. Vértesi, Recent results on the almost everywhere divergence of Lagrange interpolation, *ISNM* Vol **65** (1984), Birkhäuser, Basel, 381–91
[17] P. Erdős, Some theorems and remarks on interpolation, *Acta Sci. Mat.* (Szeged), **12** (1950), 11–17
[18] P. Erdős & G. Grünwald, Über die aritmetische Mittelwerke der Lagrangeschen Interpolationspolynome, *Studia Mat.*, **7** (1938), 82–95
[19] G. Halász, manuscript, 1988
[20] P. Erdős & P. Turán, On the role of the Lebesgue functions in the theory of the Lagrange interpolation, *Acta Mat. Acad. Sci. Hungar.*, **6** (1955), 47–66
[21] G. Halász & J. Szabados, manuscript, 1988
[22] J. Szabados, On the convergence of the derivatives of projection operators, *Analysis*, **7** (1987), 349–57
[23] J. Szabados, On the norm of certain interpolation operators, *Anal. Math.* (to appear)
[24] B. Shekhtman, On the norm of interpolating operators, manuscript
[25] P. Erdős, On some convergence properties of the interpolation polynomials, *Annals of Math.*, **44** (1943), 330–37
[26] P. Erdős, A. Kroó & J. Szabados, On convergent interpolatory polynomials, *J. Approximation Theory* (to appear)
[27] P. Erdős, Problems and results on polynomials and interpolation, in *Aspects of Contemporary Complex Analysis*, Academic Press, London, 1980, 383–91

Partitioning the quadruples of topological spaces

W. Weiss*

The partition calculus, as described in [1], can be extended to a partition calculus for topological spaces, in which the homogeneous set is required not only to be of large cardinality, but also to have some topological properties aswell. We can use similar notation; for example:

$$X \to (Y)^n_\omega$$

means that whenever the n-tuples from the space X are partitioned into countably many pieces there is a homogenenous set homeomorphic to Y. For more details about this, see [3].

Here we study limitations on this topological partition caculus. Our main result is that it is relatively consistent with the generalized continuum hypothesis (henceforth GCH) that if $X \to (Y)^4_\omega$ then Y must be discrete.

Theorem 0 *Assume GCH and $\square_\kappa$ holds for each singular cardinal κ. If X is any Hausdorff (not necessarily even regular) topological space of cardinality not exceeding the first weakly compact cardinal, then there is a partition f: $[X]^4 \to \omega$ of quadruples of X into countably many pieces such that any subspace homegeneous for f is discrete.*

We shall prove this theorem later, but first we shall place it in perspective. In [2], the following is shown.

Proposition 1 *Let ordinals have their usual order topology.*
(a) *There is a non-discrete space Y_0 such that*

$$(2^\omega)^+ \to (Y_0)^2_\omega.$$

* Research partially supported by NSERC grant A3185.

(b) *If κ is a weakly compact cardinal, then there is a non-discrete Y_κ such that for all finite n*

$$\kappa^+ \to (Y_\kappa)^n_\omega .$$

On the other hand, it is also shown in [2] that we have the following with even weaker hypotheses than we stated here.

Proposition 2 *Assume $0^\#$ does not exist.*

(a) *For any regular space X there is a partition of the triples $f: [X]^3 \to \omega$ such that any* countable *homogeneous set is discrete. So $X \to (Y)^2_\omega$ implies that Y has uncountable tightness.*

(b) *For any regular space X, there is a partition of all finite subsets of X, $f: [X]^{<\omega} \to \omega$, such that any homogeneous set is discrete. So $X \to (Y)^{<\omega}_\omega$ implies that Y is discrete.*

After the results of [2] and this article, the question remains: Does $X \to (Y)^4_\omega$ imply that Y is discrete for X of size less than the first weakly compact cardinal, without any assumptions beyond ZFC?

The structure of our Theorem 0 is patterned after proofs in [2] and [3]. Recall that a subspace Y of a topological space X is *right-separated* by a well ordering $<$ of Y if each final segment is closed in Y. Replacing 'final' by 'initial' gives the definition of *left-separated*. Given a space X, we construct a partition F of the quadruples of X, $F: [X]^4 \to \omega$, and a well ordering $<$ of X such that any subspace homogenenous for F is left-separated by $<$. We construct another partition, $G: [X]^4 \to 3$, such that any subspace homogeneous for G is right-separated by $<$. Now, any subspace homogeneous for the partition f given by

$$f(\{a, b, c, d\}) = 2^{F(\{a, b, c, d\})} 3^{G(\{a, b, c, d\})}$$

is both left- and right-separated by $<$ and hence discrete.

We first do the easy part.

Lemma 3 *Let $<$ be a well ordering of a Hausdorff space X. There is a partition $G: [X]^4 \to 3$ such that any subspace H homogeneous for G is right-separated by $<$.*

Proof Define G as follows. First, let $<'$ be a well ordering of all pairs $\langle U_0, U_1\rangle$ of disjoint open subsets of X. Now, for $a < b < c < d$, define $G(\{a, b, c, d\}) = i$ if $a \in U_i$, where $\langle U_0, U_1\rangle$ is the $<'$-least pair separating any two points of $\{a, b, c, d\}$ and $G(\{a, b, c, d\}) = 2$ if a is not in $U_0 \cup U_1$.

Now suppose that H is homogeneous for G but not right-separated by $<$; we can suppose that x is the least element of H and is a limit point of H but not a limit point of any initial segment of H. Let $\langle U_0, U_1\rangle$ be the $<'$-least pair separating any two points of H, say $a_0 \in U_0$ and $a_1 \in U_1$. Notice that $G``(H) \neq \{2\}$ because $G(\{a_0, a_1, c, d\}) \neq 2$ for any c and d in H above A_0 and a_1.

Now suppose $a_0 < a_1$. $G``(H) \neq \{0\}$ because this would mean that $x \in U_0$ and so we can find $b$, $c$ and $d$ in $H \cap U_0$ above $a_1$ and yet $G(\{a_1, b, c, d\}) \neq 0$. $G``(H) \neq \{1\}$ because this means that $x \in U_1$ and so $x \notin a_0$ and choosing c and d in $H \cap U_1$ above a_1 gives $G(\{a_0, a_1, c, d\}) = 0$. The case $a_1 < a_0$ is, of course, similar. □

Remark 4 The above proof shows that we can replace quadruples with just triples in Lemma 3.

From the basic results of the ordinary partitioning calculus we borrow the following [1].

Proposition 5 *If κ is less than the first weakly compact cardinal,*

$$\kappa^{++} \nrightarrow (\kappa)^4_\omega.$$

The rest of the proof of Theorem 0 will be by induction. We use the following inductive hypothesis, IH(κ):

'For any Hausdorff space X with $|X| \leqslant \kappa$ there is a partition $F\colon [X]^4 \to \omega$ and a well ordering $<$ of X such that any subspace of X homogeneous for F is left-separated by $<$.'

Corollary 6 IH(ω_2) *is true.*

Before continuing with the general case, we stop to prove this result for the easier case of ordinal spaces.

Lemma 7 *Suppose λ is an ordinal with no weakly compact cardinals below it. There is a partition $g_\lambda\colon [\lambda]^3 \to \omega$ of the triples such that any subspace homogeneous for g_λ is discrete.*

Proof For each limit $\alpha < \lambda$, let C_α be a closed set, cofinal in α, of regular cardinal order type. Since $|C_\alpha|$ is not weakly compact, we get a partition $h_\alpha\colon [C_\alpha]^2 \to \omega$ of the pairs such that, if H_α is homogeneous for h_α, then $|H_\alpha| < |C_\alpha|$. Define, for $\gamma < \beta < \alpha$,

$$g_\lambda(\{\gamma, \beta, \alpha\}) = \begin{cases} 0 & \gamma \cap C_\alpha = \beta \cap C_\alpha \\ & \text{or } \alpha \text{ is not a limit,} \\ h_\alpha(\{\sup(\gamma \cap C_\alpha), \sup(\beta \cap C_\alpha)\}) & \text{otherwise.} \end{cases}$$

Now suppose $H \cup \{\alpha\}$ is homogeneous for g_λ with $x < \alpha$ for all x in H and α is a limit ordinal. The set $\{\sup(\beta \cap C_\alpha) : \beta \in H\}$ has cardinality less than $|C_\alpha|$ and hence H is bounded below α. □

Lemma 8 *Suppose κ is a strongly inaccessible cardinal with no weakly compacts below it. If* IH(μ) *for all $\mu < \kappa$, then* IH(κ).

Proof Let $|X| = \kappa$. We can write X as the increasing union of $\{X_\alpha : \alpha < \kappa\}$, where each X_α is closed and $|X_\alpha| < \kappa$. We let $Y_\alpha = X_\alpha \backslash \bigcup \{X_\beta : \beta < \alpha\}$ for each α, and obtain F_α and $<_\alpha$ from IH($|Y_\alpha|$). Define $<$ on X by $x < y$ if and only if either
 (a) for some $\alpha < b$ we have $x \in Y_\alpha$ and $y \in Y_\beta$,
or
 (b) for some α we have $\{x, y\} \subset Y_\alpha$ and $x <_\alpha y$.
Define F on X such that, for $a < b < c < d$ and g_κ from Lemma 7,

$$F(\{a,b,c,d\}) = \begin{cases} F_\alpha(\{a,b,c,d\}) & \text{if } \{a,b,c,d\}) \subset Y_\alpha \text{ for some } \alpha, \\ g_\kappa(\{\alpha,\beta,\gamma\}) & \text{if } b \in Y_\alpha,\ c \in Y_\beta \text{ and } d \in Y_\gamma, \\ & \text{for some distinct } \alpha, \beta \text{ and } \gamma, \\ 0 & \text{otherwise.} \end{cases}$$

Now let $H \cup \{y\}$ be homogeneous for F with $x < y$ for all x in H. Let $y \in Y_\alpha$; note $H \subset X_\alpha$. Also, y is not in the closure of $H \cap Y_\alpha$ by the property of F_α. We finish the proof by showing that y is not in the closure of $H \backslash Y_\alpha$. If $\alpha = \beta + 1$, then $H \backslash Y_\alpha \subset X_\beta$, which is closed. If α is a limit ordinal, then the property of g_κ ensures that $\{\beta : (H \backslash Y_\alpha) \cap Y_\beta \neq \emptyset\}$ is bounded below α and, hence, again $H \backslash Y_\alpha$ is contained in some closed X_β.
□

Remark 9 Note that triples can replace quadruples in Lemma 8.

Lemma 10 *Assume that, for $\mu < \kappa$, $2^{2^\mu} \leqslant \kappa^+$. If* IH($\kappa^+$) *then* IH($\kappa^{++}$).

Proof The cardinal arithmetic allows us to write any X of size κ^{++} as the increasing union of $\{X_\alpha : \alpha < \kappa^{++}\}$, where each $|X_\alpha| = \kappa^+$ and each X_α is $< \kappa$-closed, that is, X_α contains the closure of each of its subsets of size less than κ. As in Lemma 8, let $Y_\alpha = X_\alpha \backslash \bigcup \{X_\beta : \beta < \alpha\}$ and obtain F_α and $<_\alpha$. Define $<$ and F as in Lemma 8; these will have the property that, if $H \cup \{y\}$ is homogeneous for F and $x < y$ for all x in H, then y is not in the closure of any subset of H of size less than κ. Now invoke Proposition 5 to obtain $F' : [X]^4 \to \omega$ with no homogeneous set of size κ. The partition we need for IH is F'', where

$$F''(\{a,b,c,d\}) = 2^{F(\{a,b,c,d\})} 3^{F'(\{a,b,c,d\})}. \qquad \square$$

A similar proof gives the following.

Lemma 11 *Assume that κ is a strongly inaccessible cardinal. If* IH(κ) *then* IH(κ^+).

Lemma 12 *Assume that for some singular cardinal κ we have $2^\mu < \kappa$ for all $\mu < \kappa$ and $\square_\kappa$. If* IH(μ) *for all $\mu < \kappa$, then* IH(κ^+).

Proof Let $\lambda = \mathrm{cf}(\kappa)$ and $\{\kappa_\eta\}$ be a closed sequence of cardinals cofinal in κ. We will define a matrix of subspaces $X_{\alpha,\eta}$ of X and on each $X_{\alpha,\eta}$ a well ordering $<_{\alpha,\eta}$ and a partition $F_{\alpha,\eta}: [X_{\alpha,\eta}]^4 \to \omega$ such that the following are true for all $\alpha < \kappa^+$ and $\eta < \lambda$.

(a) $X_{\alpha,\eta}$ is closed of density at most κ_η.
(b) For $\eta < \xi < \lambda$, $X_{\alpha,\eta} < X_{\alpha,\xi}$.
(c) For $\alpha < \beta < \kappa^+$, $\bigcup\{X_{\alpha,\eta} : \eta < \lambda\} \subset \bigcup\{X_{\beta,\eta} : \eta < \lambda\}$.
(d) For all $\xi < \lambda$ and $\beta < \kappa^+$, $<_{\alpha,\eta}$ and $<_{\beta,\xi}$ agree on their common domain, and they end extend each other in the following way: if

$$S = \bigcup\{X_{\gamma,\theta} : \gamma < \alpha \text{ and } \theta < \lambda\} \cup \bigcup\{X_{\alpha,\theta} : \theta < \eta\}$$

and $x \in S$ and $y \in X_{\alpha,\eta} \backslash S$, then $x <_{\alpha,\eta} y$.
(e) For all $\xi < \lambda$ and $\beta < \kappa$, $F_{\alpha,\eta}$ and $F_{\beta,\eta}$ agree on their common domains.

We enumerate X as $\{x_\beta : \beta < \kappa^+\}$ and we will ensure that for each β there is some $\eta < \lambda$ such that $x_\beta \in X_{\beta+1,\eta}$. Thus, from (b) and (c) we have that

$$X = \bigcup\{X_{\beta,\eta} : \beta < \kappa^+ \text{ and } \eta < \lambda\}.$$

Similarly we let

$$F = \bigcup\{F_{\beta,\eta} : \beta < \kappa^+ \text{ and } \eta < \lambda\}.$$

and

$$< = \bigcup\{<_{\beta,\eta} : \beta < \kappa^+ \text{ and } \eta < \lambda\}.$$

so that (d) ensures that $<$ is a well ordering of X and (e) ensures that $F: [X]^4 \to \omega$.

We must now construct the matrix so that F and $<$ will have the property required by IH(κ^+). To do this we use $\square_\kappa$, which asserts the existence of a sequence $\{C_\alpha : \alpha < \kappa^+$ and α is a limit ordinal$\}$ such that the following properties hold.

(f) C_α is closed and unbounded in α.
(g) For each C_α, $\mathrm{cf}(\alpha) < \kappa$ implies that $|C_\alpha| < \kappa$.
(h) If β is a limit point of C_α, then $C_\beta = \beta \cap C_\alpha$.

Inside each C_α choose $A_\alpha \subset C_\alpha$ to be closed cofinal and of regular cardinal order type. Note that each $|A_\alpha| < \kappa$. Since IH($|A_\alpha|^+$), $|A_\alpha|$ is not weakly compact by Proposition 1. So we get $g_\alpha : [C_\alpha]^2 \to \omega$ such that there is no homogeneous set cofinal in C_α. Similarly, for each limit ordinal $\eta \leqslant \lambda$, choose a closed cofinal $B_\eta \subset \eta$ of regular cardinal order type to help get a partition $h_\eta\colon [\eta]^2 \to \omega$ with no homogeneous set cofinal in η.

Let the construction begin: each $X_{0,\eta} = F_{0,\eta} <_{0,\eta} = \varnothing$. The remainder is done by recursion on $\beta < \kappa^+$. At successor stage $\beta+1$, check if x_β is in some $X_{\beta,\eta}$ and, if so, let each $X_{\beta+1,\eta} = X_{\beta,\eta}$, $F_{\beta+1,\eta} = F_{\beta,\eta}$ and $<_{\beta+1,\eta} = <_{\beta,\eta}$. If not, let $X_{\beta+1,\eta} = X_{\beta,\eta} \cup \{x_\beta\}$. Let $<_{\beta+1,\eta}$ be the end extension of $<_{\beta,\eta}$ by making $y <_{\beta+1,\eta} x_\beta$ for each $y \in X_{\beta,\eta}$. Let $F_{\beta+1,\eta}$ be the extension of $F_{\beta,\eta}$ such that

$$F_{\beta+1,\eta}(\{a,b,c,x_\beta\}) = h_\lambda(\{\theta_1,\theta_2\}),$$

where θ_1 is the least ordinal in λ such that $b \in X_{\beta,\theta_1}$ and θ_2 is the least ordinal in λ such that $c \in X_{\beta,\theta_2}$ (unless this ordinal is also θ_1, in which case let θ_2 be the least element of $\lambda \backslash \theta_1$). This ensures that, if $H \cup \{x_\beta\}$ is homogeneous for F with largest element x_β, then $H \subset X_{\beta,\eta}$ for some $\eta < \lambda$.

At limit stage α we use C_α. Let κ_ξ be the least cardinal in the cofinal sequence with $|C_\alpha| \leqslant \kappa_\xi$. For each $\eta < \xi$ let $X_{\alpha,\xi} = F_{\alpha,\xi} = <_{\alpha,\xi} = \varnothing$. For each $\eta \geqslant \xi$ let $X_{\alpha,\eta}$ be the closure of

$$\bigcup \{X_{\beta,\eta} : \beta < C_\alpha\}.$$

Each $X_{\alpha,\eta}$ has density at most κ_η. So for each $\eta \geqslant \xi$ we use IH($|X_{\alpha,\eta}|$) to obtain $<^*_{\alpha,\eta}$ and $F^*_{\alpha,\eta}$ on $Y_{\alpha,\eta}$ which is defined to be

$$X_{\alpha,\eta} \Big\backslash \bigcup \{X_{\beta,\theta} : \beta < \alpha \text{ and } \theta < \lambda\} \Big\backslash \bigcup \{X_{\alpha,\theta} : \theta < \eta\}.$$

We now define $<_{\alpha,\eta}$ and $F_{\alpha,\eta}$ by recursion on $\eta \geqslant \xi$.

Define $<_{\alpha,\eta}$ to be compatible with each of $<_{\beta,\theta}$ for $\beta < \alpha$ and $\theta < \lambda$ and with each $<_{\alpha,\theta}$ for $\theta < \eta$ as well as with $<^*_{\alpha,\eta}$. Furthermore, for all $x \in X_{\alpha,\eta} \backslash Y_{\alpha,\eta}$ and all $y \in Y_{\alpha,\eta}$, make $x <_{\alpha,\eta} y$.

Define $F_{\alpha,\eta}$ to be compatible with each of $F_{\beta,\theta}$ for $\beta < \alpha$ and $\theta < \lambda$ and with each $F_{\alpha,\theta}$ for $\theta < \eta$ as well as with $F^*_{\alpha,\eta}$. Furthermore, for all $y \in Y_{\alpha,\eta}$ and all

$$\{a,b,c\} \subset X_{\alpha,\eta} \Big\backslash \bigcup \{Y_{\alpha,\theta} : \theta \leqslant \eta\}$$

with

$$a <_{\alpha,\eta} b <_{\alpha,\eta} c <_{\alpha,\eta} y,$$

make

$$F_{\alpha,\eta}(\{a,b,c,y\}) = g_\alpha(\{\beta_1,\beta_2\}),$$

where β_1 is the least ordinal in C_α such that $b \in X_{\beta_1,\eta}$ and β_2 is the least ordinal in C_α such that $c \in X_{\beta_2,\eta}$ (unless this ordinal is also β_1, in which case let β_2 be the least element of $C_\alpha \backslash \beta_1$). Moreover, for all $y \in Y$ and all

$$\{a, b, c\} \subset \bigcup \{Y_{\alpha,\theta} : \theta \leq \eta\}$$

with

$$a <_{\alpha,\eta} b <_{\alpha,\eta} c <_{\alpha,\eta} y,$$

make

$$F_{\alpha,\eta}(\{a, b, c, y\}) = h_\eta(\{\theta_1, \theta_2\}),$$

where θ_1 and θ_2 are defined as above.

Now if $H \cup \{y\}$ is homogeneous for F with largest element y and $H \cap Y_{\alpha,\eta} = \emptyset$, then there are $\beta \in C_\alpha$ and $\theta < \eta$ such that

$$H \subset \bigcup \{X_{\gamma,\eta} : \gamma < \beta\} \cup \{Y_{\alpha,\rho} : \rho < \theta\}.$$

Without loss of generality β is a limit of C_α and so $C_\alpha \cap \beta = C_\beta$. So $\bigcup \{X_{\gamma,\eta} : \gamma < \beta\} \subset X_{\beta,\eta}$. Thus H is contained in the closed set $X_{\beta,\eta} \cup X_{\alpha,\theta}$. Note that in the case that $C_\alpha \backslash \beta$ has order type ω, H will be contained in the union of $X_{\alpha,\eta}$ with a finite number of $X_{\gamma,\eta}$'s.

It is now easy to verify that if H is homogeneous for F then it is left-separated by $<$. □

Remark 13 Again, in Lemma 12, we could replace quadruples with just triples. Noticing this, the reader could prove, using the methods of this article, that, under the hypotheses of Theorem 0, if X is a Hausdorff space such that, for each $Y \subset X$, $|\bar{Y}| \leq 2^{|Y|} \leq$ the first weakly compact, then there is a partition f: $[X]^3 \to \omega$ such that any homogeneous set is discrete. Nevertheless, triples cannot replace quadruples in Theorem 0, by a recent theorem of I. Juhász & S. Shelah, so that in this sense the results of this article are the best possible.

References

[1] P. Erdős, A. Hajnal, A. Maté & R. Rado, *Combinatorial Set Theory: Partition Relations for Cardinals*, North-Holland, Amsterdam, 1984

[2] A. Hajnal, I. Juhász & W. Weiss, Partitioning the pairs and triples of topological spaces, *Top. Appl.* (to appear)

[3] W. Weiss, Partitioning topological spaces, in *The Mathematics of Ramsey Theory* (eds. J. Nesetril & V. Rödl), Springer, Heidelberg (to appear)

[illegible] and ... [illegible] ... for all [illegible]

$$[illegible]$$

with

$$[illegible]$$

Take

$$[illegible]$$

[illegible] ... are determined as above.

[illegible] ... there are [illegible] ... but

$$[illegible]$$

[illegible] ... Thus [illegible] ... [illegible] ... with a finite number of [illegible]

It is now easy to verify that [illegible] ... is [illegible] ... by [illegible]

Remark [illegible] ... we could [illegible] quadrangles with [illegible] triples. [illegible] ... the reader could prove [illegible] ... of the [illegible] ... the first [illegible] ... any [illegible] ... cannot replace [illegible] ... that in this sense the results of [illegible] ... possible.

References

[1] [illegible] ... 1981.

[2] [illegible]

[3] [illegible]

Some of my favourite unsolved problems

Paul Erdős

Problems in number theory

1. This was perhaps my first serious problem (1931). Let $1 \leqslant a_1 < a_2 < \cdots < a_k$ be a sequence of integers. Assume that all 2^k numbers

$$\sum_{i=1}^{k} \epsilon_i a_i \quad (\epsilon_i = 0 \text{ or } 1)$$

are distinct. Estimate or determine $\min a_k$. I conjectured $a_k > c2^k$. In fact, I formulated the problem in the following slightly different form: let $a_1 < a_2 < \cdots < a_t \leqslant n$ be such that all 2^t sums $\sum_{i=1}^{t} \epsilon_i a_i$, with $\epsilon_i = 0$ or 1, are distinct. Determine $\max t$. I conjectured that

$$\max t < \frac{\log n}{\log 2} + c.$$

I proved that

$$t < \frac{\log n}{\log 2} + \frac{\log\log n}{\log 2} + c$$

and in 1954 Leo Moser and I proved that

$$t < \frac{\log n}{\log 2} + \frac{\log\log n}{2\log 2} + c,$$

which is still the current best upper bound (see P. Erdős, Problems and results in additive number theory, *Coll. Théorie des Nombres*, Bruxelles, pp. 125–37).

2. Let $n \equiv a_i \pmod{n_i}$ $(n_1 < n_2 < \cdots < n_k)$, be a system of congruences so that every integer n satisfies at least one of these congruences. Is it true that n_1 can be arbitrarily large? I offer \$1000 for a proof or

disproof. The current record is $n_1 = 24$. Selfridge and I conjectured that not all n_i can be odd. Is it true that if $n_1 \to \infty$ then also $\sum 1/n_i \to \infty$? Choi proved $n_1 = 20$; I don't know when and where $n_1 = 24$ was published.

3. Let $a_1 < a_2 < \cdots$ be a sequence of integers. Is it true that if $\sum 1/a_i = \infty$ then the sequence contains arbitrarily long arithmetic progressions? If true, this would imply that the primes contain arbitrarily long arithmetic progressions. I offer $3000 for a proof or disproof. It is not even known that the sequence must contain a three-term arithmetic progression.

Note that an affirmative solution would be a substantial extension of Szemerédi's theorem.

4. As usual, let us write $2 = p_1 < p_2 < \cdots$ for the sequence of consecutive primes. I proved in 1934 (On the difference of consecutive primes, *Quarterly Journal, Oxford*, **6** (1935), 124–8) that there is a constant $c > 0$ such that for infinitely many n we have

$$p_{n+1} - p_n > \frac{c \log n \log\log n}{(\log\log\log n)^2}\,. \tag{1}$$

Rankin (*Journal of the London Mathematical Society*, 1938) proved that for some $c > 0$ and infinitely many n the following inequality holds:

$$p_{n+1} - p_n > \frac{c \log n \log\log n \log\log\log\log n}{(\log\log\log n)^2}\,. \tag{2}$$

I offered (perhaps somewhat rashly) $10 000 for a proof that (2) holds for every c. The original value of c was improved by Schönhage and later by Rankin. Rankin's result was recently improved by Maier & Pomerance; their paper will appear in the Transactions of the American Mathematical Society.

5. Ricci and I proved independently that the set S of limit points of

$$\frac{p_{n+1} - p_n}{\log n}$$

has positive Lebesgue measure. Is it true that, in fact, $S = [0, \infty)$? Maier proved a considerably weaker old conjecutre of mine: the set $S \cap [c, \infty)$ has positive Lebesgue measure for every c.

6. Let $d_n = p_{n+1} - p_n$ be the difference between the n-th and $(n+1)$-st primes. Paul Turán and I conjectured (On some new questions on the distribution of prime numbers, *Bull. Amer. Math. Soc.*, **54** (1948), 271–8; see also P. Erdős, On the difference of consecutive primes, *Bull.*

Amer. Math. Soc., **54** (1948), 885–9) that for infinitely many n we have $d_n < d_{n+1} < d_{n+2}$. We could not even prove that either $d_n < d_{n+1} < d_{n+2}$ or $d_n > d_{n+1} > d_{n+2}$ has infinitely many solutions. I offer \$100 for a proof and \$25 000 for a disproof. (Of course, the conjecture is certainly true.)

7. Let $a_1 < a_2 < \cdots$ be an infinite sequence of integers. Denote by $f(n)$ the number of solution of $n = a_i + a_j$. Assume that $f(n) > 0$ for all $n > n_0$, i.e., $(a_n)_1^\infty$ is an *asymptotic basis* of order 2. Turán and I conjectured that then

$$\overline{\lim_{n\to\infty}} f(n) = \infty \tag{3}$$

and probably $\overline{\lim} f(n)/\log n > 0$. I offer \$500 for a proof of (3). Perhaps (3) and $\overline{\lim} f(n)/\log n > 0$ already follow if we only assume $a_n < cn^2$ for all n.

8. I proved that there is an asymptotic basis of order 2 for which

$$c_1 \log n < f(n) < c_2 \log n$$

(see the monograph *Sequences* by Halberstam and Roth). I conjecture that

$$\frac{f(n)}{\log n} \to C, \qquad (0 < C < \infty),$$

is not possible and I offer \$500 for a proof or disproof of this conjecture. Sárközy and I proved that

$$\frac{|f(n) - \log n|}{\sqrt{\log n}} \to 0$$

cannot hold.

9. Let $f(n) = \pm 1$. Is it true that for every c there is a d and an m so that

$$\left| \sum_{k=1}^{m} f(kd) \right| > c\,?$$

I will pay \$500 for an answer. Let $f(n) = \pm 1$ and $f(ab) = f(a)f(b)$. Is it then true that $\left|\sum_{k=1}^n f(k)\right|$ cannot be bounded?

Several further problems can be found in my papers with Sárközy, V. T. Sós and Nathanson and many more in my book with R. L. Graham (*Old and New Problems and Results in Combinatorial Number Theory*, Monographie No. 28 de l'Enseignement Mathématique, Genève, 1980, 128 pp.).

10. Is it true that

$$\sum \frac{1}{n!-1}$$

is irrational? I conjectured that

$$\sum \frac{1}{2^n-3}$$

is irrational. This assertion and its generalizations have been proved by Peter Borwein. Denote by $\omega(n)$ the number of distinct prime factors of n. Is it true that

$$\sum \frac{\omega(n)}{2^n}$$

is irrational?

11. Is it true that if $n \equiv 0 \pmod 4$ then there is a squarefree natural number θ such that $n = 2^k + \theta$? I could only prove that almost all integers $n \equiv 0 \pmod 4$ can be written in the form $2^k + \theta$.

Combinatorics and graph theory

1. This problem is due to Rado and me; I offer \$1000 for a solution. A system of sets $A_1, \ldots, A_k$ is called a *Δ-system* or a *sunflower* if

$$A_i \cap A_j = \bigcap_{i=1}^{k} A_i$$

i.e., if the intersection of any two sets is the same. Now let $B_1, B_2, \ldots, B_{f(n)}$ be n-element sets. Is it true that there is an absolute constant C such that if $f(n) > C^n$ then one can find B_i, B_j and B_k which form a Δ system? Richard Rado and I (Intersection theorems for systems of sets, *J. London Math. Soc.*, **35** (1960), 85–90) proved that

$$2^n \leqslant f(n) < 2^n n!.$$

Later Abbott and Hanson improved the lower bound to $f(n) > 10^{n/2}$ and Joel Spencer improved the upper bound to

$$f(n) < (1+\epsilon)^n n!,$$

provided $\epsilon > 0$ and n is sufficiently large. It is not even known whether, for $n > n_c$, $f(n) < n!$. (See also, P. Erdős and R. Rado, Intersection theorems for systems of sets II, *J. London Math. Soc.*, **44** (1969), 467–79.)

2. I offer \$500 for a proof of disproof of the following conjecture of Faber, Lovász and myself. Let G_i $(1 \leqslant i \leqslant n)$ be n complete graphs no two of which have an edge in common. Is it then true that $\bigcup_{i=1}^{n} G_i$ has chromatic number n?

3. Let $m = m(n)$ be the smallest integer for which there are n-element sets $A_1, A_2, \ldots, A_m$ such that $A_i \cap A_j \neq \varnothing$ for all $1 \leqslant i < j \leqslant m$, and such that for every set S with at most $n-1$ elements there is an A_i disjoint from S. (Note that the lines of a finite geometry have this property.) I conjectured with Lovász that $m(n)/n \to \infty$, but it is not even known whether $m(n) > 3n$ if n is sufficiently large. In the other direction, we could prove only that $m(n) < n^{3/2+\epsilon}$. Perhaps the following strengthening of the above conjecture also holds: for every $C > 0$ there is an $\epsilon > 0$ such that if n is sufficiently large and $\{A_i : 1 \leqslant i \leqslant Cn\}$ is a collection of intersecting n-sets (i.e. $A_i \cap A_j \neq \varnothing$ for $1 \leqslant i < j \leqslant Cn$) then there is a set S such that $|S| < n(1-\epsilon)$ and $A_i \cap S \neq \varnothing$ for all i $(1 \leqslant i \leqslant Cn)$.

4. Let $|X| = 4n$, $A_k \subset X$ and $|A_k| = 2n$ $(1 \leqslant k \leqslant f(n))$. Assume furthermore that $|A_i \cap A_j| \geqslant 2$ for all $1 \leqslant i < j \leqslant f(n)$. Is it then true that

$$f(n) \leqslant \frac{1}{2}\left\{\binom{4n}{2n} - \binom{2n}{n}^2\right\}? \tag{4}$$

This is the only unsolved problem from our paper with Chao Ko and Richard Rado (Intersection theorems for systems of finite sets, *Quarterly J. Math.*, **12** (1961), 313–20). I offer \$500 for an answer to this question.

Inequality (4), if true, is clearly best possible. Indeed, let $\{A_1, A_2, \ldots, A_{f(n)}\}$ be the collection of $2n$-subsets of $[4n] = \{1, 2, \ldots, 4n\}$ having at least $n+1$ elements in $[2n] = \{1, 2, \ldots, 2n\}$. Then $|A_i \cap A_j| \geqslant 2$ for all $1 \leqslant i < j \leqslant f(n)$ and equality holds in (4).

5. With Béla Bollobás I proved (On a Ramsey–Turán type problem, *J. Combinatorial Theory (B)*, **21** (1976), 166–8) that there is a graph of n vertices and $\frac{1}{8}n^2\{1+o(1)\}$ edges which contains no K_4 and the largest independent set of which is $o(n)$. We asked the question whether there is such a graph which has more than $\frac{1}{8}n^2$ edges and we still do not know the answer.

6. Is it true that every triangle-free graph on $5n$ vertices can be made bipartite by the omission of at most $5n^2$ edges? Is it true that every

triangle-free graph on $5n$ vertices can contain at most n^5 pentagons? Ervin Győri proved this with $1.03n^5$.

7. The following question is due to Hajnal and me. Let G be a graph of infinite chromatic number and let $n_1 < n_2 < \cdots$ be the sequence lengths of the odd cycles of G. Is it true that

$$\sum \frac{1}{n_i} = \infty ?$$

Perhaps, in fact, the n_i have positive upper density.

A. Gyárfás, J. Komlós & A. Szemerédi proved (On the distribution of cycle lengths in graphs, *J. Graph Theory*, **8** (1984), 441–62) that the set of *all* cycle lengths does have positive upper density, and some further results were proved by A. Gyárfás, H. J. Prömel, A. Szemerédi & B. Voigt, (On the sum of reciprocals of cycle lengths in sparse graphs, *Combinatorica*, **5** (1985), 41–52). For a presentation of these results, see Chapter 5 of B. Bollobás, *Extremal Graph Theory with Emphasis on Probabilistic Methods*, CBMS, No. 62, American Mathematical Society, Providence, Rhode Island, 1986.

8. In connection with the question above, Bollobás and I made the following conjecture. For a graph G, let $C_0(G)$ be the set of odd cycle lengths in G: $C_0(G) = \{2l+1 : G$ contains a $(2l+1)$-cycle$\}$. If $|C_0(G)| = k$ then G is at most $(2k+2)$-chromatic with equality if and only G contains a complete graph of order $2k+2$.

Bollobás & Shelah checked the validity of the conjecture for $k = 1$.

9. For a graph G, let $\mathrm{ex}(n;G)$ be its Turán number. In other words, $\mathrm{ex}(n;G)$ is the smallest integer for which every graph of n vertices and $\mathrm{ex}(n;G)+1$ edges contains G as a subgraph. Is it true that the number of graphs on n vertices which do not contain G as a subgraph is less than $2^{\{1+o(1)\}\mathrm{ex}(n;G)}$? Or, if $f_n(G)$ is the number of these graphs, do we have

$$\log f_n(G) = \{1+o(1)\}\log 2\,\mathrm{ex}(n;G)\,?$$

If G is not bipartite, this was proved by P. Frankl, V. Rödl and myself (*Graphs and Combinatorics*, 1988), but the bipartite case is open even for the 4-cycle C_4. Kleitman and Winston proved that

$$f_n(G) < \exp c\, n^{3/2}.$$

10. Simonovits and I conjectured that every graph on n vertices and $\mathrm{ex}(n;C_4)+1$ edges contains at least $cn^{1/2}$ distinct 4-cycles. We could

not even prove that it contains two C_4's. More generally, for which graphs G is it true that every graph on n vertices and $\mathrm{ex}(n;G)+1$ edges contains at least two Gs? Perhaps this is always true.

11. Let H be a graph and let G^n be a graph on n vertices which does not contain H as an induced subgraph. Hajnal and I asked whether there is an absolute constant $c = c(H)$ such that either G^n contains a complete graph or an independent set of n^c vertices? If H is C_4 then $\frac{1}{3} \leqslant c < \frac{1}{2}$.

12. Let G_1 and G_2 be two graphs with chromatic number $\aleph_1$. Is it true that there is a graph G whose chromatic number is 4 and which is a subgraph both of G_1 and of G_2? Perhaps there is a common subgraph of chromatic number $\aleph_0$.

13. Here is another problem we posed with Hajnal. Let $k \geqslant 3$. Is it true that every graph G of infinite chromatic number contains a subgraph of infinite chromatic number whose is girth greater than k? The finite versions of this problem are also of interest. Let $f(n;k)$ be the smallest integer for which every graph of chromatic number at least $f(n;k)$ contains a subgraph of girth at least k and chromatic number at least n. It is not at all clear that $f(n;k)$ exists. For $k = 3$ Rödl settled this question in the affirmative (see On the chromatic number of subgraphs of a given graph, *Proc. Amer. Math. Soc.*, **64** (1974) 370–71).

14. There are many unsolved questions in connection with my paper with Hajnal and Szemerédi. Let $f(n) \to \infty$ as slowly as we please. Is it true that there is a graph G of chromatic number $\aleph_0$ so that for every subgraph of n vertices we can omit $f(n)$ edges to make the graph bipartite? On the other hand, it is easy to see that if G has chromatic number $\aleph_1$ then there is a constant $c(G)$ and a subgraph of n vertices which cannot be made bipartite by the omission of fewer than cn edges. Is it true that every graph of chromatic number $\aleph_1$ has a subgraph of n vertices which cannot be made bipartite by the omission of cn edges? This should hold for every c and infinitely many n.

Is it true that if $F(n)$ tends to infinity sufficiently fast then every G of chromatic number $\aleph_1$ has for all sufficiently large n a subgraph of at most $F(n)$ vertices having chromatic number n? This is certainly false if we only assume that G has chromatic number $\aleph_0$.

See P. Erdős, A. Hajnal & E. Szemerédi, On almost bipartite large chromatic graphs, *Annals of Discrete Math.*, **12**, *Theory and Practice of Combinatorics, Articles in Honor of A. Kotzig*, (eds. A. Rosa,

G. Sabidussi & J. Turgeon), North-Holland, 1982, 117–23; see also V. Rödl, Nearly bipartite graphs with large chromatic number, *Combinatorica*, **2** (1982), 377–83.

15. Let G_1 and G_2 be two graphs of chromatic number $\aleph_1$. Is it true that they have a common subgraph G of chromatic number 4? Perhaps they have a common subgraph of chromatic number $\aleph_1$. An old theorem of Shelah, Hajnal and myself asserts that there is an n_0 such that both G_1 and G_2 contain a cycle C_n for every $n > n_0$.

16. Let G^n be a graph of n vertices which contains no $K(2,2,2)$ and whose largest independent set has $o(n)$ vertices. Is it then true that the number of edges of G^n is $o(n^2)$?

17. Let G be a bipartite graph and let $\mathrm{ex}(n;G)$ be the Turán number of G. Then Simonovits and I conjectured that $\mathrm{ex}(n;G) < cn^{3/2}$ if and only if G contains no induced subgraph every vertex of which has degree greater than 2. Neither the sufficiency nor the necessity have been settled.

18. Let Q_n be the graph of the n-dimensional cube $\{0,1\}^n$. Thus the vertices of Q_n are the 0-1 sequences of length n and two sequences are adjacent if they differ in exactly one place. I offer \$100 for a proof or disproof of the following conjecture. For every $\epsilon > 0$ there is an n_0 such that if $n \geqslant n_0$ then every subgraph of Q_n with at least $\frac{1}{2} + \epsilon e(Q_n)$ edges (i.e., with at least $(\frac{1}{2}+\epsilon)n2^{n-1}$ edges) contains a quadrilateral (C_4). Furthermore, I conjecture that if $\epsilon > 0$ and n is sufficiently large then every subgraph of Q_n with at least $\epsilon e(Q_n) + \epsilon n 2^{n-1}$ edges contains a hexagon (C_6).

Geometry

Let me start with two \$500 problems which go back to 1945 (On sets of distances of n points, *Amer. Math. Monthly*, **53** (1946), 248–50).

1. Let $x_1, \ldots, x_n$ be n distinct points in the plane. Denote by $D(x_1, \ldots, x_n)$ the number of distinct distances determined by $x_1, \ldots, x_n$. I conjectured in 1946 that $f(n) = \min D(x_1, \ldots, x_n)$ satisfies

$$f(n) > \frac{cn}{\sqrt{\log n}}. \tag{5}$$

The lattice points in the plane show that if (5) is true then it is best possible. The current record is

$$f(n) > \frac{n^{4/5}}{(\log n)^l},$$

due to Fan Chung, Szemerédi & Trotter.

I further conjecture that if $s_1 \geqslant s_2 \geqslant \cdots \geqslant s_k$ are the number of times the distances determined by our n points occur, so that

$$\sum_{i=1}^{k} s_i = \binom{n}{2},$$

then

$$\sum_{i=1}^{k} s_i^2 < cn^3(\log n)^\alpha,$$

for some $\alpha > 0$.

2. Denote by $A(x_1,\ldots,x_n)$ the number of pairs x_i, x_j whose distance is 1 and set $f(n) = \max_{x_1,\ldots,x_n} A(x_1,\ldots,x_n)$. I conjecture that

$$f(n) < n^{1+c/\log\log n}. \tag{6}$$

Again the lattice points show that, if true, (6) is best possible. In *Combinatorica*, **3**, Beck, and Spencer, Szemerédi & Trotter have papers on these problems.

It is likely that, if true, both (5) and (6) are very difficult to prove. Here are some stronger conjectures. Let again $x_1,\ldots,x_n$ be n distinct points in the plane. Is it true that for some x_i there are more than $cn/\sqrt{\log n}$ distinct distances from x_i, i.e., that there are $cn/\sqrt{\log n}$ distinct numbers among the distances $d(x_i, x_j)$ $(1 \leqslant j \leqslant n)$? In fact, perhaps, there are $c_1 n$ such points. I may 'overconjecture' here but I have no counterexample.

3. Let $x_1,\ldots,x_n$ implement $f(n)$ in the preceding problem, i.e., let $x_1,\ldots,x_n$ be such that there are exactly $f(n)$ distinct distances among the x_i. Thus for $n = 5$ the regular pentagon implements $f(n) = 2$. Is it true that for $n > n_0$ $(n_0 = 5?)$ there are always *two* and for large n *many* non-similar sets which implement $f(n)$. For $n = 5$ a colleague proved that the regular pentagon is the only set which implements $f(5) = 2$.

4. Let $x_1,\ldots,x_n$ be the vertices of a convex n-gon. I conjectured and Altman proved (*Amer. Math. Monthly*, 1963; also *Canad. Math. Bulletin*, 1972) that $D(x_1,\ldots,x_n) \geqslant \frac{1}{2}n$ (equality holds if n is odd and the polygon is regular; for even n equality is also possible). I also

conjectured that there is an x_i so that there are not less than $\frac{1}{2}n$ distinct distances from x_i. This conjecture is still open. Furthermore, Szemerédi conjectured that if $x_1,\ldots,x_n$ are such that no three are on a line then $D(x_1,\ldots,x_n) \geqslant \frac{1}{2}n$, but he only could prove this with $\frac{1}{3}n$.

5. Leo Moser and I conjectured that there is a constant $c > 0$ such that if $x_1,\ldots,x_n$ is a convex n-gon then $A(x_1,\ldots,x_n) < cn$, i.e., there are fewer than cn pairs (x_i,x_j) whose distance is 1. The current record is due to Edelsbrunner & P. Hajnal who proved that there is a convex n-gon with $2n-7$ pairs at distance 1. Füredi proved that the number of pairs at distance 1 is less than $cn \log n$. I further conjectured that in a convex n-gon there always is a vertex which has no three vertices equidistant from it. Here I again overconjectured because Danzer constructed a convex nonagon every vertex of which has three vertices equidistant from it. But I then conjectured that there is a k so that in every convex n-gon there is a vertex which has no k vertices equidistant from it. Perhaps this conjecture holds already for $k = 4$.

6. Let there be given n points in the plane in general position, i.e., no three on a line and no four on a circle. Denote by $h(n)$ the largest integer such that these points determine at least $h(n)$ distinct distances. I conjectured that $h(n)/n \to \infty$ as $n \to \infty$, but I could not even prove $h(n) \geqslant n$. In fact I conjectured a much weaker result which I could not prove either. For $n > n_0$ one cannot have n points in general position so that for every $1 \leqslant i \leqslant n-1$ the i-th distance occurs exactly i times. I. Palásti observed that for $n = 8$ such a set of 8 points exists, but it seems unlikely to me that for much larger n such a system should exists. Recently $h(n) < n^{\log 3/\log 2}$ was proved by Pach.

7. Let $f(n)$ be the largest integer for which there is a set of n points in the plane with $f(n)$ lines which contain at least $\sqrt{n}$ points. I conjectured that $f(n) < c\sqrt{n}$. Beck, Szemerédi & Trotter proved this but with a very large c and Sah showed that $f(n) > 3\sqrt{n}$. The papers of Beck, Szemerédi & Trotter appeared in *Combinatorica*, **3** and Sah's paper appeared in the volume of the Siófok conference, 1986.

For $\alpha > -1$, denote by $f_\alpha(n)$ the maximum number of lines which contain $(1+\alpha)\sqrt{n}$ points. I suspect that the function $f_\alpha(n)/\sqrt{n}$ has a discontinuity at $\alpha = 0$. This problem is still open.

8. Let $x_1,x_2,\ldots,x_n$ be n points in the plane with no five points on a line. Denote by $r(n)$ the maximal number of lines which go through four of our points. I conjectured that $r(n)/n^2 \to 0$ and I offer \$100 for

a proof or a disproof. Branko Grünbaum showed that $r(n) > cn^{3/2}$ is possible but perhaps we have $\lim r(n)/n^{3/2} < \infty$.

9. Let $x_1, \ldots, x_n$ be n points in the plane such that $d(x_i, x_j) \geqslant 1$ for all $1 \leqslant i < j \leqslant n$ and such that if $d(x_i, x_j) \neq d(x_k, x_l)$ then

$$|d(x_i, x_k) - d(x_k, x_l)| \geqslant 1,$$

i.e., if two distances differ, they differ by at least one. Is it then true that the diameter of $\{x_1, \ldots, x_n\}$ is at least cn and perhaps if $n > n_0$ it is at least $n-1$? Kanold proved that the diameter is at least $n^{3/4}$.

Miscellaneous problems

1. Let $f_n(z) = z^n + \cdots + a_n$ be a polynomial of degree n. Is it true that the length of the curve $\{z \in \mathbb{C} : |f_n(z)| = 1\}$ is maximal if $f_n(z) = z^n - 1$?

2. Let $f_n(z) = z^n + \cdots + a_n$. Denote by $E_n(f)$ the set where $|f_n(z)| \leqslant 1$. It is easy to see that $\max_{z \in E_n(f)} |f_n'(z)| \geqslant n$ and equality holds for $f_n(z) = z^n$. I conjectured that

$$\max_{z \in E_n(f)} |f_n'(z)| \leqslant \tfrac{1}{2} n^2 \qquad (7)$$

and that the maximum is assumed if $f_n(z)$ is the Chebyshev polynomial. Szabados observed that (7) is not correct as it stands but that $\frac{1}{2}n^2$ must be replaced by $\frac{1}{2}\{1 + o(1)\}n^2$. Pommeranke proved (7) with cn^2 about 20 years ago. As far as I know no further work has been done on this problem.

3. F. Herzog, G. Piranian and I have a paper (Metric properties of polynomials, *Journal d'Analyse*, **6** (1958) 125–48) where many problems on polynomials were stated, many of which were proved or disproved by Pommeranke in his papers in *Michigan Math. Journal* (1961–62). But here is a problem which is still open. Let

$$f_n(z) = \prod_{i=1}^{n} (z - z_i) \qquad (|z_i| \leqslant 1).$$

Denote by $A(f_n)$ the area of the set where $|f(z)| < 1$. Then $A_n(f) \to 0$ is possible but perhaps

$$A_n(f) > \frac{1}{n^c} \quad \text{or even} \quad A_n(f) > \frac{1}{(\log n)^c}.$$

4. Let G be a group. Assume that it has at most n elements which do not commute pairwise. Denote by $h(n)$ the smallest integer for which $G(n)$ can be covered by $h(n)$ Abelian groups. Determine or estimate $h(n)$ as well as possible. Pyber (*Journal of the London Mathematical Society*, 1987) proved that

$$(1+c_1)^n < h(n) < (1+c_2)^n,$$

for some constants c_1 and c_2. The lower bound was already known to Isaacs.

5. I have not included our many problems on set theory with Hajnal since undecidability raises its ugly head everywhere and many of our problems have been proved or disproved or shown to be undecidable (this happened most often). However, I think that the following simple problem is still open. Let α be a cardinal or ordinal number or an order type. Assume $\alpha \to (\alpha, 3)^2$. Is it then true that, for every finite n, $\alpha \to (\alpha, n)^2$ also holds. Here $\alpha \to (\alpha, n)$ is the well-known arrow symbol of Rado and myself. If G is a graph whose vertices from a set of type α then either G contains a complete graph K_n or an independent set of type α.

6. Let $|z_n| = 1$ $(1 \leqslant n < \infty)$. Put

$$f_n(z) = \prod_{k=1}^{n} (z - z_k)$$

and

$$M_n = \max_{|z|=1} |f_n(z)|.$$

Is it true that $\overline{\lim}\, M_n = \infty$? This conjecture was settled by Wagner: he proved that there is a $c > 0$ such that $M_n > (\log n)^c$ holds for infinitely many values of n. I further conjectured that $M_n > n^c$ for some $c > 0$ and infinitely many n and, in fact, for every n we have

$$\sum_{k=1}^{n} M_k > n^{1+c}. \tag{8}$$

Inequality (8), if true, may very well be difficult; so I offer \$100 for a solution.

7. Let $x_1, x_2, \ldots$ be a sequence of real numbers tending to 0. Is it true that there is a set $E \subset \mathbb{R}$ of positive measure which contains no subsequence $(y_n)_1^\infty$ similar to $(x_n)_1^\infty$? We call $(y_n)_1^\infty$ *similar to* $(x_n)_1^\infty$ if $y_n = ax_n + b$ for some $a, b \in \mathbb{R}$ and all n.